THE

STEM CELLS, CLONES, CHIMERAS,

and QUESTS FOR IMMORTALITY

SECOND TREE

ELAINE DEWAR

CARROLL & GRAF PUBLISHERS
NEW YORK

THE SECOND TREE
Stem Cells, Clones, Chimeras, and Quests for Immortality

Carroll & Graf Publishers
An Imprint of Avalon Publishing Group Inc.
245 West 17th Street
11th Floor
New York, NY 10011

AVALON
publishing group incorporated

Library of Congress Cataloging-in-Publication Data is available.

ISBN: 0-7867-1488-3

Printed in the United States of America
Distributed by Publishers Group West

To Mom and Dad

CONTENTS

PART TWO

. . . *the Lord God formed man from the dust of the earth. He blew into his nostrils the breath of life, and man became a living being.*

The Lord God planted a garden in Eden, in the east, and placed there the man whom He had formed. And from the ground, the Lord God caused to grow every tree that was pleasing to the sight and good for food, with the tree of life in the middle of the garden, and the tree of knowledge of good and bad . . .

When the woman saw that the tree was good for eating and a delight to the eyes, and that the tree was desirable as a source of wisdom, she took of its fruit and ate. She also gave some to her husband, and he ate. Then the eyes of both of them were opened and they perceived that they were naked; and they sewed together fig leaves and made themselves loincloths . . .

And the Lord God said, "Now that the man has become like one of us, knowing good and bad, what if he should stretch out his hand and take also from the tree of life and eat, and live forever!" So the Lord God banished him from the garden of Eden, to till the soil from which he was taken. He drove the man out, and stationed east of the garden of Eden the cherubim and the fiery ever-turning sword, to guard the way to the tree of life.

TANAKH, GENESIS 2 AND 3

INTRODUCTION

This book, like a clone thrown up from a lilac bush, grew from another. I was working on a project about the origins of the first people to inhabit the Americas when I was introduced to a brand new set of ideas—the use of genetics to trace the movements of people over thousands of years. Anthropologists claimed that the DNA handed down to us from our mothers and grandmothers, all the way back to Eve, can tell us where we came from.

I had learned what little I knew of biology and genetics in high school in Saskatoon, Saskatchewan, thirty-five years earlier. (I don't count a first-year university general science course that included four weeks of biology. I was so nervous at the lab test that I broke a specimen slide under a microscope. I failed it, but freed myself from my suicidal plan to go to medical school.) In high school, in my day, biology was all about divisions. We sliced the study of living things into botany and zoology. We committed to memory the classification scheme into which eighteenth-century biologists had pigeonholed life in its wondrous variety. We drew pictures of plant cells and their internal structures, and of single-celled animals and their organs. We tiptoed gingerly past Darwin's evolution theory, to avoid raising the wrath of any fundamentalists lurking among us. There was no discussion of the molecules that living things are made of, even though the field known as molecular biology had made vast strides since 1953, when James D. Watson and Francis H. Crick made a model showing how deoxyribonucleic acid (DNA), the gene-carrying molecule, is put together. In biology class, the high point of the year was when we got to dissect a frog. My bench mates and I shredded our frog's muscle and skin in what can only be described as a hormone-fueled frenzy. (What we did with the frog's eyes should never, ever be recounted.)

Genetics, invented by the monk Gregor Mendel to describe the way traits are passed to the next generation (from both parents, but randomly—some dominating, some hidden), was dealt with mainly in guidance class. We endured this class, pinned under the stern gaze of a freckled, balding and dour man who perched habitually on the front edge of his desk, banging his artificial leg against it for emphasis. We thought we were there to learn about how to be adults, but discussion of the basics, like sex, was mainly forbidden; I was kicked out for putting forth the radical notion that people might want to have sex before marriage to find out if they are compatible.

Our instructor seemed to think we needed to know human genetics, but he taught it as if it were eugenics. Genetics concerns the expressions and mutations of particular inherited characteristics, embodied as genes. Eugenics proposes that people are like corn or flies, that complex human traits are passed on in the same way as the wrinkled-skin trait is among peas. Eugenics proposes we weed out our bad traits, such as impulsiveness or violent temper, while encouraging the good, such as Christian kindness or high intelligence, through selective breeding. My guidance teacher mainly warned that inbreeding is dangerous and had led, in the isolation and ignorance of the American backwoods, to whole families of drooling, feeble-minded moral degenerates. The names of these unworthies? The Kallikaks and the Jukes, whose unfortunate pairings and unacceptable children, as well as their brushes with the law, had been recorded by the Eugenics Record Office of Cold Spring Harbor, New York, in the early part of the twentieth century.[1] Of course, by the time I took guidance, the Eugenics Record Office had long since closed in disgrace (although not before it influenced some thirty U.S. states to adopt laws regarding who was unfit to have children). The Record Office founders were soon forgotten by polite society, along with any notion that eugenics is a science.

Yet our teacher thought eugenics was an obvious extension of the work of Darwin and Mendel. He was not alone. Next door in Alberta, there were still eugenics laws on the books, long after the obedient servants of the Nazi eugenics laws were declared at Nuremberg to be perpetrators of crimes against humanity. I was certainly aware of the trial of Nazi doctors and of the judgments at Nuremberg when I was in high school, but not that the Alberta government's Eugenics Board was still overseeing the forcible (and in one case illegal) sterilization of the intellectually challenged.[2] The high point in guidance class occurred on the last day of the last year of high school, when we were freed from it forever.

I avoided thinking about genetics/eugenics for many years, until I began to research a book on Native American origins. I was then confronted with scientists' belief that one can trace a human population's movements by comparing mitochondrial DNA gathered from the living with that from the dead. Scientists doing this work assert that there are patterns of change in this DNA that are distinctive enough to show relationships of descent. This was terrain I'd never traversed before. I knew that mitochondria in a cell do something that fuels each cell's operations—I knew it because I looked it up in my daughter's high school biology textbook. But until then I'd thought that human DNA is found only in the nucleus of a cell, as part of a chromosome. I had no idea that these mitochondria are circular strings of DNA floating within the larger structure of the cell, that they also carry genes, or information, from one generation of cells to the next, just like the DNA locked in chromosomes.

Changes in the structure of these mitochondria happen mainly by mutation, not by sexual recombination, as is the case for DNA inside a cell's nucleus. (Did you know that when human sperm and egg combine, the chromosomes of both literally line up side by side, like to like, and swap pieces of themselves with each other?) Such a mutation occurs spontaneously in the order of DNA's base pairs. These are the smaller molecules that hold the larger DNA macromolecule together as if they were rungs on a ladder. These base pairs carry information: their order, or the sequence in which they are arrayed along the whole molecule, amounts to a code for the creation of proteins, which do the cell's work. When any cell divides, each chromosome, and each mitochondrion, is copied. But every time a copy is made, there is a chance for error. These smaller base-pair molecules can break away and be lost, or multiple copies can be made instead of only one. These accidental changes are then passed on from mother cell to daughter cell, and eventually from mother to child, and so on, down through the generations. Since mitochondrial DNA does not cross over when sperm meets egg, it keeps its base-pair sequence essentially unchanged over many generations, so these mutations can be used as markers to show relationships.

I was surprised to learn that mitochondrial DNA probably originated in a bacterium that somehow became incorporated into an ancient cell—the very distant ancestor of all the cells of which we are made. In other words, human beings do not descend in a straight line from one single cell: we are the fused product of two. In our most basic nature, we are what biologists call "chimeras." We are made up of trillions of cells that grow from the division of one fertilized egg cell, but each cell (except for blood cells) has two different information systems, one immense and complex, the other—the mitochondria—small and simple.

Yes, I had to look up *chimera*. The *Concise Oxford Dictionary* gives the origins of this word as being from Greek mythology, as follows:

> 1. Monster with lion's head, goat's body and serpent's tail.
> 2. Bogy, thing of hybrid character; fanciful conception, hence chimerical . . . 3. (Biol.) Organism formed by grafting etc. from tissues of different genetic origin . . .

So many questions boiled up. What did the Greeks know about the origins of life that inspired them to weave the fusion of unlikes into their founding stories? How exactly does the information system of one organism penetrate another? How is it possible that two separate systems can learn to work together so intimately? Is there a larger pattern that coordinates them? Where is that larger pattern written in our cells? Is this the ghost in the machine, what religious people call an immortal soul?

Sometimes a door opens and refuses to shut. I wanted to leave these questions behind, sealed between the covers of a finished book. But they lingered and grew until they could not be ignored. It seemed that every time I opened a newspaper I found myself reading revelationary stories about things biological. It became abundantly clear that biology is in extraordinary ferment; that the theory of evolution is under reconstruction, as are the reputations of the giants who invented the study of life as we now know it; and that the lives of the great biologists are the stuff of novels. Jean-Baptiste Lamarck, an early evolutionist who argued that the environment induces changes in individuals, which are then passed on to their descendants, had been discredited by Darwin's time. But now, at the turn of the twenty-first century, with so much new information about how environment guides development, Lamarckian thinking was enjoying a resurrection.[3] New and old articles suggested that the modest monk Mendel, cooked his results,[4] that Charles Darwin—kindly, retiring and neurotic Darwin—had made his name by clambering over the back of Alfred Wallace, to win the only battle that counts in science, the battle to be first.[5] James D. Watson and Francis H. Crick, founders of molecular biology, stood accused of taking advantage of the X-ray crystallographer Rosalind Franklin, using her work to figure out DNA's structure and beat Linus Pauling to the Nobel.[6] The same James D. Watson, now president of Cold Spring Harbor Laboratories (an organization in an indirect line of descent from the Eugenics Record Office), had become the founding director of the Human Genome Project but resigned in outrage after accusations of conflicts of interest. The Human Genome Project (which puzzled out the sequence of all the DNA in the human nucleus), though wrapped in idealism at the start, had evolved into a pitched battle between scientists around the world, pitting the standard-bearers for science capitalism against those for science as free inquiry, those who see benefit in patenting the sequences of human genes against those who believe such information belongs to everyone.[7]

Every other week *Nature* or *Science* announced the overthrow of some fundamental dogma about how cells, tissues or proteins work. Articles described the routine manufacture of new life-forms that combine genetic information across species, orders, phyla, kingdoms. The business sections of the papers recounted stories of biotech companies making initial public offerings, their scientist-executives touting this patented protein or that newly discovered disease gene. Global chemical companies lobbied governments against labeling genetically modified rice, corn or canola, which they claimed would save children in the third world from starvation and disease. Greenpeace protested against the release of these organisms into the environment, and demanded warning labels on everything. People in Europe and the United States burned test fields growing strange new Frankencrops.

What kind of journalist could hold back from such a rich stew of human perfidy and noble purpose?

This one.

Too hard, I thought, too steep a learning curve. But then the clones erupted.

In 1997, Roslin Institute, created by the British government to scientifically advance British agriculture, called in the press to make an announcement. In a shed outside its headquarters near Edinburgh, journalists were introduced to a sheep named Dolly, the first large mammal to be successfully cloned from an adult cell. The word *clone* is a slippery biological term; like *gene,* it carries several layers of meaning. Geneticists routinely write of "cloning genes," by which they usually mean capturing a small portion of DNA that codes for a particular protein. But they also use the word *cloning* to describe their methods of making myriad copies of a DNA sequence in other organisms, usually bacteria, sometimes yeast, which replicate the inserted DNA sequence as they replicate themselves. Developmental biologists use the same word, *cloning,* to describe making a copy of a cell, or a whole organism, usually by splitting an embryo at its earliest stages and making a twin.

These particular scientists at Roslin Institute, led by Keith Campbell and Ian Wilmut, had made an extremely complicated kind of clone. They took the nucleus out of a skin cell scraped off an adult sheep's udder and pushed this nucleus into a ewe's egg, from which the original nucleus had been removed. This rebuilt egg was then tricked into thinking it had been fertilized. As it began dividing, this pseudo-embryo was placed into the uterus of a ewe, removed before the implantation stage, then placed in yet another ewe and carried to term. The result was Dolly, a living animal with no direct father and with the same nuclear DNA as a ewe that was not its mother.

Roslin Institute and an associated company, PPL Therapeutics, soon called the press again to introduce Polly, a sheep cloned in the same way as Dolly but whose nuclear DNA had been modified. The nucleus from which Polly was made had been first given a human gene that allowed her to produce a human protein in her milk. Roslin Institute explained that the point of cloning was to cheaply and quickly create flocks of genetically modified identical animals, which could be milked for human pharmaceutical-grade proteins on an industrial scale. They called this "pharming."[8]

A biological dogma died with the birth of Dolly. It had once been believed, as an item of faith, that the development of an animal or plant moves inexorably from fertilization to death, never in the opposite direction. It was firmly believed that once a cell takes on its adult, most differentiated form, it can never go back to being what it started from, an embryo capable of forming any of the tissues of an organism. Although some domestic animals had been cloned before Dolly, none had been cloned from an adult cell. Dolly destroyed the idea of this impertubable vector of development.

The world's press and the public were not interested in broken dogmas: they noted that if biologists can clone a sheep or put a human gene into a sheep's embryo, they might soon put a sheep's gene into a human embryo and produce altered human beings on an industrial scale. Bad human genes could be wiped out, which might be good, and good human genes might be reproduced widely, which could be better, but freaks of all sorts could be made, which might be terrible. All our daughters could look like Sophia Loren, but legions of Hitlers could be grown from even one intact Hitlerian cell. The rich would pay fortunes to perfect their children; the poor would be left behind. Eugenics could make a comeback! Legislators in the United States, Britain, Germany, France, Canada and Australia thundered about stopping such evils before they got started. Laws were written, and some were passed. The United Nations produced a declaration. It was one thing for scientists in universities and large corporations to meddle with the molecular nature of mice and canola and corn. It was something else entirely to manipulate the stuff of humanity.

In 1998, a scientific team at the University of Wisconsin and another working out of Singapore and Australia took a few cells from early human embryos and put them in petri dishes lined with mouse cells.[9] Both teams got their funds and the human embryos from private sources. The U.S. federal government, the chief source of grants for academic scientists, would not fund human embryo research, so the team at Wisconsin got its research money from a U.S. company, Geron Corporation. Both teams found that by using the right kind of media, they could take cells from these embryos and keep them dividing in their undifferentiated state, indefinitely. The embryos were destroyed, but the embryonic stem cell lines derived from them were essentially immortal. In the right contexts, these cells could become very particular types of tissues.

Such stem cell lines had been made from the embryos of other animals, but never before from the embryos of human beings. These stem cells allowed researchers to study the earliest stages of development by interfering with it in various ways. Combined with the technology that made Dolly and Polly (called "somatic cell nuclear transfer"), it seemed likely one could soon grow any of the approximately two hundred different kinds of human tissue to order, in a dish, with genetic improvements or deletions as required. These changed stem cells could also be added to embryos and the resulting chimeras could be grown in a uterus or their cloned tissues could be transplanted into a person in need of them.

That same year, Geron Corporation announced that it had patented the gene that directs the manufacture of the enzyme telomerase.[10] Normal adult human cells will divide in a dish or in the living body only about fifty times before they stop. For thirty years, biologists had tried to figure out why. One theory proposed that the ends of each chromosome form a kind of cap called a *telomere*, which keeps the chromosome from unwinding. Because of the linear way in which the DNA molecule copies itself during cell division, the very end of the chromosome is not normally replicated. With each cell division this telomere shrinks, until the chromosome begins to unravel, like a shoelace without its plastic tip. The result: the cell stops dividing. In the 1980s, two researchers showed that there is a substance that actually puts base-pair sequences back on the telomeres of the chromosomes of a certain yeast, allowing it to keep dividing indefinitely. This substance was dubbed *telomerase*.[11] Telomerase was eventually found in human cells. Questions immediately followed: Could one fight cancer by turning off the gene that calls up the production of telomerase? Could one perpetually rejuvenate cells by turning this gene on? When Geron announced in *Science* that it had found key segments of the gene directing the manufacture of telomerase in humans, the company's stock soared.[12] The market seemed to think the company was on the verge of a cure for cancer, but more important, on the edge of a cure for death.

A few months later Geron bought the company, spun out of Roslin Institute, that controls the patents on nuclear transfer cloning. By the end of 1998 Geron held rights to the cloning of mammals by nuclear transfer, an exclusive license to grow human embryonic stem cells in immortal cells lines, and a U.S. patent on the gene that makes human telomerase. The company's very name suggested that this was all part of a grand strategy—a search for a cure for the very worst human disease, mortality.

In the summer of 2001, a group of biologists based primarily in Vancouver and expert in the life of *C. elegans,* a very interesting worm, published an article in *Genome Research* that made news.[13] They had compared genes expressed in the worm's normal state to those expressed only in its hibernation-like state, a state that can extend the worm's life almost indefinitely. They found a new and unpredicted gene expressed in great abundance in this hibernation state, a gene that they thought might interact with telomeres or telomere-like sequences. While they made no claims beyond that, the implications floated up from this work in the same way in which the subtle aroma of sandalwood drifts from an East Indian curio shop—inevitably, pleasurably, intoxicatingly. One of the senior members of the group was later pleased to tell me that he is very interested in solving this vexing problem of mortality, a subject that concerns him more and more the older he gets.

These stories found their way to the big cardboard box by my desk. They seemed connected to each other, although I couldn't say exactly why. The growing pile of clippings made me nervous. I'd done this to newspapers before and then found myself tied for years to some monstrous project, like a pirate's hostage lashed to the mast. I kept telling myself that the fact that I had saved these stories signified nothing, nothing at all. I wasn't committed; I was not yet locked into that long tunnel with no light at the end that is the true beginning of a book. I would just make a few phone calls, see a few people, see where things went.

A tiny woman, call her Ms. G., led me briskly down the hall to an elevator. (I won't tell you her full name for reasons that will soon be apparent.) Her boss, Andras Nagy, head of his own laboratory at the Samuel Lunenfeld Research Institute at Mount Sinai Hospital in Toronto, had asked her to show me how mouse chimeras are made. She had been churning them out for him and his students for most of a decade. Nagy had invented the first successful method for cloning mice, *tetraploid aggregation*, in 1993. I'd found his name on a list of those getting mouse stem cell research funding from the Canadian Institutes of Health Research. I'd never heard of the Samuel Lunenfeld Research Institute until I called Nagy, even though it is between the maternity floors at Mount Sinai, where I gave birth to both my children. I'm sure I had noticed the "research" sign in the elevator both times I went up, Lamaze pillow clutched in my arms, and both times I went down, with my babies. But I never imagined that a part of the hospital is given over to the study of the births and deaths of thousands of mice.

Not far from Nagy's lab is the office and lab of Alan Bernstein, the president of the Canadian Institutes of Health Research, which has close to three-quarters of a billion dollars in public grant money to hand out each year and a mandate to make Canadian health research the best in the world. (We'll come back to him later.) Nagy's senior colleague Janet Rossant had just been given $21.5 million by the notoriously stingy governments of Canada and

Ontario, to expand the use of the mouse in biological research. (Soon she would also be appointed the CEO of a $67 million mousery, a state-of-the-art facility to breed millions of mice for research.) The mouse is considered a great model organism—we can learn much about ourselves by studying it. Its life span is short, and it produces many offspring quickly, which makes it good for the study of genetics and development. It also has a genome (the term used now to describe the totality of an organism's DNA) that is closely comparable to ours—it has been less than a hundred million years since we diverged from a common ancestor and, as a result, vast stretches of our DNA are virtually identical. Best of all, one can do some things to mice that cannot be done to people and few will object.

I followed Ms. G. and her bouncing earrings into the elevator, which took us to another floor. I'm not going to tell you which one. Animal rights activists in the United States have done things that have made some researchers afraid. The elevator door opened to a long hall. At the far end, another door required a special access code. In a small anteroom there was a bin of little paper booties. Ms. G. showed me how to slip a bootie over one foot and step down inside the clean square marked with tape on the floor, then, while balancing like a stork, slip another bootie on the other foot and put that foot down within the square. Feet covered so we wouldn't track in dirt, we entered the mouse labs.

Behind the last locked door, the windowless hall was bright and wide. The main lab area was like most of the others I would visit over the next year. There were a number of large microscopes set at various stations on the spotless counters that girdled the room. There was a warming incubator ready to hold the embryos Ms. G. was about to alter. There were bottles of buffers, oils and specially filtered medium. Each work area had its boxes of scalpels, forceps, sterile gloves and small glass tubes set out beside Bunsen burners. When heated, the glass pipettes stretch out into long, hollow wands as thin as a hair. And, of course, there was a stand piled with small, steel-wired, labeled cages with water bottles attached to their tops, cages containing tiny mice, some white, some gray, some brown, who crawled all over each other.

There is a certain smell in a mouse lab—not unpleasant, not pleasant either. But it was something I wanted to scrub off later, after I noticed people's heads turning when I squeezed onto the subway.

"Are you squeamish?" asked Ms. G., with a warm Russian flourish of r's and vowels.

"No," I said. "I don't think so."

This was a small lie: the idea of touching a human eyeball sends shivers up and down my spine. But I am certainly not as squeamish as some. When I was a teenager, I worked in the lab in my father's medical office, helping his technician do simple things like blood counts. Once I had to help her heave a six-foot adult male back into the lab chair after she pricked his finger to squeeze out a drop of his blood. He had slid, boneless, onto the floor, in a dead faint.

"Good," said Ms. G. "We will sacrifice the mouse this way."

Before I realized what she meant, she had a mouse out of a cage, and a stiff piece of metal frame that normally held a label in her other hand. She placed the mouse belly down on the cage lid, where it squirmed. With the mouse's head held firmly by the metal frame, she yanked sharply on its tail, as if she was popping open a party favor. Just like that, the mouse died. The other mice below seemed to crawl around a whole lot faster.

This mouse and the other five she proceeded to sacrifice, had, she informed me, been "plugged" two nights earlier—mated to males. Their fertilized eggs, now embryos, would be moving down toward their uteruses. She would catch them, remove them and use them to make her chimeras. These embryos would be 2.5 days old and would therefore have formed a morula, a round ball of perhaps eight to sixteen cells surrounded by a thin shell called the zona pellucida. The morula would not yet have compacted; at that point the cells on the outside of the ball would become committed to forming, at the blastocyst stage, trophectoderm, the material which becomes the placenta. All the inner cells, from which stem cells emerge under the right conditions, would be able to become any kind of tissue, except placenta. She would extract these early embryos and remove their zonae pellucidae with a little puff of acid. Washing each embryo in warm salt solution, she would later put each one in a separate depression in a petri dish, then add to each one a genetically modified stem cell taken from another mouse's developing embryo. The stem cells are marked with inserted fish genes. Under the right light, all the cells and tissues that are daughters of these stem cells will glow green. The embryo and stem cell will stick together. Even without a protective zona pellucida, when placed back into a receptive mouse uterus they will likely develop together, as one chimeric mouse. In fact, the original embryo's cells and the foreign stem cell will compete for dominance as they divide and differentiate into specialized tissues, even as they also cooperate to form a new mouse.

Ms. G.'s boss, Nagy, and his colleague Rossant had created this method of mouse-making as a developmental study tool. The stem cells' genes can be altered with various colour markers. One can distinguish between the descendants of the stem cells and the embryo cells in the developing mouse by their colors and can demonstrate the exact sequence and pathway of each stem cell's daughter cells as the mouse grows. One can watch how the alteration—the designed mutation—affects development. This method doesn't work on all mice, though, and nobody knows why. Ms. G. had spent years working out her recipe, and, as in all recipes, the hand of the cook matters. Nobody knows why that is, either.

She took the six dead mice over to her work area. She laid out a surgical pad and placed the mice there on their backs. The first mouse's hind legs twitched feebly. I felt a pang of pity—I was worried she was about to cut open a live animal. I pointed at it.

"It's moving," I said.

"Don't worry," she said, as she snipped here and there and pulled off the mouse's skin in one smooth motion. "It's dead."

I watched her slit open the dead animal's stomach, move aside its bowel, pull up the uterus, cut twice between the ovary and the uterus to dissect out the oviduct. She held the glittering piece of tissue under the microscope and washed several embryos out with a spritz of filtered medium pushed through the tip of a needle. She examined the embryos carefully, discarding those that appeared to her eye to be less than healthy. She was unhappy with the low number of useful morulae.

Placing my eye to the scope (making damn sure I did not touch it with my clumsy hands), all I could see was a black line describing a sphere, a space, then a group of much smaller tiny balls within. That is all each morula amounted to: a thin black line outlining a sphere, grainy little balls that appeared to be hollow too. Incredible to imagine that this little bit of nothing added up to all that is needed to make a live mouse.

She moved, swiftly, quietly, from mouse to microscope, mouse to microscope. Finally she'd removed the zonae pellucidae from all the useful embryos, and placed each in its proper depression in a dish, with a stem cell added on top, just so, the whole thing protected by oil. These stem cells had been transfected with a jellyfish gene, which had insinuated itself into the stem cells' DNA. These altered cells had then been laid out in a dish and grown in a culturing medium as a stem cell line. (When the cells divide sufficiently to fill the dish, the culture is split and put into another dish. Each split is called a "passage.") I had seen these stem cells being prepared upstairs and had looked at them, too, under a microscope. There seemed to be nothing significant about them at all; they also looked like a grainy group of tiny spheres. Ms. G. placed the finished dish in the incubator.

There was a little laptop computer set up in a corner. She showed me images of the strange organs and animals that have emerged from this work. You see, she said, clicking the mouse to bring up image after colored image, the morula stage, the stem cells.

She stopped at one, which filled the screen. It was a mouse heart, pulled from the chest of a chimeric animal. The mouse had been made with two different color markers inserted into each of its two genomes. As the mouse developed, each of the original cells, with the color-marker genes in them, produced daughter cells, and the daughter cells produced more daughter cells, until, eventually, a mouse was born whose cells were all daughters of the original marked cells, either green or blue. The heart hung there in cyberspace, all its tissues vivid blue and brilliant green. The colors twisted and twined but never mixed, as if the heart had been made from two sticks of Plasticine.

What could more graphically express what I had witnessed here? The animals she'd just made had four genetic parents; a fifth would carry them to term. All of their tissues would be the product of a profound manipulation, of both cooperation and competition for a place in the developing mice. I had been told in high school that competition goes on at the species level. She had shown me that it also goes on at the level of the cell in any developing organism, and at the level of every molecule, too.

"You are interested in this?" Ms. G. asked. "Wait, I will show you something."

She disappeared down the hall to the mousery, sliding silently in her booties along the polished floor. I stood there, eyes locked on the image of the blue and green Plasticine heart.

"Here," she said. She held in front of me a cage of mice, several newborns with a mother. She took two of the furless, wrinkly little things in her hand, showing their exposed skin. They were pale, pale pink, the color of a healthy human scalp. They nibbled and nuzzled at each other as she thrust them under a lamp with a special blue filter and turned it on. The pink skin suddenly glowed a brilliant yellow-green. She put the mice back in the cage and pulled out the brown-furred adult. She put it under the same light. Its hairless ears and tail turned an amazing jewel green, the color of an oak leaf with sunlight pouring through it. A jellyfish gene, she explained, had been inserted into an embryonic mouse stem cell, which was then aggregated with an embryo, just as she had done this morning. The stem cell had found its way to the germ line—the cells that will become either egg or sperm—of a developing mouse. The offspring of that mouse all have the jellyfish gene and will pass the color of a backlit forest canopy to their descendants.

I went home on the subway that day barely able to contain my excitement. This work was terrifying; this work was magnificent. It was also, for the first time, real to me. Until this point I had been clipping news stories and reading journal articles, but now I had seen glowing, changed, living things. These animals, running in their cages in a hospital—my own hospital, a place that was so familiar and so safe—could never have developed in the natural course of things. I stopped neighbors on the street to tell them what I had seen—this green and blue heart, these bizarre, colored mice, these chimeras. They looked at me unmoved, as if I was babbling about a dream. I wanted to shout: Do you not see what is going on, do you not understand what we are becoming? We can make living things in whatever shape or color we want. We can make life reveal all of its secrets. Maybe we don't have to die!

I calmed down over dinner. I explained to my family that biology had become unrecognizable. The separate disciplines I remembered from only thirty-five years ago (botany, zoology, genetics, biochemistry, evolutionary theory, development) seemed to have merged. In just 150 years, biological science had evolved from Darwin's and Mendel's simple attempts to observe the laws governing change to this roiling stew of complexity. Such eighteenth-century classifications as the definition of a species—the capacity to breed and get viable offspring—were completely beside the point if one could make chimeras by fusing embryos of different species in a dish (as had already been done when scientists in Britain manufactured a "geep," a chimera of goat and sheep). And when the genes of fish and bacteria and mice can be combined in animals or plants, when stem cells can be reproduced in almost infinite numbers, when adult cells can be turned back to the start point—the equivalent of pouring spilled flour back into the bag—then biological science has become a kind of revelation.

This idea of biology as revelation made me read, again and again, the sad tale of Adam and Eve. Their story welds the invention of the first biotechnology (the use of leaves to cover nakedness) to the knowledge of good and evil, and both to their fall from grace. For the first time, I noticed that the story deals with two trees, not just one. They eat the fruit of the first, but it is the second tree that matters, the tree of life. This is what God sought to protect by ejecting the parents of humankind from the Garden of Eden, setting cherubim and a fiery turning sword at the gate to guard it from us, lest flawed human beings also eat of the second tree and live like God, forever.

At some point, the cherubim and the fiery turning sword must have left their posts, because biologists are definitely swarming all over this tree of life. They are grafting on new branches, meddling with the tree's molecules, stretching out their hands to eat and live forever. How did this happen? Where will it end? Who will make rules for them as they extend their dominion over all forms of life?

No journalist, especially one, like me, sliding down the dark side of fifty, could walk away from a story like that.

And so I plunged into the biologists' world. I followed their networks, learned their histories, studied their casts of mind and their political behavior. They do their work in laboratories all around the globe, from Vancouver to Cambridge, from San Francisco to New York, from London to Toronto to Haifa to Melbourne. They are a wandering tribe, like members of the craft guilds of ancient times, carrying with them their libraries of copied genes and altered cells, their reagents and study animals, their techniques and recipes, their attitudes and beliefs. They move from one jurisdiction to another as opportunities appear, as laws intrude.

Often I had to fight the fear (especially while wading through their writings) that those leading the assault on the second tree are beyond our ken. They are setting out a new view of nature so fundamentally different from the way most of us see life that it is hard to take it in. It was tempting to take their integrity and goodwill on faith, and simply consume the products of their labors. Sort of like Eve, the snake and the apple.

The trouble is that this revelationary biology—a phrase my husband coined when I was struggling to describe the way separate disciplinary streams have flowed together like rivers in flood—is also the greatest product and expression of Western democracies. Our governments pay for it and compete for its stars; our elected representatives struggle with its meaning and its implications; our markets rise and fall with its rhythms of discovery and hype; our wants call forth the creativity of its practitioners. We reward them, praise them, give them prizes and hope our children will join their number, even while we pass laws that could turn some of them into criminals. In democracies we are all responsible, we are all complicit—even when we don't understand, even when we don't know what is being done in our names. So it is up to those who do this work to pause and teach us. And it is up to the rest of us to listen, think and either give our assent or put a stop to it. Many of these biologists, as you will see, have lost their ability to speak directly to those who don't share their language and assumptions. It is the journalist's job to translate, to make bridges, to find frameworks and, most of all, to report without fear or favor on a world that is being changed under our feet.

Some of those climbing among the highest branches of the second tree do not look lovely in the harsh light of the sun. But there are others who can show us things of such power and beauty that there will never be words enough to thank them.

Part One

CHAPTER ONE

BIOLOGISTS SEIZE THE DAY

On a late November day in Ottawa, in the dreadful fall of 2001, the grass on Parliament Hill was a strangely vivid green. Though the trees were bare spines, their fallen leaves all mashed in the gutters, winter was still biding its time—waiting for a moment of weakness. I had already spent months weaving my way through library stacks and Internet sites, popping into this lab, that office. I was burdened by so many questions; for instance, what is a gene, exactly, and why did almost every article about biology trumpet a breakthrough, an imminent cure for Parkinson's, Alzheimer's, diabetes and paralysis? Where did the weird business of making clones and chimeras and this questing after immortality begin?

I climbed up the steps to the Hill, a secular pilgrim seeking the light. The Standing Committee on Health was in the middle of hearings on a proposed law that dealt with all these things. I had this notion that I could follow them and their work, the way a cross-country skier follows the leader who breaks the trail. They had researchers; they'd heard testimony from experts; they'd know so much more than I did.

The committee was examining a draft law put forward the previous May by the federal minister of health. It proposed to ban human cloning and permanent genetic change in human beings and to regulate the making of human embryonic stem cells, chimeras and laboratory-bred babies. It seemed that almost every other Western government was struggling with its own version of this bill. President George Bush had announced on August 9 that the United States would fund research only on human embryonic stem cell lines made before that date: he considered the use of human embryos for research immoral. Senators and congressmen had put contradictory bills forward, some banning all human cloning, others promoting certain forms of it. The British decided to permit the manufacture of embryos to extract stem cells for research and to use an existing agency to regulate it. Interested parties from several countries had come before the Canadian committee to cheer on the Canadian draft bill, or to object. People opposed to abortion had made it clear that they didn't like the destruction of human embryos for any purpose; some of them were MPs of the governing Liberal Party.

I got to the right room in the gothic stone pile known as the West Block, and found a chair. I was early, so in the unseasonable warmth, I amused myself by picking through the magpie nest of biological tales I'd already acquired. I had one foot in the journalist's idea of heaven and the other in hell. I'd found so many good stories (heaven) but I couldn't decide which I most wanted to tell (hell). The history of modern biology is full of bad luck and cruelty. Then there was the hubris, and the culture war between the materialists and the mystics, not to mention the flat-out pursuit of fame.

The plaque on the West Block's south wall said the building was begun in 1859, the very year Charles Darwin brought out *On the Origin of Species by Means of Natural Selection* to a ragged chorus of acclaim and derision.[1] Darwin's theory of why living things change, his understanding of how and why species differentiate from each other over time, pretty much defined cruelty. Struggling to make sense of what he had seen on his five-year voyage of discovery on the *Beagle* and of what he had learned from his correspondence with naturalists everywhere, Darwin had borrowed from Thomas Malthus's brutal social theory to frame his ideas.[2] He observed that nature displays a pattern: failures are wiped out while success breeds more. He saw that nature throws up random changes in individuals and, like a farmer, selects those creatures with desired traits. Those who acquire the right changes to fit their changing environments have more progeny, while those who don't die young and are cut off. Species slowly evolve as these changes accumulate in populations over time. Darwin's theory of natural selection eventually swept aside the dearly held Christian belief, accepted by most scientists of the day, that creation unfurls in accord with God's perfect plan. Darwin showed there is no plan. As book followed book, he also knocked humankind off creation's pinnacle and ground us into nature's dust with all the other animals. He unraveled the warm blanket woven from the belief that a loving, or at least a just, creator intentionally made each one of us. Death was Darwin's creator.

Darwin suffered more than a little bad luck as he worked out these ideas over a span of more than forty years. He had a peculiar undiagnosable illness that interfered with his work and virtually imprisoned him at his country home. Two of his children died of diseases easily treated one hundred years later.[3] And he failed to read Gregor Mendel on heredity.[4]

Darwin's ideas about how heredity works, how changes are passed on to the next generation, were his theory's soft, white underbelly, the place his critics kept poking at. He believed that these random changes that result in new species are passed on from parents to children through the medium of something he called "gemmules." He thought gemmules from both parents blend in the child, circulating in the blood. His cousin, the Cambridge-trained mathematician Francis Galton, ran a series of experiments to demonstrate this theory. Galton injected the blood of colored male rabbits into white females, then bred the females to see if the colors averaged out in their progeny. He sent numerous letters to Darwin reporting his failures.[5]

But the fact that Darwin's ideas rested on such a shaky foundation did not dissuade Galton from his conviction that Darwin's theory of change through natural selection was correct. Galton believed that through the application of scientific principles, humankind could become like nature and impose conscious selection on itself. In the 1880s, Galton used the word "eugenics" to propose a new statistical science of human attributes. He suggested in several tomes that fine English families should be encouraged to breed, in the same way that good farmers breed their best sheep—and that the poor rabble should be strongly discouraged.[6] He suggested that eugenics would be a kindness.

There exists a sentiment, for the most part quite unreasonable, against the gradual extinction of an inferior race. It rests on some confusion between the race and the individuals, as if the destruction of a race was equivalent to the destruction of a large number of men. It is nothing of the kind when the process of extinction works silently and slowly through the earlier marriage of members of the superior race, through their greater vitality under equal stress, through their better chances of getting a livelihood, or through their prepotency in mixed marriages . . .

Whenever a low race is preserved under conditions of life that exact a high level of efficiency, it must be subjected to rigorous selection. The few best specimens of that race can alone be allowed to become parents, and not many of their descendants can be allowed to live. On the other hand, if a higher race be substituted for the low one, all this terrible misery disappears. The most merciful form of what I ventured to call "eugenics" would consist in watching for the indications of superior strains or races, and in so favouring them that their progeny shall outnumber and gradually replace that of the old one.[7]

Darwin was already known worldwide in 1866 when Gregor Mendel published the results of twenty years of work backcrossing peas.[8] Mendel sent Darwin a reprint of his paper. But by then, all kinds of strangers were sending things to Darwin—even Karl Marx sent him a copy of his *Das Kapital*. Darwin didn't read Marx or Mendel; the pages of Mendel's reprint remained uncut until after Darwin's death.[9] Yet Mendel's statistical work was a neat complement to Darwin's theory. He showed the independent passage of both parents' traits to progeny, some of the traits dominant, others recessive but nevertheless still there. He showed that these recessive traits would reappear with the right crossbreeding of later generations and demonstrated that these traits (much later dubbed "genes" by William Bateson) pass from generation to generation in discernible ratios. These genes must be what Darwin's selection worked on.

But Mendel had bad luck of his own. No one took note of his paper, except for one highly respected Czech botanist, Karl von Nägeli, then at the University of Munich. Unfortunately, Nägeli, like Darwin, believed that hereditary traits are blended. He was skeptical of Mendel's methods. In order to convince Nägeli he was right, Mendel tried backcrossing a plant Nägeli worked on himself, hawk-weed, which reproduces mainly by cloning. He was not successful.[10] Mendel died in 1884, no doubt convinced that his work would never matter to the world of natural science. In 1900, thirty-five years after he published, three other scholars, one of them a former student of Nägeli's, cited Mendel's paper almost simultaneously.[11] Once discovered, Mendel's ideas were taken up immediately.

Hubris is a Greek word that means "insolent pride or presumption." Hubris seemed to grow in the minds of biologists out of all proportion to their accomplishments as the twentieth century progressed. With little more information about inheritance than Mendel's observations about peas, some biologists jumped to the conclusion that complex human traits are carried by genes, and that an individual's future could be predicted from the behavior of his or her parents. In 1911 Francis Galton died and left in his will money to create a national eugenics foundation in London. The year before, Charles Benedict Davenport, director of the Department of Experimental Evolution at Cold Spring Harbor, New York, had set up the Eugenics Record Office. Davenport intended to apply the ideas of Darwin, Mendel and Galton to better human breeding. He began to accumulate the life histories of prostitutes, criminals, immigrant Jews, "hybrids" created by the union of blacks with whites, and children of men in *Who's Who*. He believed he found in these stories evidence of complicated human traits passing to progeny, evidence that could be subjected to Mendelian mathematical analysis. Eugenics, Davenport opined, would one day be able to say which human breeding pairs would give rise to destructive traits such as "criminality and red hair." Eugenics would also predict what kind of people could, if mated, produce children of high IQ and good temper. Davenport thought these traits were carried by some actual, physical material, which probably resided somewhere in the nucleus of the cell, maybe in the chromatin (later called "chromosomes").[12] In concert with the American Breeders' Association, with generous donations from certain rich neighbors and foundations, such as Mrs. E. Harriman, widow of a railroad baron,[13] John D. Rockefeller, and the Carnegie Endowment, and with no less a great Canadian than Alexander Graham Bell as chairperson of one of its associated committees,[14] the Eugenics Record Office set out to change humankind.

Davenport's eugenics turned the moral order of the world's great religions upside down. Why help those unfortunates born with defects of mind or body? Better they should never be born. He wasn't too keen, either, on modern medicine and its newfangled notions that certain small organisms cause infections such as tuberculosis. Berating physicians for focusing entirely on germs and conditions of life as the causes of disease, Davenport wanted the emphasis back on something universally applicable, on these traits that travel through time and determine what men and women will become. Medicine, he railed, "has forgotten the fundamental fact that all men are created bound by their protoplasmic makeup and unequal in their powers and responsibilities."[15]

Baldly asserting that human babies are the world's most valuable crop, he was appalled that no one seemed to notice the plague of unfitness among them. "It is a reproach to our intelligence that we as a people, proud in other respects of our control of nature, should have to support about half a million insane, feeble-minded, epileptic, blind and deaf, 80,000 prisoners and 100,000 paupers at a cost of over 100 million dollars per year."[16] Davenport thought state and federal laws drafted on the basis of eugenics would prevent the passing on of such destructive traits and forestall any need to be kind.

Many eugenics laws were passed. In 1927, long before anyone had shown what genes are made of or which complex traits, such as intelligence, were embodied by them, the U.S. Supreme Court gave the green light to eugenics enforcement. In the case of Buck v. Bell, the court allowed Carrie Buck, the daughter of a mildly retarded woman, who was herself mildly retarded and had already given birth to an allegedly retarded child, to be sterilized against her will. Mr. Justice J. Holmes wrote for the majority about why this cruel assault on the dignity of an individual should be carried out by society: "It is better for all the world, if instead of waiting to execute degenerate offspring for crime, or to let them starve for their imbecility, society can prevent those who are manifestly unfit from continuing their kind. The principle that sustains compulsory vaccination is broad enough to cover cutting the Fallopian tubes . . . Three generations of imbeciles are enough."[17]

Though eugenics was eventually discredited, the hubris that drove it continued to thrive among biologists. It grew with each technological advance, until some biologists proposed that we can and should take charge of the evolution of all of nature. At a symposium held at Wesleyan University in 1983, biologist Barry I. Kiefer said, "We are about to reap a harvest sown at the turn of the century . . . The objective is the control of the genetic architecture of many of the individual life forms which inhabit this planet—including humans— for the benefit of our species. This is not necessarily the goal of individual scientists: it is the goal of our species in the truly biological sense. The fundamental controversy, then, is which interest group will be at the controls, not whether the objective is correct."[18]

Biologists wanted to be the interest group in charge. Since the 1970s, biologists had been founding corporations, selling shares and the products of their manipulations. By the early 1990s, some biologist-businesspeople were closing in on fixing human genes. At a conference called Engineering the Human Germline, held in Los Angeles in 1998, a number of attendees argued that since some governments were afraid to fund important research, such as genetically altering human embryos, scientists in business should take the lead and change the course of human evolution. Cold Spring Harbor Laboratory's president, James D. Watson, argued that privately funded scientists should not wait for permission from any state to delete certain disease genes from the germ lines of babies who otherwise would be born with defects. In his view, the private scientist is trustworthy while the state is dangerous:

> Consider what happened in Russia, where they essentially banned genetics because the concept of genetic inequality didn't appeal to them. Since there is genetic inequality of all

sorts, it's denying reality . . . There's an enormous amount of variation that is there to create the variations that have been necessary in the past for survival in changing environments. We have quite a high mutation rate, so many people are born with very obvious defects where their genes don't let them function as well as other people . . . Some people are going to have some guts and try germline therapy without completely knowing that it's going to work . . . The biggest ethical problem we have is not using our knowledge . . . Evolution can be just damn cruel, and to say that we've got a perfect genome and there's some sanctity? I'd like to know where that idea comes from, because it's utter silliness . . . Terms like sanctity remind me of animal rights. Who gave a dog a right? This word right gets very dangerous. We have women's rights, children's rights; it goes on forever. And then there's the right of a salamander and a frog's rights. It's carried to the absurd.[19]

No one could beat James Watson for hubris.

The culture war between the materialists and the mystics, or vitalists, was another great story, but hidden, like a river confined to an underground cavern. The materialists imagine life as the random product of interactive chemistry, as a giant, four-billion-year-old kluge—an engineer's term for a machine that starts simple but to which so many parts are added, to do so many special things, that the final product bears no resemblance to the original. Materialists believe this vast machine called "life" can be reduced by science to its molecular constituents and understood. The vitalists think about life's patterns and plasticity—like the tension between competition and cooperation, like the way cells can be ripped from one context and slammed into another and no harm done—and they wonder.[20] Is there some spark in living things that will never be pinned down, will never be reduced to something measurable and countable? The materialists have little patience for such romance. Francis Crick, a cofounder of molecular biology, is a committed materialist. As James D. Watson tells it, in 1962 Crick resigned in protest as a founding fellow of Churchill College, Cambridge, when the college decided to build a chapel. According to Watson, Crick "gave 75 guineas to the Cambridge Humanists to sponsor a prize essay on 'What shall be done with the college chapels?' The winning essay saw several different futures for them, including swimming pools."[21]

I shuttled back and forth between these two poles. On the one hand I was enthralled by the materialist view after reading physicist Erwin Schrödinger's *What Is Life?* This essay, published in 1944, convinced generations of biologists, especially the young Watson, that life can be studied at the molecular level. Schrödinger pointed out that there is only an apparent paradox separating the living from the nonliving. Life is orderly, at least until it's over, and yet molecules and atoms appear to be orderly only when in large numbers. Energetic particles, and atoms and molecules are not individually restrained by simple Newtonian rules of cause and effect. They are phenomena that exist and behave in random and probabilistic ways, whose behavior can be studied through statistical methods.[22] Living creatures, Schrödinger explained, maintain order while also being sensitive to individual molecular changes.

Schrödinger made me think about how molecules add up to cells, how cells add up to organs and organisms, how organisms add up to species—each domain of increasing complexity accumulating errors and yet still functioning, still working. I found myself with altered vision. It was like being eleven years old again, lying on my front lawn on a summer's night when the blazing prairie sky seemed like an open picture window on the universe. With my hair in the dirt, sharp pebbles jutting into my back, solidly connected to the earth, I'd nevertheless fall vertiginously off the planet and out among the stars, a tiny speck among the pinwheeling galaxies.

I walk every day with my small male dog. He hauls me around the little urban forest at the end of my street, sniffing frantically for the chemical signals left on the grass by the females who wandered through earlier. After reading Schrödinger, the huge, leaning oaks and the vast three-hundred-year-old spreading beech, which had once seemed to me to be utterly distinct and individual, appeared intimately connected to each other. Sometimes I'd sense a wiggle of air at the corner of my eye and, for an instant, imagine I could see the constant interchange between and within all these living systems—their molecules agitating, repelling, fitting together, enabling, assembling, breaking up, falling apart, reshuffling, reordering, then falling apart again.

But then my dog would bark, and I would look up and see the sun break through a blanket of cloud, see the light flow down through the canopy, the crows mob and shriek at invaders, smell the rotting duckweed on the pond. And I'd find myself thinking like a vitalist. I'd wonder about the difference between the living and the dead, about the way death immediately takes away something indescribable and indefinable from a living thing, even though the processes of decay proceed at a much slower rate. What is that something that vanishes?

The flat-out pursuit of fame was going on in every major lab in the country, in every lab on the planet. It had been going on ever since Darwin put aside his virtue and allowed his friends to arrange publication of his hastily written paper on natural selection, alongside Alfred Wallace's, though Wallace's was written first. And nobody I met was running harder than Mick Bhatia, a young stem cell researcher at the John P. Robarts Research Institute in London, Ontario. Bhatia was trying to figure out how to distinguish a stem cell from any other cell by certain physical characteristics. He had a row of champagne bottles on the shelf above his desk, set there like trophies, to remind himself of each of his lab's papers published in major journals. I had expected his lab to be a noisy place, loud with the spirited jousting of big brains and big mouths. Instead, his postdoctoral fellows and doctoral students worked away in a common room outside his door, in grim silence. Competition is so intense and the stakes so high, Bhatia explained, that he rarely tells even his former mentor, now his competitor, what he's working on.

Desire for fame, the desire to be the first, was important, but the need for money also forced this silence on researchers, particularly if the money came from corporate sources. I'd noticed that some of Bhatia's research was paid for by Geron, the U.S. corporation working on telomerase and human embryonic stem cells. Couldn't say a word about that, Bhatia told me. Why not? He'd signed a confidentiality agreement.

Bhatia had also said he wanted to work on human embryos to extract stem cells but was unwilling to go forward until the Standing Committee on Health had finished its work and Parliament had passed a law. He told me a committee of scientists working through the Canadian Institutes of Health Research was writing its own funding guidelines in a process parallel to the standing committee's. He hinted that there were rifts between researchers about what should and should not be allowed.

The members of the Standing Committee on Health finally took their places in the hearing room. They represented five Canadian political parties and many walks of life, but there was not a scientist among them (unless you count the chiropractor). The draft before them was titled *Proposals for Legislation Governing Assisted Human Reproduction,* an innocuous name for a bill that poured salt on the tail of revelationary biology.

This was Canada's second attempt to set some rules—any rules—for certain vital aspects of biological science, at least as it applies to human beings. There was nothing in Canadian law regarding even the most basic of reproductive transactions, such as the buying of human eggs. There were no rules dealing with surrogates, who carry and hand off children for a fee. In vitro fertilization clinics were run privately, not paid for by the government-funded health care system, and, like American clinics, were not regulated. These clinics were the places where human cloning experiments and germline genetic experiments were most likely to be conducted. There were also no laws forbidding the manufacture and study of human embryonic stem cells, the selection of embryos by sex or the deletion or manipulation of the genes of a human embryo. There were food and drug rules to regulate the transplantation of animal parts into humans (called *xenotransplantation,* another word stolen from the Greek, for *xeno,* meaning "foreign"), but nothing at all on the making of human-animal chimeras.

As revelationary biology gathered steam and all these things became possible, the Canadian government had simply failed to deal with them. The previous Conservative regime had appointed the Baird Royal Commission on Reproductive Technologies in 1988, but that was ten years after Robert Edwards and Patrick Steptoe had caused Louise Brown to be conceived in a petri dish, implanted in her mother's womb and successfully brought to term. By 1988, private in vitro fertilization clinics were doing a brisk business. The British government, by contrast, after extensive hearings and raucous parliamentary debate, created a regulatory body to oversee and license all such work—the Human Fertilization and Embryology Authority, which had recently been given the task of regulating the manufacture of human embryos for stem cell research. The U.S. National Institutes of Health had struck committees and designed various funding rules, but presidents George H. Bush, Bill Clinton and George W. Bush had taken their own soundings, issued moratoria, withdrawn them, reissued them, and left the field open to the mercies of the market.

The Baird Commission brought in its report in 1993, and the Liberal government introduced a bill in 1997, but the bill died when the government called an election. It took four more years to get around to this draft. In the interim, private researchers in the United States had successfully transferred human cytoplasm and mitochondria from one woman's healthy egg to another's deficient one and produced a human child with three genetic parents. A privately employed researcher in Worcester, Massachusetts, named Jose Cibelli, had scraped a cell from the inside of his cheek, pulled out its nucleus, transferred the nucleus to a cow's egg and tricked the egg into dividing as if it had been fertilized. He then stopped the experiment.

The committee's clerk took his place beside the chairwoman, Bonnie Brown, a Liberal representing the Toronto-area riding of Oakville. I'd spoken to the clerk several times trying to get a reading on who the committee was listening to and why. I was interested in whether certain stem cell biologists were coming forward, particularly a woman named Freda Miller. She had published with her colleagues, a month earlier, astonishing work on adult stem cells. Seemingly ordinary cells taken from human skin and stimulated by certain growth factors had turned into cells that looked remarkably like neurons, suggesting that one might not need to kill human embryos to get stem cells. Oh, the clerk had said, the committee had displayed a "lack of receptivity" to the evidence of the technical people. The committee members were "very wary of the vested interests."

I thought that this was an oddly indiscreet thing for a clerk to say, but then several things about this whole exercise were odd. The draft law had a preamble about how the government wanted to recognize the importance "of preserving and protecting human individuality and the integrity of the human genome."[23] What did integrity have to do with genomes? As James D. Watson liked to point out, genes change all the time through random mutation. The draft bill banned all kinds of things, chiefly the production of a human being by cloning. But it would permit under license the extraction of stem cells from human embryos for research. These prohibitions, and others regarding payment for human reproductive tissues, cells or genes and doing regulated research without a license, would be punishable as crimes.[24]

However, the strangest thing was what the draft allowed: it permitted researchers to make certain human-animal chimeras under license from a regulatory authority. The definition of a *chimera* was "a human embryo into which a cell of any non-human life form has been introduced, or a non-human embryo or fetus into which a cell of a human being or of a human embryo or fetus has been introduced."[25] In other words, a researcher with a license could put human cells into a cow's embryo, or rat cells into a human embryo. Why would a government that considered cloning an attack on human dignity allow that?

The witnesses did not touch on these matters. They were mainly lawyers who gave beautifully reasoned presentations on jurisdiction. Brown steered a careful course, asking respectful questions, but she responded energetically when a witness argued that the federal government had to take charge of IVF clinics, which were getting away with unregulated experimentation on women. The committee was aware, Brown reassured the witness, and intended to bring this whole IVF and surrogacy business to heel.

The session broke up after less than an hour. As the witnesses and members left, I caught up with Brown by the door. She appeared to be in her fifties, like me. There was something about her I immediately liked. Was it the carefully cut blond hair, the elegant figure poured into the chic black wool dress, the black leather boots with the season's correct toe? Or was it the faint scent of tobacco on her breath, a sure sign of human foibles? I've always been a sucker for human foibles.

I told her I was interested in some of the things allowed by the bill, some of the things with animals.

Are you interested in animal rights? she asked.

No, I said. I asked what work the committee had done on the making of chimeras.

She looked at me with a blank expression. She'd had this bill in front of her for months but didn't appear to know what a chimera is.

She would later admit to me that biology made her eyes glaze. She'd relied on the views of a few experts, such as Ron Worton, on the tricky science stuff. She didn't trust the IVF industry, but she thought she could trust him. He'd been up to testify twice. In fact, she thought he'd said that the destruction of human embryos to make stem cells might be forestalled by new work by some woman named Miller.

No, I said, there are some kinds of research that can only be done with embryos. A British researcher, Marilyn Monk, had just sent me a paper showing that certain genes active in the earliest stages of human embryo development are also expressed in certain cancers. You couldn't do work like that with adult stem cells.

Could I send it to her? Brown asked.

I said I could, but I thought it might be a waste of time. I had sweated over Monk's paper, and the others I was reading. The language of revelationary biology is evolving like its subject: it describes methods such as whirling things in centrifuges, running currents across gels, infecting cells with genes spliced into viruses, making chemical primers that complement DNA sequences, not to mention the unrelenting slaughter of myriad flies, mice, fish, frogs and worms. Just when I had a definition firmly in hand, some new paper would give it a twist and I'd be perplexed again.

And so I found my way to the offices of Ron Worton, Brown's reliable expert. He is the CEO and scientific director of the Health Research Institute of the Ottawa Hospital, but also a founder of the Stem Cell Network, one of the centres of excellence funded by the Canadian government. By the time I walked through his door, I knew that the Standing Committee on Health would be no source of wisdom on the biology I was grappling with. At first, I had hopes that I could lean on Ron Worton. If he was giving Bonnie Brown and her colleagues disinterested advice, why not me?

The Stem Cell Network is a publicly funded private nonprofit corporation involving the CIHR, thirteen universities and two companies. Worton had just started it with a $21 million grant to spend over four years. The network would make sure that those doing excellent work on stem cells got support and that this work would move to the market as quickly as possible. According to Worton, the Americans were envious of the whole system. In the United States, it was all cutthroat competition. The kind of collegiality represented by the Stem Cell Network was Canadian science at its best.

Worton's secretary showed me into his private office. There was an ordinary round table, some ordinary chairs. I've been in a hundred offices in Ottawa just like it. Worton sat there in front of me, in shirtsleeves, a pale man, balding, with blue eyes, glasses, a slight paunch, a hint of jowls. I've seen a hundred bureaucrats who look just like him. But his c.v. demonstrated that he is a human geneticist with a long string of honors. Among many other things, he and his colleagues cloned the gene that causes Duchenne and Becker muscular dystrophies in a series of brilliant experiments published in journals such as *Nature* between 1984 and 1998. This gene is so big that it takes sixteen hours to direct the assembly of its protein, dystrophin. He and his colleagues had demonstrated that it is dystrophin that is lacking in boys with Duchenne muscular dystrophy and is altered in Becker victims. Lack of dystrophin results in the degeneration of muscle, which is replaced with connective tissue, and eventually kills young victims, mostly male. A hum of energy was leaking out around Worton, a restlessness he could barely contain. I found myself watching his well-shod feet bounce as he talked.

He had started his series of experiments and papers with not much more than a single clue. He thought the muscular dystrophy gene had to be on the X chromosome. Women have two X chromosomes, and men have only one. Studies of families with the disease showed that while it rarely affects females, women are carriers. The big break came in 1979 when a pediatrician, Dr. Christine Verellen, "came to my lab to train on chromosome studies," he said. "She had a patient with Duchenne with a translocation of a chromosome." The patient was a woman, and when they studied her they discovered that the top of chromosome 21 had been exchanged with the top of the X chromosome. The break in the X chromosome occured in its short arm. The break on chromosome 21 was at a block of known genes. While she, unlike males, had another X chromosome, it was inactive. "It was possible that this woman's chromosome had broken right at the point where the gene responsible for Duchenne resides," Worton said.

Using a microscope, Worton and his colleagues could see the break in the active X chromosome. It was right near another gene whose function and sequence was already known. That allowed them to make a chemical probe to fish for the DNA that normally would cross this break point. They finally cloned the whole gene, made the protein it called up and proved it was the right protein. The protein part alone took two years.

Worton and two colleagues decided to enter the federal government's competition to fund centers of excellence in certain research fields. His colleagues organized the Canadian Genetic Diseases Network in 1988, when he was the head of the genetics department at Toronto's Hospital for Sick Children.[26] The network gave its twenty-two members access to about $4 million in research money. Membership was not like applying for a grant. "It was all self-selecting," Worton said with a wave of his hand. He and his colleagues in the network were all trying to clone human disease genes. It was possible to share methods, technologies and ideas because members did not compete: each was the only one working on a specific disease.

He rolled out his favorite example of the kind of sharing that makes these networks worthwhile. The University of Toronto geneticist Peter St. George-Hyslop was working on adult-onset Alzheimer's, tracking its inheritance in families and then trying to find its gene. He believed the gene was on chromosome 14, in a region thick with other genes, but he was struggling to find a method to precisely locate it. When Worton invited him to join the genetics network, he introduced St. George-Hyslop to Johanna Rommens at the Hospital for Sick Children, who had developed a technique to find a gene that was perfect for his problem. Six months later, he'd located the gene, cloned it and, at the urging of the genetics network, applied for a patent. "He beat the competition by six weeks—he got the patent." It was licensed to a major pharmaceutical company.

They had incorporated the network as a nonprofit and as a result could apply for patents. The network's share of the revenues from licensing patented work to major corporations could be used to help fund members' future research. Worton had patented some of his muscular dystrophy work, too.

"Did you get rich?" I asked.

He waved the hand again. He never earned a cent from any patents. The only reason to patent, he explained, is that no pharmaceutical company will spend what it takes to develop a drug without the protection of exclusivity. This mention of patents and drug companies made me wonder how he had enjoyed his tenure at the Hospital for Sick Children. It had been the scene of a spectacular fight between the generic drug company Apotex and thalassemia researcher Nancy Olivieri over academic freedom, and other intrigues less well known.

"I think I should write a book," Worton said. "Who got the credit, the infighting, who tried to take advantage of whom." There was an undercurrent of bitterness in this statement, peculiar to hear from one as accomplished as Worton. What had ever stood in his way, after all? Apparently, ignorance on the part of the gatekeepers, the people he had to drag money from. He'd tried to get NSERC, the Natural Sciences and Engineering Research Council of Canada, to fund a Canadian share of the Human Genome Project back in 1991, when the Americans and the British were already hard at work on it. "They looked at us as if we were nuts. Went to the Medical Research Council." The president asked Worton, What's the human genome? (A few months later, Henry Friesen became MRC president and took steps to set up a Canadian genome project, which Worton led.)

Worton had his head back, cradled by his arms. He wasn't looking at me but staring at these curdled bits of the past. He began to talk about what should have been his great moment of glory, when *Nature* published his most important paper about the muscular dystrophy gene.

"I collaborated with a group at the University of Pennsylvania who helped us with about 5 percent of the work. All the experiments published were from my lab," he said, "but we included them as coauthors. We did our press release, and in our coverage we abided by *Nature*'s rules—no press releases till the day before or of publication. They did a press conference the day before. The next day ours was old news." A headline appeared in a Toronto paper that said the gene for muscular dystrophy had been cloned by a team at the University of Pennsylvania with some help from an unnamed laboratory in Canada. The story didn't even name Toronto, let alone the Hospital for Sick Children or Worton and his colleagues. The report of the press conference Worton gave wound up as a page 12 story the next day.

"One of my technicians, who worked day and night and was very proud, she threw it across the lab and didn't come back to work for a week and a half," he said. His teeth were jammed tight together, as if by clenching his jaws he could hold in the rage that still boiled years later.

"About stem cells, the network," I gently reminded him.

"Stem cells," he said, straightening up. In the 1960s, as a PhD student at the University of Toronto in biophysics, he'd worked in the laboratory of the physicist, Jim Till, research partner of E. A. McCullough, who had been the first in the world to publish on the existence of adult stem cells. McCullough and Till had been studying how long mice could live after being irradiated. Radiation kills the bone marrow, which makes the cells that become blood. As part of these experiments, they injected irradiated mice with healthy bone marrow. When McCullough did autopsies on these mice, he noticed odd lumps of cells in their spleens. These cells turned out to be daughter cells of the injected marrow cells. They had somehow lodged in the spleen where they formed little colonies of blood cells. McCullough tried to determine what distinguished the precursors of these stem cells from other blood precursors in marrow. The process that set them on their path seemed to be random.

Long after Worton left Till's lab, he kept thinking about stem cells. But he tried first to use an adenovirus to carry the dystrophin gene into cells taken from boys with muscular dystrophy whose own genomes were deficient. The idea was to see if the patients' cells would eventually be able to produce the protein they lacked. It didn't work. (To date, no gene therapy has worked. In fact, some improperly conducted U.S. gene therapy experiments became a scandal. On September 17, 1999, eighteen-year-old Jesse Gelsinger died in a gene therapy experiment conducted at the University of Pennsylvania's Institute for Human Gene Therapy. Gelsinger suffered from ornithine transcarbamylase deficiency. In the investigations that followed, it was revealed that the experiment had not been conducted according to the rules set out for experiments on human beings. The required risk disclosure document had failed to mention that other participants had already suffered serious side effects and that monkeys had died in earlier versions of the experiment; the institute had failed to halt the trial as required or to report to the FDA after side effects appeared; the director of the lab had started his own company, which had a large financial interest in the altered virus used.[27] After the National Institutes of Health publicly reminded investigators of their obligation to report all adverse results in gene therapy trials to both the FDA, which keeps them secret, and to an NIH oversight committee, which makes them public, it received a wave of 691 adverse reports, including three deaths. Only 6 percent of these incidents had been reported at the time they occurred.[28])

In the fall of 1998, Worton read articles in *Nature* and *Science* reporting that stem cells had been found in all kinds of adult

mammalian tissues, even in brains. He began to think that a cure for
kids with muscular dystrophy might be to take such adult stem cells
from their bodies, grow those cells in culture, encourage them to
become muscle cells and transplant them back. The first problem was
to learn to distinguish stem cells from the other cells around them.

Meanwhile, early in 2000, the Canadian federal government said
it was interested in starting four more centers of excellence. Worton
canvassed his colleagues: how about a new network for stem cells?
He called Janet Rossant at the Samuel Lunenfeld Research
Institute. "She said, yeah, do it. When I asked her to lead it, she said
no." Worton's colleagues prevailed on him to do it, and the government accepted his application. It took four months to work out a
standard agreement that every partner would sign. The last step was
to get the board to ratify his science plan, which happened the week
after I interviewed him. "We want to build the experiments to control stem cell behavior totally," he said.

And he wanted this done without delay. Bonnie Brown would
have been surprised to learn that Worton had privately argued with his
colleagues, urging them not to wait for her committee to do its work,
not to wait for the government to pass legislation, not even to wait for
the CIHR to finish writing its funding guidelines, but to start studying
human embryonic stem cells right away, ahead of the Americans. He
had pointed out that there were no laws in Canada to stop them, no
laws to prevent the CIHR from funding them. But his colleagues were
"conservative Canadians," he told me. "A couple said, why go out on a
limb—wait for the guidelines, do it with impunity."

As soon as the CIHR's research guidelines were out, his Stem
Cell Network would put up $1 million for this work. Worton wanted
$15 million more, "to be competitive." His members were divided
into groups working on ethics, on plasticity and change, on stem cell
bioengineering. He also had a group on stem cell therapeutics. This
one was subdivided into the diseases the group wished to address:
factor 8 deficiency (hemophilia A), muscular dystrophy, diabetes,
Parkinson's disease.

"These are not small dreams," Worton said.

It was dark outside and I had to pull myself away if I was going
to make my train. But I had one more question: how much time had

the Standing Committee on Health spent asking questions about embryonic stem cells?

About a day and a half, he replied.

I got home late on Friday night and went over my notes that Sunday. There was no mistake: Ron Worton, a distinguished leader among Canadian scientists, the man the standing committee's chairwoman relied on, said he had urged his colleagues to move ahead of Parliament, to seize the day. (Later Worton would tell me that he and his colleagues decided to cooperate with the government game plan lest they rile parliamentarians and shoot themselves in the foot.)

I looked up the Stem Cell Network's website. Some of the researchers who were writing stem cell funding guidelines for the CIHR were also stem cell network investigators—who would get research money to do this work. The CIHR president, Alan Bernstein, a stem cell researcher, had appointed Janet Rossant as chair of the CIHR stem cell guidelines committee. Worton had made Rossant a senior investigator on his stem cell network. Francoise Baylis, a bioethicist at Dalhousie University, was also on the CIHR guidelines committee. She had helped Worton write the network's application for funding and was a stem cell network investigator, too. There wouldn't be much scope for a government-supported stem cell network if Canada's elected representatives decided that human embryonic stem cells are immoral subjects for study. No wonder Bernstein, Rossant, Worton and Baylis had all appeared before the Standing Committee on Health.

I turned on the television later that Sunday night. CNN had breaking news. A U.S. corporation, Advanced Cell Technology, led by a scientist named Michael West, had announced the first successful human cloning experiment. CNN said it had been written up in Scientific American[29] and that U.S. News & World Report had announced its own exclusive story.[30] A peer-reviewed article was available on the website of a journal I'd never heard of before, e-biomed: The Journal of Regenerative Medicine.[31] The Americans had seized the day first.

The names West and Advanced Cell Technology went round and round in my head, then clicked. Advanced Cell Technology employed Jose Cibelli, the man who'd made that human-cow chimera. And Michael West was the founder of the company I'd clipped so many stories about, Geron Corporation.

CHAPTER TWO

CLONES, ANYONE?

"The First Human Clone," screamed the cover of *U.S. News & World Report*. "The First Human Clone," echoed *Scientific American*. Everyone should have seen it coming, and some people did. That's why *U.S. News & World*'s reporter took a hotel room in the late summer in the gritty little city of Worcester, Massachusetts, across the way from the offices of Advanced Cell Technology. Jose Cibelli, research vice president of Advanced Cell Technology, had almost let the cat out of the bag when he had testified before Canada's Standing Committee on Health a few weeks earlier. I'd seen his name on the witness list and had even glanced over his testimony but didn't read it carefully until weeks later, by which point the story was on front pages around the world.

"As of today . . ." Cibelli told the committee, "we have replicates of living individuals in cattle, sheep, goats, pigs, and mice. On the data gathered with animals, it is reasonable to speculate that attempts to clone human beings at this stage are likely to succeed. However, the price we would have to pay for this is going to be very high. So some of us in the scientific community are against reproductive cloning at this time."

Cibelli explained that he and his colleagues had little interest in making a complete copy of a human being, like Dolly the sheep. He defined this as "reproductive cloning." He wanted to take the nucleus from the skin cell of a prospective patient, put it into an immature, unfertilized egg (called an *oocyte*) from which the nucleus had been removed, and get that egg to divide like an embryo. Then, instead of putting this dividing embryo into a woman's uterus so it could produce a human clone, he would make a stem cell line. Each of the cells in this line would carry a nucleus identical to the patient's. Any tissues or organs grown from any of these cells would be a perfect genetic match for that patient. He called this "therapeutic cloning."

Therapeutic cloning was, in his view, the answer to our deepest needs. It would give us new organs for transplant matched perfectly to individuals, new spinal cords, new brain tissues without the ravages of Alzheimer's or Parkinson's disease. But he saved the best pitch for last. "And there is one more advantage," he testified. "This actually was proved in our laboratory. Our data have indicated that cells generated by cloning will be completely rejuvenated, allowing us to envision not only treatment, but prevention of all age-related diseases . . . We have to remember that if we take cells from a person who is 70 years old and we do this procedure, when we bring back cells—let's imagine we are bringing bone marrow cells back—those cells will not only be 100 percent compatible with the patient, but will be completely rejuvenated. We can be thinking about humans when they reach a certain age who will have an immune system completely reset at zero."

In other words, he and his colleagues at Advanced Cell Technology had a method to extend our lives indefinitely, piece by piece—a biotech version of the fountain of youth.[1]

On the last Tuesday in November 2001, I gazed upon a serene shot of Michael West, president of Advanced Cell Technology, on the front page of my Globe and Mail. He was wearing a white, long-sleeved shirt and a deep red tie, and he was sitting in front of drawn red drapes.[2] He looked for all the world like a newly elected prime minister with a full slate of the people's business before him. The title on the story said, "The man who aims to cheat death." It included an interview with a Texan named Miller Quarles, who claimed to have given West an early shot of investment way back when. West had promised to find a method to help Quarles, already eighty-two and with three wives in his past, to live a great deal longer. (As Quarles would later tell me, he hoped to live to at least two hundred, long enough to enjoy several more spouses.)

In just two days, the coverage of the story of the first human clone had morphed from the trumpeted advance of revelationary biology to tongue-in-cheek accounts of these wacky bits. West had hinted that the real purpose of human cloning is life extension, and reporters seemed to think that was funny. Perhaps this was because the actual advance was small.[3] West and Cibelli and their colleagues had managed to take the nuclei out of skin cells from various donors (a former schoolmate of West's now in a wheelchair, among several others) and to transfer them into eggs, shorn of their nuclei, bought from young women in the Boston area. But most of these eggs failed to divide. Advanced Cell's researchers had finally taken a cumulous cell, clinging to the outside of one denucleated egg, and put it inside the egg. This was the "embryo" that divided without the application of sperm, but it stopped at the six-cell stage. If it had kept on replicating, the egg donor would have cloned herself.

Some researchers said flat out that no learned journal should have published this article.[4] There were suspicions that West and Cibelli, both members of the editorial board of e-biomed, had somehow been able to influence the decision to publish. John Gearhart, who had created the first human germ cell line in 1998 and who was also a member of the journal's editorial board, demanded to know the names of the peer reviewers. When he could not get an answer to his question, he said he was very embarrassed, very "chagrined by this publication."[5] A well-known bioethicist at the University of Pennsylvania, Glenn McGee, once a member of Advanced Cell Technology's board of ethical advisers, referred to the whole affair as "science by press release."[6]

This had a ring of truth to it. During the months leading up to Advanced Cell's announcement, the notion of human cloning seemed to have slid out of the lab and into the realm of the National Enquirer. This was the fault both of the proponents, who seemed less interested in the science than in its possibilities, and of the critics, who seemed less interested in the science than in its morals. Canadian bioethicist Timothy Caulfield, a member of the Stem Cell Network and the CIHR's guidelines committee and a proponent of avoiding criminal legislation in this area, had told

Brown's standing committee that cloning should not be banned out of the fear of creating perfect copies of individuals.[7] Such a bill would perpetrate a kind of scientific fraud, he said, a genetic determinism no scientist could possibly believe in. Multiple Hitlers could never arise from a cloned Hitlerian cell. The Hitlerian nucleus would be put in an egg from a mother other than Hitler's. The embryo would be carried in a different uterus. The resulting child would be born into completely different circumstances. All these differences would matter. But most cloning proponents ignored these facts. Most proponents seemed to be selling cloning as a way to perpetuate identity (or any body part you'd care to hold on to).

Various self-professed human cloners had given testimony before a U.S. congressional committee the previous March[8] and before a National Academy of Sciences panel on cloning held in Washington in August. Richard Seed, a physicist, said he intended to clone himself. Severino Antinori, an Italian gynecologist, had teamed up with Panayiotis Zavos, an American who specializes in infertility in males. Antinori had already enabled a woman over sixty years old to conceive, carry and give birth to a child. Zavos ran an infertility institute in Kentucky, the Institute of Andrology. Antinori and Zavos had previously announced at a press conference in Rome that they intended to use human cloning to cure absolute infertility.[9] Zavos explained to the panel that they had access to unlimited amounts of money from unnamed sources in the Middle East to clone hundreds of couples already waiting in line.[10]

I had called Antinori in Rome. The dragon woman who answered his office phone bellowed down the line at the top of her lungs that the good doctor was far, far too busy to see the likes of me. The good doctor was indeed busy: the Italian medical association was trying to revoke his medical license because human cloning is illegal in Italy.[11]

Zavos said he'd be happy to see me, but I had to understand that he had signed an agreement with the BBC, for a fee, that required him to give them exclusive access when his human cloning experiments resulted in success. He dropped broad hints that his experiments might already be taking place somewhere in the

Mediterranean (perhaps, I thought, on the island of Cyprus, where the Andrology Institute happens to have a spa-like facility). He could grant me an interview, regardless of his deal with the BBC, so long as our conversation was not exclusively about cloning. Of course, there was the matter of payment—he could not possibly spare the time simply to talk to a journalist. "What's in it for me?" he asked.

And then there were the Raelians. They too had come before the panel and the U.S. congressional subcommittee after telling the world they intended to clone people. The Raelians are a science-fiction-as-religion group based in the Eastern Townships of Quebec, not far from Montreal. They have a theme park called UFOland. A woman named Brigitte Boisselier had explained to the congressional subcommittee that she is in charge of the Raelian cloning company, Clonaid.[12] She has two doctorates, one from a university in France, the other from the University of Houston, in chemistry.[13] She said Clonaid had received expressions of interest from a hundred people who wished to offer their tissues and wombs, and to pay for the privilege. The Raelians had set up their company in the Bahamas since it was not clear whether the United States would pass a law forbidding all forms of human cloning. After a raid on Clonaid's U.S. lab, Boisselier signed an agreement with the Food and Drug Administration promising not to attempt human cloning in the United States and not to do research on human eggs there either until there is legislation permitting it or a court decision approving it. The Bahamians also told Clonaid to take a hike, and the company then got involved with a company in South Korea.[14]

If distinguished panels were willing to pay attention to Raelians, perhaps it was because the *New York Times Magazine* had carried a story about their plans.[15] Rael, the leader, had formerly been a French racing-car driver named Claude Vorilhon. In 1973, he claimed, he had an encounter on a back road in France with beings from a planet on the far side of the galaxy. They had explained to him that the whole of humanity resulted from a breeding experiment of theirs about 25,000 years ago. These advanced beings were big on cloning. Moses was one of their clones. Jesus was too. Vorilhon claimed he was taken on a voyage to their home planet, where he learned all about their abiding love for all of us and where he himself was cloned. Upon his return, Vorilhon changed his name to Rael and created his church, in which the prospect of human cloning and the worship of science and reason play central roles. The *New York Times Magazine* story did not go into all these details of the faith, but it did include a photo of Rael, a man in his early fifties with thinning brown hair pulled up high on the top of his head in a pompom. He was wearing a white jumpsuit open to the waist, with wide Flash Gordon shoulders.

Maybe people paid attention to the Raelians because the Nobel laureate Francis H. Crick once published a Rael-like theory on the origins of life on earth. In a book called *Life Itself: Its Origin and Nature,* published in 1981, Crick argued that life on this planet might not have evolved from the accidental union of a few organic molecules that somehow became self-replicating and eventually formed cells.[16] According to Crick, small life-forms could have arrived in an unmanned spaceship "sent to earth by a higher civilization which had developed elsewhere some billions of years ago." Crick called this theory "directed Panspermia."

A gang of committed followers of Rael trooped into an airless room in Toronto's Ontario Institute for Studies in Education, where we strangers sat, bored and restless, waiting for their Sunday service to get going. There was something about them. They were all under forty, perhaps even under thirty. Their skin and hair glowed with the sheen of fleeting youth in perfect health. They rubbed each other's backs and shoulders and hips as they embraced and kissed. There was a certain over-the-top intimacy among them, as if they had all, at some point, been each others' lovers, learned each others' body secrets, and this was a reunion.

This room was normally a classroom, but for this morning's service a screen had been pulled down over the whiteboard so we could watch a video. Around this screen were small plinths topped with displays of fresh flowers and fruits and scented candles. There was a little square of wooden flooring set out for an awkward and bony girl of eleven, dressed in a tutu that was far too long, who was going to dance for us. In the corner opposite the room's only door there was a pedestal draped with a square of silk. It supported a stiff poster that displayed the Raelian logo, a melted, silvery star of David. The same star appeared on the empty donation envelopes that had been set out on each chair. Right beside the door there was a display of books by Rael—for sale, of course. The Raelian's PR woman had mentioned that a new book by Rael, on cloning, would be out soon.

Those of us who'd wandered in here out of curiosity, or perhaps need, were fundamentally different from these gleaming men and women who were already Raelians. They were young and beautiful; we were old and lumpy. Take, for example, the man beside me, a grizzled, gap-toothed fellow with shoulder-length gray hair, a heavy, wrinkled shirt and a deeply aggressive attitude about paleontology and Darwin. He had the ruined complexion of one who spends most of his days on the streets holding out a cup in the hope of coins; he rambled into my ear about how Darwin was dead wrong, even as the lights dimmed and a beautiful young man stepped up to the podium to tell us how we should shut our eyes and arrange our limbs so as to properly meditate. As the young man dropped his voice down into its deepest, most sensuous register, I smelled a latecomer even before she sat down about a foot to my left. She carried with her the cloying aroma of unwashed feet and the metallic, acrid stink of extreme anxiety. In the dark, she mumbled and then shouted out questions that had no discernible point. The young man, unfazed, exhorted us to touch our own arms and legs and to feel the pleasure.

The lights came up and the video began. It told us Rael's story and then presented an ad for Rael's summer conferences. It became crystal clear that Rael was trying to extend his market reach from Quebec to other parts of Canada and that these young people were his emissaries. Their chief job appeared to be to sell his books and a one- or two-week stay at the Raelian headquarters in the Eastern Townships. There, in what looked like a summer-camp setting, one could learn all about oneself, exploring with others in the most intimate ways one's deepest nature. And would one please leave a contribution in the envelope?

Before I could escape, I had to run the gauntlet of the now anxiously smiling young men and women who had lined up to take back the envelopes and sell books at the door. I got the distinct impression they were expected to make a sales quota, but there weren't enough of us here to get them there.

I had no intention of giving them money, but I wanted to know what they thought of human cloning. I wasn't sure I would want the version Joe Cibelli and Michael West wanted to sell, the chance to extend my life indefinitely with cloned cell transplants—immortality piece by piece. Could this form of rejuvenation give one back one's youth? I had looked hard at myself in the bathroom mirror that morning and had forced myself to face my own ruin. There were the deep and permanent creases between my eyes, the bags below, the expanding network of crow's-feet; there were the sagging lines dragging down the corners of my mouth, the little dewlaps at the corners of my jaw and, of course, the threads of gray at the left temple spreading like a web over hair that was once auburn and shining and was now a faded, frizzled chestnut. This was the face I'd earned, not the one I remembered from my early twenties, not the one I trick myself into thinking I still present to the world. At that moment, the idea of fresh new skin had seemed terrific. But looking at these youths around me, I wasn't so sure I wanted to be one of them again. And surely a time would come when I'd be sick to death of my own voice inside my head, of all the habits of mind and body acquired over too many decades? Wouldn't I want to be done with it, to get it over, to recycle my stories in the ground?

A short-haired, stocky Raelian woman thrust a book at me. During the service, she had testified that the great thing about a week in the Eastern Townships with Rael was that it had helped her come to terms with her sexuality, the fact that she is a lesbian. "Do you want to be cloned?" I asked her.

Can't wait, she said.

But why would you want a copy of yourself? I asked. Your new body will take just as long as it took you to become an adult, and it will have its own experiences. It won't be you.

No, she said, that's not how it would work. Rael had explained how it would work.[7] One day, in the not too distant future, not only would she get a second, fresh body, exactly as hers had once been, quick-grown from embryo to adulthood, but she would be able to open up a porthole in her brain and pour into that nice, fresh body all of the experiences and knowledge acquired in her life to date.[18]

"But that's a nightmare," I blurted. To have all the skepticism and fear and caution learned in a long life thrust into a young body? To have an older woman's need for mental serenity denied by the pounding urges of young flesh? How could any body grow to adulthood, no matter how fast, without its own physical memories and personality? And what about the old, discarded self? Would it be left to rot in a geriatric ward, memoryless, or would it be trashed like a banana skin? Or would there be two of her, the old and the new, where once there had been only one?

"To be nineteen again," she said, "and know the things I know now? It would be fantastic."

CHAPTER THREE

HOW MANY WAYS CAN YOU MAKE A MOUSE?

Was the world pregnant with a whole new way of making human beings? Was immortality just one step away? The daily press was covering every twitch of the leading cloning figures, but it can rarely distinguish a truly important scientific advance from trivia when one first appears. Watson and Crick's description of the shape of DNA was a perfect example. When they convinced themselves that their model of the molecule was accurate, Crick told everyone in their favorite pub that they'd found the secret of life.[1] The secret of life made it into *The New York Times* in June 1953, under the title "Clue to chemistry of heredity found," a one-column story with no byline—indicating the editors hadn't much faith in its significance. Crick wrote it up himself for *Scientific American* more than a year later, as "The Structure of Hereditary Material." The magazine editors put his piece next to an article about race that used the seemingly objective voice of science to purvey drivel. On the one hand, the neighbor article discounted the notion of racial purity; on the other, it actually used the word "miscegenation" to describe America's human mixtures.[2]

Cloning had started a long time ago, without newspaper headlines. The early pioneers were the makers of mouse chimeras. They would know the history and what matters here, I thought. They would know what's real and what's not.

Andras Nagy had forgotten our appointment. I found him in a little office off his main lab in Mount Sinai Hospital. He sent me out to wait while he finished giving instructions to the woman in charge of his embryonic stem cell lines.

I leaned up against a lab table, usually called a "bench." The room was filled with row upon row of them. Shelves of experimental protocol books and glassware reached for the ceiling. Each of Nagy's fifteen PhD students and postdoctoral fellows was assigned about one square foot of personal space. Two of them, a man and a woman, were talking at the next row about a problem. He was suggesting that she should use transfection as the method to insert a gene cloned from a fish into the stem cell of a mouse. The infecting agent could be an engineered virus.

I had been stuffing myself for weeks with the history of genetics and developmental biology. I was at that tricky stage where I thought I understood basic concepts—at least until I tried to connect one to another and to imagine things in three dimensions. Transfection was one of those ideas that kept slipping away from me. It involved the use of a virus to carry genetic material from another source into a cell—transport by infection.

I knew that viruses exist in a state midway between life and everything else. Life is usually defined as something that has the capacity to make copies of itself. Viruses cannot copy themselves without help, so they enter life like thieves by forcing open the protein barriers on the outside of a cell. When the cell divides and the cellular DNA copies itself, the virus's genetic information is copied too.

Viruses are usually single strands of ribonucleic acid, or RNA, wrapped in protein (although some viruses also contain DNA.) RNA is a giant macromolecule like DNA, formed of an array of sugars and phosphates and base pairs (except it substitutes thymine for uracil, about which more later), but RNA is mainly a message-carrying substance (although it does other tricks too, about which, again, more later). In cells, RNA assembles itself on DNA and, by its shape and nucleic acid sequence, carries a complementary version of the DNA's information physically to organs called *ribosomes*. The ribosome reads the RNA like a recipe and assembles amino acids (brought to it by small transfer RNAs) in the right order to make the specified protein. In a cell, RNA is primarily an intermediary. It carries messages, but it can also *be* the message. For a long time biologists have thought the first RNA molecule assembled itself before the first DNA molecule came into existence.

But imagining all this to-ing and fro-ing, assembling and disassembling, changing and shuffling of these long, stringlike molecules made my head ache. And it still left me wondering: how could single-stranded RNA wrapped in a protein carry a piece of double-stranded DNA into a cell?

The technology of transfection arose out of the study of viruses. Biologists had studied simple viruses that prey on bacteria, called *bacteriophage*, since the 1940s. The infected bacteria are grown in large batches, allowing geneticists to study viral traits of interest. Much of this phage genetics was invented in Cambridge and Paris and Cold Spring Harbor Laboratory, where it was also taught during the summers, when the then tiny tribe of the world's geneticists gathered together to party, swim, sail and learn each other's techniques. If you want to know who drank what and who slept with whom (or tried to), read James D. Watson's memoirs, *The Double Helix* and *Genes, Girls and Gamow*.[3]

Until the Second World War, geneticists had worked mainly on larger animals or plants, such as the fruit fly (*Drosophila melanogaster*), corn and various yeasts. Early in the twentieth century, an American named Thomas Hunt Morgan showed that if one bred many generations of fruit flies, one could eventually spot physical mutations and even make maps correlating these mutations to changes on chromosomes. In 1926, Hermann J. Muller discovered that one can make mutations happen more quickly by irradiating test subjects than simply by breeding them, and that the number of mutations increases with the radiation dose. This new capacity to make mutations happen at the speed of science instead of the speed of nature helped expand the study of genetics. But it wasn't until 1944 that Canadian bacteriologist Oswald T. Avery, at the Rockefeller Institute in New York, showed which of the molecules that make up chromosomes actually change with mutation. Until then, most biologists had refused to consider DNA or its cousin, RNA, as the likely carriers of genetic information. The orthodox view was that since living things are made up of an almost infinite variety of proteins, then surely proteins must somehow carry the information that guides their own creation. Avery proved that it is DNA that carries information from one generation to the next, and his work was confirmed by two Cold Spring Harbor researchers, but not until several years later. Those geneticists who believed Avery at the time (not many did) knew they had to learn about the molecular nature of DNA if they were going to figure out how genetic information is transmitted and how that information guides the development and behavior of living things.[4]

Enter James D. Watson. In 1951, Watson, a former doctoral student of the Italian phage geneticist Salvador Luria, was sent to Europe on a postdoctoral grant to learn about DNA. Watson has famously documented in *The Double Helix* how he managed to get his grant switched from the Copenhagen lab of a biochemist who was having marital problems to the Cavendish Laboratory at Cambridge. There he met Francis Crick, a former physicist who had switched fields and was working on his doctorate on the crystallography of proteins. By the time Watson arrived at the Cavendish, Crick had also become fascinated by DNA. Watson was determined to work with Crick to solve the secret of DNA's structure.[5]

It was known by then which smaller molecules make up the macromolecule DNA, but not how they relate to each other. In Britain after the Second World War, research problems were taken up by particular groups: the question of the shape of DNA had already become the virtual property of X-ray crystallographers Maurice Wilkins and his new colleague, Rosalind Franklin. Their lab was at King's College in London. Wilkins and Franklin were taking X-ray pictures of extremely pure crystallized DNA. After making minute measurements of the ghostly images, they hoped to infer how this huge molecule is put together. Watson and Crick chased after both Wilkins and Franklin, trying to get information, trying to help, to the point of being warned by their superiors at the Cavendish to stop bungling around in others' intellectual territory. They did desist for a while, until Peter Pauling arrived at the Cavendish to do his doctorate. Peter was the son of the great Caltech biochemist Linus Pauling, already famous for describing the helical shape of certain protein molecules. Peter mentioned to Watson and Crick that his father was also working on the DNA problem. As Watson described it in *The Double Helix*, this was a spur to action. He and Crick were determined to solve the puzzle before Pauling, or at least Watson was—he imagined himself writing the paper explaining his discovery in suitably dry and understated English prose, and, of course, winning the Nobel.

Their real problem was that they couldn't persuade Rosalind Franklin, who was making spectacularly good X-rays of crystallized DNA, to share her work. They all believed that the shape of the molecule could be deduced from these X-rays, and Watson and Crick thought, from what they had seen already, that the shape was likely a helix. But what kind of helix? Only if they got precise measurements from Franklin's X-rays could they know the radius of its turning, whether there was one phosphate and sugar backbone or two, and how the nucleotides, or base pairs, were attached. Franklin, according to Watson's acid account, wasn't convinced that her images suggested a helix shape at all, and she didn't get on with any of them, least of all Maurice Wilkins, who seemed to think he was her boss. She eventually showed her images and measurements to her colleagues—not including Crick and Watson. Wilkins, who was by then convinced that Crick and Watson were right—that DNA formed a helix and that an experimental model should be built—passed her information on.

Watson and Crick eventually deduced that the DNA molecule is shaped as a double helix, not a triple one as Pauling first mistakenly published. They thought two ribbons composed of phosphates and sugar were somehow tied to each other by the four kinds of nucleic molecules—cytosine, guanine, thymine and adenine (abbreviated as A, T, G and C). But they couldn't figure out exactly how these smaller molecules fit together until Watson read a paper by the Austrian chemist Erwin Chargaff, then at Columbia. Chargaff's students had painstakingly weighed all kinds of DNA taken from all kinds of different animals and had found that no matter what, although the amounts of these molecules differed from animal to animal, there was always parity between the quantities of cytosine and guanine and also between the quantities of adenine and thymine. This suggested to Watson the ratios of Mendelian genetics. But when he and Crick tried to build a model, it turned out that only if the right form of these molecules were paired—adenine to thymine and cytosine to guanine—could they fit on the inside of a double helix in a turning radius that matched the X-ray images. With the right form of these bases in the right complementary pairings, everything fit neatly. The physical model they made, screwed together with bits of wire and balls machined to correctly signify the size, charges and positions of the atoms involved, was "too beautiful," in Watson's phrase, not to be correct.

Still, it took the imagination of a cryptographer to grasp that the sequence of these four nucleotides along the DNA molecule (or its RNA cousin) could actually be a code for the assembly of all of nature's proteins. As Watson and Crick realized, these four bases could pair in sixty-four different combinations. These combinations could then organize the assembly of the twenty amino acids in almost infinite variety. Because C only pairs with G and T only with A, if one knew the sequence of one strand of the helix, one would also know the complementary sequence of bases along its partner strand. This suggested that during cell division these helices came apart and each strand acted as a template for the assembly of a partner helix, pulling wandering nucleic molecules together in complementary order out of the cellular soup. Where once there had been one double helix, there would then be two, the daughter exactly like the mother. Watson and Crick sent a letter describing their discovery to *Nature*, where it was published in April 1953. The article was written in a style so coy that Crick felt impelled to publish a second article that said explicitly that this model of the molecule showed how DNA carries the information of heredity.

Most biologists teased out nature's secrets by doing experiments. But Crick and Watson had not done any; instead, they'd pulled together the work of others and made a model. This model suggested theories about the kind of code DNA used to organize information. Crick called one of his early theories the Central Dogma (claiming later not to have actually understood the meaning of the word "dogma"). He argued that information runs in only one direction in living things—from the DNA sequence to a messenger molecule to protein assembly. Crick thought that life directed itself along a one-way vector. He did not consider it possible that information might travel in the opposite direction. It was only a matter of time before the Central Dogma was shown to be wrong.

Cambridge's Cavendish Laboratory became the flagpole from which a new molecular image of the structure of life unfurled. Everyone who was interested in how DNA's shape embodies traits that are passed on was soon drawn there, as either a student, a graduate student or a postdoctoral fellow. Eventually, they moved on to other institutions, carrying with them these new ideas, which came to be known as "molecular biology." Watson returned to the United States, to Caltech and then to Harvard, and finally took up permanent residence as president of Cold Spring Harbor Laboratory. Crick stayed at the Cavendish, and then moved to the Medical Research Council's molecular biology lab on the outskirts of Cambridge. He too eventually went to the United States, settling in at San Diego's Salk Institute for Biological Studies. Rosalind Franklin died at thirty-seven of ovarian cancer.[6]

It seemed to me that the spread of this revelation to centers of learning around the world mimicked Darwin's theory of how a variation starts in some particular place and, eventually, if successful, conquers everywhere.

Andras Nagy waved me into a conference room. He was a slim man, wearing a checked shirt, with a graying ponytail hanging between his very wide shoulders. Little metal Santa Claus glasses hung on his sharp nose. He had a rich, *mittel* European laugh. Nagy had been born in Hungary in 1951 and had studied in Budapest to be a mathematician, but he had decided that he wanted something else in science. "I was good," he said, "but I couldn't sit all day with a pencil."

His girlfriend, later his wife, was then in medicine. "The shocking milestone was one single experience," he said. He went one day to her human anatomy class, where the students were cutting into fresh human brains. He was transfixed by the amazing patterns formed by the tissues as the scalpels sliced through them. "It was the combination of the horror and the beauty. You cut into the brain with a flat, wide knife and these beautiful patterns were there inside." He added classes in biology, biochemistry, physics and anatomy to his course load.

In 1974 he began to study the genetics of fish. Hungary, an agri-cultural nation, wanted to produce fish with the right characteristics, and to do so reliably. Nagy wondered if he could breed them without the randomizing effects of sex. His doctorate, as he described it, was on "making fish without a father." In other words, he tried to make fish clones.

How did he do it? "It was simple," he said. "We irradiated the sperm." A big dose of gamma rays made the sperm's DNA ineffec-tive, but, even so, it still started the eggs dividing. The goldfish and carp produced by this method turned out to be viable. However, since the fish carried only their mothers' DNA, "we got only females," he said.

To make changes to these clones, to get fish with the right mix of characteristics, he still needed to breed them. But two females can't breed. "So we had to reverse the sex of these females," he said. "We just fed them with testosterone after hatching. They developed testes. We had to determine the concentration and the timing of the treatment, but the sex reversal was almost 100 percent."

How bizarre, I thought, to be able to change something so basic as how fish reproduce, to change their gender by feeding them a hor-mone. I asked him how he felt about this kind of work. Did this meddling with the natural order bother him?

"I didn't see it as harmful or against any religion or laws. For me, biology is something to understand, which you do by changing things and seeing the consequences . . . It's curiosity about how things work—that's the leading thing," he said.

In 1978 he succeeded with changing the sex of his cloned fish and breeding them. But by then he wanted to do molecular biology. He wanted to understand the chemistry of the genes instructing behavior and development. He couldn't do that work in Hungary, so he got a postdoctoral position in New York at a biochemistry institute. He did two years of work on the biochemistry of the nervous system, and had a great time, but he decided it wasn't for him. When it got right down to it, he didn't think he'd ever learn to understand brain function by chopping up hunks of brain tissue, dropping them in a centrifuge homogenizer and then looking at the properties of the chemical residuals he spun out. He went back to Hungary in 1983 to a behavioral genetics lab, where he thought maybe he could take what he'd done in fish and apply it to mammals. He wanted to know if it was possible to get a mouse egg to become an embryo without any contribution from the DNA of a father.

He did get the mouse eggs to start dividing when he put them in a 7 percent ethanol solution. (He used to make jokes to the women in the lab that they'd better be careful about what they drank.) The embryos developed hearts and brains, but they died.

"Mammals are different from fish," he said wryly. "The basic difference is they can't do without a father or a mother—you need both genomes in a cell. The reason is you get a different message from the mother and the father. Some are expressed only from the male or the female source. The lack of that message kills the embryo. If there is no mother, the embryo develops very bad—there's no embryo," he said. If there is no father? "You get a good embryo, but a poor placenta."[7]

It turned out that about one hundred genes involved in the development of mice are imprinted, which means they set to work only when inherited from the parent of the right gender. Two groups discovered this imprinting process in mammals at about the same time, just after Nagy applied for a grant to make fatherless mice. "Both groups showed this phenomenon exists in the mammalian system. We were lucky," he said with a wicked grin. "We promised fatherless clones in 1983. Imprinting came up in 1984, but we already had the money."

And so he decided to use the money to study mouse embryology. He and his colleague used tourist visas to go around to the top mouse labs in Europe. He went first to see Anne McLaren in Cambridge. McLaren had shown, many years earlier, how to take an embryo from one mouse and implant it into the uterus of another, inventing techniques that made the study of mouse embryology possible. She'd been studying development with chimeric embryos for years. McLaren told Nagy never to waste a single embryo because something could be learned from every one. It was obvious that Nagy considered McLaren among the most significant leaders of this field. He had a digital recording of her stored in his computer and asked me whether I would like to see her. He ran back to his office and brought me his laptop, opened it up and showed McLaren to me on the screen, as if she were some sort of precious stone he kept in a shiny box. He had recorded the moment when she received a prize for lifetime achievement from the Society for Developmental Biology, based in the United States. I watched a tiny, dark-haired woman with a plummy accent give a gracious and funny acceptance speech.

After his travels, he came back to Hungary with an idea. While he couldn't make a fatherless mouse, maybe he could induce a mouse egg to develop partway to a live mouse. And what if he then joined a normal embryo to this dividing fatherless egg? He wondered if some cells from the fatherless egg would integrate into the developing two-parent embryo and, if they did, what tissues they could make. He wanted to make a new kind of chimera.

He pulled up on his computer the image of an ancient Mesopotamian chimera cast in bronze—a ferocious lion-like beast with a long tail. "The mythical idea goes back to 600 BC," he said. (Actually, it seems to go back a great deal further than that: the early Egyptian gods are all chimeras of one form or another.)

So who made the first chimeras? I asked.

"Doing this in mice goes back to 1961, and a man named Andrzej Kristoph Tarkowski. He's still around, a good friend of Anne's," Nagy replied. Tarkowski, then at Cambridge, had taken a pigmented mouse's embryo and an albino mouse's embryo, burned off their zonae pellucidae, and fused them together. The embryos attached to each other and grew as a single embryo, producing a live mouse with both albino and pigmented skin. "A single embryonic structure with two types of cells," said Nagy.

This whole idea of sticking embryos together, of adding two to get one, seemed to me so irrational that I couldn't imagine why anyone had even thought of trying it. What kind of mind would dream up such an experiment?

"It was just for curiosity," Nagy answered. "But it was a great tool. We use it like crazy. You get a chessboard animal. The genetic material is not mixed . . . So we can follow these cells, where they go [in the developing mouse] with genetic reporters . . . So we did this with the fatherless and the normal mouse embryos to see which organs need a father. Do you know what we need the male for?" he asked. "Guess." He was laughing again at the ridiculous comedy of life, at the "of course" of these things. His eyes danced, waiting for me to catch up. I was too slow.

"You need the father for muscle," he sang. He looked at me as if he was disappointed, as if to say, how could you, a woman, miss something so obvious? The fatherless embryo's cells made the heart or the eggs or the brain cells of the chimeras, but the muscles were made by the two-parent part of the embryo, the part that had male parental genes. "You don't need it for important parts, just muscle," Nagy said.

He published reports of his experiments in important journals. But he wanted to learn more molecular genetic techniques, so in 1988 he wrote to Janet Rossant in Toronto and sent her his third paper in this area. He wanted to learn from Rossant how to genetically alter mouse embryonic stem cells. She wrote him right back and said he should come to her lab as a visiting scientist for three months. She paid for his flight and his support.

In the 1970s, in California, Paul Berg, Herbert Boyer, Stanley Cohen and their colleagues had figured out how to snip out particular DNA sequences from viruses, bacteria and even frogs, insert them into circular bacterial DNA called *plasmids* and then grow these altered plasmids in bacterial colonies. This recombinant engineering, or "gene splicing," as they called it, could soon be done in animals and plants. In 1981, two teams reported at the same time that they had established mouse embryonic stem cell lines. If kept in the proper conditions, these embryonic stem cells reproduced indefinitely without differentiating. In the late 1980s, the capacity to make stem cells and the capacity to change their genes with recombination techniques came together. If one altered a stem cell's DNA, one could precisely change the next generation of stem cells cloned from it. What if one added these altered stem cells to a mouse embryo and made a chimera? Instead of having to breed several generations of mice, hoping for a desired change to emerge, one could make the change in stem cells, make chimeras that will transmit the change through the germ line, and be in no doubt as to why it occurred. Mouse genetics, biochemistry, molecular biology and developmental biology began to merge.

In 1988, when Nagy arrived in Toronto, Rossant's lab was already doing gene targeting, knocking out specific genes in mouse stem cells to see what happens without them. I signaled Nagy to stop and explain. I couldn't make head or tail of this knockout business. I wasn't sure what it meant, or how it is done.[8]

"You cut the gene out with a special vector—a piece of DNA introduced into the cell."

I still didn't get it.

"It's like a pair of pants," he said, rubbing his leg to enhance his metaphor. "You make an ugly patch, put it on the pants, and it finds its way to replace what is underneath. If you want it to attach to the pocket, you create one that is similar to the pocket. It binds to that pocket and replaces the area. That is called 'DNA sequence homology'"—literally, knowledge of the sameness of DNA sequences.

I couldn't see what making a pocket patch had to do with the word "knockout." It didn't sound like a gene was being cut out; it sounded more like another gene was being sent in to cover the gene of interest, to keep the original gene from expressing itself.

Give me this idea again, I said, one step at a time.

"Well, say you have some DNA that is known—you know it codes for a protein. You know where it is, you can sequence it and clone it into plasmids . . . You design this patch so its two arms are homologous to the gene of interest, and you put nonsense in between."

"What do you mean, 'nonsense'?"

He meant a sequence of DNA that didn't call up any protein. "It codes for nothing. It replaces an essential part of the gene."

Actually, part of this nonsense sequence of DNA often does code for something—usually it's a gene that conveys resistance to a particular drug. Getting a changed gene into a stem cell is done by transfection, which is a random process. Not all the stem cells will be transfected, and not all their genomes will be changed. So how could one find the changed cells among all the others? If one exposed a population of stem cells to that drug, only the properly altered cells with drug resistance would keep on reproducing. The others would die. "This is how you find your targeted cell," Nagy said. "The resistance. It's a simple idea."

Not so simple, I thought—clever. Some five thousand genes had been knocked out in the mouse using such methods.

Nagy wished to add these methods to the making of chimeras. Tarkowski had shown that a tetraploid chimera (containing double the usual number of chromosomes) made at the two-cell stage of development by fusion will be compromised enough to make only a partial embryo. Usually it produces a healthy placenta, not a living mouse. Nagy and Rossant wondered if mouse embryonic stem cells would integrate with such compromised embryos. What if it made only the placenta and they could get a live mouse created from only the embryonic stem cell lineage? By 1993 Nagy and Rossant had made mice with this method. The placentas came from the embryos with extra chromosomes, but the tissues of the live mice came only from the stem cells. The resulting mice were normal: they could breed, passing on the stem cells' alterations to their offspring.[9] These were also the first cloned mice ever made, a major milestone in the mouse world but ignored elsewhere.

"It was our Dolly," Nagy said. "It was definitely a clone of the embryonic stem cell line. It was not recognized as a clone because it was not from a somatic [adult] source."

This method of making a mouse directly from stem cells provided a new tool to study mouse development. Nagy and Rossant could knock out one gene after another in stem cells and see what developed, or failed to, when the stem cells integrated into an embryo. Since then, they have created many different ways to make chimeric embryos, gaining spectacular control over the making of mutants to answer very specific questions about development.

In 1994 Nagy closed his lab in Hungary and took up permanent residence at the Lunenfeld, returning to Hungary only to direct the work of his doctoral students. He now planned to move quickly into the study of human embryonic stem cells. But, I said, if he used human embryonic stem cells in the same way he'd used them in the mouse, he'd be doing human cloning, wouldn't he?

He looked at me over his glasses, then said he thought cloning of human beings would more likely be done like Dolly, with nuclear transfer, something achieved in the mouse only in 1998. The study of human embryonic stem cells would be important mostly in the development of cell therapies.

I thought about this. Nagy was making the same distinction Jose Cibelli had set out, like a plate of cookies and milk, before the Standing Committee on Health: reproductive cloning versus therapeutic cloning. Yet cloning, as Nagy had clearly shown, is cloning, whether one is copying a gene or a stem cell or a transferred nucleus. He'd stuck mouse embryonic stem cells to fused, compromised embryos, and he'd made live and healthy mice, even after twelve passages of those stem cells in petri dishes. It seemed to me possible that if he put human embryonic stem cells beside a compromised human embryo and put that construct in a woman's womb, he might eventually get a live cloned person, in the same way he'd got cloned mice. And what might happen if he stuck an adult human stem cell beside such an embryo? Would he be able to get a live human clone out of that too?

Would you try reproductive human cloning? I pressed.

No, Nagy insisted, he would not.

Why?

Because it is unethical. "It took almost 400 tries to get Dolly," he said, meaning 400 transfers of nucleus to egg to make an embryo before they got one live sheep. In fact, it had taken only 277 tries to make Dolly, and 112 to make Polly. Ian Wilmut, of Roslin Institute, had argued that this nuclear transfer method was actually a more efficient method than others to get changed genes into sheep,[10] but Nagy's aggregation method of making mouse clones seemed more efficient still.

"The techniques are not there," Nagy said. "It's not a safe procedure. You'd be likely to end with a monster or a defective individual. It's really shocking. Dolly was 400 tries!"

Even though the mousery was many floors away, there was a faint whiff of mouse wafting through the hallway of the Lunenfeld. It reminded me of stale graham crackers. I walked along past the emergency showers, past canisters of gas, past the beat-up yellow lockers, and came upon a woman pulling papers out of a file cabinet. It was the end of a long day, and my appointment with Janet Rossant was for five. I was in a hurry, but there was something about this woman, something intense that screamed "I am not a file clerk" and made me stop.

"You must be looking for me," she said.

"Are you Janet Rossant?"

She was, though she did not closely resemble her picture, which had just appeared in The *Toronto Star*,[11] along with her comments on a paper published in *Nature* by Joy Richman, a British Columbia researcher. Richman had managed to meddle with the signals for facial development in chicken embryos; the chicks had developed with two beaks. Rossant's photo made her look large nosed and slightly witchy, but in the flesh she was small, with a fine bone structure, a perfectly lovely nose, very thin and veined hands and a body so trim it was almost frail. Her dark hair was ribboned with gray, and her eyes, behind her round glasses, were a soft gray-blue. She radiated impatience, as if time spent with a journalist was time down the drain.

We walked to her office. There was a student at work at one of the two desks in the room, who kept right on as we talked. Rossant was at first careful, almost suspicious. What did a journalist want with her? She answered questions about her background with blunt speed. She grew up in Chatham, Kent, in the U.K., where she went to a good school. It was her female biology teacher who opened the door on her adult life, taking Rossant and her other top students on field trips up to London and Oxford and down to Cambridge, so they could imagine themselves at the great universities there. "No one in my family had been to university," Rossant said, "which was not so unusual in Britain." Her dad worked in a furniture store; her mother was a typist. "She typed my thesis for me," she said, pointing to it on the shelf behind her.

Rossant had the choice of Oxford and Cambridge, but Oxford "just felt better." She entered in 1969 and dipped her toe in various sciences, but went on in zoology with a scholarship. It was a small department with some professors who were, or would be, known worldwide. "One chatted with Nobel laureates," was how she described the atmosphere. Richard Dawkins (eventually the author of a controversial book on evolution, called *The Selfish Gene*) was an "upstart" professor.[12] Rossant was interested in developmental biology, which was taught by John Gurdon, another star, the first person to clone a vertebrate organism—a frog. Gurdon persuaded her to study mammalian development, which was then still a small field.

Why small? I asked.

"It was very hard to work with mouse embryos," she explained. "I could count on the fingers of one hand the people doing mammalian developmental biology."

It was hard not only because the mouse embryo is small, but because there weren't many tools to investigate it with. The first mouse chimera had been made only a few years earlier. She pulled down a textbook to remind herself of the dates. "Yes, 1961, Tarkowski and Beatrice Mintz," she read. Tarkowski had isolated blastomeres, cells found at an early stage in the mouse embryo, in 1959. In 1961 he and Mintz, independently, put two embryos together to make the first chimeras (so two people had the same weird idea at the same time). Just before Rossant started at Oxford, Richard Gardiner, then a doctoral student at Cambridge, figured out how to inject teratocarcinoma cells (cancer cells that can be grown in a dish) into an embryo to see what kinds of cells they'd make. Although there was a big group working on frogs at Oxford, the mouse attracted Rossant more, because of its closer evolutionary relationship to humans, and also because "I thought it better to go into a pioneering area."

In 1972 she joined Gardiner at Cambridge, where Tarkowski was working. She was Gardiner's first doctoral student. They were all housed at the Marshall Lab on Downing Street, a main road in Cambridge. "It was a terrible space," she recalled, with a laugh. "It was the ground floor of a building with no air-conditioning. There was dust and fumes and exhaust. We came in every day, and every day you'd wipe layers of muck off the counters . . . The other thing was, everybody smoked . . . Tarkowski did all the manipulations, cutting eggs in half with a glass needle, by hand, dropping ash . . . We had microscopes and glass needles and we were the leading edge of mammalian development because we could manipulate this tiddly embryo. I was trying to look at the commitment of cells in the early embryo. We knew what the inner cell mass and the trophoblast [the outer layer of embryo cells] gave rise to. I wanted to know when they were committed . . . The blastocyst [the early stage of the embryo] was the size of a grain of dust."

By "committed," she meant the point at which the cells had started on an irrevocable path toward becoming a particular tissue. The belief in this unchangeable commitment was fundamental among all developmental biologists, like Crick's belief in the one-way flow of genetic information. Rossant reached up onto her shelf to pull down her leather-bound thesis, with its yellowing pages. She smoothed the cover with her hands and carefully turned each page to show her precise but artless drawings of embryos. Since mouse embryos were so difficult to handle, her conclusions were drawn from very few—she had only four controls and six others to work with, and she had squeezed from them information sufficient for four papers and one doctorate. She had used radioactive labels to mark the cells. The questions she asked in this thirty-year-old thesis, she was still asking now: "How does the embryo decide to make cell types? How does a round structure get a head and a tail, a front and a back? . . . If we move cells around, how does [the body] come to have an axis?"

She had my full attention: these were vitalist questions and they gnawed at me. They suggested there is some metaprogram that guides all the rest, something that can't be explained just by the embryos' genes. Four or six sets of different parental genes can be put together in an aggregated embryo and yet one animal will emerge. Why?

In a review article written for a recent special issue of the *International Journal of Developmental Biology* honoring the lifetime contributions of Anne McLaren, Rossant and Nagy had described the strength of this larger pattern in mammals. Embryos seem to withstand almost any assault researchers have cared to throw at them, starting with Tarkowski and Mintz:

> Those original experiments attested to the highly regulative nature of the mammalian preimplantation embryo: the embryo aggregate started at twice the normal size, but ended up as a normal-sized mouse with no abnormalities in patterning. Since that time, experimental manipulation of the early mouse embryo has been refined and modified in many different ways; cells can be removed, reoriented, added back at different stages; cells from sources other than the embryo can be added; embryos can be reconstituted from their constituent cell

lineages; DNA, RNA and protein can be added; and still the embryo manages to survive and compensate for the indignities to which it has been exposed.[13]

Barbara McClintock, who pioneered the genetics of corn, mainly at Cold Spring Harbor, had jousted with these questions since the 1940s. Long before Watson met up with Crick, she had observed and recorded the breakage of chromosomes in the cell divisions of her growing corn, as well as these chromosomes' cycles of repair. Breaks tended to occur at the ends of the chromosomes, which led her to suppose that some genes were lost while others were preferentially passed on because of their position within the chromosome. She also showed that genes, or genetic information, seemed to move around in the corn genome. While particular genes produce particular effects, she found they can also come under the sway of other genetic structures that she called "controlling elements." Some of these elements, called *transposons,* physically move from chromosome to chromosome over time.

At first McClintock's findings were ignored. Then she was treated as a mystic. But then regulators, which synthesize repressors and operators, which influence the expression of genes, and which also move around, were found in bacteria by French biologists—Jacques Monod, François Jacob and André Lwoff. They got the Nobel in 1965. In 1983, McClintock got one too.[14] She came late to glory because she insisted on saying things that disturbed materialists. She said that their molecular descriptions did not really resolve fundamental questions about heredity and development so much as restate them from another perspective. "The point is we forget we're all one thing. It's all one. And we're just breaking it up into little things," McClintock said.[15]

These resilient patterns were discovered early in the study of the mouse embryo. "Experiments in the 1960s and 1970s showed . . . you can take it apart and put it together, it'll be fine," said Rossant. "Richard Gardiner shows if you leave it alone, one side of the blastocyst will form the front or the back end of the embryo. So how come [this pattern] reestablishes if you take it apart?"

I wondered where she thought this larger program was written.

"I think there is information the embryo uses to establish

asymmetry," she said. "The back end comes from mesoderm . . . It uses signals we don't understand that relate to the axis of the blastocyst. It has to be reconstituted by every cell . . . But how much is genetics, from the embryo's genome, or is carried in by the mother? The egg carries information within that determines how development will occur. The egg lays down a metapattern."

It seemed to me this was no answer at all. It just presented a new question: how does the egg do that?

Rossant followed Richard Gardiner back to Oxford (although Cambridge granted her doctorate). She did a postdoc there, married a Canadian, Alex Bain, and arrived in Ontario looking for a job. Bain was hired at McMaster University in Hamilton, Ontario. In 1977 Rossant became an assistant professor at Brock University in nearby St. Catharines, and soon had two children.

At Brock, she continued to work on early embryos, but there weren't a lot of resources to get down to the molecular domain. So she began to collaborate with Verne Chapman at Roswell Park Cancer Institute in Buffalo. They looked at DNA interactions, specifically at methylation, the suppression of gene expression by a methyl molecule that moves in and sits on the gene. These methyl molecules prevent the assembly of RNA messages. Methylation and its opposite, demethylation, manipulate expression of the right genes, at the right level, at the right time in the development sequence. The more Rossant spoke of this, the more I formed an image in my mind of hands playing on a great church organ, of small fingers covering and uncovering keys, building a musical phrase.

Learning about methylation opened up a whole new under-standing of the rhythms of information exchange going on in embryos. Rossant was particularly interested in imprinting—in those genes that behave differently according to whether they come from the mother or the father.

In 1983 she got a fellowship that allowed her to stop teaching and focus on research for two years. That's when she met Alan Bernstein, now the head of CIHR. Bernstein was working on retroviruses, which work their way into cells and actually write their own message on

DNA (just one of the many instances in which the Central Dogma proved wrong). He thought he could use them to alter genes in patients with genetic flaws. Rossant wondered if retroviruses could be used to take foreign genes into mouse embryos. Together, they showed that although one could get the retrovirus in, the foreign gene would not integrate into the mouse's genome unless it carried its own promoter—a regulating sequence that helps a gene to express itself.

In 1985 Rossant was appointed to the Lunenfeld and to the University of Toronto, and she and Bernstein set up the first transgenic mouse lab in the city. By this time, biologists were beginning to realize that there was a lot of similarity in gene sequences between species widely separated on the tree of life. McClintock had been making that point since the early 1970s. Darwin, according to Francis Galton, had had a similar insight. "It is shown by Mr. Darwin," Galton had written, "in his great theory of *The Origin of Species,* that all forms of organic life are in some sense convertible into one another, for all have, according to his views, sprung from common ancestry."[16] McClintock had taken this idea much further. According to her biographer, Nathaniel Comfort, McClintock had said that "any organism can make any other,"[17] by which she meant that the basic genetic components of all species are so similar that differences might be the result of nothing more than a change in the position of a control element between chromosomes.

Work in the fruit fly had convinced geneticists that one class of genes regulates a whole sequence of development. "We knew the genes that regulated in *Drosophila* . . . encoded transcription factors that turned on cascades of genes," said Rossant. Her new colleague at the Lunenfeld, Alex Joyner, had previously been in the laboratory next door to Tom Kornberg at the University of California at San Francisco. Kornberg had cloned a fly gene, which he called *engrailed,* that controls segmentation of the fly's body. Joyner cloned two similar *engrailed* genes in the mouse, and found that they are expressed in the brain early in development. Biologists began to think that these classes of genes must have been conserved across time and myriad species because they do something vital.

"Everybody went to meetings," Rossant said. "The mice people would say there are genes conserved from the fly to the mouse—the fly people said, now show what they do." Unlike the fly people, the mouse people were unable to show what their gene sequences actually did because they had no method to turn genes on and off in developing mice. This is when Rossant and Joyner started down the long path to the knockout mouse. One of their graduate students went to study at a lab that was using the then brand-new polymerase chain reaction (PCR). This technique takes advantage of the fact that the two strands of a DNA molecule will come apart if heated to the right temperature and will make two identical daughter strands when allowed to cool in the appropriate chemical bath. Through this method of heating and cooling, one could make many copies of the tiny DNA fragments recovered from cells. These multiplied fragments could then be more easily sequenced. Once one knew the sequence of a gene, one could make a nonsense sequence to sit on top of it and shut it off. The nonsense gene could be carried into a mouse stem cell attached to an infecting agent.

"We were the second group to publish on the ability to knock out any gene in a mouse embryonic stem cell. We published in 1989 in *Nature* . . . That changed what everyone could do in using embryonic stem cells and targeting mutations to analyze gene function."

Andras Nagy had arrived at her lab at just the right time: techniques and ideas he and Rossant had developed independently came together perfectly. Rossant could change the stem cells' genes; Nagy could make a new kind of chimera. Together they found a way to get those cells to make a whole cloned mouse.

"So is that what you've got all that money for—that $21.5 million?"

She waved her hands, as if to divert attention from such a huge heap of good fortune, then grabbed at her hair as if she'd had to explain this one time too many to jealous colleagues. It wasn't all for her, she said. It came from a government-business partnership called Genome Canada, and it was for several projects, of which hers was one. "It's not our intention to mutate all the genes. But other centers worldwide are doing this too. Together, the long range goal is to do them all."

But there must be a goal beyond that, I thought. It must include doing this work in humans. Only so much could be learned about human genes by knocking out mouse genes; to understand human development, one would have to knock out human development genes one by one in human embryonic stem cells, put the cells in embryos and see what developed. In fact, one of Rossant's colleagues and a former protégée of Anne McLaren's, embryologist Marilyn Monk, had been doing gene expression work on human embryos in the U.K. for years. Monk had written about why she moved from studying microbes to mice to men in the same special journal issue devoted to McLaren that had included Nagy and Rossant's contribution:

> Mouse development is an excellent model for the study of human embryology since for the first week of development mouse and human embryos are very similar. So the many years of study in the mouse and my training in Anne's Unit have paved the way for similar studies in the human over the last decade at the Institute of Child Health. Mouse and man may have the vast majority of their genes in common but we now know that there are many differences in gene expression and its regulation between the mouse and the human . . . It is essential to study the human embryo directly.[18]

I could see that this was inevitable, but I didn't know what to think of it. On the one hand, turning embryos into research subjects made the hair stand up on the back of my neck. On the other, how else could one ever understand human development and human disease? The materialist in me asked, is there really any difference between studying a two-cell human embryo and studying a human cadaver? Well, yes. The embryo is alive. And then there's the matter of informed consent: an adult can make a gift of his or her future dead body to science; no one can ask an embryo anything.

"I think everybody who works in these areas is very aware of the implications of translating what we do in mice to humans," said Rossant. "I felt studies in mice were essential to understand how human genes function. At the same time I wouldn't support using these techniques on humans. I'm totally opposed to genetic alterations in humans."

Her vehemence surprised me. She is a senior investigator in Worton's Stem Cell Network, and she chaired the CIHR's human embryonic stem cell guideline committee.

She recognized that there is no intrinsic difference between selecting for certain genetic traits in people and the selection farmers make for certain traits in plants and animals, something that has been going on for millennia. Take Holstein cows, for example. "They can hardly walk," she said. "It's a milk machine. That's selection. Whether it's bad, per se, I'd say no. But when you get to humans, I say it's categorically a bad thing." Selection for traits in humans is eugenics. She wanted no part of eugenics.

I asked her what she thought of James D. Watson's suggestion that private scientists should try to get rid of human disease genes by making permanent genetic changes to embryos. Selection against certain traits is also eugenics.

"To start with, that's nuts," she said. The furthest she would consider going is what she called preimplantation diagnosis—examination of a single cell taken from an early embryo for signs of a known disease gene. If one was concerned about a child inheriting Duchenne muscular dystrophy, for example, she thought it was all right to determine an embryo's sex in order to avoid a male child likely to get it. "That is deemed acceptable," she said.

The more I examined her argument, the more it seemed beset with contradictions. On the one hand, Rossant didn't object to making chickens with two beaks and mice that glow in blue light, or to inserting artificial chromosomes into plants, as a woman named Pruess in Chicago had just done; all these genomes could be changed, but not ours. In effect, Rossant was putting people back up on the pedestal of creation that Darwin had knocked us off of. Yet she had no problem studying human embryonic stem cells, which can be extracted only by destroying the embryo. She wouldn't alter genes, but she would destroy embryos.

" . . . to alter genes is unnecessary and unacceptable," she was saying with finality.

"When will the CIHR guidelines for human stem cell research be ready?"

"January."

Aha, I thought. Her committee would scoop Bonnie Brown's.

CHAPTER FOUR

UNRAVELING MORTALITY

My suspicions that the pursuit of human cloning was only hype vanished beneath a pile of mouse articles. What one can do in the mouse is being tried in the human. So I set out along another branch of the second tree, following the trail of telomerase, the enzyme that puts DNA sequences back on the ends of chromosomes, the enzyme with the flavor of immortality.

I went to the airport in Toronto to catch the first flight by a foreign carrier to Washington DC's Ronald Reagan National Airport since September 11. I kept telling myself that life had returned to normal, but feathery tendrils of fear attached to simple things such as walking through the terminal. The tools of terror had now expanded from commandeered airplanes to life-forms, specifically to anthrax. Weapons-grade anthrax, sent by mail by person or persons unknown, had killed in Miami and New York and New Jersey. The discovery of anthrax spores in a letter to Tom Daschle, the leader of the Democrats in the U.S. Senate, had shut down mail centers and congressional offices in Washington.

The gate was in a part of the airport I'd never seen before. All the gates nearby were empty, as if someone had tried to remove us, like a possible source of infection, from normal traffic. I was searched twice by large men carrying stubby, gray guns and wearing thick body armor. I thought that this frightening atmosphere would keep many eminent persons away from the annual meeting of the American Society for Cell Biology. I couldn't have been more wrong. As dawn broke the next day, a seemingly endless line of people snaked around the vast perimeter of the Washington Convention Center. We each had to clear security and were allowed to enter through only one unlocked door. Why so many here so early? The professor in line in front of me explained it this way: the cell is the unit of life, and so cell biologists get up earlier than everybody else to go about their work figuring it out.

Close to eight thousand brave souls had registered, up from four thousand only five years before. Hundreds of companies had set up booths in the exhibitors' hall. The program required 350 glossy pages. A separate book containing abstracts of papers and posters was 523 pages of tiny print split in two columns. The society's journal presented articles that defied my usually elastic capacity to extract meaning from just about anything. A few choice titles: "Na,K-ATPase Activity Is Required for Formation of Tight Junctions, Desmosomes, and Induction of Polarity in Epithelial Cells" and "Golgi-to-Endoplasmic Reticulum (ER) Retrograde Traffic in Yeasts Requires Dsl1p, a Component of the ER Target Site That Interacts with a COPI Coat Subunit" and "APC2 Cullin Protein and APC11 RING Protein Comprise the Minimal Ubiquitin Ligase Module of the Anaphase-Promoting Complex." The only one that really spoke to me was "Repair of Chromosome Ends after Telomere Loss in *Saccharomyces*,"[1] though I wasn't sure what *Saccharomyces* were (they're yeasts that reproduce without sex—a process called *budding*). At least I'd heard of telomeres.

The main purpose of my trip was to meet Carol Greider, a professor at Johns Hopkins University whose telomere work appeared in yet another glossy society book of forty-two *Landmark Papers in Cell Biology*, offered for sale outside the press room. Along with her doctoral supervisor Elizabeth Blackburn, a woman the society had invited to give its E. B. Wilson Lecture that night, Greider had kicked open the door on telomerase. Greider and Blackburn had already won the prestigious Gairdner Award, given in Toronto for major contributions to medicine. The Gairdner Foundation advertises itself as a Nobel predictor: one in four Gairdner laureates go on to win the Nobel. While no Nobel had yet been forthcoming, Greider and Blackburn's work had certainly encouraged Michael West to create Geron Corporation and launch it on an anti-aging path. Greider had served on Geron's board of scientific advisers.

Telomeres and their functions played a role in the debate over the safety of nuclear transfer cloning. Dolly was cloned from the mammary cell of an adult sheep, a sheep dead long before Dolly was born. Some wondered if the telomeres on the chromosomes in Dolly's cells would be shorter than normal since the skin cell from which she was cloned was the product of many cell divisions, each one supposedly making the telomeres shorter. Would Dolly age prematurely? Wouldn't anything cloned from an adult cell die young? Roslin Institute researchers had let it be known that Dolly's telomeres were shorter than average. Yet Jose Cibelli had explained to the Standing Committee on Health that telomere length is actually rewritten to the start point by cloning. He maintained that dropping an old nucleus into an immature egg made chromosomes new again. If Cibelli was right, all we need to do to extend our lives indefinitely is to clone ourselves sequentially. Bingo—youth could be forever.

Greider had suggested we meet by the notice board the first morning of the convention. There I stood, at the appointed time, waiting for a woman looking as if she was looking for me. Many women appeared, but they all went up and down the rows of notice boards looking for messages, not for me. A dreary hour passed by. I pinned a note to the board telling Greider how to find me.

There was nothing for it but to go and listen to the reading of papers. Upstairs there were three large rooms normally separated by movable walls, which had been opened for the occasion. One lecturer, supported by large video screens, was mumbling through a presentation before an audience of thousands. It was all about apoptosis, the name for programmed cell death. The fact that healthy cells are programmed to die had been demonstrated in the early 1980s by Sydney Brenner, John Sulston and Bob Horvitz, who initiated the study of the worm C. elegans.

The talk on apoptosis ended as I found a chair in the dark. Then a woman with a heavy accent, whose name I didn't hear, began to talk about polarity in certain kinds of skin cells. Everything she said was new to me, so I found myself frantically making notes, trying to keep up with her. The experiments of her group showed that some cells, when exposed to magnetic beads, reverse their polarity, but that this capacity is somehow lost in tumors. As far as she could see, this matter of polarity had something to do with how cells orient themselves in the three-dimensional structures that are living bodies.

There were big pictures on the screen of these cells she was talking about. They were dead cells, stained and sectioned and laid out on microscope trays so they could be studied. They lay there on a seemingly flat plane, but as she spoke it became obvious, like a slap of cold water in the face, that the way cells are normally studied cannot reveal how they actually work: there is no such thing as a flat plane in a live body. All living things have to orient themselves in three dimensions.

She and her colleagues wondered if the chemical signals to which polarized cells respond could be used to change tumor cells back into normal cells. Taking cells from the human breast and growing them in culture on flat plates, she noticed after fifty passages that the cells finally lost their capacity to hold onto their structures and then became malignant. But how did they know what kind of environment they were in?

The room was beginning to empty. I couldn't understand it. Where were they all going?

She explained that her grad students found that malignant cells can be reverted to normal by being placed in a correct three-dimensional environment. This allowed them to signal appropriately, which they could not do on a flat plate. "The pathways," she explained, "are not connected the same way. This is a fundamental problem in biology . . . Form and function are connected dynamically and reciprocally."

She was talking about that unwritten pattern too, I thought—the one that keeps asserting itself in the mouse embryo even after foreign cells are shoved beside it, even when two embryos are fused together to make one. In her experiments, it had taken fifty passages to make that pattern break down, but even then, when the deranged cells were placed in the right context, they reverted to normal.

One of her students then took six different agents known to kill cells in a dish and asked, would these agents cause a different response in three dimensions? "All these agents kill in two dimensions," she said. "In three dimensions these cells are resistant to all."

Further, she found that if the cells' polarity was reversed, and if they were placed in a three-dimensional context, they also did not die. "It boggles my mind," she said. "These findings have implications for dormancy, for drug resistance and for chemotherapy."

In other words, studies to figure out whether something is a carcinogen done on cells in flat plates would produce different results from studies on cells grown in a suspension, in a ball, in a state in which up and down and left and right matter. I found myself breathing shallowly as I struggled to make my notes. Why were people around me walking out on her?

Hers was the last paper I grasped that morning. It wasn't just that the rest of the speakers labored in fields in which I could not work up even the slightest interest. It was also that they used a minimum of English words and a maximum of code to signify genes or proteins. The first woman had been talking about a new idea. She had been forced to resort to English; no jargon was at hand. Those who followed her were reporting on well-established areas, so they used myriad dialects to convey their findings. At first I was embarrassed by my own deficiencies, but I was bemused to learn later that many of the professionals around me couldn't follow either. A protein expert from South Africa complained over lunch that he had grasped only two out of the hundreds of poster displays on proteins he examined. Two stem cell researchers said they were equally stymied. So, finally, I got it: this convention only seemed to be about communication. In fact, researchers were hiding their gems in plain sight.

I went downstairs and checked the notice board. Carol Greider had come just after I'd left; there'd been some sort of problem on the train from Baltimore. She would meet me tomorrow. Same time, same place.

Outside a small meeting room, a sign advertised a seminar to discuss "Biological Weapons, Obligations and Opportunities." I stopped. Opportunities? What opportunities? I took a seat. Tara O'Toole, director of the Center for Civilian Biodefense Studies at Johns Hopkins, noted that a congressional report had found that "biological weapons are the most strategic threat to the U.S. in this century . . . Only nuclear is in the same category."

"One hundred kilograms of anthrax properly prepared and upwind of the District of Columbia would have the same fatalities as a one-megaton hydrogen bomb," she said. There had been enough anthrax spores in the one letter sent to Tom Daschle to cause "200,000 lethalities." Biological weapons were ideal for rogue states and third world tin-pot dictators and terrorists of all stripes. "Not only can they be built without a large infrastructure, they can be hidden . . . And, third, they appeal to the asymmetric nature [of world power], a way around the U.S. hegemony. And they can be wielded even by individuals."

In O'Toole's view, therefore, the open nature of biological inquiry had to be curtailed. The literature provides instructions on how terrorists might put anthrax into aerosol form. Biological scientists didn't seem to understand that national security takes precedence over intellectual freedom. However, in danger she also saw opportunity: if biology were done within the defense infrastructure, it could acquire awesome budgets.

Beside the main lecture hall, a wall of doors stood open to reveal a vast area given over to poster displays. They were pinned up on room dividers, set up in rows organized like streets. I found the poster of a Canadian researcher, Richard Oko of Queen's University, which dealt with how mammals' eggs can be induced to divide without fertilization, by the simple application of a protein called PT 32, produced only by the testes.

I couldn't find the author, so I kept on walking and found myself in the corporate domain. Each booth was decorated with monitors, displays, logos, samples, offered free coffees, cheeses, nice-looking young men and women—anything to get a researcher with a capital budget through the cardboard-framed front doors. The booth for Cold Spring Harbor Laboratory displayed videotapes and books for sale. One video machine played a speech by James D. Watson, and I stopped to listen. He was describing how he and Crick discovered the shape of DNA. The tape was new; he was an older version of the weedy, grinning youth who winked at me from the pages of *The Double Helix*.

Cold Spring Harbor's booth was close to that of Celera Inc., the publicly traded company formed by Craig Venter and his investors to sequence the entire genomes of the fly, the mouse and the human being. Venter and Watson had been at war for most of the last decade over whether the Human Genome Project should be a private or a public endeavor. Celera's booth was decorated with a colored poster showing its version of the sequence of all twenty-two human chromosomes run together end to end. Little brackets and parentheses and colored patches distinguished zones with known genes. There was no video of a talking Venter aimed at the talking Watson, which would have been appropriate given the insults they'd lobbed at each other.

I came at last to a booth advertising the products of a North American human tissue network. A brochure said that researchers need only fill in a form to join the network. They could then request the kinds of human tissue they need to study. The network offered chunks of placenta, discarded tumors, tissues whacked off corpses during autopsies, all prepared "according to your protocol for a very low price." Private companies could get these tissues too, but they had to pay three times the academic rate for the same thing.[2]

Jesus, I muttered, to no one in particular. I buttonholed the salesman. You actually sell the chunks of organs and flesh that surgeons pull out of their patients? I asked him. Who gets the money? What share does the patient get, or the patient's estate?

There was some sort of general consent form for patients to sign, the salesman said vaguely, but he was unable to find a copy of it for me.

To zip back and forth between the sales floor and the lecture halls was like moving between two worlds, or maybe two moieties of the same world. I walked into a session on the organelles in cells. First I learned that cells are not just squishy bags of organic chemicals suspended in water. They have an architecture, a skeleton, a physical structure that keeps on repeating itself with every cell division. Within the various sectors of the cell there are organs with curious names like the Golgi apparatus, the endoplasmic reticulum, vacuoles.

I'm sure I sketched them in biology class, with only the vaguest idea of their functions. It seemed that the questions being asked about them now, more than one hundred years after they were first seen through microscopes, were still very basic. What happens to these cellular organs as a cell divides? Do they simply keep dividing, getting smaller and smaller until the cell dies, or do their molecules break apart entirely at each division and somehow come back together again, reassembling like Tinker Toys in each daughter cell? What would instruct or guide molecules to keep making the same structures? I sat and listened to researcher after researcher describing dynamic processes using terms I strained to follow. Some of the researchers worked for universities or private research institutions. Others worked for big companies or little ones they'd started themselves. No one seemed to distinguish here between academics working in the public interest and private researchers working for a corporate interest, often their own.

In the big room upstairs, late in the day, there was an award ceremony. The American Society for Cell Biology had decided to hand out its eighth public service advocacy award, and actor Christopher Reeves, the best-known quadriplegic in the world, was the recipient. He had tirelessly lobbied President Bush to permit federal funding of human embryonic stem cell and germ cell research, morally difficult research that might get people like him out of their wheelchairs. He had appeared before Congress to argue against restrictive bills on cloning; he had dissed the president for his August 9 funding decision; he had appeared on *Larry King Live* to make his points. There were big screens set all around, so everyone in the room could see in close-up the man who had been Superman. The room filled right up and spilled over.

The moderator made a long speech about Reeves's history, before and after his 1995 accident. Reeves had always been politically active; he'd been a member of Amnesty International and the Natural Resources Defense Council and had demonstrated in Chile against the junta. In 1999 he set up and became chairman of the Christopher Reeves Paralysis Foundation to seek cures for disabilities like his. A video rolled with images of him as he once was—a pilot, a skier, a sailor, a horseback rider—and then as he became—a man who can't raise his hands, let alone an arm and leg, and who is barely able to turn his head. Somehow, over the years since his accident, the muscles that once defined his eyes, cheeks, lips and nose had all shape-shifted. A man once beautiful now displayed his suffering. His voice rang out over these images as the lights came up: "We must, we can, we will" cure spinal cord injuries. Under cover of darkness, he'd been transported onto the podium and strapped into his chair with its high back. Someone stood beside him to wipe spittle off his chin. He had a respirator hooked up to his throat.

Everyone stood up to applaud.

He plunged right into the politics of cloning and stem cells. "We are in a race for public opinion," Reeves wheezed. "CNN had Representative Weldon on. No one was invited to challenge him. He said outrageous things—that adult stem cells are doing what we need and we don't know [whether] embryonic stem cells will have any therapeutic effect on human beings. It almost made me get out of the chair." He laid out the debate over whether therapeutic cloning should be allowed but reproductive cloning banned. He had no time for the niceties of ethical arguments.

"In the clinic, they say, therapeutic cloning might turn to reproductive cloning. Well, doctors are regulated . . . Extremism in this country [comes] from the religious right. We have to stop calling it 'therapeutic cloning.' These are buzz words that terrify a lot of people. Weldon used those words . . . to ban cloning in the House." The fight had now moved to the Senate. To win in the Senate some words had to change. "What has to happen is we call it 'nucleus transplant.'"

Senator Arlen Spector was on the right side, Reeves's side, biology's side. The Democrats were fired up. The public would come on board if things were just properly explained. He reminded his audience that a grassroots movement "put AIDS on the map. Today, the NIH funds AIDS to the tune of $1.8 billion a year. I urge you to remember, communicate with the public, educate the politicians, giving them the cover so they feel safe. It's gonna be critical. If there's a ban, not all of us can go overseas to get a treatment. The race is on."

It was a strange speech, garbled but weirdly inspiring, as if his terrible suffering enabled him to float above his own interest, as if his dreadful condition allowed these biologists to feel they were advancing a greater good while advancing their own interests. When he was done, people leaped to their feet to show their gratitude and respect. None of them seemed troubled that he assumed that these human embryonic stem cells would cure him though no researchers had demonstrated any such thing.

Reeves called for questions. A man came to a microphone set out in an aisle. We are outnumbered, he said. How can we be better organized, what can we do?

I wanted to stand up on my chair and yell, try speaking English! But I restrained myself.

Reeves answered that there was an umbrella group of the "diseases" who were going down to lobby senators. "Before August 9 we counted sixty-three senators in favor of stem cell research. They need reinforcement; they need to know from experts like yourselves that it's safe . . . The number who could be helped is 100 million, with Parkinson's or ALS, we all know somebody . . . We gotta create a grassroots movement here."

Another question: "Those of us who engage with those opposed encounter a yuck factor in doing work with the human embryo."

Reeves counseled that changing the words could fix that. "The best way is never, ever use the word 'embryo,'" he said. To deal with concerns about having to kill embryos to get stem cells all they had to do was to explain it this way: "There's a room on fire. [On one side] a petri dish has little cells inside that some say is life. [On] the other side is a policeman in a wheelchair. You can only save one of them. Which one will you save? This puts the question of saving life into proper perspective. It's about saving the lives of the living. It must be allowed to go forward if we consider ourselves a compassionate society, and they're just gonna have to get over it."[3]

I trudged back to my hotel, loaded like a pack mule with bags of books and pamphlets and press handouts, my right hand numb from taking too many notes. It seemed to me that biologists functioned in three domains: academia, the business world, politics. They were like any other organized form of life, learning to evade and avoid or to invade and overwhelm that which impedes their growth. And they didn't seem to worry when knowledge in one sphere contradicted knowledge in another.

Carol Greider did not show up at the message board the next morning. I was beginning to think that I had braved the plane, the guns and the fear of spores that go bang in the lungs for nothing.

A panel discussion on the practice of science was underway in one of the small rooms downstairs. It was called "Fuzzy Borders" and was said to deal with the interactions between academic researchers and industry.

"When I was training," said the moderator, "industry was the evil empire. There's been rapid changes in the past two years."

He listed the issues: conflicts of interest between academics and their universities over intellectual property—what should be patented, who should get what share, who decides. How could one keep the openness required by science and protect vital information as demanded by business?

The conversation soon whipped from sponsored research agreements to the separation of one company's interests from another's in the same lab to reach-through agreements (permitting companies to profit from any research that they helped fund originally, even if the product is made by someone else). A panelist from the University of California at San Francisco made it clear that his university no longer waits for business to knock on its doors but had set up its own incubator enterprise, spinning off its own products through its own companies. However, he said, UCSF wasn't doing this for the money. "I think [it's about] a social contract," he said. U.S. scientists at universities get taxpayer support because society believes this benefits society as a whole. The 1981 Bayh-Dole Act requires all American universities receiving federal funds to make maximum efforts to turn new inventions into actual products. "We have a social contract to get this from the bench to the bedside," he said.

A much older man grabbed a microphone set out for the audience.

"I'm from NIH," he said. "I agree there is an implied social contract. However, it doesn't follow that the university has to link to biotech or to industry. Why couldn't you say [the social contract] is met by generating new knowledge anyone can draw from?"

Save for the person from UCSF, the panelists responded as if the NIH man had stood upwind and released a loud fart: he was politely ignored.

A dark morning of light drizzle. It was my last chance to meet Carol Greider before I had to catch my flight. I arrived at the main convention center, where a tripod recorded the names of donor corporations that had sponsored this convention, ranked in the order of their largesse. Finally, Greider walked toward me. How did I know her? She had a name tag hung around her neck and a mop of streaky blond California-surfer-girl corkscrew hair bouncing at her shoulders. She wore a black leather jacket, black corduroy pants and a sloppy hot pink sweater. She led me across the street to Starbucks.

My watch ticked through precious minutes while she ordered coffee. Finally she sat down. In the context of the younger women chattering all around us, she was obviously a senior scholar, someone with her own laboratory at Johns Hopkins University School of Medicine, her own postdoctoral fellows, doctoral candidates, technicians, big grants and big prizes. (She is now the director of the university's Department of Molecular Biology and Genetics, and a professor of oncology.) She seemed forthright, but there was an undercurrent of wariness in her manner.

The daughter of a physics professor at the University of California at Davis, Greider always knew she was going to university, just like everybody she went to school with. She won a Regents Scholarship to the University of California at Santa Barbara, where she did her undergraduate biology degree. She found her research direction when she applied to graduate school. She was interviewed by faculty at various universities, but then she met Elizabeth Blackburn at the University of California at Berkeley. Blackburn had been pursuing questions on the nature of telomeres since the 1970s. Originally from Australia, Blackburn had done her doctorate in biochemistry with Frederick Sanger in the Laboratory of

Molecular Biology on the outskirts of Cambridge, as he was working out methods to learn the sequence of base pairs in small fragments of DNA. In her postdoctoral work at Yale under Joe Gall, Blackburn had worked out her own sequencing method, which in 1978 she had used to winkle out the sequence on the ends of linear RNA found in tetrahymena, an organism with a nucleus.[4] She was trying to understand how these linear RNAs maintained themselves that way, and she found that their end sequences were simple repeats. Blackburn's questions about these ends, or telomeres, appealed to Greider. Blackburn asked, why do these linear RNAs hold together in tetrahymena when the telomeres on their ends should shrink with each cell division? She proposed that there was some kind of unusual enzyme that kept putting the telomere sequence back on the ends.

When Greider arrived at her lab, Blackburn already had an NIH grant to search for the enzyme. They worked with extracts from tetrahymena cells. Blackburn had already established that tetrahymena telomeric sequences could be put into a yeast, where they would also stabilize yeast chromosomes, so she was pretty sure that whatever did this had been conserved over evolution, that there was something that made these sequences and put them in the right place.

"After nine months we got the first hint that there was something there," Greider said. "Then it took another year of characterizing it. We published the paper discussing this in 1985."

Their first publication, in *Cell*, demonstrated the existence of the enzyme through a process of elimination. The enzyme itself hadn't been found, but their experiments showed that its existence was the only credible explanation for how telomeres were able to stop shrinking.[5]

The existence of the enzyme led to the next questions: what is it and how does it work? Enzymes catalyze, or speed up the chemical reactions in living cells by as much as one thousand times, without actually becoming a part of the reaction. But enzymes are proteins, and proteins don't usually direct the assembly of DNA or RNA sequences. However, David Baltimore had demonstrated that some kinds of transfer RNAs can rewrite DNA sequences; these RNAs are called *reverse transcriptases*. Greider and Blackburn soon found an RNA component within this telomerase substance that could act as a template for the formation of repeating sequences. Whatever else it was, telomerase was in part an RNA. They published in *Nature* in 1989.[6]

Greider had got her doctorate in molecular biology in 1987. She stayed at Blackburn's lab for a while, then applied to Cold Spring Harbor Laboratory to be a postdoctoral fellow. As part of the interview, she gave a seminar to the faculty. Two weeks later she was offered an appointment. Obviously someone at Cold Spring Harbor thought her work was red hot.

"Did you run into Jim Watson there?" I asked.

"Watson was the one who hired me," Greider said. "He made all the decisions . . . He was amazing on the details. He was really interested in my project, in part because he first pointed it out." Watson's last published research paper, in 1972 in *Nature New Biology*, had shown why someone needed to solve this puzzle about chromosome stability.[7] Watson must have been very interested, because not only was Greider hired, she didn't even have to find her own money—normally at Cold Spring Harbor Laboratory, postdocs were expected to bring in grants to support themselves. "Everything there is soft money," Greider said. "You get salary and operating costs from grants. But for the first two years they funded [me] as a fellow. Then a graduate student wanted to join me. They said write an NIH grant and get promoted to staff."

The year before Greider and Blackburn published their findings about the RNA component of this substance called "telomerase," Bob Moyzis identified the human telomere sequence.[8] Greider decided to study human telomeres with a Canadian researcher named Calvin Harley, then at McMaster University. She had met him when she was still a student in Blackburn's lab. Greider's boyfriend, Bruce Futcher shared half a lab room with Harley. Greider would go up to Hamilton, Ontario, on long weekends and at Christmas to see Futcher, and she ended up spending a lot of time in the lab with Harley. Harley was interested in why human cells fail to keep dividing after fifty passages in a dish, why they eventually succumb to senescence, and whether telomere shrinkage might have something to do with it. "Cal and I thought it would be interesting to see if senescence had to do with telomeres. I called him and we collaborated long distance."

Harley grew human fibroblasts (cells found in connective tissue) in dishes and sent Greider samples of their DNA as they went through various passages. Her job was to test the DNA and see if the human telomere sequences shrank with cell division. They did.

"The first experiment showed [the telomeres] shortened with cell changes," she said. "We published in *Nature* in 1990, Harley, Futcher and Greider."[9] They drew no conclusions about what this might mean for human aging, but only two years later, Harley, Greider and others, including Silvia Bacchetti of McMaster, did suggest one. In an article that appeared in the *EMBO Journal,* they measured telomere length, telomerase activity and chromosome rearrangements in human cells grown in culture. They grew embryonic kidney cells through many passages, and they also took some of these cells and transformed them with the addition of either a virus or an oncogene that causes malignant growth. Some of these transformed cells died, but those that did not die became immortal. In the mortal cells, the researchers found telomeres shortened by about sixty-five base pairs with every generation until the cultures became senescent. These mortal cells and the transformed cells that died often had chromosomes rearranged into circles, as if their ends had stuck to their bodies. In immortal cells, by contrast, chromosomes rarely formed these rings, and only immortal cells had telomerase activity. The first line of their abstract was extraordinary: "Loss of telomeric DNA during cell proliferation may play a role in aging and cancer."[10]

So, I asked, when did Geron get into this game?

She looked at me for a moment, as if trying to decide how much she wanted to say, then rolled the story back to that point in 1989 when she wrote her first grant application at Cold Spring Harbor Laboratory for this project with Harley and Futcher. The peer review committee that evaluated her application included Woodring (Woody) Wright of the University of Texas Southwestern, a researcher exploring aging. Michael West, who went on to form Geron, worked with him. Wright, she said, wrote a review that led the NIH to turn down her application. He "trashed it. [But] he thought it was so interesting, he started working on it," she said.

I put down my pen for a moment. I wanted to make sure I understood her and asked which grant application this was.

"This was the NIH grant to support looking at shortening in human cells, to see if it led to senescence. I was the principal investigator."

This is obviously not the way things are supposed to work when peers review grant applications. The denial of her first application meant she had to wait months to apply again, and months in a new area is a long time. Someone else did publish in *Nature* that same year on human colorectal cell telomeres that shrink with aging.[11] This is a scientist's worst nightmare—that they reveal their hot idea in a grant proposal and the reviewers turn it down but then run with it themselves. So I asked her directly: was she making that accusation against Wright?

She said that when she ran into Wright at Geron, she'd asked him why he had criticized the application. "He told me we didn't know the aging literature well enough," she said. "He stopped it. It put us back nine months." The second time around they got the grant.

(Later, I asked Wright if he'd voted against the grant so he could get ahead of Greider. He denied it. He was already working in the area, he said, and in any case his first paper on the subject didn't come out until 1994.)

Biotech companies routinely market themselves to prospective investors by creating scientific advisory boards composed of people with top publications in the right area. A good board with the right scientific advisers can make the difference between getting a lot of investment and getting none. Greider and Harley had important publications in this brand-new area and were among the very few studying human telomeres. If Geron wanted to make a name for itself in this area, it needed their names on its board. Greider and Harley were appointed to its scientific advisory board.

"Did they pay you?"

"Typically, I think it was $30,000 a year. That's what everyone gets." As I would later learn, that is certainly not what everyone gets. In fact, it was two times what another adviser got for being on the same board.

Greider first met Geron's founder, Michael West, after she got a call out of the blue. People she didn't know said they wanted to come out to Long Island and take her to lunch. They turned out to be West and Alex Barkas. Barkas, the chairman of the board of the brand-new Geron, is a venture capitalist, and at the time was a partner in the Palo Alto firm of Kleiner, Perkins, Caulfield & Byers, the hottest venture capital company in the United States. Greider said she should have known how much they needed her from the very first meeting. Cold Spring Harbor is a reasonable train ride but a very expensive taxi ride from New York; "Alex Barkas and Mike flew from Texas and took a cab from New York and had it wait while we had lunch. Nobody does that."

She joined Geron's advisory board in late 1991. Geron soon set up its offices in Menlo Park, California, the heart of the action for the biotech business. Calvin Harley joined the company in 1993, eventually becoming its scientific director. In time, Geron went public. Its stock price went up and down like a roller coaster, especially when Geron trumpeted a paper on telomerase.

So, I said, where did all this life extension and immortality stuff come in? She looked at me with an expression of distaste, as if to say, how could anyone be so stupid as to think her work could extend human life, or that she'd ever been interested in such nonsense?

"No," she said. "There was no extension of life for humans—it was not of interest at all."

But of course it was, I said. Michael West got early money from a Texan named Miller Quarles because West promised that the company would search out ways to greatly extend human life. I knew that was true—I'd already talked to Quarles.

Greider said she'd never heard of Miller Quarles.

She had her head cocked to one side, as if she'd been through this whole immorality business a hundred times before and was sick of it. "I wanted to know if telomeres play a role in tumor growth," she said. "I was always talking about cell senescence."

"You mean if I asked you if your work could be used to extend human life—"

"I'd say you're insane," she snapped. "To tell you the truth, I didn't take Mike West very seriously. But there were other things in the company. And Cal is a serious scientist. That kept me involved for a while."

"When did you get out?"

"I went off the board in 1997," she said. "It was a conflict of ideals, I guess. I wanted to publish data. They wanted to put out a PR line . . . They did the experiment that if you put telomeres back in a cell, you extend the life span, but it was not my way of doing things."[12]

When she joined Geron's scientific advisory board, she was paid cash, but she was also given warrants to buy shares that vested. These warrants could be exchanged for tradable shares only after a named date. Greider left the advisory board in the fall of 1997. At around the same time she also left Cold Spring Harbor for a position at Johns Hopkins. She wanted nothing more to do with Geron, and she sold her shares as soon as they vested. "I didn't want to be invested in biotech," she explained. "It was too much of a conflict of interest."

A friend of hers, a policy wonk she'd known since her days at Berkeley who worked for the Clinton administration, had asked her to serve on the National Bioethics Advisory Commission. In 1998 it began to examine the ethics of human embryonic stem cell research for President Clinton, who was trying to decide whether the United States should fund this research. Greider had attended several meetings of the commission on this subject when "this package arrives in the mail." When she opened it, she found warrants to purchase Geron shares that she didn't realize were still owing to her. Geron had invested in 1995 in the work of James Thomson, the first to grow a human embryonic stem cell line. So there she was, a member of the commission studying the ethics of stem cell research with a financial interest in Geron that inadvertently she had not disclosed. She thought she could just put the warrants in a blind trust, but when she declared them to a lawyer for the commission, she was told she needed to recuse herself from the remaining meetings. And so she did.

As we threw our empty cups in the garbage, I tried again. I'd chased her down because I thought she was hard on the trail of immortality. I thought her work had shed light on the mechanics of mortality, the most basic human condition. I thought the paper she published in 1992 with Silvia Bacchetti and others said as much. "So none of this will lead to making humans immortal?" I asked, plaintively I'm sure.

"We said cell senescence. We never said it will change the life span of humans."

But surely there was some connection? The people who worried that Dolly the sheep would live a shorter life because of shorter telomeres hadn't just made it up?

She waved her hand as if to dismiss this as so much fluff.

I persisted, asking why people would keep suggesting that telomeres have something to do with aging if there is nothing to it. "People like to hear that," she replied. "There's not a single paper that [shows] it will have anything to do with life span. . . . We did a side experiment . . . We found strains [of mice] with tenfold different [telomere] lengths." There was no difference in the life spans of these mice that could be correlated to the lengths of their telomeres. "We published that three years ago."

Greider, like Elizabeth Blackburn, her former supervisor, was interested in manipulating telomeres to block cancers, not in immortality. And since Mike West had left Geron, the company had also moved away from aging and was focusing instead on stem cells and cancer. Greider made a wry face about the way the meaning of her work had been abused, wrinkling her nose as if she smelled scalded milk.

The day after I got home, I called up Geron's website. According to the press releases posted there, she was right, the company was focusing on stem cells and on the use of telomerase to fight cancer, not old age. But the company had not completely buried its past. It still had an interest in the degenerative diseases of aging and some of those still on the scientific advisory board were human-aging specialists, such as Jerry Shay and Woodring Wright. There was one other person on the list whom I had not expected to see: James D. Watson, Nobel laureate and president of Cold Spring Harbor Laboratory. Though he has since stepped down, he had joined the board when Geron was formed.

CHAPTER FIVE

A LITTLE SKIN MAGIC

The train rumbled along beside Lake Ontario. It was mid-December and the grass in Toronto was still green, but as we cut north through the sedimentary ramparts guarding Kingston, everything changed. The color drained abruptly from the world, disappearing under the cover of pure white snow. When I stepped onto the platform in Montreal, the air was cold enough that it hurt to breathe.

Cold air is a dose of salts: it goes right up the nose to the brain and turns romance to ice. I was suddenly filled with doubt. If I'd been wrong about the meaning of Carol Greider's work, could I be wrong about Freda Miller's adult stem cells, too?

Miller and her colleagues had published that they could isolate, from the second layer of the skin of juvenile and adult mice, certain stem cells that have the potential to turn into various tissues. Dubbed SKPs (for skin-derived precursors), these adult stem cells, when grown in flasks and then in plates in the right chemical contexts, transformed into muscle, fat and neural cells that produced the right sorts of chemical signals. Miller then found cells in human skin that would do the same thing.

The article was fascinating for various reasons. It was carefully bold. The paper's first author, Jean Toma, and the last, Miller, and the rest of their colleagues (including Miller's husband, David Kaplan) were not the first to show that stem cells found in adult tissues are capable of making cells different from the tissues where they're found.[1] But Miller's experiments were the first to show that adult human stem cells can be so versatile, and so the article was political dynamite: it offered strong support for those who argue that there is no need to extract stem cells from embryos in order to cure Parkinson's and diabetes and to help Superman walk. The research undercut Reeves's position, both morally and scientifically. At the end of the article Miller and colleagues wrote, "fetal tissue is the current tissue source for human neural and embryonic stem cells, raising important ethical issues. Moreover, the use of human fetal tissue involves heterologous transplantation, and the requisite accompanying immuno-suppression is particularly problematic in individuals with long-term neural problems, such as spinal-cord injury or Parkinson's disease."[2] If Miller and her colleagues could figure out how to make these adult cells develop appropriately, Christopher Reeves might be able to mine his own skin for cells that could give rise to neurons to repair his spinal cord, without having to cross any moral frontiers.

In the few months since the publication, Miller had been on CBC Radio's national science show and had been written up as a celebrity in the newspapers. But she had not given a paper at the American Society for Cell Biology. And although I went to as many sessions as I could on stem cells, no one had mentioned her work. Was there a flaw in it somewhere?

Miller's laboratory was in a brand-new brick and glass addition to the Montreal Neurological Institute's Penfield Pavilion, halfway up the mountain on the McGill University campus. A quotation from the great neurosurgeon Wilder Penfield had been etched in a bronze plaque set into the building stone at eye level: "The question for neurology is the nature of mankind."

The more papers I read, the more the idea of human nature residing in the brain seemed a quaint relic of pre-Darwinian thinking. The difference between the genes of man and chimp is said to be a simple matter of the double copying of one gene important in brain development.[3] If this assertion is correct, one doubled gene is the origin of human nature, and of all of human history, our gods, our wars, our cultures and our moral conviction that all the world is ours to dominate and subdue. If this doubled gene is responsible for these vast differences between man and chimp, surely the nature of humanity can be found not just in neurons, but in every single human cell?

Miller's work raised vitalist questions. What makes a human cell decide to become one sort of tissue rather than another? Are stem cells a different kind of cell from all the rest, imbued with something that allows them to decide their own destiny? Or is every cell a prisoner of circumstance, constrained by forces of gravity, electrical charge and magnetic polarity? If we can answer these questions, can we control the development of any cell?

Freda Miller's size was completely at odds with her voice, which is rich and deep—as melodious as Adrienne Clarkson's, as lively as the late Barbara Frum's. It carried a layer of warmth that made it completely remarkable. If I closed my eyes, she was a woman in her ripe forties, a woman of imposing height and physical authority, a Shakespearean sonnet of a woman. Eyes wide open, she was a tiny little thing, with street-kid peroxide blond hair cut in a mullet with long bangs. Her thick black eyebrows stretched straight above her eyes like accents denoting a hard vowel sound. Intensely blue eyes, black-rimmed with thick lashes, stared out from chic brown-and-red-flecked glasses, deflecting attention from a tiny chin. At forty-six, she was very personable and pleasing but with a certain feisty core.

Miller grew up in Calgary, Alberta, the oldest of six children; her father tended vending machines for a living. She was a gifted student who attended the neighborhood vocational high school. "It was hard-core working class," she said. So how did she find her way here, to one of the most important neurology centers in the world? In high school, she explained, she had two key teachers. She'd always imagined that she would grow up to be a writer, but it was the late 1960s and her English teacher did not want to teach literature or pass on the discipline of writing, just to "sit and hold hands and commune. It was so awful." Her other key teacher taught chemistry and told the students, "'Forget everything you learned about science. We'll just do experiments.' It was amazing." So one door shut while another opened.

Her parents moved to Saskatoon, so she did her undergraduate work at the University of Saskatchewan. Not surprisingly, she studied chemistry and was thrilled by biochemistry, the study of the giant molecules found in living things. "I loved how big molecules interacted. I liked math, it was problem solving. How do you put A and B together?" The thing she most liked about biochemistry was its visual element: "You have a 3-D picture in your head."

But she hated biology and avoided it until graduate school, where she became interested in "how big molecules like DNA work, and then how it makes the cell what it is." At the University of Calgary, where she did her doctorate, she became interested in how genes shape organisms in development.

I asked what she knew then about how cells differentiate.

She knew that fertilized eggs, or early embryonic cells, are capable of making any kind of tissue in the body, but that once they have differentiated, they can't go back again. "I was taught that to make a human and a rat you start with germ cells and embryonic cells and these make everything, and to make a liver you progressively restrict the cell until you end up with a liver cell, and it can't be anything else."

This dogma was under assault by the time Miller did her doctorate, along with the idea that all genes are arranged, side by side, like pearls on a string. By 1980 genes were known to be organized differently in animals with nuclei (eukaryotes) than in organisms without nuclei (prokaryotes, such as bacteria). The genes in bacterial DNA are set side by side, but in eukaryotes, only parts of the gene—called *exons*—actually code for a protein, while other stretches of DNA on both sides of these exons, sequences called *introns,* do not. While assembling themselves on genes, messenger RNAs edit out these introns, arranging themselves to carry only the exon's make-this-protein message. It was also known that segments of DNA that permit the expression of these exons, called *promoters,* are read first by an RNA but might not even be on the same chromosome as the gene they turn on.

By the early 1980s, the dogma of development as a progression of limitations had been undermined. When Miller was a post-doctoral fellow in 1985, "people could take a muscle and a liver cell and fuse them, and lose the liver nucleus." She began to ask whether "the muscle cell only makes muscle things, or whether something in the environment of the cell could convince it to make liver things."

"We thought that the DNA was actually modified by enzymes, that they actually physically modify it . . . Once silenced, it never turned on again," she said. But experiments were showing that something else was possible. In the mid-1980s, researchers discovered that a protein transcription factor could turn a fibroblast cell into a muscle cell. Not everything about development was irreversible.

Miller was interested in the brain, though not in the questions Wilder Penfield asked—about the relationship between particular cells and feelings of consciousness—but in how the brain develops and repairs itself. In particular, she wondered how early embryo cells decide to become one of the several types of neural tissue: neurons or glial cells. "And in one form or another," she said, "I've been following that ever since."

She was accepted as a postdoctoral candidate at the Scripps Research Institute in La Jolla, California, in 1985, in the laboratory then run by Floyd Bloom, who went on to become editor-in-chief of *Science,* the major peer-reviewed general science magazine in the United States. Miller wanted to know why the same genes express themselves differently at different points of development. She used a technique called "subtractive hybridization" to find genes expressing at higher levels (producing more of a particular protein) at one stage of development than at another. She isolated the messenger RNAs that assemble themselves on active exons, from which she could infer what the exon sequence must have been. The answers she got opened new questions. The *Drosophila* experts had found master switches that turn on cascades of genes; she wanted to know what caused these master switches to flip over. The methods she was using imposed a cause-and-effect order on systems that are anything but linear.

She was particularly interested in axons—the long arms of neurons—and in the gene that causes axons to be made. She drew me a picture of an axon. A neuron is a weirdly shaped cell with one end that looks like a starfish, but with another extremely long arm, which she called a process. "A neuron is a complex cell," she said. "There's no nice rounded shape. There are long processes to make connections, one neuron to another. In the spinal cord there is a cell that sends a long process all the way to the muscle in the foot. In a giraffe it's meters long. The gene we identified builds that long process . . . This gene is the member of a family of genes that is part of the structural organization of the cell. It is a tubulin, a neuron-specific one made to build axons. So then we used that as a tool to ask, what makes a precursor into a neuron?"

In her three years at Scripps, Miller also got more insight than she wanted to into the signaling pathways used by her colleagues to orient themselves at the top of the science heap. She had arrived there as an idealist who considered science a pure form of inquiry. Then her group collaborated with another group, which had located a gene they thought was associated with Alzheimer's disease, and she soon found herself surrounded by colleagues jockeying ferociously for authorship position. She almost quit science.

Was it the thought of getting rich, or being famous that drove them? I asked.

"It's all about the fame," she replied, "the Nobel, being known by peers."

In 1988 Miller left the Scripps for a job at the University of Alberta. She was offered her own lab along with research money from the Alberta Heritage Fund. There she soon allied herself with a group interested in how the nervous system develops. She was asked to join one of the Canadian government's first generation of elite science networks, this one set up to study prospects for neural regeneration.

Some of Miller's colleagues in Edmonton were expert in growth factors, various chemicals that encourage cell differentiation. Miller was soon able to show that a particular gene she was interested in turned on as cells made the transition from precursor to neuron. She then found the controlling region that turned the gene on, and located the promoter. Miller connected this promoter sequence to a reporter—a gene from another species that produces a distinctive blue color in the tissue when the promoter is working. In genetically altered mouse cells, when the promoter turned on, the cells turned blue.

In 1993 Sam Weiss of the University of Calgary discovered the existence of stem cells in the adult rat brain. Miller and Weiss had been graduate students together, so she heard about it long before he published. The discovery of such cells lurking in a developed brain of a rat was a shock. Until then, everyone doing neurobiology in humans thought that no new brain cells were made after the age of three and that brain cells begin to die in humans by the age of thirty. "He showed stem cells in the adult rat brain that make neurons," said Miller. What was found in the rat would likely be found in man.

At first Miller was very skeptical that such stem cells could ever be used in humans for therapy. For one thing, she saw big ethical problems in trying to get human neuron precursors from the sources then available since they would have to come from the brains and spinal cords of aborted fetuses. But she did work on the problem of how to keep such neural cells growing indefinitely in a dish. That's when she began to ask another question: if stem cells can be found in the brain of a mammal, maybe they can also be found in other parts of the adult body?

The first area she and her colleagues searched was the layer of cells in the lining of a mouse's nose. Miller's thinking was that since one always breathes things that are not good, these cells must die off, and there had to be some precursor nerve cells in a mouse's nose that can be recruited to replace dead cells. What if these stem cells were not restricted to making olfactory neurons? What if they could make all kinds of neurons? In 1996 she and her colleagues set out to find these cells and see if they could turn them into brain cells. And, she said, "It worked." Her tone of voice was very matter-of-fact, but she was sitting in her chair in a very peculiar way, as if she wanted to push away from her desk, and maybe from me. Something went wrong here, I thought.

"We contacted *Science*," she said. "They said please submit." She paused.

"And then an unfortunate thing happened. I got the nastiest review I got back ever in my life." Her mobile mouth had tightened right up. She could still quote from the review five years later. "'What are these mysterious cells? Impossible. Who are they kidding?' The other review was good."

But one out of two wasn't good enough, and *Science* wouldn't publish. The postdoc in her lab who'd developed the experiments, Andrew Gloster, had a job offer in Saskatchewan. She realized that extra experiments needed to be done to convince skeptics, such as the reviewer who had made such obnoxious remarks. "He was to do the extra experiments. We realized it was so radical we had to show it ten ways from Sunday. It's politics."

How is it politics? I asked.

"*Science* and *Nature* and *Neuron* have editors," she said. "At *Science*, the editors can accept even if the reviewers have issues." But these editors, she explained, behave differently according to the status of the persons they deal with.

"Somebody with a big name submits a paper and it gets one good and one bad review. The big shot convinces the editor to take it anyway," said Miller. "So, first, Canadians aren't as plugged in and don't reach superstar status like a big shot at Harvard. And with women, it's harder to do that. We won't phone and bully and cajole. And Canadians are not as aggressive as American scientists. The Canadian way is to do everything they ask."

She was certainly correct in describing how some scientists bully and cajole the editors running big journals. Sir John Maddox, the editor of *Nature* from 1966 to 1973 and again from 1980 to 1995, has commented on how relations between scientists and editors declined between his first term and his last. In a preface to the expanded edition of Horace Freeland Judson's *The Eighth Day of Creation,* a history of the early days of molecular biology, Maddox wrote,

> By 1980 authors would call to ask whether their manuscript had arrived, whether it had been sent to referees, why we had declined to publish it and why it was *Nature*'s consistent practice to rely on referees whose intelligence was below par, whose judgment had been warped by self-interest, whose charitable instincts had been blunted by cynicism and whose parentage must even be in doubt . . . These endless telephone conversations from 1980 onwards were a telling sign that good manners had ceased to matter.
>
> Sadly, that is only the tip of an iceberg. By 1980, secretiveness had become commonplace. Authors had taken to sending long lists of names to whom, please, manuscripts should not be

sent for review, with the explanation, to authors always self-sufficient, that the listed names were those of people working on the same problem . . .

Then there was outright fraud. The first case that came *Nature's* way, in 1980, was that of an ingenious Jordanian in Houston whose technique was to steal manuscripts from other people's mailboxes, retype them, replace the authors' names with those of himself and two or three distinguished people (not at Houston) . . . Latterly, the spate of accusations has been attenuated, but the competitiveness is still there.

Indeed, the competition has even intensified, with the recognition that there are useful things to do with molecular biology, and money to be made as well.[4]

Miller and her colleagues knew that unless they moved fast to do new experiments to answer lingering questions, they would lose their window of opportunity. Her postdoc left for his new job, and though she asked him to do these experiments in his new lab, he didn't do them. But she filed a patent application.

Since she was still a member of the Neural Regeneration Network, the network provided her with a patent lawyer. The matter of who would own the patent was tricky. McGill has the right to file patents arising from the work of any of its professors, but, according to Miller, McGill wasn't protecting its patents because it was too expensive. There are lawyers' fees, fees to file the application in every appropriate jurisdiction and then annual fees that must be paid to maintain the patent until the end of its life. If she hadn't been a member of the Neural Regeneration Network, the patent opportunity would have evaporated, just like the *Science* publication.

It was in conversations with the lawyers, when she had to explain why they had decided to look for stem cells in the cells inside a mouse's nose, that she followed her own logic and arrived at the next set of experiments. She realized that sensory receptor neurons also exist in skin, and that skin cells, like nose cells, die and slough off all the time. Surely there were stem cells in skin to replace those nerve cells that died? If stem cells in the nose could be induced to make neurons, why not stem cells in skin?

The Neural Regeneration Network did not get its federal grant renewed and ceased to exist, but one of its former administrators called Miller one day. He had a new job in a venture capital company interested in investing in basic medical research, and he asked her how she was doing with that stem cell stuff. When she explained she was waiting for her former postdoc to do more experiments, he told her not to wait anymore but to do it herself. He said that his company would find a way to fund it.

I had noticed that at the bottom of her recent article, she had acknowledged both CIHR for its research grant and a private company called Aegera Therapeutics. She explained she is a founder of Aegera Therapeutics, a company in Montreal that works with growth factors. She had patented early work on certain growth factors and assigned them to the company in return for shares. Aegera employs thirty people and has already raised its first round of outside capital, which significantly diluted the value of her shares. "They pay me consulting money and fund some research projects," she said. The venture capital company invested in Aegera, which in turn entered into a contract with her for $150,000 to do the experiments searching for stem cells in skin.[5]

She and her senior research associate, Jean Toma, used the same techniques to find stem cells in skin that they had tried earlier with the cells from the nose of the mouse: they grew the cells in suspension. "The hallmark of the cell," Miller explained, "is that when you grow it in suspension it grows in these clusters. They respond to growth factors."

She and Toma took samples from the skin of juvenile and adult mice. They broke it up, digested it with enzymes and put it in tissue-culture flasks with a growth factor called EGF (epithelial growth factor) and another called FGF (fibroblast growth factor). "Most cells stick to the bottom or the sides of the flask and die," Miller explained. "But some clumped together in spheres." Toma pulled these spheres of cells out of the flask, purified them and eventually got a pure population of floating clusters of cells; individual cells taken from these clusters also proliferated and made more clusters. "They looked a little like olfactory spheres," said Miller. "They were more like grape clusters. Then we asked, do these cells express neural stem cell proteins? They did."

When they put these cells on a sticky substrate in a dish, after several passages the cells began to differentiate. They produced markers normally seen in various forms of neural cells, including Schwann cells, glial cells and astrocytes. "They have long processes and express proteins specific to neurons," she said.

The researchers determined by elimination that the cells that clustered in spheres arose only from the dermis, the second layer of skin, not from the epidermis. Further, they discovered, they weren't restricted to developing as neural cells. When they exposed these cells to different growth factors or serums, they got different kinds of cells. They exposed them to rat serum and generated a small population of smooth muscle cells. They put them in fetal bovine serum and this time they turned into fat cells. They could distinguish these cells from other stem cells, such as those from bone marrow, by the different chemical markers they produced, markers consistent with the shapes they assumed: cells that looked like neural or fat cells made the proteins of neural or fat cells. These SKPs seemed to be able to develop into cells of two different lineages. They were also able to keep passaging these SKP cells for months. They are functionally immortal.

And then they tried the same experiments with human skin; they got their skin samples from the small pieces of scalp tissue that are removed when neurosurgery is performed.

As Miller told this part of her story, she appeared elated and anxious at the same time. She knew exactly what people believed was possible when she began these experiments and how her findings contradicted it. As with the mouse cells, the human skin cells spread out in the flask; some stuck to its edges and died, but some formed into clusters. When exposed to the right growth factors, they produced neural markers and looked like neurons. "It caused me more sleepless nights," she said. "I could never believe it. I kept asking, where's the flaw? Or, I'd say, it won't repeat again . . . They were much more multipotent than expected."

In the language of cloning and development, biologists describe cells as *totipotent* when they can produce any of the cells of the embryo, including the placenta. Only fertilized eggs cells were supposed to be totipotent, but that changed when Dolly was born. *Pluripotent* cells can make almost all cell types except for placenta.

Miller's use of the word *multipotent* suggested these cells might differentiate into several of the lineages that descend from the fertilized egg, but it also suggested limits.

"Now the question is how far can they go," she said. "Well, we've worked on how plastic they are. I can tell you . . . they can also make bone and cartilage cells with different growth factors." She did a group of experiments in which they broke up the clusters of cells. Daughter cells pulled from the cluster could give rise to different cell types. "At least 60 percent of those in spheres are multipotent," Miller said.

Her next goal is to try to differentiate them until they become mature neurons. After that, the next big problem will be to find a way to distinguish these cells from others in the second layer of skin. "At that level, they are little round cells with big nuclei—they look like nothing, but . . . they are bags of potential."

Miller liked to think of this potential as something that had once been broadly available but had been progressively restricted as mammals evolved from an amphibious ancestor. A newt with a limb cut off will generate a replacement limb; the cells for the new limb are recruited from the dermis, the second layer of newt skin. Similarly, an earthworm, cut in half, will regenerate as two worms, and the cells recruited to grow the other half come from the connective tissue, which is similar to the dermis in the worm. So, she said, it wasn't so crazy to find such primitive stem cells in the dermis of mammals, including humans.

So why can't mammals regenerate their limbs? I asked.

"Maybe in mammals you can't afford to have that ability anymore." Why not?

"The best speculation is those cells could be dangerous. Embryonic stem cells can make a teratoma [a cancerous cell]. You don't want that popping up. We are long lived and exposed to ultraviolet radiation. Maybe there's a trade-off. The animals that regenerate, like newts, are short lived."

Where are you getting human skin from now? I asked.

She had arranged to use skin left over after cosmetic surgery, and also foreskins taken from circumcised babies.

Circumcision among Jews began as a religious practice that signifies an ancient contract between God and man: man will obey, and

God will not require a human life in exchange for favor. How strange, then, that a piece of foreskin might enable us to regenerate nerves, and maybe organs and even limbs—to keep mortality at bay. Just that summer I had attended a bris, the ritual circumcision of the newborn son of one of my young cousins. It was performed by a doctor who is also a moyel, religiously trained to do this ritual. My tiny eight-day-old cousin was laid on a pillow on his grandmother's dining-room table. A drop of wine was placed on his mouth, and the foreskin on the end of his penis was ripped away with a scalpel. While everyone cooed over the wailing baby and made certain his mother was bearing up under the strain, I asked the moyel what he did with the foreskin. He said that the foreskin is supposed to be buried in the ground in a secret place by the father of the child. In other words, the burial to come is foreshadowed by this painful beginning; death is part of the bargain; life is a circle.

Miller doesn't worry anymore that there is an error somewhere, or that her work will not be believed. The NIH has asked her to apply for grants and has told her that it would not require her to have a U.S. partner because her work is unique. "No one in the U.S. is doing this," she said. "They are desperate to fund human adult stem cell research."

Well, why isn't anyone in the U.S. doing it? I asked.

Because it's really hard, she said. Keeping such cells alive in culture in a flask, growing and dividing, as hers have for more than a year, is not easy, and is very difficult to teach to others. "This is like gardening," she said. "You and I can do the same thing, you can have a beautiful garden and mine is dead."

So when would she try to transplant these differentiated cells into a human being?

Not any time soon, she said. Some researchers think brain tumors start in these precursors cells. "What if I transpose them to the head, and they make bone?" she said. "We've gotta put them on a journey to becoming neurons and put that in. That's a problem. We have to learn how to produce the cell we want. We're close, as a field. I'm amazed at the progress." There was a person in the Stem Cell Network, working out of the University of Toronto, who was developing what she called a "bioreactor," a big flask to grow cells.

He was already moving forward to the next step: growing them in clinical quantities. When she was ready, he'd be there.

This talk of flasks, newts and recipes, and, most of all, of vaulting ambitions, brought the three witches in *Macbeth* to mind. In act IV, they are bent over a bubbling cauldron, making their potion for construing the future, getting ready to push Macbeth to the last notch of hubris and bring on his doom:

> Second Witch: Fillet of a fenny snake.
> In the cauldron boil and bake:
> Eye of newt and toe of frog
> Wool of bat and tongue of dog.
> Adder's fork and blind-worm's sting,
> Lizard's leg and howlet's wing.
> For a charm of powerful trouble
> Like a hell-broth boil and bubble.

Eventually Macbeth arrives, demanding answers to his burning questions. He would allow castles to tumble, oceans to overwhelm all navigation, "even till destruction sicken," if it would help him learn what he wants to know. And he is not content to hear the answers from the witches' lips: he wants the goods straight from their masters. The recipe to conjure such apparitions requires even more foul things to be added to the broth.

> First Witch: Pour in sow's blood, that hath eaten
> Her nine farrow: grease that's sweaten
> From the murderer's gibbet
> Throw into the flames.

"Did you like magic when you were a kid?" I asked Miller.

"Totally," she replied.

"Isn't this like magic? You do something, you get a result, you don't know why, but it works, so you repeat it?"

"I tell people we know nothing," she said.

CHAPTER SIX

FLIES, WORMS AND RUNNING-DOG GENETICISTS

I flew to Vancouver at the turn of the new year. The plane descended into a warm Pacific gloom, through ever-thickening layers of cloud and fog. I was hoping for the flash of sun on snowcapped peaks or the brilliant shimmer of a sun-pocked sea, omens of an approaching state of clarity. Clarity makes me happy. What I got was murk. Result: misery.

Biologists' curiosity had produced such peculiarities—chimeric green mice, grape clusters of potential transformed in goopy cauldrons. Could one call tinkering with recipes, as Miller described her method, or learning by doing, as another eminent biologist described it, science? Wasn't it more along the lines of alchemy? Science is not supposed to be like this, I kept telling myself. It is done by reduction, simplification, specialization. Yet biology's separate streams of learning had come together in a flood, like the torrents of water slopping out of the buckets under the dubious command of the sorcerer's apprentice. When I tried to snag one branch of the tree of life, down came all of it—cell biology, biochemistry, developmental embryology, cloning, genetics, genomics.

Freda Miller might have liked magic as a child, but I didn't. I watched like a prosecutor, hoping the magician would slip up and expose his manipulations to the light. Magic, I used to explain to my credulous friends, is just some machine you didn't notice. I remained a relentlessly rational materialist until my early teens, when I almost died of asthma. Actually, I think I did die. The white light was there; I followed it and flew up. I saw my poor body sprawled below me, head in a spew of vomit. Whatever and wherever I was, I was not in that helpless flesh on the linoleum floor under the bright, clinical light of my mother's kitchen. But my father injected me with adrenaline he'd kept in the fridge, just in case. Suddenly I was back, using my own muscles to turn my head, fearing my own frantic heart, hearing my lungs sob and whistle for breath, disgusted by my urine-drenched jeans and acid-seared throat. After that, my certainty about material explanations receded. How could I account for a persona that seemed to exist independently of my body, of my material self? Ms. Rational, the one in search of the machine, the one with the explanations, the one so pleased by tight fits between theories and actual measurements, had to acknowledge that there was also this other, ephemeral self that hovered above. This Other One has developed many curious and irrational habits over the years: she looks for omens, reads horoscopes, notices coincidences in life and thinks of them as part of a larger plan, is somehow connected to the pain of the distant child, wakes up in the night with foreboding just before the phone rings. This self is rarely happy.

For a long time, I thought any talent I had as a reporter came from Ms. Rational, since she finds the patterns that loop things together. But gradually, I came to see that the Other One also makes a contribution: she feels the hidden story beneath the veneer of a proffered biography and digs down to bring it up to the light. I had been trying for months to keep the Other One under the reliable control of Ms. Rational. How could she grapple with these nouveau Darwinians messing around with creation? But by the time I got to Vancouver, Ms. Rational was at her wit's end. Biologists' explorations of the material realm had exposed epiphenomena as ephemeral as the Other One herself.

And there was another problem. I had never had such trouble getting people to talk to me. Well, I shouldn't say never. There have been sources who would speak only off the record because they were afraid of retribution. There have been people I couldn't get to because they were dead. There have been important people who, for various reasons, thought it imprudent to speak to me, did so anyway and wished I were dead. But eminent biologists weren't afraid, weren't dead and kept turning me down. It took a while to figure out why. I thought they would make more time for an author than for a reporter, because an author can take more care to study their work than a daily hack can. But it turned out that I was calling media-slick people who know that books are read by few, whereas television or newspaper reports are seen by many. Media appearances can be used to support appeals for tenure, lead to government appointments, help sell shares. Fame, in other words, is power, and power brings forth money.

This concern with fame had exposed itself right away. It had seemed to me that if I wanted to understand the story of revelationary biology, I should start with its living founders. I had phoned Cold Spring Harbor Laboratory and asked to speak to James D. Watson, but various conversations and e-mail exchanges had got me nowhere. Watson was far too busy to talk to me. He was writing a book, planning promotion for another book, giving interviews to the BBC, which was creating a documentary series for the fiftieth anniversary of the model of the DNA molecule, and so on. Did I not know that everybody wanted to see him, all the time? I tried every stratagem I could think of. I said I thought Cold Spring Harbor Laboratory is the most important institution in North America for the dissemination of new ideas in biology. (I admit I didn't say I thought Cambridge mattered more.) I said I wanted to attend a conference, which was true. Cold Spring Harbor is in the business of selling conferences, among other things, so they liked that. They suggested that if I came to a conference, I might bump into Watson since he often wanders around loose on the laboratory's campus. I looked up their conferences on their website: I'd have to shell out many hundreds of U.S. dollars to get in the door, too much money for a maybe.

I began to hunt for people who could introduce me to Watson, and I learned, by the serendipitous act of reading the proxy statement of a company in which my daughter owns one share, that I have friends, a couple, who know him very well. Watson is a director of a publicly traded company they helped found. "Darling," said the wife, "he's like a rock star, everybody wants him, we're trying to figure out how to get him up here for the Gairdner Awards." How about his private number? I asked. "I never give out my friends' numbers to lawyers or journalists," said the husband. This goaded me to follow other investigative lines, about which more later.

Watson had always been a man in search of fame; he had even posed for *Vogue* magazine.[1] He had written one self-revelatory book, *The Double Helix,* and was about to publish another. But Francis Crick had not sought out celebrity. Surely he would be reassured that I was writing a book, that no anonymous editor would cut his valuable words by the column inch or the sound bite?

I called Crick's secretary at the Salk Institute. She was charming, but explained that Crick had just had major surgery and, besides, he never, ever talked to journalists. If I insisted, I could send him an e-mail request and she would pass it on. I did, and I got an e-mail right back signed Francis Crick. He said I could write to him again, closer to the moment when I might want to come see him, and he would consider my request. I'm on a roll, I crowed, he'll consider it. But when I did write to him again asking for an appointment, he replied that nothing I was interested in was of interest to him; therefore, he would not see me. Crushed, I aimed lower— much lower. I tried for the CEO of Nexia, a little company that is genetically modifying goats to produce spider silk. After various phone calls and e-mails, I got a very sharp no. Where would one find the time for science if one talked to all the press who called? And yet Nexia made time for the *New York Times Magazine, CBS News* and the *Toronto Star.*

I was on the verge of despair when I couldn't get even Vancouver scientist Steven Jones to take my call. He was first author of that article about the semi-immortal worm *Caenorhabditis elegans* that had caught my eye in the summer. Jones and his colleagues had studied the genes expressed when this worm enters a state of suspended animation, a state that can vastly extend its life span. One of those genes had not been previously known; it expressed itself twenty times more in the dormant state than any other and seemed to have something to do with telomeres.[2] As the first author listed, Jones could not be the most important guy in the world—in biomedical publishing, the first author is the grunt; the last author is the one with the name on the laboratory door. Nevertheless, the assistant said Jones was far too busy writing grant applications to talk to me. There was another man named David Baillie I could try if I really insisted.

Since Baillie's name was neither first nor last on the credits, I deduced that he was dispensable enough to be thrown at stray journalists. Wrong. Baillie holds a prestigious Canada Research Chair in Genomics at Simon Fraser University, a position funded by the government of Canada so that the best researchers will stay at home.[3] When I called him, I learned that he also knows well many of the biological wunderkinder who had passed through Cambridge, and he had been there himself, working with the perennial Nobel bridesmaid Sydney Brenner. He had done post-doctoral studies under Francis Crick. He knew James Watson. He knew Frederick Sanger, not to mention Sir John Sulston, just knighted for his services to Britain on the Human Genome Project. Baillie had been working on genetics, genomics and the worm from the early days. His life in science spanned the period in which the study of biology changed entirely. Baillie would know why all these disciplines began to merge, how to open all the doors shut to me. He talked for an hour on the phone and actually said he'd be pleased to see me.

I could have cried with relief. Baillie, I thought—Ms. Rational thrust firmly to the background—you will be my Virgil, you'll guide me through this Hell of Fame.

The rain came down as a droozle of fine mist. There was a bitter taste of salted mildew in the air, and there was so little sunlight it could have been dawn or twilight; in fact, it was early afternoon. I had to use my cell phone to find my way through the cobbled maze of streets beside Vancouver's False Creek. I walked by concrete townhouse after concrete townhouse, with the seawall on my right, as directed. Finally his voice said, do you see the red door? Yes, I said. Look up, he said, look way up.

David Baillie was standing in the greenhouse window of his second-floor living room. He waved, came down and led me in past the little front office, through the long, dark hall and up the stairs. Right across from his front window, bobbing in the water, were count-less rows of sailboats. On the other side of the creek's wide mouth, countless condo towers snuggled tight to each other. As the afternoon faded, their lights turned on and brilliant neon hues and skied across the water, moving streaks of red, blue, orange, green, yellow.

Baillie was tall and square shouldered. His hair was pepper and salt, gray over brown, thick but dead straight, with eruptions of cowlicks in strange places. He had a mustache of the same color and density. He wore aviator-style glasses, a sweater vest over a long-sleeved shirt, black pants, socks, no shoes. I wasn't certain how old he was; he seemed old enough to be kindly, young enough to have hidden sharp edges. Early fifties, I thought, like me.

Actually, he was in his late fifties, born in 1944 in Britain to a Canadian-soldier father and a British mother. He had to think hard to come up with anyone in his family who had attained any higher education before him. His father, a carpenter, built the small suburban Burnaby house Baillie grew up in. He also built a shortwave radio and, from his basement, roamed the airwaves with the likes of Hussein, king of Jordan, and Thor Heyerdahl aboard the *Kon-Tiki*.

By age ten Baillie knew he wanted to be a scientist. He learned about biology by messing around in ponds and using a little microscope to look for paramecia and amoebas. He was the only person for miles around who read *Scientific American;* he picked it up at a local newsstand that brought in only one copy every month—his. He was contemptuous of organized religion, and it wasn't for lack of exposure. While his father was an atheist, his mother, an Anglican, sent him to five different denominations' Sunday schools over the years, where he learned that "they all called each other idiots." When it came time to go to the U.K. to do postdoctoral studies and he had to be sworn in as a Canadian citizen to get his passport, he would not swear on a Bible. They told him to bring any book he believed in; he swore on *Fundamentals of Genetics.*

Baillie was the only one of his science-nerd friends who took biology in high school. His friends' attitude was, if you couldn't put a number to it, it wasn't much of a science. When he studied Mendel and the statistics of inheritance, Baillie realized that you could put numbers to biology, but he was more captivated by the classification system, which demonstrated the evolutionary links between species. He was amazed that you could recognize by their structures that a butterfly, a beetle and a silkworm are all related organisms. "This was a marvel to me," he said, and that they are also grouped with crabs "made it even more remarkable."

Physics and biology labs were like night and day. "A person trained in physics can't accept that biological solutions are never perfect," he said. "An organism can't do something it has no genes to do. It can only do the best it can with what it has. There is no optimal solution . . . Every change has to keep working—that is the rule of biology. You have to stay alive and change the thing you started with."

He first learned about the chemistry of biology when he entered the University of British Columbia as an undergraduate. He found a book in the bookstore called *Design and Function at the Threshold of Life,* by geneticist Heinz Fraenkel-Conrat, which described the tobacco mosaic virus. This virus's genome is nothing more elaborate than a string of RNA. As Fraenkel-Conrat laid it out, this RNA specifies the protein that folds and covers it in such a manner that it can move into a tobacco plant and multiply. "It allowed the organism and its information to get into a new organism and do it again," Baillie said.

But Fraenkel-Conrat didn't know exactly how the RNA specified the protein because the genetic code was not yet known. "Crick wrote a paper in the 1950s that was based on an elegant model but he got it all wrong," Baillie laughed.[4] Baillie had no awe of Crick because he had a living knowledge of biology, of the confusion and false certainties that had hindered all those trying to puzzle out how genes work. I realized he might even tolerate it if I asked him the really stupid, ignorant questions that bedeviled me. I leaped right in.

"What exactly is a chromosome?" I asked him. "I mean, what is it made of? Is it just one string of DNA, or is it also protein, and how does that work exactly?"

Chromosomes are half and half, he said—half DNA, half proteins called *histones.*

Why are genetics papers written so badly? Why are there so many names for genes? There was no rhyme or reason to the way my questions tumbled out. He looked startled, appalled.

The phone rang before he could answer me. It was his colleague Don Riddle, visiting from the University of Missouri at Columbia.

They had agreed to meet for a drink, but Baillie ominously informed him that he and I were going to take a lot longer than he'd thought. He handed me the phone, suggesting I make an appointment with Riddle too. Riddle, he explained, helped develop the use of the worm *C. elegans* as a genetic model organism. Riddle had been Sydney Brenner's postdoc, and he had done amazing things. I made a lunch date with Riddle.

I made Baillie start at the beginning—what biologists thought they knew about genes, for example, when he started school. It wasn't until Baillie's third year at the University of British Columbia that he got to take a genetics course. "All my biology professors were old farts," he said. But not this smart young genetics professor. He was David Suzuki, now a Canadian television star and environmental evangelist, then a twenty-eight-year-old expert in the genetics of the fruit fly. Suzuki started the class by explaining that the gene was just a concept, a bundle of information. Then he moved on to how genes could be mapped on chromosomes, then to the idea that genes interact, then to the chemistry of all of it. "Molecular genetics, DNA, RNA, proteins—we got none of that in biochemistry till years later."

In the early 1960s, biologists who studied the molecular nature of genes were a tiny new breed.. Biochemists didn't talk to them much; biochemists were concerned with the nature of the proteins made by living creatures, not with how living creatures make proteins. Many rabbits died so that infinitesimal amounts of proteins could be recovered, weighed, reduced to constituent amino acids. Molecular biologists, on the other hand, were trying to work out the information system that directed protein construction. "You could count them on one hand, and most of them were geneticists like Watson and Brenner, and people from physics like Crick," said Baillie. They saw genetics as the chemical transfer of information about development. "Biochemists didn't work on the chemistry of information," Baillie said. "They didn't think of how proteins get made, just what they did."

Baillie loved the genetics class. It also dealt with the mechanics of chromosomes, how they open out like an unraveling string in the early stage of cell division, literally pulled apart by protein structures in the nucleus (called *spindles*), then snap back into a

condensed state. It was in Suzuki's class that Baillie first heard that
John Gurdon had cloned a frog. "Dave said to me, he said, 'Before
you die they'll clone humans and you'll have to decide whether it's
moral or ethical.'"

Specializing in zoology, Baillie got to make use of a small study
house set aside by the university for its forty honors biology students.
Only four or five of these students shared Baillie's new interest in
molecular biology and genetics. All the streams of revelationary biology that he would follow in his career first came together in that
room: developmental biology and physiology, genetics, evolutionary
theory. He and his fellow students talked across the disciplines; their
professors didn't. "In a third-year embryology class, we were talking
of the cell. The professor said, 'Some people think development is
guided from the nucleus, the DNA, but that's nonsense.'" Baillie
laughed and frowned as he remembered that narrowness. David
Suzuki was a complete contrast. "He was excited, which was rare.
He knew what he was talking about."

In 1965 Suzuki got a big grant and asked Baillie if he wanted to
work in the genetics lab over the summer. "Dave's thing was, what
does heterochromatin do?"

Baillie explained that these are areas of the chromosome that
take up a colored stain introduced by a researcher at different stages
of cellular reproduction. One part of the chromosome stains beautifully in mitosis (the process of cell division). Another part stains when
the chromosomes are condensed. These stainable areas are found
near the center part of a chromosome, although some are also found
near the ends. Suzuki had a theory that heterochromatin holds the
genes that control how cells divide. Orthodox science is done by piling up evidence high enough that one can draw reasonable conclusions, or by disproof—finding evidence that shows a theory must be
wrong. "Our task was to disprove this theory," said Baillie.

I asked him what he knew about chromosomes back then.

He could tell me more about what they didn't know. "We didn't
know a chromosome was one string of DNA," he said. "In 1969 it was
shown. We knew they carried units of inheritance, genes, arranged
linearly on them. This had been proven in fungus and yeasts. Mouse
genetics was way behind, but it looked like it might be true." Various

bacteria, such as *E. coli,* and various viruses that prey on bacteria, such as T2, T4 and lambda, were being studied by Sydney Brenner, Francis Crick and a few others.

It was known that messenger RNAs deliver DNA's protein-making instructions to strange little RNA/protein organs in the cell, called *ribosomes.* But no one knew which specific arrangement of the base pairs coded for which particular amino acid, or how the ribosome knew when to start and stop the protein's assembly.[5]

In Suzuki's lab, Baillie was introduced to the primary manipulations used by geneticists to answer such questions. Suzuki reasoned that if they could breed enough generations of flies, eventually they could find a mutant in which the genes for cell division failed to function. Then they could try to trace those genes to specific places on specific chromosomes. The trick was to find a mutation that would not cause immediate death, one that would allow the mutant fly to be kept alive only under specific laboratory conditions. This technique of breeding to find such conditionally lethal mutations had been used by geneticists for generations. For example, bacterial mutants had been bred that would die without an auxiliary supply of food. Some mutations would be lethal only at specific temperatures because some proteins become unstable at temperatures higher than 37 degrees Celsius. So a fruit fly with such a temperature-sensitive mutation will live at one temperature but die at another.

In the summer of 1965, Baillie and his lab mates set out to use a new mutation-making process developed at Caltech. They made many flies with strange mutations, most of which were lethal, but some of which, to their joy, were sensitive to temperature. "These were the first ever found in a multicellular organism in an organized way," said Baillie. The use of the descendants of these mutated temperature-dependant flies "became a growth industry in *Drosophila* for the next thirty years."

The mid-1960s was a very fruitful time for geneticists. The leading edge was moving so fast that even a smart and connected scientist like Suzuki couldn't keep up. In 1965, Baillie recalled, there was an unproved theory about how DNA's messages are coded. Brenner and Crick thought that three base pairs, or a triplet, codes for one amino acid, and that some amino acids might be specified by more than one such triplet; they worked out that this triplet code had to be read from a particular start point along the DNA strand, and if the start point is pushed just one base pair out of whack, nonsense results and no protein is assembled. The stretch of base pairs between this start and an end point became known as the "reading frame."[6] Still, no one knew which triplet called up which amino acid. One day in 1966, Baillie went with Suzuki to hear a public lecture given by a biochemist, Har Gobind Khorana. Khorana had once taught at UBC and was well remembered as a man of spectacular talents. He'd been born in India in a village so poor that his first schoolroom was a patch of grass under a tree, his first writing pad river clay; from there on it was scholarships all the way to the U.K. On the way into Khorana's lecture, Suzuki turned to Baillie and said, Dave, one day we'll have the genetic code. "We walked in and Khorana was putting up a slide of the genetic code," said Baillie.

Khorana was soon rewarded with a Nobel. Using biochemical methods, manufacturing simple RNAs of known order in the lab, Khorana and his colleagues showed which particular triplet called up which particular amino acid. Later, Crick, Brenner and others identified the triplet that signifies the end of the protein chain and the triplet signifying the beginning. With the code broken, the amino acid sequence of a protein could be inferred from a gene sequence and vice versa. Genes could be studied not just as concepts, but as specific arrangements of molecules. Molecular biologists could decipher these messages and even make their own.[7]

Baillie did his master's with Suzuki and then moved to the University of Connecticut for his doctorate. He studied with Art Chovnick, who was working on the positions of mutations on genes and whether messages were the same in multicelled organisms as in bacteria—in other words, whether there was continuity of the meaning of this genetic code across great evolutionary distance. (There is, and there isn't. More on this later.)

Now Baillie answered one of my first questions: why genes have names. While genes have a standard letter and number nomenclature, mutated genes are often known by nicknames. It was Thomas Hunt Morgan, the first geneticist to work with fruit flies, who started it. He called a gene that turned the eyes of the fly from red to white the *white* gene. At first, Baillie said, these mutants were mostly named after the absence of a function that made the flies look strange. "*White* is a white-eyed fly without eye pigment . . . *Legless* would not have legs, the gene required to make legs." By the time Baillie's generation of geneticists came along, the nicknames were often silly. When Baillie found a dominant mutation that was temperature-sensitive in the fly, he called it *DTS,* the initials of his teacher, David T. Suzuki. Suzuki found a mutant fly that couldn't move, which he dubbed *shits.* This produced a nifty, scandalized intake of breath from any lecture audience. "So this cute naming started with young running-dog geneticists of the '6os and '70s," said Baillie. "And students thought it was how to get ahead in the world . . . It was a thing people did."

Baillie's work under Chovnick involved finding the gene that orders up an enzyme that breaks down specific amino acids. Baillie wanted to know if parts of the gene were separated in sexual recombination. Each gene has two forms, one contributed by the male and the other by the female; each form is called an *allele.* He was trying to follow the male and female alleles through various generations to find out if there were regulators of this gene. It was known that promoters turn some genes in viruses and bacteria on, but it was not then known whether the same is true in multicelled organisms like flies. It turned out they are and aren't the same.

Higher organisms have very long DNA molecules, with more than enough DNA to make half a million genes, yet they don't display nearly that much genetic complexity. "There was this weird thing, a clear problem of too few genes or way too much DNA," said Baillie. "We wanted to know, what is that excess DNA doing?" This extra DNA was often referred to as "junk," although no one really knew whether or not it did some sort of job. Thirty years later, *Scientific American* dubbed the notion of junk DNA the biggest mistake in biology.

Baillie became interested in working with the bacteriophage virus T4. He liked it because, unlike with fruit flies, which take two weeks from hatching to produce a new generation, he could do genetic experiments with viruses overnight. Since Sydney Brenner had done pioneering work with this virus, Baillie decided he'd like to do his postdoctoral work under Brenner. But by the time he asked his colleague, Joe Speyer, who knew Brenner, how he could make contact, Brenner had moved on to a new organism, a nematode, or tiny worm, called *C. elegans.*

"They are self-breeding hermaphrodites," Baillie said, adding that there are also a few males that fertilize the odd hermaphrodite, making it possible to do genetic crosses. The generation time is three days. Brenner had apparently already bred a hundred mutants. By contrast, geneticists working on fruit flies had managed to find only a few hundred useful mutants in forty years. This high number of useful mutants suggested that the worm might be ideal for genetic study.

Baillie was fascinated by the question of how and why genes locate in particular places on chromosomes. He did not think location was random. He saw that recombination, the swap of bits and pieces of genes between chromosomes at fertilization, was "precise breakage at exactly the same location . . . The sheer wonder of recombination is that it's very, very precise." In other words, there is something about the way chromosomes are organized that determines that this exchange of genes or chunks of genes will recur in the same place over many generations and be similar across many species.

I was now bold enough to ask the stupidest question of all, considering the reams of material I had been reading in which the word "gene" frequently appeared. I still couldn't really explain to myself what a gene is. Some people said that a gene is just a sequence of base pairs, also called nucleotides, along a string of DNA that specifies a protein; others insisted that a gene is a group of nucleotides regulated by others that have to be activated before the gene can express. I couldn't figure out what the word "express" means, either. Sometimes I thought of it as genes talking and then falling silent, or genes somehow extruding from a string of DNA, exposing themselves to the cell and then withdrawing, or like stops on a calliope being pushed down so that a cap opens and steam and sound emerge up above. How could a gene like the muscular dystrophy gene take sixteen hours to express itself? Where were these larger patterns of organization laid out?

"What's a gene?" I said, timorously.

Baillie made it clear that my confusion is widely shared. "We're still, up to now, very sloppy with the word 'gene,'" he replied. "One definition is it makes a protein. But others code a gene whose product is an RNA, for example the ribosomal RNA genes." Different species of RNA molecules have been found in cells over the past forty years. There are hundreds of different transfer RNAs. A new class of RNAs had just been discovered, RNAs that regulate how long a ribosome can work at making a protein. Other RNAs have been found that don't behave like RNAs but like enzymes—catalysts that aid a chemical reaction. The concept of the gene is like a balloon: it keeps expanding to contain the burgeoning complexity of information about the creation and destruction actually going on in cells.

Getting a handle on complexity was why Sydney Brenner had fixed on the worm as a study subject. Baillie insisted we go downstairs to his office so he could print off a copy of a letter Brenner had written on this subject (the letter is available on his colleague Don Riddle's *C. elegans* website).[8] Brenner had laid out his ideas about how to parse complexity in a letter to Max Perutz, then the head of the Medical Research Council at Cambridge, in June 1963, when Baillie was just beginning to study genetics, three years before the genetic code was published. The letter displays Brenner's enormous optimism, and also his remarkable insight and lucidity:

> Dear Max:
> . . . It is now widely realized that nearly all the "classical" problems of molecular biology have either been solved or will be solved in the next decade. The entry of large numbers of American and other biochemists into the field will ensure that all the chemical details of replication and transcription will be elucidated. Because of this, I have long felt that the future of molecular biology lies in the extension of research to other fields of biology, notably development and the nervous system . . . Molecular biology succeeded in its analysis of genetic mechanisms partly because geneticists had generated the idea of one gene—one enzyme, and the apparently complicated expressions of genes in terms of eye color, wing length and so on could be reduced to simple units which were capable of

being analyzed. Molecular biology succeeded also because there were simple model systems such as phages which exhibited all the essential features of higher organisms so far as replication and expression of the genetic material were concerned, and which simplified the experimental work considerably. And, of course, there were the central ideas about DNA and protein structure.

In the study of development and the nervous system, there is nothing approaching these ideas at the present time . . . The experimental approach I would like to follow is to attempt to define the unitary steps of development using the techniques of genetic analysis. At present, we are producing and analyzing conditional lethal mutants of bacteria . . . Our success with bacteria has suggested to me that we could use the same approach to study the specification and control of more complex processes in cells of higher organisms . . . I would like to tame a small metazoan organism to study development directly.

At first, Brenner thought that the metazoan to study was *C. briggsae,* a close cousin of *C. elegans.* But *C. elegans* had much to recommend it. These are smooth-skinned, transparent, unsegmented worms only 1 millimeter long, tapered at both ends, free living or parasitic, existing both on land and in water. They have only 959 somatic cells when fully developed. As Riddle's website explains: "*C. elegans* is about as primitive an organism that exists which nonetheless shares many of the essential biological characteristics that are central problems of human biology . . . It has a nervous system with a 'brain' (the circumpharyngeal nerve ring). It exhibits behavior and is even capable of rudimentary learning . . . After reproduction it gradually ages, loses vigor and finally dies . . . (We must, alas, assume that the greatest biological enigma of all, consciousness, is absent from *C. elegans*—although this remains to be demonstrated!)"[9]

Baillie wrote to Brenner and asked if he'd take him on as a post-doctoral fellow. Brenner wrote back to say he wasn't taking postdocs. Baillie tried again. And again. Finally Brenner agreed to meet with him at a conference at the Woods Hole Research Center in

Massachusetts in 1970. Afterward Brenner told Baillie he could come and work with him on the worm, but he had to be supervised by someone else—Francis Crick.

In 1971 Baillie presented himself to the Medical Research Council's Molecular Biology Laboratory, located near a hospital three miles from the center of Cambridge. He was not allowed to work on Brenner's worms for a whole year because Brenner was protecting young colleagues from wasting their time on a model organism that might not pan out. Baillie found himself in a strange new world in which mere brilliance was hardly worth noticing. "There were eight Nobels in the building," he explained. He was irritated by the class distinctions and stuffiness of English academe, but also beguiled by the beauty of Cambridge and its traditions. He was in a hothouse, an atmosphere of smoldering intensity. There was the working-class British postdoc with a big chip on his shoulder, who felt he didn't belong. "Sydney was a communist—he could talk to him," laughed Baillie. While Baillie could talk easily with Francis Crick, others were afraid of him. "He was a thinker; he had a reputation for telling you that you were talking nonsense." Crick's sharpness was completely different from the demeanor of Frederick Sanger, the biochemist, a man of marvelous humility. "He walked out of the lab the day he retired [at 65]. He apparently never strove for Nobel prizes. He just won them."

Finally, Brenner allowed Baillie into the worm lab. Two and a half years flew by, and Baillie was offered a staff position at the Medical Research Council and took it. He had arrived at Cambridge, the center of biological study, at a turning point—just as biochemistry, genetics and developmental biology came together in an explosive mixture and politics was thrown into the brew.

Biologists were no strangers to politics. Some of the great ones in the West, such as Hermann Muller, had been communists and eugenicists. Some joined the Party in the 1920s and 1930s; some rethought their positions in 1939, or in 1947 at the start of the cold war. In the West, after World War II, science was primarily funded by the state, so all research scientists learned to keep a wary eye on politicians. Physicists learned during and after the war that receiving large amounts of public money for research has drawbacks, including close scrutiny of political attitudes, the necessity for security clearances and some control of publication. Baillie's generation of biologists agitated against the war in Vietnam, but, until 1971, only those few biologists who worked on U.S., Canadian and British germ warfare research felt the heavy hand of government restrictions. For everybody else, curiosity and openness were normal. Physicists turned themselves into geneticists and turned their hands to the manufacture of batches of bacteria and viruses. No one asked about their qualifications or methods of containment, no government laid down rules or cared—until Paul Berg discovered how to make DNA hybrids, thus making the cloning of DNA possible.

In 1971, Berg, a biochemist at Stanford University, figured out a laborious process to merge the DNA of a monkey virus with that of a bacteriophage. Berg's method of making hybrids was slow, but it worked.

The next year, Herbert Boyer at the University of California at San Francisco and a graduate student named Robert Helling, along with Stanley Cohen of Stanford and Cohen's technician Annie Chang, figured out easier ways to make hybrids. They, like Berg, used restriction enzymes to cut DNA into fragments. Over the four billion years of life on the planet, bacteria had developed these enzymes as defenses against viral invaders. Restriction enzymes recognize the foreign viral genes that enter bacteria. The enzymes cut the viral genes up, disabling them. Each enzyme makes a characteristic cut between the same nucleotides; some will always cut between cytosine and guanine, others will cut in different places. Boyer, for example, found that the DNA of the virus lambda was cut in five places by a restriction enzyme called Eco Rɪ. Then Stanley Cohen found a particular circular piece of bacterial DNA, a plasmid, that was cut in only one place by Eco Rɪ. Cohen also found that the two ends of this cut plasmid rejoined spontaneously. It was already known that this plasmid could be inserted into E. coli bacteria and that it carried the genetic instruction for resistance to an antibiotic, which made it an important genetic study tool. Cohen and Boyer and their colleagues showed that hybrid plasmids could move into a bacterial cell and be copied every time the bacteria made a copy of itself. This process became known as "cloning." There was no longer any need for Berg's complex methods to fuse hybrids. Anyone in a lab could cut DNA precisely with restriction enzymes, put it together with a piece of DNA cut out of something else, put the hybrid plasmids in bacterial cells in a dish and watch as the new hybrids multiplied along with the bacteria.[10]

In the meantime, in the summer of 1972, Paul Berg sent a student to Cold Spring Harbor Laboratory. She described to her study group a series of experiments Berg planned that involved putting hybridized molecules of the DNA of the monkey virus SV40 and the phage Lambda into E. coli bacteria. The Cold Spring Harbor instructor was appalled: SV40 virus was known to cause cancer in monkeys; it carried what are called oncogenes. If Berg transplanted an oncogene from SV40 into E. coli bacteria, which easily infects humans, could the new hybrid E. coli infect a human being with cancer? Could cancer became an infectious disease like the common cold? What if this hybrid escaped from Berg's lab? He called Berg to protest.[11] Berg didn't think there was any danger, but he started talking to colleagues about it at a conference and held off doing the experiment.

And then came the most exciting experiment. Cohen and his colleagues inserted a DNA fragment from a frog into a cut plasmid and put the resulting hybrid into E. coli bacteria, where it multiplied. They were able to find frog RNA in the bacteria, indicating that the frog DNA was transcribed. Cohen and Boyer filed patents on their work.[12]

After the frog experiment, many realized this new technology could be very dangerous and that something had to be done to regulate it. Berg organized a small meeting at MIT in the spring of 1974, sponsored by the National Academy of Sciences. James Watson, by then president of Cold Spring Harbor Laboratory, was one of the participants.[13] They decided to hold a larger conference at Asilomar, California, to thrash out how to handle this new technology, to make rules for themselves before governments stepped in. They also wrote an open letter to Science and Nature, and the National Academy of Sciences calling for the deferral of certain "dangerous" experiments. In Britain, where two people had recently died of smallpox in a failure of containment at a laboratory, the Medical Research Council stopped all such recombination experiments. Sydney Brenner was asked by a government committee led by Lord Ashby to report on this new kind of science, these new methods of manipulation. The son of an illiterate shoemaker who had immigrated to South Africa, Brenner had been taken under Francis Crick's wing not long after he arrived in Britain, and by this point moved among the power brokers of British science. His friends included Lord Victor Rothschild, then the head of the MRC. In September 1974 Brenner reported to his political masters that this work demonstrated that one could take genes from one species and put them into any other species, even those widely separated by evolution. He recommended it be pursued but warned that there were unknowable dangers. Berg invited Brenner to be a member of the Asilomar organizing committee.

The Asilomar conference was held in February 1975. Some 130 invited scientists came from around the world. Journalists (including one writing for *Rolling Stone*)[14] and several lawyers attended. Some of the attendees imagined that this conference was a crucial, historical moment. They thought that by moving to regulate themselves, they would pass a moral test failed by the physicists working on the Manhattan Project. But others disagreed. Watson, for one, who had at first agreed that self-regulation and containment measures were needed, now said he was dead set against any controls on this work. The Asilomar conference was on the verge of breakup with no rules developed when a lawyer addressed the scientists. He explained that they and their institutions would be liable for anything that went wrong in their labs, and that they'd better not leave until they'd worked something out. With the prospect of litigation and financial ruin to guide their moral development, the scientists agreed to let a committee write some rules. The resolutions that emerged from Asilomar set the stage for a National Institutes of Health committee that would rule on the safety of planned federally funded experiments in the United States. The conference marked the beginning of a public struggle between biologists, who wanted full freedom to inquire, and everyone who was frightened by this work and who wanted to rein it in.

Baillie remembered Brenner getting ready for that first meeting. "Sydney said there was a conference about what to do about cloning of DNA," said Baillie. "He said the fox was in charge of the chicken coop. They'll do it, and stop others."

The placid seawater outside the window had gone jet black. Yes, Baillie said, he'd finish the story another day, but after I had lunched with his friend Riddle. Riddle had plumbed the mystery of the long life of *C. elegans* in its hibernating state. It was Riddle who had made the study of decision-making by a lowly worm his life's work.

It was raining lightly when I walked down Broadway the next day looking for the Vietnamese restaurant Riddle had suggested. There were two on the same block, and he hadn't been able to remember its name; which was it? I was walking back and forth between them when I spotted a tall man with a furled umbrella in his left hand. He was the right age to have shared a postdoctoral bench with Baillie, and he moved with perfect deliberation and equanimity, as if he was the kind of person who would always have all his research notes in order, who would always set down his thoughts in a clear hand and never leap to unwarranted conclusions. I leaped to my own and followed him through the restaurant door.

Riddle had a long, open face, thinning reddish hair, freckled skin, glasses. He ordered the pho beef and so did I. It came almost immediately, a huge bowl of soup with thick noodles and chunks of tasteless, gristly gray stuff pretending to be meat. It didn't matter— I forgot to eat as Riddle described his work. He had spent the last twenty-eight years solving one genetic puzzle: how does a worm with only three hundred neurons decide whether or not to develop into an adult?

Riddle grew up in northern California, son of an accountant and a stay-at-home mother, the younger brother of an industrial chemist whose example he intended to follow. In 1965 he entered the University of California at Davis, which had a very good chemistry department. He didn't take a biology course until his third year. That course was one of those experiences, he said, that opened a world— the world of molecular biology. In this class he learned about DNA and how it copies itself, and about how gene activity is regulated by other structures of DNA.

He'd been aimed at a nice secure job in industry, but he fell instead under the thrall of the phage and bacteria growing in the genetics lab. He did his doctorate at the University of California at Berkeley under John Roth, who was working on gene regulation in the bacterium Salmonella. This was a step up in genetic complexity from the T4 virus, which was estimated to have only about two hundred genes. Salmonella seemed to have about three thousand. He was interested in using a chemical mutagen to study these genes, and he knew of a DNA intercalating agent that had been discovered in Philadelphia.

I had to ask: What is an intercalating agent?

An intercalating agent is a chemical whose molecules insert themselves into another molecule, Riddle explained. DNA repels water. The interaction between the base pairs allows them to stack in a stable way: they exclude water from the ladder they form inside the double helix. This intercalating chemical also repels water, and it can squeeze itself between the base pairs and stretch the DNA helix out. This creates mutations by forcing errors in the way the DNA copies itself. "The mutations were single base-pair additions or deletions, shifting the reading frame of the genetic code." As Crick and Brenner had determined, there is a start point where transcription of any genetic message begins. A shift of just one base pair moves the reading frame to the wrong place. Any messenger RNA assembling itself on a stretch of this mutated DNA would become a message unreadable by a ribosome and would cause the synthesis of the protein to stop.

Not only are there chemicals capable of causing such frame-shift mutations, but life is so wonderful in its variety that some bacteria and viruses have developed mutations on their own that can fix this problem—they can shift the reading frame back to normal, by adding or subtracting a base pair. Brenner and Crick had used such naturally occurring repair mutants of the T4 virus to deduce that the genetic code must be in triplet form.[15] Riddle embarked on the careful manufacture of single base-pair mutations in Salmonella using this intercalating chemical, and he found strains of Salmonella that could fix this problem, called revertants. "Some I discovered were of a type with a second mutation that restored the frame." These mutations suppressed the first mutation. "This," Riddle said, "was new."

A group studying with Sydney Brenner had described other nonsense suppressors, as they were called, in the bacterium *E. coli.* They corrected substitutions of the wrong base pair in a sequence. Riddle found six different RNA genes in Salmonella that could be mutated to suppress the shift in the reading frame. These mutated genes suppressed the effects of additions or deletions in a sequence of base pairs.

He got his doctorate in 1971, did a one-year postdoc at Santa Barbara with John Carbon, then arranged to work with a Cambridge biochemist named John Smith. But when he got to the Molecular Biology Laboratory in 1973, Smith took a sabbatical at Caltech. Riddle worked on his own, but his project did not go well. It was there that David Baillie befriended him. They talked about what Baillie was doing with *C. elegans,* and wondered if the worm had any nonsense suppressors. Riddle had brought with him from California some of the intercalating mutagen he'd used on Salmonella. He did experiments to see if he could detect frame-shift suppressor muta-tions in *C. elegans.* He found no such thing, but it got him working with the worm and convinced him that it could answer interesting questions, even though it was little known and risky.

What Riddle meant by "risk" was that since Brenner had still published nothing on his work, grant reviewers wouldn't know about it and it might be difficult to get research money. Though Brenner had already taught Baillie, Bob Horvitz, Bob Waterston and several other North Americans, and though John Sulston had been working with Brenner since 1968, none of them would publish before Brenner was ready. Brenner controlled the stock of worms, doing all the genetic crosses, recording and scoring each one. His first paper on the worm didn't come out until 1974.

Still, Riddle realized he'd found an area to which he could com-mit his life. "*C. elegans* was still a black box," he explained. By this he meant that almost nothing was known about its genetic structure. Sulston, who was then a staff scientist with the MRC, was still working out the cell lineage of the worm, tracing out the entire development sequence of all 959 cells, staring into a microscope and drawing each cell and all its descendants. Riddle asked Brenner if he could work on the genetics of the dauer larvae.

Dauer, he explained to me, is a German word for "enduring." The *dauer state* is the arrested state of development the worm uses to wait out poor environmental conditions into which it may be hatched. "It doesn't feed; it doesn't develop; it lives for months." The question Riddle hoped to answer was, how can an animal with only 959 cells decide to develop or not? The answer had to lie in its genes. He set out on a hunt for a development switch that makes the worm either go into a dauer state or develop into an adult. He thought it could be a case study in developmental regulation as well as in how an animal controls development in response to environmental stimuli.

First, he isolated some *C. elegans* mutants that could not go into the dauer state and others that always formed dauer states no matter what the conditions. He could select mutants that formed dauer larva when they shouldn't and revert them to ones that would grow. They were almost all suppressor mutants, and they suppressed each other in particular patterns.

He talked to Sydney Brenner about what he'd found. Brenner laid out the genetic methods that could now be applied to parse the worm's development decisions more carefully, and he explained the concept later called "epistatic ordering." Epistatic genes suppress the functioning of other genes. He suggested to Riddle that he could make mutation that lacked the genes to make the signals that normally set off the different processes of development. He could work one type of mutation against another to figure out which turned on first in the steps of development.

Riddle pulled out a set of salt and pepper shakers, a bottle of hot sauce and a mustard bottle to show me what he meant. He placed them in a row. If he tipped the first one, all four tipped. If he tipped only the last one, the first three were unaffected. If he tipped the middle two, the last was affected as well, but not the first. He could determine the order in which genes expressed by making various kinds of mutants. "If the mutation is early [in development], the other steps shut down . . . So Sydney told me, if this works out, you'll be very lucky."

And he was lucky. He found two pathways involved, one regulating a growth factor (called TGF-beta), the other regulating insulin production. The human growth factor had been described by Anita Roberts (and others) at NIH, whom he met in 1983 at a course in Cold Spring Harbor. By then, the methods for making DNA hybrids and growing them in bacteria had a become a major tool for studying genes. One could cut genes out of DNA with restriction enzymes and make myriad copies of them in bacteria. One could then use other methods (which we will come to later) to learn the exact sequences of these genes, the exact order of their nucleotides. By then, geneticists and biochemists were coming in large numbers to Cold Spring Harbor to learn these techniques of cloning. "We were students in the molecular cloning course at Cold Spring Harbor Lab. We had to tell [the group] why we were taking the course. Anita said she had found this important factor and needed to learn how to clone its gene. I said I need to clone genes in the dauer pathway, and little did we know that one encoded a TGF-beta."

Meanwhile, Riddle's first graduate student, Jim Golden, now at Texas A&M, had described how two environmental controls affect the worm's development trigger. The worm eats bacteria and is able to sense how much bacteria lies around it and how fast the food source is multiplying. In addition, each worm senses the presence of its fellows, its competitors. Riddle and his students were learning that the worm's decision arises from a conversation between genes and environment.

"The worm hatches from the egg," said Riddle. "In the first eleven hours, it exists as a first-stage larva, at the end of which it decides to dauer or develop into an adult. It assesses the environment, whether there's enough food to support its reproduction if it grows to an adult, and the next one is the rate of consumption of that food. How much food, how many eaters." If there are too many eaters or too little food, "they shift metabolism to accumulate fat reserves . . . and then become dauer larvae."

This struck me as absolutely phenomenal: a tiny animal, with so few cells, in the first hours of life decides whether to compete or to wait for a better day. How could this be described as anything other than problem solving, and what did it mean that an essentially brainless thing could solve such problems? This capacity for foresight in a sightless creature is the kind of thing that makes creationists insist that there is a Plan and a Planner behind it.

"So," I said, getting right to the heart of it, "do you believe in God, a creator?"

He sat back in his chair. "I wonder if there's a God," he said. "I'm not an atheist. I'm agnostic, I guess. I don't want to take a profound position one way or another."

But think about this fantastic little creature, I babbled.

"I feel the same way," he said, flushing with pleasure that even a worm novice like me could recognize the miraculous qualities of this organism, the questions you could ask about development by selecting different mutants and watching what they do.

"The animal spends eleven hours assessing the environment," he continued. "When it reaches the first molt, it makes the decision. How does it integrate its experience? It has to remember . . ."

"Well, what in hell does that mean?" I said. "They calculate? They think?"

He looked at me with disappointment, as if he thought I'd already understood that this is all about chemistry, about tipping points. "It's a threshold argument," he said. "It works if there's enough food and low pheromone. TGF-Beta is sent out over time and interacts with the cells' receptors until it reaches a point where it acts to throw the switch. If it doesn't get to the threshold, the larva forms a dauer."

So what is activated in the central nervous system that in turn activates the switch? Riddle cloned, or copied, more genes and found one for a nuclear hormone receptor, part of a large family of genes that control transcriptional responses to hormone cues. "There's an implication of a hormone that mediates the developmental switch," he said. "We have another gene that we think is responsible for biosynthesis of the hormone. We have not yet identified the hormone."

It was like unraveling knotted balls of string: these hormone interactions were complex enough, but there was also the insulin pathway to consider. High levels of food activate it, and the worms will grow. Insufficient food? They form a dauer. Riddle had found ways to make these two separate pathways conflict through genetic manipulations and had learned that the worm needs both to function normally.

And so we came to Steven Jones's article, the reason I'd wanted to come to Vancouver in the first place. Riddle, one of its authors, had proposed that they try to use a new technology to read genes in the dauer state as they are being expressed. The SAGE system (serial analysis of gene expression) was invented by geneticists to characterize all the transcriptions produced by all the genes of a yeast cell. The question was, could this system also work on C. elegans, an animal with 19,000 genes, more than three times as many as yeasts?

It was simpler to test the SAGE system on a worm in the dauer state than on one in its normal condition because it was assumed that fewer genes are turned on in the dauer stage than in a developing animal. This experiment was started not long after the worm's entire genome was sequenced and published in 1998. They found genes expressed in the worm's dauer state that were not predicted by those who had analyzed the worm's entire sequence. "We are working hard to find out what the functions of those genes are," said Riddle. They were working on the new gene that, when the animal is in a dauer state, is twenty times more active than in growth stages.

So, I said, are you interested in applying this work to human beings and their mortality?

He did not blink. He did not pause. "Longevity is an issue," he said. "We're looking now at aging adults." Worms that enter the dauer state tend to live longer than those that don't. "That brought us to aging research," he said. "The insulin pathway also controls adult longevity . . . We are working on comparing those genes to human homologs." In other words, he was comparing worm genes to human genes that look similar and might act in similar ways.

"Dauer larvae live a long time," he said, carefully. "They can live as long as three months versus two weeks of normal life . . . So the model is that there may be genes for longevity expressed in the dauer."

In fact, ideas about the nature of aging and death had changed because of the work in C. elegans. Cynthia Kenyon at UCSF and Gary Ruvkun at Harvard had shown that aging is a "regulated process with signals that control how long cells survive," Riddle said. Work in rodents had demonstrated that a 30 percent calorie reduction increases life spans by about the same amount. "There could be points of pharmacological intervention; one could mimic calorie restriction in the body without the pain and suffering of the diet and live 30 percent longer." In other words, one could make an anti-aging pill.

One could make a fortune with something like that, I said. And yet I couldn't imagine Riddle chasing down some dream pill offering long life and thinness in one gulp. He seemed too staid to be a modern-day Ponce de León. "Are you really interested in extending this work beyond the worm?" I asked.

"I think I'm cognizant of the potential uses in other organisms and find it interesting," he said. "It brings up ethical and moral considerations. Is it desirable to make people live longer? I suspect that most would say I don't care about you, but it's fine for me."

I found David Baillie the next day, in a Second Cup around the corner from the BC Cancer Research Centre, where many of his colleagues do their work. We sat in its wing chairs for hours. Unlike Riddle, Baillie was not very interested in aging or mortality. In his view, the length of an animal's life is the result of a trade-off, and death has a function. "I understand it as a necessary penalty to have other things work," he said. "We have a high metabolism, consequently we have to die. We can't afford also to compete with our progeny."

In 1974, just as revelationary biology was taking off, Baillie left Cambridge. His mother had been diagnosed with a brain tumor and had only six months to live. He came back to Vancouver, puttered in the lab of a former Brenner postdoc at UBC, then got his own job at Simon Fraser University after his mother died. He found that he didn't want to go back to the U.K., even though Cambridge was a breeding ground for exotic ideas and Vancouver definitely was not. His job at Simon Fraser was teaching developmental biology, but he couldn't really talk to his colleagues about parsing development genetically. "If I said then, it comes out of the genes, they'd run out of the room screaming. At that time they didn't think of it, they didn't talk of where the controls are. The geneticists did."

But gradually, Sydney Brenner's students created various North American centers for the study of the worm. In 1979 fifty people were working on *C. elegans*. By 1987 four hundred attended a *C. elegans* society annual meeting at Cold Spring Harbor. At that meeting someone proposed that they should sequence the entire genome of *C. elegans* before anyone tried to sequence the entire human genome, a project being talked about at the time. John Sulston and Alan Coulson, both from the MRC, and Bob Waterston of Washington University in St. Louis had already put together a physical map of the sequences of the few known genes of the worm, which they displayed at Cold Spring Harbor. Those interested in sequencing the whole worm genome were particularly concerned to get Jim Watson behind the effort. "I think it was perceived that if you could convince Jim it was important, there was a good chance the National Institutes of Health could be brought onside," said Baillie. Which meant money.

Sulston and Coulson's map showed random bits of worm DNA that had been cloned into bacteria, which they called *cosmids,* each about 40,000 base pairs long. They calculated they would need to make 250,000 such clones to get enough samples of the entire worm genome to make an accurate sequence. All they needed to get it done was money. Sulston, Coulson, Waterston and Baillie's wife and colleague, Ann Rose, huddled over these figures so they could pitch it to Watson, who soon took a proposal forward to the NIH.

Baillie shut down all the sequencing work his own lab was doing after the human and *C. elegans* genome sequencing projects were launched in 1989. He wanted to work ahead of the sequencers, to figure out methods to make sense of the data they would provide. "I was interested in how we'd find where important genes and regulatory elements are. At best, it's by comparison between two animals that are about 50 to 100 million years apart."

Evolutionary theory predicts that two animals of different species that descend from a common ancestor will share elements, and, as Baillie put it, "what remains the same is required to be the same. Genetic material is changing all the time. Your egg is different from yourself in a few thousand nucleotides. Across thousands of generations there are millions of changes. Areas that are not important can differentiate. The important must stay the same." He began to look at the genes of *C. briggsae,* the first nematode that had caught Sydney Brenner's eye, a species that last shared a common ancestor with *C. elegans* 50 to 100 million years ago. "They look identical . . . They are as far apart as lemurs and people are . . . About 30 to 50 percent of the nucleotides have diverged." He wanted to know if the order in which genes fall on chromosomes is conserved or different, and which genes need to be close to each other. "The chromosome order appears to be the same, but the number of chromosomes is different."[16]

By this point we had sat so long in the Second Cup that one of the staff brought us a small plate of candies—chimeric candies, one half orange, the other half chocolate. I had saved the question that mattered most to me until I was so hyped by caffeine that I couldn't suppress it any longer. So, I said, what about chimeras, what do you know about them?

He knew that Beatrice Mintz had made a four-parent mouse in 1961.

Can you explain it to me, I asked. Can you tell me how a functioning mouse can develop from four parents, from two different sets of genetic instructions? Where's the larger pattern written that allows this to work?

He diagnosed my condition immediately. "You believe in souls," he said, with a dismissive wave of his hand. He was not about to trifle with nonmaterial explanations for very real material events. Nature, on its own, had made much weirder fusions than Beatrice Mintz's chimeric mouse. "The lichens are mixtures of fungi and algae," he said. "These are two different phyla. They live together as a multicellular organism," he said. "We like to think genes make the person, but, in fact, a person is above all a conglomeration of cells. The cells have cooperated, integrated together. It's a population, like an ant colony." His wife liked to compare a human being to an ant colony. "Why is the ant colony different from you and me? Your cells die to let others live. To protect your liver, your eyes and you. So there's not much difference between you and an ant colony. And, of course, she's right."

The idea of comparing human beings to ant colonies disturbed me.

"It only makes sense in a nonteleological universe," he was saying, by which he meant that there is no end point at which nature is aimed. There is no glorious creature, human or otherwise, at the pinnacle of creation. There's just a string of random successful accidents: whatever works, lives and leaves offspring. Whatever works hangs around.

I left Baillie and wandered up the street, tired and bemused. What a vision of humankind these biologists were making. We are an aggregation of cells. We have to die so we can move fast, so our progeny can have enough food to reproduce—without death, we'd have no kids! In a way the story of Adam and Eve in the garden also expressed this trade-off. Didn't God sentence Eve to the bearing of children in pain and suffering, and Adam to a life of sweat and toil as the price of knowledge? We're all chimeras, bogeys, monsters, I thought. We can climb the second tree, we can live much longer and be very thin, but the lesson of the worm is that we'll have to give up something vital, make some Faustian bargain. No children. No Big Macs. The lesson of the worm is that life and death decisions can be made by Rube Goldberg machines.

Metaphors have a way of wrapping up one's thinking into nice, neat bundles and shutting out alternative ways of seeing things. Crick's one-way information metaphor, his Central Dogma, had trapped everyone for years into thinking of the gene as a signal sender but never a signal receiver. What of this image of the human being as an agglomeration of cooperating cells, like an ant colony?

I tried to imagine it: my nose cells working hard to find sugar (preferably in the form of milk chocolate); my skin cells sloughing off, dying to defend the future of the cells below; other cells shuttling about from one organ to another, looking for an invader to throw themselves away on. It seemed to me that this metaphor missed something important—our sense of individual identity.

If genes don't encode all the mechanisms of birth, development and death, if each embryo's cells compete to live and make offspring yet each descendant group of cells survives through cooperation and altruism, regardless of Baillie's dismissal of soul, surely there had to be something that directs these vital and contradictory ways of being, something that is sovereign over this anarchy? If that is not

THE SECOND TREE • 139

the genes, what is it? And if it is not written anywhere, how do we account for it?

I found myself remembering Thomas Hobbes's image of society as described in *The Leviathan; or the Matter, Form and Power of a Commonwealth, Ecclesiastical and Civil,* published in 1651. It describes Hobbes' theories of political power and duty—how sovereignty is a property of the individuals in society, which they give to the king. The state is a living entity that emerges from this contract. I had a paperback edition when I was at school. There was a cover illustration of this society, this Leviathan, as a large man with a hat, the whole man formed of myriad tiny men and women, all furiously going about their business—some at war, some at work, eating, cooperating, fighting—all giving their power to govern themselves into the hands of a sovereign monarch, who orchestrates the whole. I liked Hobbes's metaphor better than the ant colony. The notion of a human being as a cellular society made room for Freda Miller's strangely plastic, opportunistic cells, which can be recruited to new tasks. It caught the anarchy of individual cellular endeavor, the discipline created by the need to get along and the sense of a directing intelligence. Like the life of any individual person in any society, or any individual cell in the body, the Leviathan too had a beginning and an end. It arose as a product of interaction.

My cousin's car pulled up at the curb and I climbed in. His father, my uncle, was in hospital. He was old, very old; his entire elimination system was in rebellion. A surgeon had just cut out most of his bowel, and it didn't look as if he was going to make it.

Soon I was sitting in the hospital waiting room with two cousins and my aunt. There, stamped on each of us, was a particular version of our shared family face. Each one of us bore the awkward family ears, which sag a little more with each passing year. Each one laughed the raucous family laugh; each one was burdened by a family ailment—acid reflux, arthritis, hypoactive thyroid. If I am a colony, I thought, what are we together, a sickly imperium?

My uncle lay on his back in the next room, either unconscious or so deeply asleep he might as well have been. His newly made bowel opening glistened like a wet mouth. There was a welter of tubes running here and there. He was the most awful color of beige, as if he'd been dipped in a vat of plastic. His eyes were half open, half shut; his chest barely rose and fell.

I approached his bed on tiptoe, saddened by all the changes time had made in him. I remember him best from when he was the age I am now, still tall and strong, hair still thick, skin still brown from a life outdoors. In those days, he liked more than anything else to make things. He had climbed a mountain with building supplies strapped on his back and had built himself a cabin. He had come one summer to visit me and had built my kids a tree house, which became an under-the-tree house, since there was no tree big enough to carry it, but which they loved nonetheless. There was no one else in the world quite like him.

Was he following the light? Had any part of him already left his physical body, removed itself to a place free of pain? Was that private self hiding somewhere in the corner of the room, watching me, watching him?

I touched his hand to tell him I was there. His skin was still warm. The Leviathan soldiered on.

CHAPTER SEVEN

THE PIG FARM OF DR. MOREAU

One day a picture of a man and his cow appeared in the *Globe and Mail*. Ralph Warren, a veterinarian from Port Hope, Ontario, told the reporter, Wallace Immen, that his champion Holstein cow, Margo, had been cloned after her death by a company in Wisconsin. The cow in the picture was a two-year-old clone named Margo II. A resurrected cow! Warren told Immen that Margo I had been a great producer, with such a sweet temperament that he took her for walks on a leash. Margo II was as sweet as Margo I. She had also given birth to a calf of her own and was a prodigious producer of milk, but Warren threw it all away because some people might be disturbed to drink the milk of a cloned cow.[1]

What was the word to describe Margo II's relationship to Margo I? Daughter? Sister? Resurrected Other? And if one could resurrect a cow, could the return of a dead human being through cloning be far behind? Margo II had been cloned from a cumulous cell taken from the outside of one of Margo I's eggs, which had been taken out of the ovaries cut from Margo I's dead body. This cell had been dropped into another cow's denucleated egg. The reconstructed egg was stimulated until it started to divide as an embryo. Margo II had been delivered from a surrogate by Caesarean section and she also developed pneumonia, but these were small complications, exactly the kinds of things that had happened when in vitro fertilization techniques were being perfected for cows twenty years earlier. Warren told me he thought it would be easier to solve such problems in humans because women's eggs are much bigger than a cow's and therefore easier to manipulate. Infigen, the firm that performed this service for Warren, had done it for nothing because cloning a dead cow from her retrieved eggs was an interesting challenge. But the company was already in the business of cloning prize farm animals for large fees.

It was not surprising that such breathtaking biological experiments had so quickly become a business. Commercialization of revelationary biology went into high gear after Stanley Cohen and Herbert Boyer made their first hybrid DNAs. In 1976 Boyer founded Genentech Inc., which genetically engineered bacteria to produce human insulin.[2] Twenty-five years later, the *Atlantic Monthly* carried a story detailing how major research universities had become so deeply embroiled in business, much of it biotech, that they no longer functioned as places of free inquiry.[3] Even David Baillie, former member of Students for a Democratic Society and a lifelong socialist, had appeared at our second meeting wearing a vest emblazoned with the logo of a company in which he holds an interest. He had been given shares in three companies started by students.

Infigen's website showed it had already cloned a small herd of cattle and was on the verge of the commercial cloning of pigs. Now that was astonishing. Pigs had been declared impossible to clone only two years earlier, in March 2000, at a private meeting held at Cold Spring Harbor Laboratory's Banbury Conference Center. Of course, on the same morning this dogma was announced, it was overturned: Keith Campbell, the man most responsible for cloning Dolly the sheep, had stood on the Banbury Center's glorious lawn along with a colleague, Alan Colman, to tell a press conference that they had successfully cloned a pig for PPL Therapeutics, a company affiliated with Roslin Institute. Since then, PPL's and Infigen's cloning methods had developed so fast that Infigen's website described a deal to genetically alter and clone cows and another to clone genetically altered miniature pigs whose organs could be used for transplant into humans.

By early January 2002, Infigen and its partners had an article in press with *Science* claiming the birth of the first cloned knockout pigs, engineered to suppress a gene that results in a sugar that causes pig cells to stick to each other. This sugar would cause a disastrous immune response in a human being. But the day before the *Science* issue arrived on the newsstands, PPL Therapeutics scooped it,

THE SECOND TREE • 143

announcing the birth of its own cloned knockout pigs. Its share price zoomed up 46 percent on the London Stock Exchange.[4]

The plane flew over Wisconsin's dairyland—low, gently rolling hills, dust brown soil without a hint of snow cover. There were lines and crosshatchings of trees and bush growing in the gullies. From this height, it looked as if the earth was covered by a tanned, wrinkled skin, the cracks sprouting unshaved hair.

Infigen was in the country several miles from the Madison airport, its offices in a large building set back from the roadside behind a large sign that said American Breeders Services–Global. Signs on another long, shacklike building off to the left, buried deep in the ground like a bunker, said Keep Out, Biohazard.

I had made arrangements to attend the farrowing of a pig pregnant with clones. I did not tell Infigen's PR person exactly why I thought it so important to witness such an event for myself, but it had to do with the testimony of some senior scientists to the National Academy of Sciences panel examining the safety of cloning humans the previous August. Rudolf Jaenisch, of MIT, had said that almost all animals born of cloning had something wrong with them—shortened telomeres, too-large-at-birth syndrome, deformities, tumors.[5] Dolly the sheep had just been presented to the world press again because she had developed arthritis. The suggestion was that these were side effects of cloning. And yet here, at Infigen, cloned pigs and cattle seemed to be rolling out of the lab like so many Model T's off the line. It had occurred to me that perhaps Infigen and its competitors were hiding their monstrosities. So I wanted to look at each piglet as it emerged.

I sat at a boardroom table waiting for Michael Bishop, Infigen's president and the director of its science program. The room was absolutely plain, with a view of the rolling hills. A young woman stuck her head in; she was the veterinarian responsible for the farrowing pigs and the calving cows. She didn't want to shake hands—she said she had manure on hers. She gave her first name, which begins with the letter M, and a last name that brought to mind a certain mythical stable, but she did not want me to print either one. She never did say why, but I figured it out soon enough: fear of religious nuts. If they shoot abortionists, why not someone meddling with the farm animals most representative of human dominion over creation? She informed me that the pig I'd come to see farrow (a commercial job for a farmer who wanted many copies of a prized boar) would not be ready until after my departure. She'd arranged instead to show me how she implants cloned embryos into gilts.[6]

Michael Bishop loomed into the room. Dr. M. fell silent. In fact, the whole place seemed to become unnaturally quiet, as if that was the prudent thing to do in Bishop's presence. He was a huge man, his vast upper body covered in a tieless lilac shirt. His hair was very short and very black, like a small wool cap stretched over a globe. His face was round as a melon, his hands meaty, and his thighs too wide to allow him comfort in a normal chair. I could see him running a cattle spread. He reminded me of more than one rancher I knew in northern Saskatchewan, men who bulked up as big as their bulls, as if they needed the mass to cow their animals into submission.

Bishop sat down and began to tell me his story, and a fine, melo-dramatic business story it was, too, with all the usual ins and outs, hirings and firings, changes of ownership, strategies and defeats that any business reporter would long to hear. He had that business voice, too: it oozes out always at the same timber, never with too much affect, not too low, not too high. This voice comes to a man after countless dog-and-pony shows, where he lays out his story over and over again to prospective investors or explains himself to shareholders.

Michael Bishop's father died when he was in grade 11, so he had to run the family farm while he finished high school, after which his mother sold the place so that he and his younger brothers wouldn't be saddled with it. "A defining moment," is what he called the tragedy of his father's early death. He went to North Dakota State and graduated in 1976 with a degree in animal science and a minor in economics. He got a job managing a purebred Polled Hereford cattle farm in Ohio for an independent oil contractor with ties to Amoco. He made money, invested it in silver as the Getty brothers were trying to corner the market, and made enough to buy himself a nice ranch in North Dakota. His place was so far removed from city life that it was "a 123-mile round trip to the nearest Big Mac." He got married, and by his third season on the ranch he was at another defining moment.

In 1981 inflation was at its peak. Bishop had a choice between taking on more than a million dollars in debt and figuring out what he really wanted to do with the rest of his life. He was sitting at the table, stewing over his projections and business models, when an item came up on *Good Morning America* about a transgenic pig. He thought that was fascinating. He decided to "get rid of the iron, debt, pieces of land," and go back to school. He went to Ohio State. At that point, molecular knowledge of the cattle genome was way behind the work in the mouse or the worm or the fly, where markers for some genes were known. "There were less than one hundred markers for cattle," he said. He worked on tracking the gene for a bovine growth hormone, IGF-1, but the main thrust of his doctorate was gene mapping: he located the gene on its chromosome.

To learn molecular genetics, "I rebooted myself," he said. This paid off. He was hired, before his PhD was granted, by the United States Department of Agriculture, which sent him to Zurich, Switzerland for more training. The USDA wanted to find markers and to map useful genes in domestic livestock, and to do it before any other country succeeded. In Zurich, Bishop learned a strategy to break the huge bovine genome up into much smaller segments, clone those segments into bacterial libraries and map these sequences. From Zurich, he went directly to Clay Center, Nebraska, to a 30,000-acre spread in the middle of nowhere with refurbished ammunition bunkers left over from a World War II army base that had become the USDA Agricultural Research Center. He was a member of a team of five, and they had DNA from selected families of cattle, but their first task was to fill the brand-new, empty lab with equipment. "In eighteen months we put together the first cattle and pig maps," said Bishop. The work was essentially secret. His bosses told his unit to keep very quiet about what they were doing and to limit the academic meetings until they were done. "We were told we were in a race and we didn't go anywhere." They finally unveiled their work at a symposium in August 1993 in Seattle, five months before they published in *Genetics*.

He came to Wisconsin in 1994 to work for American Breeders Services, then still a subsidiary of the conglomerate W. R. Grace, to initiate a program for DNA testing of cattle. But in 1995 ABS was bought by an investment company called Ardshiel—a leveraged buyout firm composed of a few smart guys in New York.

At that time, there was a skeleton crew at ABS working on bovine cloning. In 1982 ABS had invested in Neal First, a professor at the University of Wisconsin who was trying to create a commercially viable method of cloning cattle. He successfully cloned cattle in 1987 by nuclear transfer between eight-cell and sixteen-cell embryos.[7] In 1990 W. R. Grace had bought Granada, a cattle company in Texas that had its own cloning research program. Granada's team had been led by the brilliant embryologist Steen Willadsen, who cloned the first lambs by splitting embryos when he worked in Cambridge. Willadsen also made interspecies chimeras—fusing a sheep with a goat to get a live "geep," or "shoat." In 1985, in *Nature*, Willadsen described how he made the first cloned sheep with nuclear transfer. He did not apply for a patent on this process, and he used a similar method to clone cattle embryos for Granada.[8]

But by the time Bishop got to ABS, the cattle-cloning project had been downsized because it was too expensive. A lot of key people had already left. Bishop focused first on genetic marker–assisted selection of Holstein bulls. But the second thing on his list was to develop a commercial method to clone cattle and pigs. He hired back one of the young technicians who had left, and asked the team to stop going to scientific meetings. Why? Going to meetings just leads to sharing of ideas. "Sharing is good," he said, "but you can give away secrets, and that's death."

Keeping information to oneself for strategic reasons is the defining difference between science for curiosity and science done for business. Science driven by curiosity requires publishing the news as quickly as possible. Science done for business means keeping one's brilliant innovations to oneself until one has a patent application filed and the prospect of a market monopoly.

Bishop and his team had this idea that one could rewrite a differentiated cell to the start point of development by transferring its nucleus into an egg. The nucleus that made the first such cloned bull, Gene, was taken from a cell of a fetal genital ridge, the area of a fetus that will eventually become the ovaries or testes. Bishop and his colleagues treated these fetal germ cells with inhibitory factors to keep them from differentiating further, then with a growth factor to stimulate their proliferation in culture. A nucleus was removed and transferred to a denucleated egg. The egg and the nucleus were fused with a zap of electric current, and the egg began to divide like an embryo. The research was going well, but the company was in turmoil. The CEO had been fired and a hatchet man ran through the company, accosting people in their offices, asking "what the fuck do you do?"—looking for deadwood to fire.

The room we were sitting in was named after Gene, the first nuclear transfer–cloned bull, who was born on February 6, 1997. "Dolly was announced February 23," Bishop said. The news of Dolly's birth stirred up a media hurricane. Everyone focused on the fact that if a sheep had been cloned, human beings were next. "I was late that morning, taping the news media . . . I had the CFO and the CEO come in to the conference room. I showed them the video and said, 'Gene is the equivalent of Dolly.' They turned white. Human cloning! We're in a conservative business in agriculture. The people who buy are farmers, maybe religious. And here's what my company is doing. And this management team with no scientific background whatsoever, they are freaking out . . ."

If Bishop announced the birth of Gene right on the heels of Dolly, who knew what would happen? He had his research program all laid out: he would make genetic changes in cells, put those changed nuclei into eggs and get very specific kinds of clones. "Our owners in New York couldn't spell DNA and were totally blown away . . . I gave them champagne on a beer budget. The whole research budget was less than $500,000 a year. We have patents that go back to 1986, they originated here . . . We have all that intellectual property, plus what we added, the activation protocols. The activation of an embryo and how to fuse it was created here—that was the twist, what was different from Dolly." Bishop had figured out how to maintain his reconstructed embryos in culture, unlike Roslin Institute. "We did it in an incubator," said Bishop. "We were right at an industrial process right then."

"We decided not to announce Gene," he said, after taking a deep breath. Instead, they developed a plan demonstrating that cloning cattle could be a viable business. But there was a problem. Roslin Institute had applied for a patent, and its animal had been born first. Bishop and a member of Ardshiel flew to Edinburgh to meet with Ian Wilmut and Roslin's other principals. "They told us they'd applied for a patent. We said we'd filed too. We were there to head off war, to talk about working together on cattle breeding."

But they couldn't come to an agreement, so Bishop and his colleagues called the press out to introduce the world to Gene. "It wasn't as wild as Dolly—it was all national news. NBC carried it live. I needed media training." He told the media what they wanted to know, but nobody asked the question he was dreading: what happened to the other animals they tried to make? Gene was a twin, and Gene's twin had died.

Several more clones died over the next year, during which time Infigen was also spun out of ABS as a separate company. The new CEO thought they should make a deal with PPL Therapeutics, Roslin's affiliate. Instead, the company made an agreement with a Netherlands company called Pharming NV. "We did a deal with Pharming to clone cows for producing human therapeutic proteins in milk . . . So that was the first deal. It brought in $3.5 million up-front funding for research and development as well as annual research funding plus milestones." But then Pharming got into financial trouble, and it was a year before another live, healthy clone was born. Infigen did another joint venture deal, with Novartis, which owned a company called Imutran, in 1998—"In pigs, for God's sake."

No one had yet managed to clone a pig; many thought it was impossible. Bishop was also trying to clone deer, elk, dogs and cats. The Novartis-Imutran pig-cloning deal put another $3 million in Infigen's pockets, most of it in the first year. But just as Pharming had got into financial trouble, so did Imutran. Pig cloning was aimed at transplanting pig organs into humans, but in 2000 an article was published showing that pigs have retroviruses in their genomes that don't harm the pig but might very well jump to humans if pig organs are transplanted. Novartis had been working on xenotransplantation for a long time.[9] Imutran was disbanded.

Infigen's first cloned pigs farrowed in July 2000, a few months after PPL announced success. By October, Infigen had its first transgenic litter of cloned pigs. By Bishop's count, Infigen had since cloned 75 or 80 pigs, 160 head of cattle and 1 sheep. He was now cloning bulls and cows, sows and boars to order, asking up to $50,000 an animal. His company had no debt. Ms. Rational put it together, and out popped the question: when were they doing an initial public offering (IPO) on a public exchange?

"We don't need to make an IPO," he said.

I sat there stone faced, knowing this was nonsense. IPOs are not just for raising capital, they also satisfy the greed of early investors. Bishop admitted that his owners would have liked to take out an IPO in 2000, but had not been able to get an offering ready before the market for dotcoms and biotech stocks collapsed.

Patents? I wondered.

"We've got patents on bovine and porcine cloning, on oocyte activation and stem cells in Europe, and mammalian cloning in New Zealand."

But not in the United States?

He replied that their U.S. patents were for nonhuman cloning using nuclear transfer from all kinds of cells, and they were interfered with by U.S. patents granted to Advanced Cell Technology and Geron through Geron's ownership of the company spun out of Roslin Institute. (Within months, the U.S. Patent Office would rule that Geron holds the senior patent, putting Infigen in the difficult position of having to prove that it invented nuclear transfer cloning first.) In other words, Bishop was trying to sort out whose patent would prevail, and he would be in court for a long time.[10]

Infigen had already sued Advanced Cell Technology once, because the company used Infigen's patented processes without a license. The 1999 court settlement forced Advanced Cell Technology to pay Infigen in return for a license to use the patents "in the field for human cell therapy using nuclear transfer derived cells."[11] The settlement was onerous for Advanced Cell Technology: not only did the company have to pay for this license over the course of several years, but Infigen was granted a first security interest in all Advanced Cell Technology's hard assets. Advanced Cell could not even license out its own patents in any way that might adversely affect its capacity to pay Infigen.[12] Bishop didn't want to say anything about it because the following Monday, Infigen's lawyers would be in a courtroom in Worcester, Massachusetts, arguing that Advanced Cell Technology had misinterpreted this settlement. As he dropped me at my hotel, I wondered: how many lawsuits can a small company carry? And what happens to this kind of alchemical science if the company goes bankrupt? Who takes care of the clones?

Dr. M. arrived at the door of my hotel, as planned, at 7:45 a.m. sharp. The sky was still black and there was a bitter‧wind, but for January in Wisconsin it was a balmy day. She drove her beat-up little car over a series of paved country roads until we arrived at our destination, the University of Wisconsin's pig barn, which has its own surgical suite.

"We scrub in but not out," she said, as she led me through the cinderblock entry. I had no idea what she was talking about. I followed her as she walked through a set of doors into a shower room. She handed me a basket of hospital greens, and another with cotton underpants, sports bras and socks. Then she started pulling off her clothes. It's a weird thing to watch a woman you don't know kick off her jeans and reveal that she doesn't bother with underwear.

The shower had two curtains and two entrances, one on the dirty side of the room, the other on the clean side. There was a thin stink of pigshit in the air—a smell that I found difficult but not impossible to deal with, like the mousery at Mount Sinai but more acrid. I pulled off my clothes, walked into the shower and sluiced down my hair, which then hung damp on my back. Reeking of anti-bacterial soap, I stepped carefully through the other side of the shower to the clean room. Dr. M. had short hair, so she was already pulling the supplied cotton clothes on and swinging through the door. She pointed to a bench with various sizes of boots and I found a pair, spattered with pig manure, that I could push my feet into. Odd, I thought, that pig manure is considered clean but I'm not.

We walked down a short hall to the surgical suite. I was handed a surgical cap, to keep my hair from falling into incisions, and a mask, to keep my germs to myself. She pushed open a set of swinging doors and a young man named Garrett, no more than twenty, was there waiting for her. So was the patient. The pig was a young gilt, a female who had just reached sexual maturity, about eight months of age. She was in estrus, receptive to being fertilized. Someone had painted a slash of green on her back to identify the exact status of her cycle. She was trapped in a peculiar pen with a metal floor and metal mesh sides, which looked vaguely like a grain bin, wider at the top, narrow at her feet. It was elevated above the level of my waist; a ramp connected this pen to a door behind which the rest of the pigs in the barn went about their business. She couldn't move in any direction, forward, back, sideways. If I'd been trapped like that, I would have been wild. She turned her head as best she could to look at us. She had little bright eyes and thick white eyelashes. She was curiously dainty, almost pretty.

Dr. M. walked right up to her, talking in the unnaturally high voice I use when I'm introduced to babies who don't know me. "How ya doin'?" she cooed. She took the pig's ear in her hand, and with the other grasped a needle. She gave the pig a shot and the pig bellowed and squealed. "That's just the pig defense mechanism," she said to me as I jumped back. "It's okay," she cooed to the pig. The pig began to relax, though her eyes shuttled back and forth between us, panicky, asking—demanding—what's going on? What are you doing to me?

I was unnerved: this pig obviously had some sense of self, a capacity to feel violation. Dr. M. was now searching the pig's ear for a vein she could use to infuse a sedative. Pigs' veins reach the skin surface only in the ear; the rest are buried inches deep in fat. She found it and the pig was going, going, gone.

Garrett rolled in a strange apparatus, two long polished stainless steel beams on rollers with an empty V between them. The beams could be spread wider, raised and lowered; the whole thing functioned as a movable surgical table. The two of them pushed and shoved until it lay parallel to the pig's pen, then they lowered the side of the pen and the pig rolled out onto her back on the two beams— the operating table—exposing her white-haired belly and teats to the air, trotters curled on her chest. Garrett swabbed the pig's belly with detergent to clean off any manure, then they pushed the table through a pair of swinging doors into the operating room—not a sterile room, but a clean one. When the pig and the table were properly positioned under the lights and over the hydraulic system, and the table had been brought to the right position, they tied her forelegs across her chest and her back legs straight out and down. This was done so that if she woke up in the middle of surgery, no one would get hurt. Dr. M. then stuck two tubes into the pig's nostrils and turned on an anesthetic gas that was once used on people. Pigs, she explained, can breathe through their mouths, but they don't like to. The seemingly unconscious pig winced as Dr. M. pushed the tube higher into the second nostril. "Sorry, sorry," Dr. M. cooed. While she went to scrub up, Garrett covered the pig's feet with little plastic booties so she couldn't kick manure into the incision.

A technician came in from another room carrying small petri dishes and a plastic pipette. On the previous Wednesday, he had taken eggs from the ovaries of a freshly killed pig obtained from a slaughterhouse, sucked out their nuclei and pushed into each of these eggs a nucleus taken from the cell of a boar they wished to clone. He had zapped the constructs with a jolt of current to fuse them. These reconstructed eggs had then been allowed to divide in the right growth medium until they had become an embryonic society of four to eight cells. He placed the petri dishes beside the microscope set up at the back of the room. Then he helped Dr. M. tie her gown and put on her surgical gloves.

I stood beside the unconscious, trussed-up pig. Her head was lower than her feet, a matter of convenience for Dr. M. A pig's uterus and ovaries are all pushed together under the intestines, and if the pig is dipped in this fashion, the intestines slide forward, leaving a space where the surgeon can easily grasp the uterus and find the oviducts. The pig's belly had by now been washed several times more with antibacterial detergent and alcohol and then covered, save for a small square, with surgical draping. Dr. M. made an incision through the layers of skin, then used her hands like a spade to plow through the two-inch-thick layer of fat below. This caused the fat to divide without tearing. It is better to push than to cut, she explained; it is easier for the pig to heal. When she arrived at the thin layer of connective tissue below the fat layer, she used her knife again, then pushed more fat aside. Finally, she grasped the uterus in her hands and pulled it up.

It was like a thick sausage, long and twined with what looked like muscle. Yes, she said, the uterus *is* a muscular organ; it has to be to push those piglets out. There was no blood to speak of as she moved her hands along the uterus in search of the oviducts leading to the ovaries. The ovaries looked like little red balloons tied together as a bouquet. She found the right point, pierced a duct and then called for the technician, who had a pipette into which he had sucked several reconstructed eggs. She guided the tube on the end of the pipette into the oviduct opening and told him to release. He took his thumb off the top of the pipette, ending the suction, letting the dividing eggs drop from the tube into the duct and from there, hopefully, into the uterus to implant. And that was it. The whole procedure took less than five minutes. The technician now poured saltwater solution into the area around the uterus as Dr. M. stuffed it back in place. She sewed up the connective tissue, then sewed up the skin. Operation over. The close took three times as long as the opening.

Dr. M. was very pleased with the results they were getting in pig cloning: they got healthy pigs from 25 percent of the reconstructed eggs they made. Every one they implanted produced a healthy cloned animal, a phenomenal success rate, way above the norm for human in vitro fertilization techniques.

After the incision was closed, the pig was untied, the tubes removed from her snout and the table was rolled out of the operating room and into recovery. This room was divided into three pens, each with a thin rubber mat covering the clean concrete floor. The pig was rolled off the table by lowering one of its beams. She lay motionless, on her side on the mat, and the gate was shut.

The next pig was shoved into the elevated pen. Dr. M. went to scrub up all over again. As she performed exactly the same operation on the second pig and closed her up, I could hear the first pig groaning in the recovery room. Within twenty-four hours, Dr. M. said, both pigs would be on their feet, carrying the reconstructed eggs of a pig killed in a slaughterhouse, now behaving and dividing like embryos, motherhood on the way.

Are these gilts mothers? Or something else? Don't we need some new vocabulary to describe these relationships? Up until recently, biological motherhood meant giving a child 50 percent of one's genetic inheritance and carrying it until birth. If these pigs carrying cloned fetuses are not mothers, what are they? "Surrogate" surely doesn't cover what they do. How could anyone argue that the fetuses they carry are genetically identical to the boar, just chips off the old block? They are made from the nucleus of the boar, but also from the eggs of a dead female. The resulting embryos carry the nuclear genetic information of the dead boar, but also the mitochondrial genetic information of the dead female. Doesn't that make these fetuses chimeras, not true clones? While the embryos develop for the first few hours under the direction of the genes in the reconstructed egg, eventually their placentas attach to the surrogates. A few embryonic cells will course through the veins of the surrogates, and vice versa. These developing fetuses have to adjust to the cells of the surrogates, and their immune systems allow each others' cells to cohabit without going to a state of war. These exchanged cells will also make the resulting piglet a chimera different from the boar and from the dead female that provided the egg. Each piglet will become a unique blend all its own.

So the surrogates aren't mothers, and the piglets aren't precise reproductions of the boar. Not exactly a resurrection, but definitely a life after death.

The sun was well up, but it was a gloomy sort of a day. Dr. M.'s car carried me up and over the smooth hills and, finally, pulled into a long lane leading to a white farmhouse. No signs identified this farm as being in any way connected with Infigen and its cloned menagerie. Behind the farmhouse there was a very long, low barn, half hidden in a hill, a barn leased by Infigen to harbor its cloned pigs and their progeny. I opened the car door and got out, and was hit with a wave of pig stench so strong I staggered.

"God," I said, "awful."

"Leave your coat in the car," Dr. M. said. "If you wear it in the barn, it'll absorb the smell and you'll never get it out."

I stripped it off as fast as I could. Shivering, we walked to the barn entry, which was a locked door at the bottom of four concrete steps. In we went. The stink went up an order of magnitude to gale force. It was cloying, yet sharp; it was thick; it was everywhere, enough to make me want to retch. Jesus, I said, do we smell this bad to them?

"This is great," said Dr. M. "This is a great barn, state of the art." She gave me paper booties to wear over my shoes, then we entered a small office where a group of video monitors kept watch on all the pigs in the room beyond. The animals could be observed without actually going in and disturbing them.

"I like to watch them," said Dr. M. "You know, the secret life of pigs. They're just like us. They'd eat all the time if they could, and they get up from time to time to check the food bin just to make sure they got it all."

The images were black and white and grainy, but I could make out the backs of pigs in pens in a long, long room. After Dr. M. had satisfied herself that all was well, we walked down a narrow hall. The pigs were behind closed doors. I remembered, as we walked, my husband's various pig stories. As a boy, he had worked for various farmers on acreages near the small town where he grew up. His stories were all about how dangerous pigs are, how clever and how malevolent; they have been known to kill and eat farmers stupid enough to fall down in a pen in front of them.

I was thinking these thoughts as we walked through a door into the pigs' room. The smell inside was intolerable. I wanted to wrench off my socks and stuff them in my nose or run away, neither of which were options. The pigs were in pens raised up off the floor. This was to prevent the shit from cloned pigs finding its way, somehow, into the water table. Who knew what dangers lurked in their feces? Hunks of pig shit clung to the grated bottoms of the pens, and it was smeared on the pigs' feet, on their bellies, on their snouts, on the low pen walls. The walls were so low, in fact, that it seemed to me that any pig, with the exception of the tiny piglets squealing at their mothers' teats, could heave itself up and over and be on me in a flash.

There was a pathway down the middle of the room, between two rows of pens. Dr. M. walked over to check out the pigs about to farrow their clones. They were in separate, narrow pens that would give the newborn piglets access to them but protect the piglets from being crushed if their surrogates rolled over. Both pigs heaved themselves upright as Dr. M. came close. As I walked behind her, I could feel many sets of piggy eyes watching me. They were curious, and I could feel their curiosity build as I went down the row. Some of them were chewing away at rusty, clanking metal chains hanging from the ceiling. All of them stared as I moved by, as if they knew I was a stranger.

They do that to keep amused, to keep from being bored, Dr. M. said, pointing at the chains.

She wanted to show me a litter of cloned piglets a few weeks old. There were two adorable, black clones with gray markings, none of the markings the same. In the next pen there was a group of six who ran around or lay on their bellies or wrestled each other, a wiggling mass of trotters and snouts. I could see they were all approximately the same size, but there was a runt among them, just as there would be with any other litter. These were clones from a prize boar who had earned his owner $400,000 over his short life through the sale of his semen. There was nothing that struck me as unusual, misshapen, strange, awkward or deformed about any of the animals, and most of the pigs in the room, about thirty, had been cloned. In some cases you could tell the clones at a glance because the piglets were a totally different color from the surrogates that suckled them. Some of the sows had themselves been cloned; some of the piglets were clones of clones.

Dr. M. checked the pregnant females closely. All eyes were still on us. A better reporter would have asked a million questions, but all I wanted to do was get out of there. The smell was a physical presence, like a cloud of furious wasps circling and hissing, another form of pig defense skating over to offense. And why wouldn't they be angry? These conditions—living on a platform, locked in day and night—seemed horrible. Dr. M. saw the look on my face. These are great conditions, she insisted again. It's air-conditioned, everything's the best.

We walked back toward the hall door. I asked her, as casually as

I could, if these pigs ever tried to bolt, ever tried to jump their low fences and get outside. They lived their lives in such physical restriction, meddled with, operated on, never allowed to touch the cement floor, never mind the dirt and grass outside. I'd never allow a dog to live like this, I thought. And what was their eventual reward? The slaughterhouse. Surely they wanted out.

"Only occasionally. They're not exactly athletes," Dr. M. laughed.

We walked back down the narrow hallway to the office. Then we both saw it at the same time: there was a strange motion across one of the video screens, a swift, rushing, unfocused movement, as if the camera had made a swish pan, except it hadn't. Someone or something was loose in there.

"Hey," I said, "did you see that?"

Oh, she'd seen it, all right. Her face, freckled and a little wind burned, had gone quite pale. She had her eyes locked on the screens, as if she was frozen in place.

What in hell, I thought. What's she afraid of?

"Did a pig get loose?" I asked, thinking that that was something worth being frightened of, wondering how long it would take one angry pig, say a 400-pound sow, to bash her way down the hall and into this room.

Dr. M. was trying to maneuver the monitor to give her a clearer view. There was nothing to be seen in the main room but pigs and piglets. Then another wash of movement appeared in the video monitor focused on the narrow hall. Our eyes locked on that screen. A big hulking something was moving down that hall, moving toward us.

It was man shaped—big-man shaped. She watched it, rigid. What, I was thinking, a break-in? That's what she's afraid of—some nut, some kook, not the pigs.

The door into the hall swung open. A thick, red-jacketed arm, followed by a thick red-jacketed body, came in. It was Michael Bishop. I sat down with relief. And so did she.

I was in Michael Bishop's SUV driving down another country road. He wanted to show me a newly born bull calf, another clone only a few days old. I had my coat on. My clothes stank of the pig barn, and now this stink was burrowing its way into my skin: my hands and face reeked of the pig barn; my hair reeked of the pig barn.

Bishop rolled down the window as he drove.

He turned down a road, past a great pink barn. That was built by Pharming NV, he said, which had set up a state-of-the-art operation for the genetically altered cloned cows who were going to produce pharmaceutical-grade human proteins in their milk. He called it "the pristine palace." The barn was completely severed from the landscape and its watershed. It had cost $5 million to complete. It was a shower-in/shower-out facility, with climate control. Pharming NV needed to prove to the U.S. Food and Drug Administration that these human proteins could be collected from the cows' milk and be so clean, so pure, they could be marketed as drugs. Pharming had blown a bundle. There were sixty genetically modified cloned cows in there.

Bishop turned up a dirt road to another farm. Again, no sign showed its connection to Infigen. There was a small farmhouse and, off to the left, a large shed. He pulled up alongside the shed and we walked in. It was gloomy where we entered, but on the far end double doors stood wide open to the sun. A large cow lay on her side in a thin scatter of straw, a bovine Madonna. Not too far away, in a fenced area with a good view of the cow, the barnyard and the fields beyond, a furry bull calf mooed piteously. He was calling for his mother, except she wasn't his mother, she was his surrogate. Bishop walked over first to the cow, which had been delivered by Caesarean section. She mooed in a vaguely threatening way. He patted her down. The little calf pulled himself shakily up on his three-day-old legs so he cold get a better look at us. He was the cloned version of a prime bull whose semen had earned $20 million for its owner before the bull died. Infigen was paid $50,000 to clone him, and the company will also earn a royalty from any of the clone's earnings. Bishop walked over, opened a gate and laid his hands gently on the calf's flanks. He felt the calf's legs, took hold of his muzzle, smoothed his hands over it, looked into the calf's limpid, quivering eyes. "He's beautiful," he said, as if to reassure himself.

Michael Bishop and I walked through the back door of my hotel. We were going to have lunch with a visiting scientist who had come to town bearing a certain rare enzyme for Infigen's use. I could smell myself with every step. I pleaded for a few moments to wash up. I went in search of a bathroom, walking by the open door of a meeting room. There appeared to be a Christian revival meeting going on in there. People were testifying. Hey, I wanted to shout, you want revelations? I could show you one or two. I stood at the sink and scrubbed my hands, my face, my hands again, my face. But when I lifted my arms to pull down a paper towel, the stink of pig still billowed forth with every motion.

Later I sat with Bishop at a large, round table, talking about what is known and what is not known about why cloning works. Bishop launched into how completely hit-and-miss the whole business of cloning still is, the equivalent of the Egyptians building their first pyramids by trial and error only to have them collapse before they got the thing right.

"It's push and shove," said Bishop. "You grab a cell out of a population of cells. You don't know if it's differentiated or stem cells, whether it's in G0 [a quiet state of the cell] or in G1 [the phase after cells split into two]. Don't know if the cell is truly quiescent . . . That is a black-box situation." In other words, cloning is an endeavor as ill defined at this point as the genetics of C. elegans when Don Riddle started out, as magical a process as the manufacture of Freda Miller's stem cells. Does it work because cloners inadvertently grabbed a stem cell to clone from, or does it work because any nucleus will be reprogrammed to the start point of development after insertion into an egg?

Ian Wilmut, operating on the first theory, had spent years trying to isolate sheep stem cells with no success.[13] His colleague Keith Campbell had believed that any cell can be rewritten. Did that mean the whole idea of a stem cell as something special is just another dogma destined to fall, an extension of the belief that cells cannot revert to the total plasticity of the embryo cell they descend from? I told Bishop about the cells Miller was able to grow in culture and tweak into becoming neuron precursors. I told him these cells can be distinguished from others around them only because they will grow in a ball in the right culture medium.

He sat up in his chair. Dramatically, he asked for Miller's name and number. He wondered out loud if she had somehow infringed on an Infigen patent.

How could that be? I asked. She's published this, she has a patent pending.

Well, said Bishop, Infigen researchers can get this kind of cell from an adult cow too. "We've checked for rearrangement—there's no change, no change in telomeres. It's part of our patent. Check her patent," he said.

I didn't know what to make of this. Could he make these cells differentiate?

"We can trigger stems into muscle and into heart-type cells," he said.

At this point the visiting scientist arrived, a pediatric geneticist I will call Dr. E. because, like Dr. M., he didn't want his name to appear in print. He was a small fellow, diffident, blond, polite, who works for a company formed by molecular biologists from Yale. Their IPO brought in $500 million even though the company had no products and no quick prospect of revenues. Dr. E. wouldn't tell me which enzyme he had brought to Infigen or what Infigen would do with it. Instead, he wanted to tell me about how bioethicists have invaded the lives of researchers like him to such a degree that they are inhibiting important work.

I found it hard to pay attention. Why would he bring a human enzyme to a company that does animal cloning? This question went round and round in my mind as he unfurled his long tale of bioethical interference. About halfway through his story, the penny dropped. I'd been thinking in the old categories, that there is an unbridgeable divide between animal and human. But are we not all accumulations of cells derived from a common ancestor? Molecular biology has torn down the frail boundary between animals and humans. Human genes have been put in cows and sheep; genes obnoxious to humans have been knocked out of pigs. The products of these manufactured animals will eventually be consumed by human patients. What is done to them will be done to us. How is it, then, that moral arguments are made for and against human cloning but no one worries about cloning animals?

Having seen the healthy piglets, I could no longer understand why every government wanted to outlaw human cloning. Regulate it, sure, but ban it? Where exactly were the moral Rubicons here? No human embryo would be destroyed to make a human clone. Which embryos had died to make Gene or the piglets? Embryos are created when sperm meets egg; Gene had been made by taking the nucleus of a cell and dropping it into an egg. The egg, not an embryo, began to divide after egg and nucleus were fused with an electric charge and bathed with certain growth factors. The reconstructed egg behaved like an embryo, but it was not identical to an embryo. Since no embryo died to make Gene or the bull calf or all these piglets, and since Gene and the bull calf and the piglets were not identical to any other creature on the planet, what moral objection could there be to any of this?

"But why not human cloning?" I asked them. This was a question to which they were not accustomed, and their answers revealed that they too had forgotten the distinction between activated egg and embryo.

"My board is totally against us tampering with a human embryo," said Bishop. "In any way, shape or form."

"There are enough nuts out there, blowing up airplanes with a cause," said Dr. E. "Abortion activists have killed physicians."

"There would be a penalty if you did human cloning in the U.S.," Bishop added.

He was so nervous about being associated with human cloning in any way that he had refused to appear on camera to comment about Advanced Cell Technology's announcement of its so-called cloned human embryo, which wasn't an embryo but a reconstructed egg. "CNN wanted me live . . . We didn't want a part of it. We have a press release that says we're against human reproductive cloning."

Was he also against the creation of stem cell lines from embryos? This is true cloning. Embryos are destroyed to make embryonic stem cell lines; the cells of these lines are true clones of the original embryo. For the first time, I saw things clearly: the transfer of a nucleus to an egg to make a living being is *not* true cloning, but the creation of embryonic stem cell lines *is* cloning and raises real moral issues. Reconstructing eggs does not. Biologists were their own worst enemies, trapping all of us in their sloppy use of language. They applied the same word to technically and morally different procedures.

"We would not condone cloning human beings," said Bishop. "Every species is different. We're so successful in pigs. We need to improve in cattle. Humans are a lot different. I'd be concerned about offspring; they might live, but with severe complications. I'd hate to consign someone to a life they couldn't live fully or functionally. We're too early to consider it. We might be successful, but more by accident than by design."

But that risk hadn't stopped Edwards and Steptoe from fertilizing egg and sperm in a dish and putting the resulting embryo into a woman, who gave birth to Louise Brown. Very few studies had been done until recently on the health of children born this way.

Bishop pushed his large body away from the table and stared at me. "If you want a story," he said, "dig. I've had described to me things that are in effect nuclear transfer." A technician who had worked in an assisted human reproduction lab had come to work for him bearing tales of the things going on in certain clinics. He'd heard about manipulations that amounted to nuclear transfer between one human egg and another. "It's nuclear transfer in terms of risk and fact," he said. "The politicians don't understand what they read. No reporter has ever asked me about that. Nobody asked, is there a form of nuclear transfer used in vitro right now? I'd say yes."

He'd seen pictures on television of the equipment in use in an IVF clinic in southern California. He saw micromanipulators. "Why, for God's sake?" he asked. "Because you're working inside the egg with a suction pump . . . We're already cloning people, just not calling them clones. Who's gonna know . . . How would you know unless you compared the DNA?"

He was leaning on the table, giving me the full force of his views about his competitor at Advanced Cell Technology, a company he'd sued, a company whose very assets were now pledged to Infigen as security for future license payments.

"Mike West," he said, "this past summer, became a dad of triplets."

He stopped. I waited. He said nothing. I thought, I'm supposed to add this up and get four. So I added it up.

"So are you saying Mike West cloned his kids?"

"Would he?" he asked me. "Those babies don't look like Mike as he is today, or the mother . . ."

"Well, of course not," I said, "they're babies."

"I'm asking you," he said. "The man wants to find ways to stay in the limelight. Would he do that?"

"Are you really saying he cloned his kids?" I asked.

"He got triplets," Bishop said. "It's very rare. And how come him out of the randomness. How come him?"

CHAPTER EIGHT

RENDEZVOUS AT THE FOUNTAIN OF YOUTH

I had picked up so many tales about Michael West that if stories had wings, I would have been carried straight to Worcester, Massachusetts, without the inconvenience of planes and airports. Some of the most interesting tales came from the Houston oilman Miller Quarles who gave West money because West had undertaken to solve the problem of old age.

Quarles, then eighty-seven, was a story himself. He had two companies: the Miller Quarles Company, which explores for oil, and a nonprofit that supports longevity research. He had told reporters,[1] and later told me, that sometime in 1989 he'd made up his mind to spend the rest of his life "to find the fountain of youth or a cure for aging." He had money, determination and an MA from Caltech, where he had been taught by Linus Pauling. He'd founded the Curing Old Age Disease Society. He had told his dentist about it, and the dentist had joined the board. The dentist had a female patient who mentioned that her husband knew a terrific cytologist named Michael West. West's number had passed from the patient to the dentist and then to Quarles, who called West and told him of his quest. West mentioned that he was going to a cryogenics meeting, where they could meet and talk. When they got together, Quarles told West he was in a hurry to defeat old age, to find a cure. "I told him I wanted to have it by the year 2000," Quarles said.

According to Quarles, West soon offered him the opportunity to invest in something that could regrow skin. Quarles put some oilmen together in a room to listen to West's presentation, but they weren't interested in investing. But Quarles told West that if he started a company to cure old age by the year 2000, he'd give him the equivalent in stock of $50,000. "He called me a month or two later and said, I think I found a longevity gene in a chromosome. He said if the offer's good, I'll take it." That call came at the end of 1989 or the beginning of 1990, and the result was Geron.

West then found some venture capitalists in New York, Quarles said, who backed the company with a further $10 million and moved it to California. By 1997 the venture capitalists owned 77 percent of Geron and the management had changed—and so had the company's goals. Geron began to avoid the scientific controversy of saying it was trying to cure old age. "They publicly announced they were not trying to cure aging but the diseases of aging," Quarles said. "Mike had started the company with the promise to do it by 2000," he said. "He quit. He joined Advanced Cell. I sold all my stock in Geron and invested in Advanced Cell Technology." (He has since purchased Geron stock again.)

In 1990 Quarles also decided to offer a $100,000 award through his nonprofit to anyone who came up with a cure by the year 2000. He told me he had already given the prize away—to Michael West. "Mike knows, but it's not well known that he knows, how to extend the human life span," said Quarles. "He found the fountain of youth before 2000 and tested it on animals. I had to offer him the $100,000. Mike refused the money because it had not been tested on humans, so I put it in Advanced Cell Technology shares."

West and his colleague's method was of no use to Quarles himself, but it would help the next generation. It involved cloning, so they were holding back on it because people in the U.S. Congress were trying to pass a law making all human cloning illegal and punishable by ten years in jail and a fine of $1 million. "The minute that law is definitely defeated," Quarles told me, "he can clone a baby that will probably have double the life span."

Worcester, Massachusetts, is set in winding valleys that run between seven hills. "Just like Rome," said the taxi driver, without a hint of irony. He ferried me from the echoing airport in the worst cab I've seen anywhere, and that includes Tangier, Athens under the junta, and even Piauí, one of the poorest states in Brazil. There were no seat belts. Stuffing spilled out of ripped upholstery; busted springs erupted into my hip. I hung onto the door handle until it came away in my hands. As we drove through downtown, with its crumbling warehouses and mean streets, the driver explained that Worcester had once got its living from various kinds of mills but that now the economy was reliant on its schools of higher education. There were fifteen. He rhymed them off, starting with Holy Cross, running through Amherst and ending with the University of Massachusetts, near whose campus he found the building he was looking for. Advanced Cell Technology's offices were at 1 Innovation Way, in a brand-new redbrick overlooking the unnervingly decrepit stack of a hospital. In the lobby, beside a large staircase, some decorator had placed a neon double helix to light the way. I found Advanced Cell Technology off to the right, behind locked doors.

I was put in a boardroom to wait for Michael West, who soon appeared with his laptop and set it up on the table. He was of average height, perhaps five foot ten, with a slim build. There was something awkward about his shoulders, which pushed forward on the top of his spine. His hair was auburn and thinning, his eyes large and light-filled. As he talked, I found myself watching his fingers, which splayed and played with each other, or thrummed on the tabletop in awkward little drumrolls, or flung themselves behind his neck, where they laced together and then fit like a cap over his head. Sometimes he used them to hide his eyes. He was far better dressed than any other scientist I had met or would meet over the course of the next year. He wore a sports shirt with long sleeves in a fine herringbone check, with a shimmer that said silk blend; this collaborated with well-cut wool pants of similar but not identical fine check, the kind of expensive outfit one can see in downtown law offices on casual Fridays.

Where to begin? Should I ask him right off if he'd cloned his children, or wait till later? I thought I'd begin with something less controversial, such as, why had he published articles on cloning a human embryo when no actual embryo was cloned?

West had a whole list of reasons. On the one hand, he said, Joannie Fischer, the U.S. News & World Report journalist, had heard they were doing experiments with human eggs. What if 60 Minutes heard about it too and showed up, cameras running, at his door? Then there was his desire to be transparent, the way Bob Edwards was when he published each stage of his slow march toward the birth of Louise Brown. He also wanted to advance the public debate so that no crafty politician could attach an anti-cloning amendment to some tax bill and sneak it through. And, finally, there was the problem of the competition. There was an Israeli scientist named Itskovitz who helped James Thomson derive the world's first embryonic stem cell line. The Israelis were working on this too. "I didn't know where they were at," he said. "Not a clue. And we're proud of this work. This is Jose Cibelli's dream for years. He wants to be the achiever, the guy who gets it done."

Done first, I thought.

Now West took over: he had a spiel, which included a short history of the mythology and science of aging and death, plus a set of images to illustrate it. He turned his laptop to face me. The computer screen showed a female in profile at four different stages of life: as a baby girl, as a grown but very young woman, as an older woman of my age, and finally as a very old woman. With advancing years, the pretty features, so full and smooth in the child, began to thin, then wrinkle and finally fold in on themselves. This graphic display of my current position on the trajectory toward death upset me so much I barged in on his presentation. Screw the shmooze, I thought, who has time for another dog-and-pony show? "I want your story," I said, without so much as a "please."

He shifted in his chair, surprised by my peremptory manner, but he complied. He was born in Niles, Michigan, in April 1953. He had one sister. His dad ran a semi truck dealership. His mother was a housewife. Both parents were dead. "I was born with a test tube in my head," he said, by which he meant he was a fool for science. He had a lab in the garage. Like David Baillie, he liked chemistry, physics, electronics. He thought he'd do science for the rest of his life, but in his senior year of high school, "I began to think about the immortality of the soul," he said. "The late '60s were interesting times. ESP and things were being studied scientifically, and I approached the supernatural with the same eagerness as physical sciences. What stood at the center was the Bible."

His family was not religious, although his father was an occasional Methodist. So West became religious enough for all of them rolled together. "It was the Jesus-T-shirt-passing-out-spiritual-tracts-at-two-in-the-morning-outside-a-bar," he said. His hands had crept down to cover his eyes.

"I burned my science books . . . I was such a committed Christian. I knew my vice, where I could be in trouble in my relationship with God. . . I dismantled my lab and burned it. The religious thing—my parents were gravely concerned. I was taking it to extremes. I'd take a Bible while Dad and I were fishing . . . The track I was taking was the intellectual side of Christianity . . . What occurred to me was that the best defense of Christianity is the argument from design."

West was referring to creationist theory. The argument from design proposes that the intricacy and beauty of creation cannot be an accident but is obviously the product of infinite care. If life is designed, there must be a Designer. This is a very bad argument. It is circular, for one thing. A friend of mine had recently succumbed. "Elaine, I believe in God," he'd blurted, over lunch. Since he had been a determined materialist as long as I'd known him, which was over thirty years, I was surprised. Since when? I asked. "Since I went snorkeling," he said, "on this coral reef, and there was this amazing, swirling variety, all these different fish, all these living things. There has to be a God."

West decided to spend his life using science to demonstrate the existence of God, as "an evangelical tool." He went to Rensselaer Polytechnic Institute in upper New York State. He started in physics but switched to psychology and graduated with a bachelor of science degree in 1976. However, he realized that he needed to learn much more in order to demonstrate the truth of his Truth. "Michael was to study the fossil record," he said, speaking of himself in the third person, as if that other Michael West was someone he no longer wanted to know.

He did an MA in biology at Andrews University, a Seventh-Day Adventist school in Michigan, where he studied under Richard Ritland. "His life's work was to harmonize the paleontological record with the Bible. Very devoted Christian. Just like me, but one generation earlier." Ritland earned his doctorate at Harvard.

But certain questions plagued West. The worst were those raised by mixing blood cells from various animals—"comparative serology" he called it. "If you mix the blood of animals, they are more reactive the more widely separated by evolution," said West. "Small response human-to-monkey, a large response human-to-platypus." Didn't this escalating rejection reaction demonstrate the very opposite of what he was trying to prove? Didn't it suggest that animals derived recently from a common ancestor are more alike than those further removed? If God made man and frog at the same time, he asked, "why wouldn't God make human blood like a frog's?" Eventually, West had to conclude he'd been wrong. "I had to admit the earth is old, and there was evolution, and that humans evolved."

He came to this conclusion in 1980 (although he didn't finish his MA until 1982). His change of mind seemed to coincide with the loss of his father, who died of a heart attack in 1980 when West was twenty-seven. West had to take over the dealership in Michigan, but "didn't want to run a truck garage the rest of my life. I orchestrated a clever transaction. I sold the business for a significant amount of money, hundreds of thousands of dollars. I thought I could retire and live off the interest. I was free to do what I wanted."

I asked him when the idea of making humans immortal entered the picture. I thought I knew the answer, that it had to do with the sudden death of his father. I thought his death might have destroyed West's faith in Christ and renewed it in science.

He slumped in his chair with his hands over his eyes again. "It's hard to explain."

One day he was watching television and saw a man being interviewed about a disease that leads children to age very rapidly. The disease is apparently passed down in families. "I latched on to that—something in DNA controls aging. I realized that was huge . . . I realized the most significant problem of the human condition was the biology of aging. It's right up there with memory. It's a big mystery, never cracked, but it's at the heart of the human condition. Where it crystallized was, I was out for lunch. Silverbrook Cemetery was across the street. I'm not sure if this was before or after my dad was in there. I'm pretty sure it was before. I looked across the street. I'm a person with vision . . . I could see the day when all the people I cared for would be these names written on the tombs of the cemetery. I realized aging and death were not something I could tolerate."

For a moment I thought he was joking, and I almost laughed. No one likes aging and death, but they are inevitable, so what does toleration have to do with it? But West was completely serious.

"Emerson said, 'I taught one doctrine, the infinitude of the private man, the infinite value of the individual person.' The idea that we are like ants and it's okay to sacrifice a few to the whole—I never believed that."

So, it wasn't just aging you wanted to defeat, it was death? I asked.

"Yeah," he said. "Aging is a death sentence . . . I believed it was the most significant problem of the human condition and the most important scientific challenge I could think of. I recognized this was where I'd apply the rest of my life."

Early in 1982 West went to see Sam Goldstein, a professor at the University of Arkansas' medical school at Little Rock. Goldstein worked on the molecular biology of aging, and West wanted to get his doctorate under Goldstein's tutelage. "The day I interviewed with him was the day of the Arkansas creation trial," said West. The state of Arkansas had passed a law mandating the teaching of creationism as if it is a science. The American Civil Liberties Union sued the state on the grounds that this law was unconstitutional since it forced religion into the schools in the guise of science and the U.S. Constitution requires the separation of church and state. West walked into the courthouse to watch. The courtroom was packed with the leading creationists and with evolutionary theorists such as Stephen Jay Gould, who had come to testify.

"In the pews next to me was Stephen Jay Gould," said West. "Duane Gish was on the stand testifying to the scientific value of creationism to the judge. He was an acquaintance of mine. I'm talking to Gould—I said Gish is a biochemist by training and he is leaning on others for the scholarly evidence. He said, 'He's dug his own grave, he can lay in it.'" Gould testified against the theory of creationism and in favor of the science of evolution (though he was known as one of Darwin's more formidable critics).[2]

West considered the trial a segue in his life, marking the point where he left religion behind and moved on to the study of the aging of cells. But segues are smooth: his change of direction was more like an about-face, and things did not go well. His doctoral supervisor, Goldstein, had argued and published that DNA changes with aging. "I disproved Sam's work. It was really unexpected . . . He made some mistakes," said West.

Goldstein had to publish a retraction in *Nature,* and he wanted West to coauthor it. But West did not want his first publication to be an apology for error, so he left Goldstein's lab without a doctorate and moved on to the Baylor College of Medicine in Houston. He started over in virology and was awarded his doctorate in 1989. Then he decided to go to medical school. It wasn't that he wanted to practice medicine, but he did want to study the specifics of human aging and death. He entered the University of Texas Southwestern medical school in Dallas. There, he also worked in the laboratory of Woodring Wright and Jerry Shay. "Woody is interested in aging," he said. "They gave me an empty room to do experiments. They made me senior research scientist. I wrote my own grants [applications], while I'm in medical school."

He turned again to the laptop, where a quotation appeared on the screen. "Life maintains the appearance of immortality in the constant succession of similar individuals, but individuals pass away," it read. "The attempt to answer why," West said, portentously, "is the heart of religion."

Oh, I don't think so, I replied. In Judaism, questions about immortality start and stop with the Adam and Eve story. Judaism wastes no time on why.

"Paul said, 'As in Adam all die, so in Christ shall all be made alive,'" said West.

"So you're talking about Christianity," I said, "not all religion."

But he really didn't want to talk about Christianity or Judaism. He pointed at the screen. Egyptian images appeared. These referred to the story of the resurrection of the deity Osiris, slain by his enemy, cut into fourteen pieces that are scattered across the earth, restored to life by his wife, Isis, who gathers the pieces together. Osiris was the Egyptian god of the afterlife, to whom one must appeal to live forever. He was "a fertility god, associated with grain," West said, "personified as a god of the eternal renewal of life, the constant succession of similar individuals." On he went with these tidbits from Egypt and the age of the pharaohs, and then to Greece and its mystery cults. "The Greeks said *zoe*—immortal renewal—and, springing from it, *bios*—mortal life . . . They thought life had a spiritual origin, an immortal seed, or epigenesis." These concepts passed on to the ancient Christians.

From a discourse about the origin of the ritual of communion, we leaped to the nineteenth century. The focus also shifted from immortal life to the cause of death. A new image flashed up on the screen, a bearded gentleman. "August Weismann," West said. "Absolutely brilliant and precocious." Weismann believed we are all the result of single-celled animals that proliferated without limit, immortal cells. Eventually, some of these cells evolved into a ball and formed a primitive body. Here, in the body, in complexity, was the origin of death. Cells in bodies die in every generation, but "there is a germ cell line that continues; the lineage is immortal." In 1881, according to West, Weismann predicted that these cells that make up bodies, called *somatic cells,* would be shown to have a finite ability to divide, "and that's why the individual grows old."

But soon Weismann's theory was turned upside down. Alexis Carrel, a surgeon at Rockefeller University, grew chicken-heart cells in culture in 1912. The cell line continued to grow indefinitely,[3]—for thirty-four years. "They didn't grow old. He published annual updates," said West. "The media ate it up. There were rumors he made a giant, beating heart. He convinced the world we don't grow old because the cells grow old." Carrel argued that aging "is an attribute of the multicellular body as a whole."[4]

For many years thereafter, the idea that cells could live indefinitely if removed from the body was accepted as fact. But in 1960 Leonard Hayflick noticed that his human cell lines kept dying out after forty or fifty passages. Hayflick and a colleague showed that these cells eventually reach a state they called *senescence.* Then in 1965 Hayflick showed that cells taken from older people divide fewer times than cells taken from younger ones.[5] "Weismann was right," said West.

In the early 1970s the old discipline of cell biology began to merge with the new molecular genetics. "Alexey Olovnikov in Russia heard a lecture in Moscow about Hayflick," said West. "He went to the Moscow train station. He saw a train come in, and had a flash of insight." The way West told this story, the train reminded Olovnikov of the way a DNA strand is copied. People did not get on at the very back of the train, but near the back. Similarly, molecules assembling on a DNA strand to copy it would not be able to copy the whole thing. Inevitably, after many cell divisions, a DNA strand would shrink. Olovnikov wondered if this might explain the Hayflick limit.[6] "A little is lost each time," said West. "Olovnikov wrote in 1971 in Russian, in 1973 in English, his theory of telomeres . . . He also proposed that a special molecule keeps the immortal germ line from aging—it rebuilds the chromosomes."

Those studying cellular aging were familiar with Olovnikov's paper, published in English shortly after James D. Watson's paper about the difficulty of copying the ends of linear DNA strands. There were also indications that these telomeric sequences differ from animal to animal: it was certainly obvious that life spans vary. "The Galapagos tortoise lives three hundred years and his cells go two times [as long as] a human's cells in the dish," said West. "There were reasons to think there was something to this idea of the problem of DNA copying having something to do with aging."

More images on the computer, this time the ancient symbols for man, woman, infinity.

By this point I was growing impatient. West wrapped his ideas in a crust of mythology, peppered with the leavings of old theories. He seemed to think they would become palatable if served up first by Egyptians, Greeks and ancient Christians, as if to say, this work comes out of a long tradition, the fountain of youth is an old idea we are merely rediscovering; as if he wanted to capture my imagination first, and belief would inevitably follow. I was impatient because my imagination had been triggered long since, and I had been toying with my own what-ifs for some time. What if the ancient Egyptians knew much more about biology than we imagine? What if there was an important, but hidden, reason for their constant use of chimeric images? What if the Greeks picked up a garbled version from the

Egyptians? What if their myths encoded something important, something true about the nature of life? And what if some of these ideas were transferred to Jews during the long sojourn in Egypt? For two years I had sat in a synagogue on a hard bench every Saturday morning and on every major holiday, as part of the preparation for my daughters' bat mitzvahs. I had carefully read in English translation the portions of the Torah being sung to the congregation in Hebrew. The sections that caught my attention were the descriptions of God in Exodus and Leviticus. God appears at certain moments of crisis in the guise of a grumbling, dark, dangerous cloud hovering in front of the children of Israel. Leviticus 16 opens by explaining that two sons of the chief priest Aaron, Moses' brother, had already died because they got too close to the shrine, the center of the tent of meeting where God's ark resides and over which God hovers. After this event, God made rules for all time. Aaron or his successor could come into this shrine only once a year, to make a sin offering. He had to wash himself and put on the correct linen vestments before he entered. Even then, he had to protect himself by burning a cloud of aromatic incense thick enough to prevent his actually seeing the cover over the ark, and after he was finished purging this shrine with the sprinkled blood of a bull and goat offering, he had to remove his clothes and wash. Even the person designated to guide the community's sin offering to the wilderness (a second goat, which the priest also dedicated inside the shrine) was ordered to remove his clothes and wash them and himself before returning to camp. God ordered that the bull and the first goat "shall be taken outside the camp, and their hides, flesh and dung shall be consumed in fire. He who burned them shall wash his clothes and bathe his body in water; after that he may reenter the camp."[7] God sounded like an alien with a megaphone and a smoke machine, and his instructions sounded uncannily like methods to reduce contact with and get rid of a bio-hazard. On the other hand, maybe God thought unwashed men in desert camps smelled bad.

"Weismann says we are just a production machine, the disposable soma is like the fluff of the dandelion to carry the seed," West said. "We grow old and die when we fulfill our function. Animals, and plants, they just live long enough to replicate themselves."

Now there were images on the screen of butterflies and daisies.

I objected to Weismann's theory on the grounds that many animals and plants live longer than one reproductive cycle. Apparently I was not alone in finding it wanting. "Lots of insects live underground, crawl out, reproduce and die the next day," answered West, but that defense made it an unpopular theory. "Philosophically it reduced humans to reinforce the evolutionary model. The [old] idea was, we were the crown and glory. Scientists said we were a reproduction machine selected for by natural selection. To this day," West said, "you can go to leading scientists and say, did you realize we are made from cells that have no ancestors, that the lineage has proliferated indefinitely? The typical scientist will say, 'Never thought about it.'"

And so, said West, the question arose, "Could we transfer immortality to somatic cells?"

He did not say exactly who this question occurred to besides him, but in 1985, when he was working on his doctorate, he heard of the work of Howard Cooke, who had chopped up DNA segments of known sequence near telomeres of the X and Y chromosomes and measured the length of a marker.[8] According to West, Cooke found that in human beings, reproductive cells' telomeres are long and stable. "I looked at that and thought, holy cow—that's what Olovnikov predicted!"

West seemed to suggest that his interest in telomeres had been ignited by Olovnikov and Cooke. What about Greider and Blackburn, who published on the mysterious enzyme putting sequences back on tetrahymena RNA the same year?

"Carol Greider, who studied with Blackburn, first described a protein in single-cell tetrahymena that could add it on to the DNA. She called it *telomerase*," he said. There were several other authors who published along these lines within the next few years too, but none showed that telomere-copying problems caused cellular aging. Nevertheless, West had long since jumped to the conclusion that this was the truth. "I believed it was true. Couldn't prove it."

And because he was convinced, he decided to start a company based on the search for an immortality gene, something that could put telomeric sequences back on the ends of aging chromosomes. He explained that in his third year in medical school, he changed the name of a company he still owned to Geron, and he set out in 1987 to find backers. Then, he said, he found a backer who sent him to California, where he convinced Kleiner, Perkins, Caulfield & Byers that he had a great idea; they put up funds and Geron took off. West never mentioned Miller Quarles in this account. And Geron's website says the company was incorporated in 1990, not 1987. He also made no mention of Carol Greider's grant application coming to Woodring Wright's lab, where West was working in 1989, nor of his pursuit of Greider to sit on his scientific advisory board. I asked about Miller Quarles first.

He sat back. "Miller. There was a guy in a dentist office—I know a guy—I needed angel finance for Geron. I was burning money to fund this company, to meet venture capitalists. It costs money. Medical school is expensive. Support my wife, tuition, my food—I needed a few hundred thousand. I wanted to build a company. My specific strategy was to treat age-related diseases. I needed angels to put money at risk on a wild and crazy guy . . ."

The wild and crazy guy he referred to was himself, Michael West. Among the people he went to see was a man who brought in a Nobel laureate to review West's ideas. "He turns to the scientist and says, 'What do you think?' The scientist says, 'Baloney' . . . I'm in the room. He says, 'We'll never figure out aging.'"

West went through the rest of his savings very quickly. Finally, he found a small group "who believed in me," said West. "The first guy was Bill Ryan, an MD and biotech analyst at Smith Barney in New York. He said he never heard a more interesting, exciting story. He wrote me a cheque for $25,000." West incorporated Geron in Delaware with the goal "to . . . find the biology of aging and immortality gene, and it's all based on theory. I believed it. Bill and I knew it was risk as high as selling swampland in Florida."

This was when Miller Quarles came in: he was West's second investor. Then West got $250,000 from a group of angels. When Kleiner, Perkins wanted to do a deal, "it was a feeding frenzy." Kleiner, Perkins offered shares in Geron in a private placement to venture capital companies. "It was way oversubscribed. Here are guys who need $5,000 for plane tickets and now [we're] turning away millions."

In the first private placement of shares, they brought in $7.5 million, tripling the company's value. "I was thrilled. I saw it needing maybe $20 to $30 million before the IPO. Money was not a problem. The major thrust was telomeres: when cells immortalized, they stabilized the telomeres." He was referring to a paper published in 1994, whose first author was Calvin Harley; Carol Greider and Silvia Bacchetti worked on it as well.[9]

Part of Geron's appeal to investors was that West had managed to get James Watson to join the scientific advisory board. West explained that Watson was interested "in the telomeres thing . . . He had hired Carol Greider at Cold Spring Harbor. He spent a lot of time in her lab, where I talked to him."

"You mean she introduced you?"

No, he didn't mean that. The Smith Barney biotech analyst was familiar with Cold Spring Harbor, and he called Jim Watson. "He calls Jim and says, you oughta hear about this company. Jim says okay. I meet with him. It was very entertaining. He's very interesting. The guy has awesome concentration—he heard every word I was saying [even though] every five minutes he was interrupted . . .

"He says, 'I'll do this,'" West continued. "'I doubt that we'll get anything in my lifetime to help me, but I'll do it anyway.' He recognized it was important, but also recognized his own mortality."

It was time to ask him about Carol Greider.

"She was interested in isolating a gene, purifying it from tetrahymena," he said. "We paid money to Cold Spring Harbor and got the license to whatever she discovered. I hoped she'd get the protein part and we could find the human [gene]. It didn't go well." He explained that the company spent millions of dollars trying to isolate and clone the portion of the gene that calls up the catalytic action of telomerase. While Greider was working with tetrahymena, a graduate student of Nobel Prize–winner Thomas Cech was working on the same thing in another organism. By this point, 1996, Geron had gone public, West's wife had left him, and the woman who would become his second wife was one of four team leaders directing the company's efforts. Cech contacted Geron to say that a student had found a sequence for a piece of the gene in a protozoan, euploides, and that he thought it was similar to the reverse transcriptase HIV. "The University of Colorado sent the data under a nondisclosure agreement to several people. We licensed it through an auction," said West. "We were very aggressive. We worked a deal. Colorado filed a patent. We wanted the human version of this gene."

They all searched for similarities among the human DNA segments being randomly sequenced by various labs working on the Human Genome Project. These sequences were posted in a public depository called GenBank, which could be accessed and searched via the Internet. They were looking for a human DNA sequence that was similar to the euploides fragment found by Cech's student— similar, but not identical. "And one night, at two in the morning, Cech's graduate student types in their sequence on that day's information and up comes a sequence from St. Louis." Washington University was one of the main Human Genome Project sequencing centers in the United States. The sequence found bore a striking similarity to Cech's student's sequence. "We ran twenty-four hours; we mobilized the whole company. It was a massive effort to get the rest of the gene. We got the whole sequence. It took a little over two weeks to get the whole gene."

They ran tests to show that this gene called up an RNA that was active in immortalized cells but not in ordinary cells. "We had it. We wrote it up. We published it in *Science*. The article appeared in August 1997."[10]

But there was something strange about his version of this story. The lead author and the last author credited on this article were not Geron staff, whose names appeared in the middle, but Cech's student, Toru Nakamura, and Cech himself. "Six days after we published," continued West, "Bob Weinberg, a cancer researcher at MIT, published it in *Cell*. We got the patent. It cost us $30 million. And we got it six days before him and one grad student."

The next step was to see what would happen if one added the product of this gene to an ordinary human cell: would it extend that cell's life span? The first cells they tried this on, West said, were cells scraped from Leonard Hayflick's leg. Hayflick began the experiment in front of a French television crew. "They wanted him to cut the skin off himself. It was the most bizarre thing. I said, Len, are you crazy? Here's the camera on his leg." Hayflick's skin cells were grown in culture, said West. "I transfected his cells with telomerase. We sent them to Woodring Wright and Jerry Shay. We published in 1998 that it worked. It extended the life span. Now, just barely," said West, holding up a cautionary finger. "It was statistically significant, but just . . . We were so anxious to be first, we went with the statistical data . . . We were writing a version of the paper before we had the data."

This was the article Carol Greider so disdained. And in spite of West's description of his central role in this work, his name does not appear on the paper. Leonard Hayflick is thanked for reading it, but not for providing cells.[11]

The publication of this article coincided with West's exit from Geron. He explained his departure as the result of his growing impatience with the way the CEO ran the company. The CEO, he said, came from a much larger company, and he was trying to run Geron, "this struggling biotech," as if it was a large pharmaceutical. West spent way too much time going to retreats and writing mission statements and far too little time, for his taste, working on aging. And so, he said, one day he just sold his shares and packed it in.

Geron had been a publicly traded company since 1996. Anyone owning a publicly traded stock must not trade on material information (defined vaguely as information likely to change the value of the shares) until that information is made available to the entire marketplace. Officers of companies publicly traded in the United States, and all others deemed to be insiders, are required to report to the Securities and Exchange Commission (SEC) all their transactions in their company's securities within ten days of making them. Leaks of material information can quickly move the price of a company's shares. When Geron and its associates were about to publish on the sequence of the catalytic component of the human telomerase gene in Science in August 1997, the share price of the company jumped up just before the magazine was issued. In January 1998, just before the publication of the cell life-extension article in Science, the Geron share price bounced up again—44 percent in one day. It was a phenomenon that Science itself covered on

January 23, 1998, in an article entitled "'Fountain of Youth' Lifts Biotech Stock."[12]

The article on cell life-extension had been submitted to *Science* on December 1, 1997. Two months earlier, in October, West had acquired options to purchase 15,000 Geron shares. The option price was $10.13 per share. On December 4, West submitted a report of this transaction to the U.S. Securities and Exchange Commission, stating that he held a total of 168,900 options (but apparently no shares). On December 22, the article was officially accepted by *Science*. On Monday, January 12, 1998, the article's findings and its imminent publication in *Science* were described in an embargoed press advisory sent out by the Alliance for Aging Research, a nonprofit lobby group. Geron's CEO at the time, Ron Eastman, and various representatives from large pharmaceutical companies were members of the Alliance's board. The Alliance's press advisory described the findings, named Geron and gave notice of a press conference to discuss the forthcoming publication in *Science,* to be held on Thursday at 1:30 p.m., and warning that the information could not be released until 4:00 p.m. on Thursday, the end of the trading day before the magazine would appear. But after being asked by Geron's lawyers to move their press conference back to 4:00 p.m. on Thursday, the Alliance sent out another press advisory with a changed press-conference time. Somehow, it was distributed by a newswire to investors as well as reporters. By Monday afternoon, the advisory was also published by a database service. As Geron officials were making a presentation at a biotech exposition, the share price jumped to over $14 per share, and *Science* had to release the article early.

On January 29, West sold 25,000 Geron shares for $13, and on January 30, he sold another 15,000 shares, also for $13. West, vice president for new technologies at Geron, filed an insider-trading report with the Securities and Exchange Commission, recording these transactions. The SEC received it on February 10.[13] This filing showed that the number of securities then owned by West was 96,358; the number of options he held had declined from the 168,900 he said he owned on December 4, 1997, to zero. (There was no statement as to how these options had been disposed of.) Three days later, Geron issued a press release stating that West had left the company.

West mentioned none of this to me—I pieced it together after our meeting. Instead, he explained that part of the reason for his departure was that his thinking had by this time moved far beyond extending the life of a cell to extending the life of tissues and organs. And he backtracked a little to explain. "Since 1995 I'm thinking, could we get embryonic stem cells from humans, which are telomerase positive, and make anything from the body?"

First, he signed up Roger Pederson, then at the University of California at San Francisco, who ran an IVF clinic and was working on mouse embryonic stem cells. Then he heard that James Thomson of the University of Wisconsin was about to publish an article on stem cells in primates. "How did you hear?" I asked.

"One of the reviewers . . . talked," West said. "I heard he'd publish a paper on monkey embryonic stem cells. The next day I was on the plane to Madison." He did a deal with Thomson and the University of Wisconsin's Alumni Research Foundation. In exchange for a license, Geron agreed to fund Thomson's attempt to establish a human embryonic stem cell line. West also learned that John Gearhart, at Johns Hopkins, had access to human abortuses of less than ten weeks of age. Gearhart too was funded by Geron to try to establish human embryonic germ cell lines. Thomson succeeded in 1998, followed by Gearhart.

I asked West how he justified the destruction of embryos for stem cells in the light of the new work on adult stem cells. He looked very displeased. He said he was tired of all this talk about how adult stem cells could do everything, make every kind of tissue. "It's just absurd. The Greek Orthodox Church said adult stem cells are superior. The pope says it. What do they know about science? People need to recognize areas where they are expert and where they are not, to respect the truth. The adult stem cell is not a totipotent cell."

West stood up. He said he needed a break and he stalked out of the room. Of course he needs a break, I thought. His determination to end death had forced him, a former Jesus freak, to invest in killing embryos. He was gone for a long time.

When West slid back into his chair, I said, "So, we were on cloning." I meant the cloning of embryonic stem cells, but he jumped at the opportunity to discuss the work he had just published.

"What Jose [Cibelli] and I saw in cloning was not just an identical animal from a somatic cell . . . not much of a miracle . . . Where I saw this, and Jose as well, is a time machine for a cell, taking a cell back to the embryonic state."

By the time he heard about Cibelli's work, he had left Geron and joined a new company called Origen Therapeutics, which was "developing transgenic technology in commercial poultry." He heard about Cibelli at a conference he and his wife, Karen Chapman, attended in Australia. Cibelli had tested the idea that a senescent adult cow nucleus could be made young again by nuclear transfer, by dropping an old nucleus into an egg. "He put it in a mother, he grew a fetus, he took cells out of the fetus, put it in a dish . . . He showed the cells got their life span back. I leaned over to Karen and said, oh my God."

Better yet, West learned that the nuclear transfer patent application Roslin filed did not directly cover humans. He thought he could apply this nuclear transfer technology to human cells without having to buy a license from Geron. He had scouted Roslin Institute's spinout company, Roslin Bio-Med, for purchase by Geron, and he'd thought Ian Wilmut, whose name was first on the Dolly article, had led the cloning work. Now he heard that Keith Campbell was the man who'd made it all happen, and it was Campbell who had had this theory that putting an old nucleus into an egg could somehow rewrite the nucleus's program to its youngest state. Campbell no longer worked for Roslin. "Geron had bought Roslin [Bio-Med] for $45 million, and not for the pig-cloning patents," West said.

In 1999 West visited Cibelli, who was already working at Advanced Cell Technology. The company had been created by professors and students at the University of Massachusetts, but it was then owned by a company called Avian Farms. "I approached the owners, a chicken company in Maine. They had no CEO [at Advanced Cell]. I said, why don't you hire me as the CEO? They called me back two weeks later and said, when can you start? I bought the company a year later."

In 2000 Advanced Cell's scientists, working with others (including Vancouver's Terry Fox Laboratory's telomere specialist, Peter Lansdorp), published the results of certain experiments in *Science*. They took cells from a bovine fetus, transfected them with a reporter gene and grew the cells in a dish until they were senescent. Then they transferred the nuclei from about 1,800 of these senescent cells into denucleated cows' eggs; the eggs were activated and began to divide like embryos. They were then put into the uteruses of receptive cows. After various failures, six cloned calves were delivered by Caesarean section. Cells from these calves that showed the reporter gene were grown in culture and passaged more than ninety times, double the normal life span in the dish. These cells were also compared with the cells of normal calves of the same age: the cells taken from the products of nuclear transfer showed much higher expression of genes associated with the early stages of development than cells from normal calves, and these cloned calves' telomeres were longer than normal. In other words, ancient cells had been made new again and their life span doubled by cloning.[14] "It worked beautifully," said West. "We got this beautiful symmetry between telomerase and cloning."

He took a moment to gloat. Geron, he reminded me, had made a big deal about Dolly's telomeres being shorter than they should have been. "We did this careful study in cattle showing the telomeres rebuilt . . . These animals were cloned from senescent cells. The first one we called Persephone . . . If you can do it in a cell, you can do it in thousands of cells."

"So that's why Miller Quarles offered you his prize?"

"Yeah," he said, "that's why Miller gave me more money . . . We're not immortalizing cells, but rebuilding their life span."

"So this is a kind of immortality in pieces," I said.

But this was going too far. "What you need to realize," West said, forcefully, "is that even though I'm controversial, I'm [an] immensely practical guy. I'm not talking of people [being] immortal, I'm talking immortal cells. This is the scientific term. If I in my lifetime can make one solid contribution, I'm thrilled. My mother visits the lab when I'm with Sam Goldstein. She's laughing and giggling. She says, 'I have this image; you're an old gray-haired guy still trying to figure this out.' We have come close to making therapeutic cloning work. Can we give back a young immune system? That's doable."

Take, for example, the case of a twenty-one-year-old steer named Chance. They took a frozen sample of his skin and transferred a nucleus to an egg. "This animal, Second Chance, was made from a geriatric organ."

But are these clones normal, or are they deformed? I asked.

"We published a paper in *Science* about how normal they were. The media created the story that [these] animals are all abnormal and it's not true. Rick Weiss at the *Washington Post,* he called around. 'Cloned cattle are dropping like flies' is the title. Not true. Not even close to true. Pregnancies we lose a lot. Animals born, about 80 percent are absolutely normal. The problem we see is related to in vitro fertilization, not cloning." The most common problem was that cloned fetuses were too large and needed to be delivered by Caesarean section, which might be a result of the procedures involved in leaving embryos dividing in culture rather than of the cloning itself. Advanced Cell Technology had a subsidiary called Cyagra, which cloned animals for a fee, like Infigen does. Its price, however, was significantly lower, about $19,000 per cow.

West turned back to his computer screen. Up popped a series of images of bovine embryos and of tissues coaxed from bovine embryonic stem cells, including follicles, cartilage, bone, neurons and beating heart muscle. They'd even managed to make something that was like a primitive version of a kidney. "Adult stem cells don't do this," West insisted.

Yet, I said.

"Where do you see an adult stem cell forming intestine, follicles, developing eyes?" he demanded, outraged.

So, about cloning people, I asked. If he'd got this far—getting cow cells to double their life span through cloning, getting tissues and organs developing from cloned embryonic stem cells, surely cloned people could not be far behind.

"This is wrong," he said, like Michael Bishop's echo. Yet his colleague, Robert Lanza, Advanced Cell's medical director, had been quoted in a story in the same issue of *Science* that carried the cloned cattle article, saying, "there's a real possibility that cloned animals

might live as much as 50 percent longer than their normal counterparts—up to 180 to 200 years in the case of humans—an idea that is
going to raise an eyebrow or two."[5] Miller Quarles hadn't offered
West his prize because he could make cows live longer.

"I don't think we should clone people," West said again.

I asked why not.

Clones tend to become oversized fetuses, he said, and he was
concerned that carrying such a clone might kill a woman. But why
would it kill a woman, I wondered, if it didn't kill cows or sheep?

"I believe in the sanctity of life . . . I don't think we should sacrifice a mouse unless it's for a damn good reason."

But when I pressed him on whether there should be a law
against cloning human beings, he hedged. He said he thought there
might turn out to be a constitutional argument that people who are
otherwise completely infertile might have a right to reproduce by
cloning. "We say ax murderers have a right to reproduce, but those
restricted to cloning can't? You hear these arguments, but what do I
know about constitutional law."

Nothing, I thought, and yet he was trying to suggest a constitutional framework for a political fight he obviously thought was coming. Somebody was going to make a human being this way and
announce the birth. And West would be pleased because, eventually,
he would want to clone people from altered eggs in order to radically
extend human life.

Perhaps that was why he now launched into an argument in
favor of doing human embryonic stem cell research. He didn't need
to—the whole point of his current work is that it is done with eggs
that are not fertilized. He appealed to the natural order. He suggested that many embryos die in the normal course of things.
Blastocysts, the proper term for embryos more advanced than a
morula, get no guarantees of life. "About half the time these blastocysts will implant," he was saying. "Half don't. They die, or they go
down the toilet or something. The Christian church says . . . life
begins at conception. The orthodox lambasted us. [But] there is
nothing in the Old or New Testament about this. Nothing. When
the angel Gabriel came to Mary and said, 'That which is conceived
in you' . . . I know Greek. The Greek word in the sentence is no way

related to fertilized eggs—it's 'that which is gathered together in you.' This is—it's very reminiscent of Galileo. The Bible didn't say the earth is the center of the Universe . . ."

"So what's wrong with their position?"

"What's wrong is it ignores the facts of biology," he said. "A fertilized egg can form an embryo where the inner cell mass forms two clumps. One in three hundred embryos you get identical twins. It implants around day 14. Then you get the primitive streak, the first step to a human being; it's a first sketch line on the canvas." The primitive streak marks the differentiation of those cells whose descendants will become the nervous system, the cells that will carry sensations and organize thinking and emotion. West pointed to his computer screen again. It showed different kinds of twinning, blobs of cells with two streaks instead of one.

"Here, with two inner cell masses, there are two lines drawn. Same genetics. You get identical twins. It can occur differently. You can get two primitive streaks forming on the same inner cell mass. The result is the same—identical twins. In the first one you get two membranes, the same chorion, but each one has its own amniotic sac. Some twins share the same amnion; that means twinning occurred late, about two weeks into development . . . You can induce twinning by dividing the cells." Blastocysts, he explained, could implant in the uterus before "individuation."

The same appeal to nature was made by his opponents— ethicists and biologists working with the Catholic Church and other pro-life organizations. I'd met Dianne Irving, formerly a biologist employed by NIH, now an ethicist teaching in the School of Philosophy of the Catholic University of America in Washington, DC, at an Alliance For Life conference in Guelph, Ontario. I'd gone to the conference to do the journalist's equivalent of due diligence— making certain I wasn't being unfair by ignoring someone's arguments out of prejudice. I am prejudiced on the subject of abortion. I'm a feminist; I have thought since my teens that abortion is something a woman has a right to. I expected to hear in Guelph the badly constructed arguments of the perpetually marginalized, and I did, but I also heard positions that were consistent, particularly Irving's critique of the half-truths purveyed about embryology by proponents

of cloning and stem cell research. Later she sent me an open letter
by one of her collaborators, C. Ward Kischer, emeritus professor of
anatomy at University of Arizona College of Medicine. Kischer's
letter was entitled "The Corruption of the Science of Human
Embryology."[16] Both Irving and Kischer assert that there is no such
thing as a pre-embryo, or any embryological process known to
science called "individuation." Kischer argued that these words are
rhetorical devices invented to make the case for human embryo
research and to allow the killing of human embryos. In fact, Kischer
argued, there are no hard-and-fast rules about when twinning can
occur; some twins are formed after fourteen days. Many of these fail
to properly divide and become Siamese twins, or die. For Irving and
Kischer, the scientific way to look at these phenomena is to
acknowledge that embryos develop over time into adults who will
eventually die. From the moment of fertilization, all that is required
to be a human being through all stages of development, up to and
including death, is in place: the developmental program that unfolds
over a whole life span doesn't start at fourteen days after fertilization,
or at birth. They argued that human life is an indivisible continuum
from start point to end point. No matter where one is on that
continuum, one is a human being.

The Nuremberg Code forbids experimentation on human beings
without their consent, so one cannot, in conscience, they argue,
experiment on embryos or deliberately kill embryos. Irving and
Kischer also argue that the concept of individuation and the term
pre-embryo falsely divide that which is one into two—true human
beings on the one hand and, on the other, something that is fair
game for research. This renaming, they argued, was all part of a new
eugenics, the first step down a slippery slope. Irving supported her
version of this thesis by pointing out that in Maryland, a law had
been proposed holding that the mentally disabled cannot be considered persons with full rights and therefore can become research subjects without the same protections as the rest of us. All forms of
cloning, in Irving's view, are wrong; all the legislative assemblies
around the world trying to stop nuclear transfer cloning had utterly
failed to understand biology and therefore to capture the myriad and
iniquitous other ways to make living beings.

What Kischer and Irving don't say is that the manufacture of all these living, breathing clones and chimeras, these bizarre anomalies outside the normal development pathway, make a lie of the Catholic dogma about life beginning only at fertilization, along with old scientific certainties about the inevitable vector of development. Try as they might, neither the Catholic Church nor development theorists had been able to shove life, that four-billion-year-old kluge, into a box and keep it there.

West now asserted that human beings can't exist until embryos become individuals, which he sometimes defined as when they implant, sometimes as when they develop a primitive streak. If embryos are disaggregated before the appearance of the streak or implantation, where, he asked, is the harm? Oddly, he missed the opportunity to argue that since nuclear transfer does not involve fertilization, no embryo has been created, so no embryo is harmed. He seemed not to appreciate the devastated ruin that nuclear transfer makes of his opponents' theories of the start point of life. His cloned cows are alive, yet they were not conceived. What could the Catholic Church say about a person made in the same way—that such an individual's life had no beginning and therefore no human being existed?

Unfortunately, it followed from the argument West did make that one could do ethical abortions only up to about fourteen days past the creation of an embryo. Prior to the appearance of the streak, one would just be flushing out a colony of cells; after, one would be killing an individual.

Somehow, I didn't think that view would sell well politically: every woman who might need an abortion would rise up and smite it down. West's argument highlighted the wisdom of *Roe v. Wade*. In 1973 the U.S. Supreme Court's majority decided that definitions of the beginning of life are contradictory and irrelevant to the question of abortion. The court pointed out that state laws against abortion were less than a century old. It pointed out that the Catholic Church had changed its dogma on when life begins: originally, it accepted the Augustinian notion that ensoulment occurs late in development and that this marks the start point of an individual's life; in the nineteenth century, the church opted instead for fertilization as the start point.

The court wondered about the practical question of where the state's interest in the health of the woman and the fetus intersects with the woman's privacy and liberty. It made a distinction between potential human beings and persons. The court said that there is no person until birth, so no person with rights can be said to exist until the fetus is born.[17] *Roe v. Wade* imposed no rules on nature, and thus left room to maneuver in any situation nature might throw up. West should have stuck with *Roe v. Wade*. But he didn't.

He pointed at the image of a tiny human embryo on the screen. "They say killing this is murder," he said. But what if he took that embryo apart, tweaked some of its cells into becoming a kidney and took that kidney out of its sustaining serum bath? "Do we give the kidney a funeral?" he asked. "No. It's not a person. The individual begins with the pregnancy . . . This is offensive to some people, but instructive. If these primitive streaks diverge, you get conjoined twins. Common sense can tell you, this is telling you when human life began. It draws the line for you—it's the primitive streak."

"All this opens thorny questions," West said, "the genetic engineering of persons." Some American ethicists were forging arguments in favor of engineered human enhancements. West could see using genetic engineering to get rid of known disease genes, such as the one that conveys Tay-Sachs, but this idea of enhancement bothered him. This idea of hurrying the evolution of humans to higher levels made him worry about hubris.

That's rich, I thought. The man who would defeat death worries about hubris. Wouldn't extending the life span of a human immune system be an enhancement? Wouldn't dropping a whole nucleus into a new setting be a more radical change than adding or deleting a gene?

"Hubris I worry about all the time," he said. "People are arrogant. The enhancement part I worry about. I can't think of a single enhancement I know to be safe . . . I'm not against using technology to make the world a better place—my concern is hubris . . . What bothers me here is so many assume that scientists are hubristic."

"They are," I said.

"There's a huge difference between medical researchers and physicians," he said. "Many cancer physicians have control over life

190 • ELAINE DEWAR

and death and do tend to be more hubristic. But academic researchers, in the majority, are very thoughtful people."

So where did he fit along this hubris continuum? I wondered. He is neither cancer physician nor academic: he is a science entrepreneur. The money he helped raise gave academics the wherewithal to create the first human embryonic and germ stem cell lines. Now he was trying to raise more to turn human eggs into embryos, and embryos into lines, and mortality into a more malleable character in the human play.

How many kids do you have? I asked him.

"I have triplets and one other."

Well, I said. I've been asked to ask you this. One of your competitors suggested that your triplets could be clones. Are they?

At first, West sat in silence. Then the blank expression on his face shifted; he looked like a man who thought that he'd heard everything, that there were no surprises left in the world. Then he must have realized what this question implied—that he had secretly achieved a goal far beyond the reach of his competitors.

"They're not clones," he said. But with a rictus stretching his cheeks that no one would confuse with a smile, he added, "You made my day."

He repeated that several times.

CHAPTER NINE

VESTED ETHICS

I t was dark, dinnertime. I walked my dog in the private park at the end of my street, actually a private gated neighborhood set around a pond dredged between the banks of a disappeared creek. My dog snuffled his way from one yellow-stained snowbank to another while I shamelessly watched the soap operas playing out in the lit windows. I passed the great house in the park's center. That took a while—it's so big, it even has its own paneled ballroom. It was once owned by an old woman living off the dregs of a family fortune. One day a thief broke in. He must have assumed a house so huge was like a sign on a bank saying Free Money. When he didn't find any, he tied up the old woman and left her in the basement. The next owner was a different breed: she had smart new money. She spent a fortune—new kitchen, new pool, big parties. One summer night, the heart-lifting paean of a hired tenor floated from her veranda, across her acres of lawn, all the way to my house. But then the house went up for sale, at a discount. We heard that the woman took off to the Caribbean, one step ahead of the Mounties. The only signs of the money she might have had were the mounds of poop her trophy dogs deposited on the ballroom floor.

Money, I decided, as my dog yanked to go faster, is like an intercalating chemical. It moves in and stretches the bonds holding desire to reality; it shifts frames, shifts meaning. It also obscures, hides and deceives. For months, as I listened to biologists tell their stories, I had used money and its origins as a kind of correlative for scientists' reliability and motives. When I began, I thought commercial scientists were biased and their results suspect because they worked for or owned a company. I thought academics were more objective and their results more believable because they competed for public funds. The two Michaels—Bishop and West—made me think again. I began to understand that these distinctions among biologists have become as meaningless as the species boundary, as obfuscatory as the word *embryo* when applied to a reconstructed egg.

Just like their transfers of cells, nuclei and genes from one setting to another, biologists constantly move between public and private domains, or inhabit both at the same time. West's company, for example, had been awarded $1.9 million in NIH grants for various research projects.[1] Everybody had something to protect—shares, companies, grants, patents, important positions—vested interests. How could money's origins, then, say anything about biologists' motives? What did it matter whether greed or altruism moved Michael West or David Baillie or James D. Watson to make inquiries?

Yet their complex interests intrigued me. I couldn't shake the idea that the origin of money colors motive, that motive is a reading frame for moral stories. What did morals have to do with biology? Nothing and everything. Science is supposed to be about free inquiry, but moral motives had shaped the political discourse about revelationary biology for thirty years. And biologists had constructed this reading frame themselves.

Since the Asilomar conference in 1975, biologists had been hard at work staving off government regulation. Their desire to interfere with the processes of life, mixed with their determination to be free of a cage of law, made it essential that researchers be seen as innocent, trustworthy and motivated by the common good. The common good came to be defined as the defeat of the worst human diseases and conditions, such as diabetes, quadriplegia, Alzheimer's, cancer—Michael West helped add aging and mortality to the list. But biological methods and findings also shook beliefs about humankind's relationship to God and about our place in nature. Most democratic governments were reluctant to put public money into the most morally and therefore politically troubling experiments and left them to private enterprise and private conscience, knowing full well that private enterprise is driven not by conscience but by competition for profit and by avoidance of loss. Wise executives, seeking to limit liability, soon called on ethics specialists to pronounce on rules, to burnish appearances. Wise executives hired ethicists to sit on corporate advisory boards to say what should and should not be done before stepping into morally muddy waters. What better way to demonstrate pure motives than to say that an ethical advisory committee says everything's okay? The marketplace of ideas soon offered a full range of ethical opinion for biologists to choose from; one could pick the ethics one paid for.

As if to forestall criticisms before anyone might make them, in 2002 *Scientific American* printed Dartmouth ethicist Ron Green's argument in favor of therapeutic cloning right alongside West and his colleagues' article describing their so-called first human clone. In their introduction to Green's article, the magazine's editors themselves declared Advanced Cell's worthy motive: Advanced Cell "aims to generate replacement tissues to treat a range of diseases," they wrote. They also pointed out that Advanced Cell had assembled "a board of outside ethicists to weigh the moral implications of therapeutic cloning research." This board had considered major ethical questions before Advanced Cell went forward with "cloning the first human embryo."

Green, unlike *Scientific American*'s editors, was very careful with his use of language; he did not refer to Advanced Cell's organisms as "embryos." But he set a favorable moral reading frame around Advanced Cell's work. He acknowledged that there was debate as to whether the entity made by the company is an embryo, and whether the moral status of a cloned organism is the same as that of an embryo: "Some would argue that the organism produced in human therapeutic cloning experiments is the equivalent of any ordinary human embryo and merits the same degree of respect and protection," he wrote. "Most members of our advisory board did not agree."[2] The new type of biological entity that Advanced Cell produced is no embryo because the egg was never fertilized, he explained. The ethics board preferred the phrase "activated egg" to describe this new creation.

Green's argument seemed to imply that a true human embryo deserves protection and respect, while these new entities deserve neither. He did acknowledge that some believe that these "organisms" should not be created in order to be destroyed, but then suggested that some of those who see an activated egg as the equivalent of an embryo might also find it acceptable to extract stem cells from embryos left over from IVF, since they will die anyway. He said that his ethics board had spent the bulk of its time working out protocols for recruiting women and paying for their eggs, and deciding whether children should be allowed to donate their skin cells. In other words, he did not explicitly answer here the real questions: Can this new kind of human organism, if put in a uterus and brought to term, become a human being? And if it can, is there a moral difference between this organism and an embryo? (He told me later he is not troubled by any of these distinctions as he believes embryos have minimal moral weight and therefore doing research on them is permissable.)

In print Green maintained the usual separation between what he called "therapeutic cloning" and "reproductive cloning." He defined therapeutic cloning as the cloning of a nucleus in a reconstructed egg, excluding from this definition the cloning of stem cells from fertilized embryos, though that is also therapeutic cloning. He suggested that even if some rogue scientist tried to make people from reconstructed eggs, this should not preclude activating eggs to make tissues to save lives, nor should it prevent research that might make such human reproduction safer. In other words, he did not close off either the use of true embryos to derive stem cells and tissues or the creation of people out of these new kinds of multipotent organisms. (though he told me later he is opposed to reproductive cloning "at this time"). "We believe we have managed to give Advanced Cell Technology a firm ethical base for its therapeutic cloning research program. After researchers derive stem cells from cloned human activated eggs, ethicists will need to determine at what point it will be safe to try to transplant such cells back into volunteer donors. The task ahead for ethics boards like ours is demanding. The reward is assisting at the cutting edge of medical knowledge."[3]

Ron Green, the chairman of Advanced Cell Technology's ethics advisory board, is also director of Dartmouth College's Institute for the Study of Applied and Professional Ethics and a professor of religion. He wrote a book on human embryo research and was on the panel of the NIH that in 1994 recommended that the U.S. government fund it. He had also been for one year a part-time director of the office of Genome Ethics at NIH's National Human Genome Research Institute. He told me he had invited West to speak at a 1999 conference on stem cell research, where West told him about Advanced Cell Technology's goals. Green volunteered to help, but didn't hear from West again until August 2000. Then he was asked to pull together an ethics advisory board for the experiments West planned to do with human eggs. Green convened a meeting of those interested in providing ethical advice to Advanced Cell. They included Ann Kiessling, an expert on human reproductive biology at Harvard, as well as ethicist Glenn McGee of the University of Pennsylvania (who soon resigned). "The first thing we did was conclude we should not be significantly paid," Green told me. "I said nothing but expenses. Others thought that was derisory. We decided on the NIH per diem for each of four meetings annually. We felt we had to hold our heads high as an independent board."

Independence is the gold standard required of those serving on institutional review boards in the United States, those organizations that must rule on the propriety of the protocols of human subject experiments if experimenters are to receive public funds. Independence usually means that the members of these boards have no professional, personal or financial interest—no vested interest—in the results of the experiments and will therefore put the protection of human subjects first.

Green explained that taking the NIH per diem of $200 was a sacrifice. He normally bills out his services as an ethicist to corporations at $4,500 per day. He said that Arthur Caplan, a well-known bioethicist at the University of Pennsylvania, is paid much more. (Caplan later said he doubted if he had ever received a daily consulting fee as high as Green's.) I calculated that if Green worked as an ethical consultant at his usual rate for two hundred days a year, he'd earn $900,000 for his learned opinions, more than three times the pay of the president of the United States. His day rate was right up there with disaster PR consultants, but he told me he only averages up to ten days of consulting a year.

Green was proud of his work for Advanced Cell. He thought his ethics advisory board had helped West's science team speed up the pace of the experiments. The Israelis, who had been identified as the chief competition, had hinted to Advanced Cell that they were held back in their work by the small number of excess human eggs they could get from infertility clinics. Advanced Cell had considered asking couples involved in IVF to donate extra eggs, but the ethics advisory board decided against that. Asking a woman who is a client of an IVF clinic to donate eggs in exchange for free treatment, which normally costs between $8,000 and $10,000 per cycle, would put unjust pressure on poor women to join the research program and, if her IVF treatment failed, she might worry that the scientists got her best eggs. Advanced Cell Technology, it ruled, should buy eggs from healthy women. Green thought this was perfectly appropriate. In his capacity as a member of an institutional review board at Dartmouth that reviews experiments done there on human beings, Green had learned that young men are paid $1,000 for certain kinds of experiments involving the insertion of catheters into their hearts—a risky business—and that, in general, human subjects are often paid to participate in experiments. "It strikes me it is paternalistic and sexist to say women can't be paid," he said. Payment was a fair exchange for risks taken. Women giving up eggs face real risks. They have to take medication to produce more than the normal number of eggs; these drugs can lead to serious medical problems. Eggs are removed with a needle inserted through the vagina, and organs can be punctured. But aside from all that, he said, women are routinely paid to donate eggs, about $3,000 to $5,000 per cycle. He pointed out that Advanced Cell could have offered $20,000 per cycle, but this would have made Advanced Cell Technology "the thousand-pound gorilla," utterly disrupting the normal marketplace for women's eggs in the Boston area. So the company settled on paying each woman about $4,000 per cycle for eggs.

To find egg suppliers, Advanced Cell advertised for eggs for research. The ethics advisers created strict rules about who could and who could not be a supplier. Women had to be between twenty-one and thirty-four years of age with at least one child. "We screened participants. Medical and psychological tests, consent materials, an independent monitor to make sure it's a free consent. Many of these women want to contribute to scientific research . . . Sixty women have responded."

How many were turned away? I asked. "A significant number," he said.

As the experiment rolled forward, Green said, an ethical glitch occurred: the scientific team couldn't get the eggs to divide after nuclear transfer. One of the successful methods used to clone mice involved dropping a tiny cumulous cell, which clings to the outside of an egg cell, into the denucleated egg instead of just a nucleus. The scientific team wanted to try that method. If they succeeded, it would mean that the woman who supplied the egg would actually clone herself.

"I got a telephone call on a Friday morning from Michael West to ask if the board would approve the use of a cumulous cell from the woman as the nuclear source . . . I was distressed," he said. "She had not consented to that. Now she was asked to clone herself."

I was surprised at his distress. What possible difference could it make? If it was acceptable to the woman to have one of her eggs denucleated, somebody else's nucleus dropped in, the resulting construct zapped with an electric current and then, if they all got lucky, turned into an immortal cell line, why would she care if one of her own cumulous cells was dropped into her own egg instead? In fact, this issue of fully informed consent, which so concerned him, seemed to me trivial compared to the main issue: is it ethical to ask any woman to take a physical risk to sell her eggs and allow them to be transformed into embryos that would become, in effect, permanent research subjects? Green said I missed the point. The woman ran the risk of her DNA proliferating indefinitely—a risk to her genetic privacy. Also there was the chance a rogue scientist could divert "one of those cells to repro- ductive purposes. They face the risk they might have a cloned child in the world." Green said, "I see nothing morally wrong with the use of or the destruction of the early embryo. The issue of exposing women to risk on ethical and experimental grounds, this is not an issue that keeps me awake at night . . . In this country people do phase one drug toxicity tests all the time. I don't see why women should be exempted . . . Women who contributed eggs said they wished to make a contribution to science. I don't see why they should be prevented from doing that."

In many ways, Advanced Cell's ethical advisory board's decisions were similar to the stem cell research guidelines prepared by an expert panel in a 1999 report to the American Association for the Advancement of Science. Green had been a member of that working group too. However, that report said that neither men nor women, as individuals or as couples, should be paid to produce embryos.[4]

I asked how much he had made from this. He didn't like my sug- gestion that he did any of this work for money. "The grand sum of $800 a year, not including the hour I spent talking to you. I don't need the money," he said sharply.

Ann Kiessling had also helped form Advanced Cell Technology's ethics advisory board. Kiessling is an associate professor in the Department of Surgery at Harvard Medical School, as well as being director of the Reproductive Biology Laboratory at Beth Israel Deaconess Medical Center in Boston. She also directs Duncan Holly Biomedical, a clinical laboratory in Somerville, near Boston. She specializes in helping people with HIV get pregnant. Green told me she played a vital role, yet her name did not appear on Advanced Cell's website list of ethics advisory board members.

Kiessling explained that she had become involved with Advanced Cell Technology because she knew Jose Cibelli. For many years, she has run a seminar called the Egg Group. The purpose of this group is to allow those interested in embryology in the U.S. Northeast to meet four times a year, over pizza and beer, and share what they are learning. The group is interested in many species' eggs, but it is particularly important for those interested in human eggs. "The idea is that so few people are working on human eggs . . ."

Research on human eggs and embryos, she explained, has not been funded by the U.S. government since Congress refused to allow public money to be spent in this morally equivocal area. "There are 50,000 or 60,000 women a year going through in vitro fertilization, and no federally funded science," she said angrily.[5] "We thought Clinton would allow it. But the College of Cardinals got to Clinton's officials. To do a project on fertilized eggs there is a body of people to review projects in Britain, but nothing in the United States. Infertility treatment in the U.S. is cowboy medicine, people getting wealthy on the backs of couples wanting children."

Jose Cibelli had attended Kiessling's group for years. In the fall of 1999, after Michael West became CEO of Advanced Cell Technology, West and Cibelli called on Kiessling. They told her they needed help finding human eggs for nuclear transfer experiments. She explained to them that they should find an institutional review board that could oversee the program first because research on human beings (as distinct from treatment) must be approved by such a board. They talked about getting eggs from infertility clinics in the Boston area, but "they're all private so they had no review boards," said Kiessling. Then she thought maybe Harvard's review board would help, but no. "They didn't want to give an IR [institutional review] to a private company," she said. She thought she might be able to take the project to Deaconess Hospital, where she has an appointment. "The lawyers created a private board for the IVF lab. I thought it was a great place for a review of the Advanced Cell project. I thought I could get people. I tried awhile. I hadn't the time and didn't get enthusiasm."

In the summer of 2000, West and Cibelli came back to see her. Since she had not been able to find them an independent review board, West thought maybe they could just find some individuals, set up a conference call and get one going. "Jose left; Mike and I were chewing the fat," said Kiessling. West explained to her that Advanced Cell was under pressure to get these experiments done quickly. The company was running out of money; there were competitors out there, and the company needed to do the experiments to be able to get patents, otherwise, no investors. "The real concern was they were running out of funds," said Kiessling. West also mentioned to her that Ron Green could be the leader of a review board. Kiessling asked a former colleague with an appointment at Boston University if she'd join an ethics board for Advanced Cell. This colleague said it depended on who else was on the board. Kiessling mentioned Green's name, and the colleague, Judith Bernstein, checked him out. Bernstein said that Green was sound and she agreed to join.

Kiessling remembered well her first meeting with Green. There'd been an accident on the turnpike between Boston and Worcester—a tipped-over hearse had stopped a whole funeral procession. When she finally got to Advanced Cell's offices, she'd asked Green why he wanted to be on this ethics board. He told her Dartmouth had hired him to take a stand on this issue. "He and I and Judith [Bernstein] hammered out the details to recruit women. I'm very proud of how this works."

They took out ads in the *Boston Globe* seeking women of the right age with at least one child. They didn't get much of a response, so they advertised in free community papers. "The idea was to make this public, and no recruiting of friends," said Kiessling.

"Well, what friends could have been recruited?" I asked. Then I slapped my forehead—of course, Michael West's wife could have been recruited.

"Jose's wife or Mike's were not recruited," she said. "Jose's wife really wanted to. We thought about it. We reviewed her, but we decided not to. We couldn't stay objective enough . . . The advisory board thought, how will we convince the planet that all the eggs went to stem cell research and not to clone somebody? We thought we'd have to account for every egg—to the press and Congress."

The result was an elaborate system to make sure no one was alone with eggs. There was a twenty-four-hour video camera installed in the lab. "You need two keys to get into the lab. The incubator is locked. This is all to speak to the hysteria in the United States about cloning a person." There was a Chinese wall put up between the team extracting the eggs and the investigators reconstructing them. Advanced Cell had only numbers. The medical team extracting the eggs had only names.

But there was no such wall between the ethics board devising these rules and the people doing the experiment, as is the case with a properly constituted institutional review board. Michael West was on this board, though he had no vote. Kiessling worked with the board as well as on the experiment (though she was not permitted to vote on any part of the experiment that involved her).

Where was the egg removal done? I asked her. I had a hunch it had been done at Kiessling's private clinical laboratory.

"It's done in a private lab in Somerville," she said.

"Duncan Holly Biomedical?"

"Yes, Duncan Holly."

So, I asked, do you actually do this work?

"I just oversee it," she replied. She explained that she was the only one involved who was an expert on human reproduction; they needed her expertise. "When Mike first contacted me to be on the ethics board, they wanted to pay me. I said you can't pay for that. It's a conflict of interest. He said one was usually offered $1,000 per day."

But it wasn't the money that created the conflict here. It was the fact of being on both sides of the issue, of having a professional interest in the experiments while also devising the ethical conditions under which the experiments would be done.

I told her what Green's normal daily rate is to provide ethical advice.

"Wow," she said. "At some level that's distressing to hear." On the other hand, she recognized that companies must pay for consulting services, though she also thought that the idea of lucrative payments for ethical consulting seemed odd. "For ethics, it's very curious."

There'd been a debate, Green had said, about whether Kiessling's name should appear on the ethics advisory board list. Kiessling had been so helpful that they wanted to include it, but they didn't want anyone on the ethics board to have a professional interest in doing the experiments. Kiessling's name did appear on the scientific publication in *e-biomed* describing the experiments. But in spite of the fact that she'd helped staff the ethics board and develop its policies, Ann Kiessling's name did not appear on the list.

It was Superbowl Sunday afternoon. My member of Parliament, Carolyn Bennett, had sent an invitation to her constituents to a public meeting on the Standing Committee on Health's revisions to Bill C-13, the proposed law concerning assisted human reproduction.

The committee's revisions to the bill had focused on curtailing vested interests. The committee seemed to believe that the assisted human reproduction industry traded in human flesh as if it was property.[6] It proposed that payment for the expense of supplying sperm, eggs and surrogate services and payment of some professional fees in support of assisted reproduction become a crime. It also suggested rules to allow the children produced by such methods to find their genetic parents if they wanted to, all part of the committee's efforts to make the conception of children an act devoid of the taint of commerce.

The committee also proposed forbidding the construction of human clones, even nuclear-transferred eggs, for any purpose. However, it would allow research on leftover embryos and their stem cells—provided they were not made for the purpose of research. Researchers would be restricted to the frozen leftovers from IVF clinics or stem cell lines imported from abroad, and they would have to demonstrate that the use of these embryos was the only way to answer an important scientific question. This research would be done only under license from a regulatory authority, regardless of whether it was done by a private or a public institution. If indicted for performing a prohibited activity, a researcher would face a fine of up to $500,000, up to ten years in jail, or both.[7]

As soon as these proposed revisions became public, those with interests began to howl. Doctors and lawyers assisting the infertile for fees were not amused. Stories about the murky business of procuring eggs and surrogates appeared in the *Globe and Mail*.[8]

My MP's meeting was held in a YWCA in a midtown neighborhood on a sunny winter afternoon. I took a chair by the big windows and basked in the weak sunshine, watching giant icicles drool from every drainspout. Panelists included Bennett, committee member Judy Sgro, a doctor with his own private infertility practice, a York University ethicist who declared herself infertile in tones of utter anguish, and Alan Bernstein, the president of the Canadian Institutes of Health Research. The audience included several young men and women spooning cubed potatoes and wrinkled peas into the mouths of their small children. These were infertile couples who had had their children through IVF procedures offered by the doctor on the panel, and they were here to tear a strip off any member of the committee foolish enough to stand before them.

Bennett, who was a family physician before she was elected, made the introductions. I found my eye drawn to Alan Bernstein, who sat at the dead center of the head table. He wore baggy black pants and a black turtleneck, and his face was all drooping lines, crossed by a thick, black Zapata mustache. He looked more like a down-on-his-luck film producer than the head of a government-funded agency. All I knew about him at that moment was that he was the first appointed president of the CIHR and that he used to direct the Lunenfeld Research Institute at Mount Sinai, where he'd worked with Janet Rossant. Bennett served him up as one of the most brilliant scientists of our day, world renowned in stem cells, a person of such limpid intelligence that he could even make her understand the complex findings of this emerging science, which of course, she implied, she could not make head or tail of by herself.

Bernstein opened with the money: he said he has $600 million a year in federal funds to give to 30,000 Canadian researchers. He laid out the origins of the CIHR and stressed that part of its mandate "is to advise government on coming issues such as stem cells." He then explained that he had convened a special working group a year earlier on stem cell research, chaired by Janet Rossant and including Anne McLaren of the United Kingdom, "who got the Japan Prize two weeks ago for $400,000 U.S.," and, naturally, ethicists. The group had worked out how this research should be done, under what ethical rules. Its report on how the CIHR would fund all forms of stem cell research had just been finalized and would be made public in another month, at which point, he said, research money would be made available. He gave a short lecture on stem cells. He avoided the use of the word "embryo," but he hit the idea of the worthy motive, the glittering promise of curing the big diseases, over and over.

He blamed the committee's refusal to permit nuclear transfer cloning on Advanced Cell Technology and its announcement of the first cloned human embryo. "The committee was affected by that and came down harder than they would have if not for Advanced Cell Technology's hype," he said. The hype was all about raising money for the company, whereas his guidelines committee had examined the science and ethics of this research and had prepared strict rules. He was proud to say that his rules were the same as the government's proposals, and would place Canada midway between Britain and the United States. He concluded, "The guidelines will come into place next month. All researchers who get federal funds will be bound, and I hope the charities [that fund research] will come along as well . . . I think that's all."

The other members of the panel seemed concerned mainly with those sections of the bill that would outlaw commercial transactions in the creation of children. Infertility lawyers in the audience expressed their outrage. Infertile men and women held up their children and spluttered with fury that MPs would try to stop them from paying for eggs, sperm, surrogates, doctors and lawyers—whatever it took to get kids with some of their own genes.

If they were so laid low by infertility, I wondered, why did they not ask why human reproductive cloning is a crime? And why didn't Judy Sgro ask Bernstein how his committee's guidelines could come into effect the following month when embryonic stem cell research would be a crime under Bill C-13 unless done under license issued by a regulatory authority? No such authority could exist until the government passed a bill, which it could not do by the following month, or even the following year.

After my visit with Michael West, I got a copy of the National Academy of Sciences' report *Scientific and Medical Aspects of Human Reproductive Cloning*, written by the panel of experts who had heard evidence from witnesses the previous summer.[9] The panel included leading developmental biologists such as Anne McLaren and Brigid Hogan, as well as Canadian medical geneticist Judith Hall. It also included one medical ethicist. The panel had inquired into whether or not reproductive cloning is safe in animals, and whether it is therefore a justified experimental risk to try to clone a human being.

The panel reported that reproductive cloning in animals often goes awry, though it was unable to say why, and so it found that there is an unquantifiable risk to women who might carry clones. The panel also found some of the techniques involved in reproductive cloning and therapeutic cloning to be the same. It described how assisted reproduction is unregulated in the United States and Canada, and how there is no oversight required by law of private scientific investigations conducted on women and their eggs if they are defined as part of a process of "innovative therapy" instead of as research. The panel feared that researchers would investigate with human cloning but call it therapy, and argued that anyone trying to reproductively clone a human being without reference to prior studies in animals, which showed it could be dangerous, would be in contravention of the Nuremberg Code.[10]

However, the panel pointed out that if experiments were done as part of a proper systematic program of research, they would conform to the Nuremberg Code and would also fall under the umbrella of U.S. federal human subject research rules. Private companies doing this work would then be forced to have their experiments evaluated by review boards independent of the investigator. But the panel saw that determined people who wished to evade oversight could do it, so a voluntary ban on human reproductive cloning would not work. It therefore called for a legally enforceable ban on human reproductive cloning, to be reviewed in five years after a broad public debate involving ethicists.

The panel decided that there is no reason to ban embryonic stem cell research. Declaring that all stem cell research falls outside rules for experiments on humans beings, the panel called for it to be funded by the government so that it would come under U.S. federal research rules. The panel also found a good phrase to rename therapeutic cloning: they called it "nuclear transplantation to produce stem cells."[11] One could say this phrase in public without hinting of the life and death of embryos or other troubling embryo-like organisms.

The report, in sum, read like a positioning document developed by a lobby. Some of the panel's members were researchers who would want to use these technologies, people with a vested interest in the result. None had a greater interest than Irving Weissman, the panel chairman. A professor at Stanford University School of Medicine, Weissman is also a founder of two companies interested in embryonic stem cell research, including one called StemCells Inc.

Less than a year after this report was published, Weissman and Stanford University announced the creation of a private research institute to study human genetic flaws by turning genes on and off in human embryonic stem cells. The institute is funded by a $12 million anonymous donation. Weissman, the institute's first chairman, explained that if human embryonic stem cells could not be obtained from government-approved sources—the stem cell lines made before August 9, 2001—he was quite prepared to harvest eggs from women and construct them into embryos so he could derive cell lines for his research purposes.[12]

CHAPTER TEN

DR. WATSON, I PRESUME?

I trudged up New York's Madison Avenue. I had just spent several hours wandering through the halls of the Metropolitan Museum of Art, whose echoing main floor was devoted to ancient imagery of death as expressed in ceramic and stone. There were huge urns painted with black and red scenes of Greek youth stabbing each other with sword and spear. There were wonderful Egyptian sphinxes, and Greek funerary chimeras plucked from the tops of ancient stelae. These chimeras had women's or men's heads and upper bodies fused to lion torsos. They had wings, often spread forward as if to cover the dead below. They were beautiful and naive and puzzling.

Why did chimeric unions of species symbolize death or guard the dead? Was it because death transforms, death recycles? Or was it an early insight that life arose from more than one source and that everyone would become something else in the fullness of time? Why did I enjoy peering at these images so much more than reading papers produced by biologists? What would happen to the drive to make art, stories, songs and make them heard, if one actually became immortal and lost the need to be known before it's too late?

I arrived at this oh-so-deep question at a street corner. The Armani store was on the other side. The light was red. A woman in a long, well-cut melton coat stood waiting for the light to turn green. She had blond hair as shimmery as stainless steel, cut with such perfection that it spread like wings in the wind but fell just so. As the light changed, I noticed something familiar about her and the way she moved—duck-footed, shoulders swaying—into the street. She turned her head to the side and her profile gave the game away.

She was Candice Bergen, best known for her sitcom character Murphy Brown.

The doorman at Armani saw her coming and swept her in as if she was Milady come to trade secrets with the cardinal. Fame has such muscles as could move mountains, I thought, as the door swung shut in my face. And fame's power was something James D. Watson certainly knew all about.

I was in New York to get a good look at Watson—as he is now. Fifty years ago he was a geek, or so he describes himself in *The Double Helix*. The pictures of him back then displayed an unruly British schoolboy haircut and bad teeth, the polar opposite of his cat-smooth intellectual partner, Francis Crick.

Watson's latest book had just been published. It was called *Genes, Girls and Gamow*, and the title pretty much summed it up. It was a way-too-candid account of his life as a science bachelor, after he and Crick had made their model of the double helix but before his marriage. It was the kind of autobiography that would make some of its named living characters (such as his first serious girlfriend) desperate for revenge and the dead ones (such as Linus Pauling) try to heave themselves up out of their graves. It could have been fun and insightful, but Watson had spent far too many words describing travel itineraries and his numerous pretensions, particularly his poor-boy determination to become an honorary British gentleman of science. Barbara Ehrenreich had just given it an awful review in the *New York Times Book Review*.[1] She had ridiculed the simplistic reductionism of the early days of molecular biology. She took advantage of the ample evidence Watson's book provided to note the rampaging sexism of the period. She also opined that the Human Genome Project was much overhyped and that Watson and Crick's models and theories would soon be supplanted by those of biologists capable of handling both self-reflection and complexity, work that might possibly be done by "a 'girl.'"

I had tried to get an interview with Watson about his book and, once again, had failed. His publisher, Knopf, is related to my own publisher since both are owned by Bertelsmann. I had got Watson's promotion schedule, then phoned Knopf and asked for an appointment. I was given one for the day I requested and told to call back to confirm. So I did, but then, with some asperity, the publicity person demanded to know which media outlet I represented. I said I was writing a book for a related company. No appointment.

Frustration has a way of pushing one on, as Watson would be the first to attest. I had bought a ticket to a public talk he was to give at the Young Men's and Young Women's Hebrew Association on 92nd Street, part of the YMHA's public education series by great men of science and a chance for him to autograph and sell books. I had also assembled a file on Watson's Cold Spring Harbor Laboratory. I had begun to think of it not just as a renowned center of learning, but also as an important organization balking at presenting its president to a journalist's legitimate inquiry. He had just appeared on the Question and Answer page of the *New York Times Magazine*, but he wouldn't talk to me. Why not? I had questions; for example, where does Cold Spring Harbor Laboratory's money come from?

Michael Bishop of Infigen had started me down this path. He had mentioned being invited to a special meeting at the Banbury Center, an adjunct of the Cold Spring Harbor Laboratory, in March 2000, to discuss cloning. Only a handful of people were there, he'd said, and they had come from as far away as Japan. Only invited guests could attend such Banbury events; everything said was off the record, unless permission was later sought and given, and no press conferences were allowed. Yet the Banbury Center and the cloning conference had been used as the backdrop to the announcement of the first cloned pigs.

On the Cold Spring Harbor Laboratory website, I read that the Banbury Center puts on thirty-six such hush-hush meetings every year. These are not like learned society gatherings, where papers are read in public, nor like Cold Spring Harbor Laboratory's famous courses or conferences, where hundreds come and pay for the privilege. These were organized by three top scientists invited by the center to do so, or by top persons in public policy, and all participants' expenses were paid by the center. The list of sponsors included the usual suspects—important corporate names in medicine and pharmaceuticals. Various costs were covered by "foundations, federal grants and by company support for individual meetings."

I wrote an e-mail to the center's director, Jan Witkowski, and asked to attend one of these Banbury conferences as a fly on the wall—present but silent. I was very curious about private science meetings where unpublished results are unveiled and arguments that might be used to influence public opinion can be polished. I explained that I thought science is supposed to be a public process. Witkowski's e-mail responses boiled down to "not a chance." One silent journalist witnessing thirty-six to forty scientists at work at the Banbury Center was one journalist too many. But Witkowski did say I could come and see him and we could talk about my misapprehensions about "the role of meetings and other forms of communication in science. I think," he wrote, "that you are confusing science as an enterprise with the ways in which science is done."[2]

"Enterprise" is a word that more often refers to business than to scientific inquiry. Did he consider biological science a business? I wondered. Were Cold Spring Harbor Laboratory and the Banbury Center set up as businesses? I soon learned that, in fact, the laboratory is a 501(c)3 U.S. tax-exempt organization, a registered charity that must fit a legal definition of charitable purpose. It is required to make certain IRS filings available when a member of the public asks to see them. I asked one of Cold Spring Harbor's PR people where I could read financial statements and its 990s, as the IRS filings are called. They're on the website, I was told. But the financial reports were bare-bones, and the 990 returns weren't on the website. When I pointed this out, I was told I could read the 990 returns when I came down to visit.

So I dug around and found some of the laboratory's 990s through an independent website that posts the filings of a number of charities, and I later received more from the IRS. These showed that three related charities are intertwined at Cold Spring Harbor—the Cold Spring Harbor Laboratory, the Cold Spring Harbor Laboratory Association, and Robertson Research Fund Inc. There was no filing by the Banbury Center, and a call to the IRS confirmed that it does not exist so far as the IRS is concerned. The money to support the Banbury Center seems to come from both Cold Spring Harbor Laboratory and Robertson Research Fund Inc., which have interlocking boards of trustees and common purposes. Robertson Research Fund Inc. received its tax-free status in 1973. In its filing for the year 2000, it showed assets of $110,740,919 and income of

$13,177,869. Its purposes were listed as supporting the biological research of the laboratory ($3,300,000 in 2000) and support for the maintenance and upkeep of the Banbury Conference Center ($451,195 in 2000).

The Cold Spring Harbor Laboratory was granted its tax-exempt status in 1964. It spent $68,815,279 in 2000, and its net assets and fund balances totaled $190 million. It can confer undergraduate degrees, and it had just been granted the right by New York State to confer graduate degrees (its graduate school is named after Watson). The laboratory gave the Banbury Center $1,165,989 in 2000. Witkowski, the center's director, is listed as an employee of the laboratory. The returns also showed that the laboratory is primarily dependent on government grants for its existence. Of the $61,548,750 the laboratory received in 2000 as gifts, more than half came as grants from government. The purposes of the laboratory, as described in the 990s, fit within the charitable definitions of education and the betterment of the whole community. The purpose of the Banbury Center, as described in the laboratory's 1997 return, was given as organizing meetings, but the return did not say that the meetings are private and off the record.

The Cold Spring Harbor Laboratory Association had won its tax-free status in 1936, back when the Eugenics Record Office still existed. The association had a little over $300,000 dollars in assets, and revenues of under $28,000. It spent less than $20,000 a year organizing lectures to educate the public.

Only Bruce Stillman, chief executive officer of Cold Spring Harbor Laboratory, sat on all three boards. Watson, the president of the laboratory, sat on its board and also on the board of the association, but not on the board of Robertson Research Fund Inc. In 2000, the laboratory paid Watson more than any of its other employees, officers or trustees: he earned $342,000 that year, and $76,251 in fringe benefits.[3]

All three organizations declared that they spent nothing on lobbying or on advocacy to get laws changed. While that might well be true of the organizations, it could not possibly be true of the laboratory's president, James Watson. As his new book made clear, Watson had taken a leading role as far back as the early 1960s, trying to advise presidents, tweak government policy, lobby Congress against the imposition of inconvenient laws and for more government spending on biological science. In the later half of his career, he had earned a fine living from the laboratory while he did this—from a charity that gets its money primarily from governments and from private donations. Watson lived in public and private worlds at the same time, eating from both public and private tables. His career was a prime example of science as enterprise, which was beginning to remind me of other forms of modern capitalism—even cowboy capitalism.

In the 92nd Street Y, names of big contributors are rendered on plaques in gold-leaf letters. They are mainly famous business names, like Schiff, Simon, Tisch, Bronfman—all people who changed the face of New York City, yet whose fame is as nothing compared to Darwin and Watson, who earned their immortality on battlefields of the mind. Watson's story is almost a complement to Darwin's. Charles Darwin was the well-born son of an upper-class British doctor who invested shrewdly and married a Wedgwood. As Janet Browne so brilliantly explains in her two-volume work on Darwin's life,[4] Darwin also became an entrepreneur: he doubled his considerable patrimony by investing in railroads and mortgages. His books sold well and made money. Watson, on the other hand, started with little. While he is reticent about his family circumstances in his two autobiographical books, it is known that his family was not well off. But he too did well from publishing: The Double Helix is still in print more than thirty-five years after it was first published. Similarly, his textbook on molecular biology has had a large readership and a long shelf life. Like Darwin, Watson spent formative years in Cambridge and made a moral choice at a turning point in his career without which he would never have been famous.

Darwin's choice concerned what to do when the amateur botanist Alfred Wallace innocently sent him a paper outlining his theory of the origin of species and asked him to pass it on to Charles Lyell. Darwin realized that the paper was very similar to his own unpublished views. Should he throw in the towel, letting Wallace have the primacy he'd earned, or let his friends fix things for him? For the rest of his days, Darwin maintained cordial relations with Wallace, even though Wallace was not a gentleman like Darwin.

Watson's moral choice concerned whether to make use of Rosalind Franklin's X-rays of DNA, which he was shown by her colleague at King's College, London, and which convinced him that DNA is shaped as a helix. Details of Franklin's accomplishments were handed over to Watson and Crick, not by Franklin but through a report on the progress of her King's College lab that was drawn up for an outside committee evaluating the work.[5] Watson and Crick were shown a copy of this report because a member of their own lab was on the evaluation committee. They could not have built their model without it, yet, unlike Darwin and his polite relations with Wallace, when Watson wrote of Franklin in *The Double Helix,* he mocked and disparaged her. (Crick on the other hand, became a close friend, and Franklin stayed with the Cricks in the last stage of her fatal illness.) Watson's book was published after she died, so she could not even defend herself. In *Genes, Girls and Gamow,* he dismissed Franklin's contribution, and that of her colleague Maurice Wilkins, too, though Wilkins shared the Nobel with Crick and Watson: "No one could then have anticipated that in less than a month, Francis and I alone would have found the answer and one so perfect that the experimental evidence in its favor from King's almost seemed an unnecessary accompaniment to a graceful composition put together in heaven. Our writing of the tiny manuscript for *Nature* that would announce the double helix seemed even then an historic occasion."[6]

I scrunched down into my plush seat in the auditorium, waiting for Watson to appear. People scrambled to their places all around me. My old friend who knew Watson well had warned me that I was in for no treat. "He's a terrible speaker, darling, but they flock to hear him," she'd said.

The stage was empty save for a podium, two chairs, a table and an azure drapery. Robert Krulwich, the interlocutor for the evening, was introduced as a special correspondent with ABC, a regular on *Nightline* and co-host of a prime-time special in 1999 called "Brave New World." He took his place in an armchair. Then Watson came out from behind the curtain. His hair had thinned to a white fringe, wisping across a bald scalp. He had filled out, but he moved like a gangling young man, ungainly as a stork but vigorous. He flapped his arms as he walked, hiked his pants up too high. He wore a checked sports jacket, a shirt and tie, brown leather lace-up shoes. His hands described little circles and rubbing motions on the arms of his chair. Strange slushy, slurping noises issued from his mouth and nose, along with strangled honks and snorts. Maybe this is why Krulwich tried to play the interview for laughs. It took Watson a long time to catch on; every joking question from Krulwich was answered with almost lugubrious honesty. He was the straight man's straight man.

Krulwich began with Watson's early days. He had grown up in Chicago, with a mother who encouraged the finer things, a father who took him bird-watching at a nearby state park, and a sister. Watson went to the University of Chicago at fifteen—not because he was drop-dead brilliant, but because at the end of the Depression, the university opened its doors to high school students to encourage enrollment.

"Chicago was not a polite place," Watson said. "[It was about] logic, and thinking clearly. You were to learn ideas and say why they were great ideas or read great books and have lofty goals." Watson studied biology because he was interested in the lives of birds. At seventeen, he read Schrödinger's book. "He said the essence of life is information," said Watson. "Not the enzymes that do the work, but the script. It must be physical. The idea was, you're going to explain life through chemistry."

"Did you close the book with the question in the air and say, this is what I'm gonna do?" Krulwich asked.

"I wanted to understand life," said Watson, without a hint of grandeur. Not so long before, he said, many people believed in vitalism, that there was some spark that made the difference between life and nonlife, that this was a gift from God. "That's a totally untestable idea," he said. "So science, it's a method, using logic . . . so it was satisfying to see the script."

"So only seven years later you found the answer to a question of that size," said Krulwich, with a hint of an edge.

"The brain is very good in the late teens," said Watson, oblivious to the undercurrent. And later, more modestly, he said, "I never thought I was very bright. I had keen curiosity. I couldn't draw. No musical talent. A very good memory."

And so, as the interview progressed, his story emerged, punctuated by his odd snorts. He finished his degree at eighteen and applied to graduate school at Caltech, but even though he was Phi Beta Kappa, they turned him down. Because he was interested in genetics, he applied to Indiana State, where Hermann Muller, the eugenicist and one-time Marxist who'd shown mutations can be induced by radiation, was teaching. Muller had just won the Nobel. There was, Watson said, a good protein chemist there too, Jewish . . .

"Salvador Luria," Krulwich offered.

"He was an Italian doctor," explained Watson. Luria had been interested in physics, had spent time in Enrico Fermi's circle and had developed some knowledge of viruses. Luria had left Italy to get away from the Fascist regime and the restrictions imposed on Jews. He had fled to France, then to the United States. Watson was his first graduate student. Then Renato Dulbecco arrived. "Both got Nobel prizes."

"Probably your elevator man got it," said Krulwich. The audience roared, but Watson appeared not to notice.

"You get the Nobel if you are working on an important problem," Watson responded. Luria, he explained, had a reputation for arrogance. "He'd say some of the research at Indiana is worthless. These people tell you the truth fast. It's difficult for polite people to know what's up." And then he added this: "Almost all the scientists were young and Jewish."

"There's Jews everywhere in this building," said Krulwich, in a stage whisper, making a joke, but also trying to warn Watson that something about this harping on Jewishness was getting a little dicey.

"You wanted to make people feel if [they were] not Jewish, they may have a chance," said Watson. Luria, Watson explained, was very disdainful of chemists. He felt physicists were brighter than chemists.

Krulwich wanted to speed the story up. He jumped to Watson's days at the Cavendish in Cambridge. Watson inveigled a position as a postdoc there after being sent on a government grant to Denmark to learn something about the chemistry of DNA. "Then you go to this magical place," Krulwich said, "the Cavendish Lab, and there's Max Perutz, John Kendrew, Crick, and they all win the Nobel Prize."

"It was the best-equipped lab in the world," said Watson, "and there were more bright people who understood X-ray crystallography. Bright people like to go to a place which is bright." There was an unhurried pace at the Cavendish in those days, too, he said, which made it a good place to work on a difficult problem because one didn't have to get an answer by next year.

"And your boss [Lawrence Bragg] got the Nobel Prize at twenty-five," said Krulwich.

"It was important," said Watson. "Bragg was in a position of power. The English were poor, but he could get resources."

Watson and Crick got along almost immediately, although, according to Watson, many people found Crick's laugh and demeanor almost impossible to tolerate. Watson had opened the first chapter of *The Double Helix* with, "I have never seen Francis Crick in a modest mood." But Watson had no problem with Crick's idiosyncrasies. They shared an interest in trying to work out the structure of DNA, which both believed to be the macromolecule that embodies genetic information. While Crick was supposed to be working on a boring protein problem for his PhD, Watson was supposed to be learning X-ray crystallography under Lawrence Bragg. Instead, Watson spent most of his time in the library, reading. Watson and Crick got together for tea, and for lunch, and they talked. "And Francis had a certain style," said Watson. "He'd tell you who was boring; [he was] filled with gossip. His wife was, you know, arty. Francis liked the demimonde. I'd be thrilled to meet the prime minister. Francis, oh God—no interest in power. The only petition he ever signed was when the Stones got busted, for the legalization of pot . . . Francis was a character out of Shaw. He had no use for middle-class morality."

Krulwich led him quickly through his and Crick's discovery of the molecule's shape, but Krulwich wasn't really interested in the facts of the science. He wanted to know more about the social lives of scientists, as described in both of Watson's autobiographical books. He wanted to know, for example, if Watson's characterization of Francis and Odile Crick's marriage and its strict division of spheres of interest by gender was typical of the period. He wanted to know if Watson considered it important for a scientist to be able to share his work with a spouse.

"It depends," said Watson. "It's asking too much of people. I wrote a lot of science in my letters to the girl who didn't marry me. Don't try and inflict your occupation on somebody."

Finally Krulwich got where he really wanted to go. "Rosalind Franklin was very adept at taking pictures," he said. "You said she would have been the most famous scientist who ever lived if she'd found the structure of DNA." Krulwich then read out one of Watson's descriptions of Franklin from *The Double Helix*. The scene he selected was of Franklin giving a report of her findings at a seminar. "There was not a trace of warmth or frivolity in her words," wrote Watson of her performance. "And yet I could not regard her as totally uninteresting. Momentarily I wondered how she would look if she took off her glasses and did something novel with her hair."[7]

Krulwich asked if Watson went through this kind of double vision normally when faced with a woman scientist.

"Sure I do," said Watson. "They're women and scientists." But he said the real problem between him and Franklin was that Franklin "was not warm." In any case, Franklin wasn't really focused on him.

"She took an instant dislike to Francis," said Watson. "I was a sidekick of him." And, more to the point, she didn't think that DNA was shaped as a helix, as Watson did. She and Wilkins had fallen out over that before Watson had arrived on the scene, and Watson was unable to budge her, he said, even though he told her her arguments were lousy. She didn't want to build models ahead of the data, either, but Watson and Crick knew that Pauling was already doing it, had done it and, even though he'd gotten it wrong, was trying again. Watson was determined to be first. To him, Franklin's and Wilkins's images of the B form of DNA suggested helices with backbones on the outside. Eventually, he and Crick put two and two together. The result, according to Watson, was a theory of the molecule's shape so "beautiful" it had to exist. "It's obvious," Watson said, with a particularly forceful snort, "within two seconds . . . The only one who didn't like it was Barry Commoner . . . It was his Communist past, nature versus nurture being controlled by genes. To me, the alternative was God, and that was terrible."

I found myself leaning forward in my chair wondering why God was the only alternative to genetic information carried by DNA.

"It was either something like this, or religion," said Watson, which was not a much more careful formulation of the same idea.

"There's almost a religious belief here that something beautiful is true," said Krulwich. "Why was it Watson and Crick, not Crick and Watson?"

"Francis remembers we flipped a coin," said Watson.

How about patenting? asked Krulwich.

"When I came to Cold Spring Harbor, Leo Szilard asked that. But you can only [patent] something new and useful. There was nothing useful about it. Then the suggestion was, we copyright it."

"You'd be a really rich guy," said Krulwich.

He *is* a really rich guy, I mumbled to myself.

"I know," said Watson. "In those days people didn't see science as a direct route to wealth."

Watson said he wrote this new book because he wanted people to be able to read George Gamow's letters. Gamow, a physicist, had sent letters to Watson and his other genetics colleagues describing his theories about how RNA relates to DNA, how information is carried from gene to cell. The letters, which are reproduced in the book, are childish yet interesting and strange, with many stickmen drawn on them to illustrate points. Watson described Gamow, a Russian who had fled to the United States in the 1930s, as a brilliant person who also drank too much, was unhappy "and felt a desperate need to abuse people . . . He was never a social success."

From this, Krulwich moved to the use Watson had made of people who were not as well known as Gamow. Krulwich put this gently; without naming names, he said that perhaps Watson had made some of these people "more famous than they might have wanted to be." He was clearly referring to Christa Mayr, Watson's first serious girlfriend, and no doubt to the awful scene recounted in the book by Watson in which he tried to make love with her in a

British country home (belonging, he told Krulwich, to a left-wing Gosford Park of a family). The reader learns he fumbled Sex 101.

Yet Watson seemed to think Krulwich meant Rosalind Franklin, who seemed to be alive in his mind and still shrieking at him fifty years after she threw him out of her office, and forty years after her death. There is a Yiddish expression for the pain inflicted by guilt. My grandmother used to say, *Auf un ganif brent der hittle*, which, loosely translated, means "the thief, his head burns."[8] "If I hadn't written *The Double Helix*, she'd have been forgotten," Watson now cried. And he went on to add, later, that even if Franklin had not died before the Nobel was awarded, she would not have got it anyway because the best X-ray image of crystallized DNA had been made by Wilkins, not her. (Why then had he chosen to reproduce her X-ray picture, taken in 1952, in *The Double Helix?*) Since the Nobel goes to a maximum of three people, the Nobel jury, in Watson's view, would have given the prize to him and Crick alone if Franklin had lived. "Because I wrote about her using language of fifty years ago, it was sexism," Watson cried. "A biography came out, and I was the bad guy."

"What about Christa," asked Krulwich. "What does she think about this?"

"I guess, why this attention?" said Watson. "I like her. She kept my letters. The moment she married someone else, I burned hers."

Krulwich asked, "Are you a deist?"

"I just assume [a god] doesn't exist," said Watson. "I have no religious feeling. If God created the universe and left us alone . . . I don't know what happened in the beginning so I leave it alone. I don't think evolution had any help," he added with a flip of his hand.

"Do you believe stem cells are important, and why?"

Watson assumed Krulwich was asking about embryonic stem cells. Watson wanted it understood that the difficulty over embryonic stem cells is just religion at work, and he didn't think religion would interfere for long. He thought that if there is a chance to make these cells useful in the treatment of disease—if someone could cure his wife of Alzheimer's, for example, through embryonic stem cells—then, he said, "I can't imagine any American president keeping the benefit from citizens."

"And cloning?"

"The first clone won't upset the world," he said. "Most women won't want to have children that way. If the techniques work perfectly and no arthritis at thirty, you could say, who is harmed? Well, if thirty people look the same in a small village, it would be a mess. It doesn't morally upset me . . ." He segued from there to genetic enhancement. "For people at the bottom, they might like it. Their children compete with the best. It's an unpopular thought, but your rank in society relates to what your genes are. It's unpopular but true."

Watson had expressed these views before, and in public, but so many ethics papers had been published emphasizing the danger of the notion that genes are destiny that I was shocked. He didn't seem to care how it looks when the president of Cold Spring Harbor Laboratory, the former site of the Eugenics Record Office, says in public that social "rank" is associated with genes.

Didn't you write an article against cloning? Krulwich asked. Watson had, back in the early 1970s.

"Yeah, but then science said it can't happen," said Watson. "Then it did happen . . . But if there is no risk, you know, it becomes a question of who you clone and under what circumstances."

"Any more Nobel Prizes in the future?" asked Krulwich.

"I'm not sure," said Watson. "[My] only hope for the Nobel Prize is in literature, and many would say the current book will finish that."

Krulwich showed no mercy. He leaped upon him with Ehrenreich's unfavorable review. "Is the genome project overhyped?" he asked.

Watson pushed his back into his chair, gripping its arms with both hands. It looked as if he'd been waiting for this one and intended to bash it out of the park. "I think Ehrenreich's education at Reed left something to be desired," he said, loudly. "She's an intelligent woman, but clearly jealous. People pay attention to science more than to social critics. She's a good writer, but [the review] just revealed she is ignorant."

Oh my God, I thought.

From one female critic, he soon returned to the first—Franklin.

"People say we were on her shoulders," said Watson, but he refused to acknowledge any debt to Franklin at all. "The great pictures were taken by Maurice."

Jan Witkowski stood waiting at the end of the Long Island Railroad platform at Huntington. He was tall and seemed to tuck himself carefully into a small social space, like a gallant diplomat. His well-groomed head dipped with a questioning tilt as I moved toward him. He wore a full-length black leather coat, gloves and a scarf. The railway platform around him was surrounded by piles of industrial equipment, but very soon we were traveling in his Volvo through a green velvet landscape into which grand mansions were set like Fabergé eggs. We drove over lovely, narrow roads, flowed up hill and down dale under the chapel architecture of beach and oak.

As he drove, I couldn't get Watson's performance of the previous night out of my mind. Watson had answered whatever he was asked, without sparing himself. He had barked and snorted and slurped his views and stories and opinions without fear to a large room full of strangers. He deflected criticism that no one made. He seemed bemused by his failure to do anything else in science equal to his description of the shape of DNA (leaving me to wonder if answering a big question too early in life is a kind of curse, especially when the circumstances are morally bracketed in one's colleagues' minds).

He had also made me see a side of myself I didn't like. I understood, now, that I had gone into that auditorium not as a journalist should—open, impartial—but disliking Watson, a man I'd never met. He'd refused to see me, and the character he presented as himself in his books was so arrogant, disdainful, hubristic, callow and sexist that it obscured his virtues. So what did I think of him now, this founder-exemplar of the whole field? I was still astonished by his arrogance, by the simplicity of his political ideas. Did he really believe that genes equal destiny, that the poor could rectify their genes faster than the rich could enhance theirs, and that this could change the balance of social power? But I liked his courage.

I liked his struggle to be honest about himself, even though he failed. Who doesn't fail at that?

We arrived at last at the entry to an estate so big the main house could not be seen from the road. The Banbury Center is in the middle of 55 acres of rolling sward and woods, all of it given as a gift to Cold Spring Harbor Laboratory. The grand main house is a lodge for invited guests. The meetings are held in the renovated former garage. Witkowski's office was tucked into a tiny space under low eaves, with its own lovely view of hills and trees.

The estate had been donated by Charles Robertson, for whom Robertson Research Fund Inc. is also named. Robertson studied at Princeton, became a lawyer and then, when he was out of work, met and eventually married Marie Hoffman whose family founded the Atlantic and Pacific Tea Company. His original goal was to fund science research to be done on these grounds, but Watson, by then director of Cold Spring Harbor Laboratory, explained that the nearby village would never allow it and that the estate would be more useful as a center for small meetings. What Watson wants, he gets: Robertson handed over the estate in 1975. The buildings were refurbished and were opened for use by the Banbury Center at a ceremony in May 1977. Francis Crick did the honors. Crick, who had threatened to sue Watson for libel over *The Double Helix* because Watson had described himself and his fellows as being motivated by ambition, had by then made up with Watson.[9]

The laboratory began in 1890 as the Brooklyn Institute of Arts and Sciences summer camp. Charles Davenport arrived in 1898 and, shortly after the rediscovery of Mendel's work in 1900, became a passionate follower of Mendel's theory, publishing his first genetics paper just two years later. In 1910, Davenport persuaded the widow of E. H. Harriman,[10] to establish the Eugenics Record Office at Cold Spring Harbor, where the Carnegie Institution already had a station for the study of "experimental evolution." The Eugenics Record Office was first located in the Carnegie Institution's building, then moved to a building of its own. But there was no real division between the three organizations. Harry Laughlin was the Eugenics Record Office's superintendent; it was his involvement in legislation to restrict immigration on eugenics grounds in the 1920s that drove many geneticists away from eugenics. The office was shut in 1936, but the Carnegie Institution's Genetics Department, the descendant of its experimental evolution station, continued, along with the biological laboratory, on the Cold Spring Harbor campus.

Drosophila geneticist Milislav Demerec became the director of both labs in 1941, and Barbara McClintock, the corn geneticist, was hired in 1942.[11] Watson spent his first summer here in 1948. But by the late 1950s, Cold Spring Harbor was on its financial scuppers. One of many crises arose when the Carnegie Institution withdrew support for its Genetics Department. John Cairns was hired as director in 1963 to revitalize the place and Cold Spring Harbor Laboratory was set up as a legal entity in 1965. In those days, researchers were mainly dependent on federal grants, and there wasn't much federal money around. The lab had been publishing reports arising from its symposia since 1933. "The 1966 one sold incredibly well," said Witkowski. "It was the *Genetic Code* volume. It single-handedly kept the lab afloat that year." With all these crises came "machinations," to use Witkowski's word, among the trustees. In 1968, Cairns resigned.

A few years earlier, to resolve an earlier financial disaster, Bentley Glass, a *Drosophila* expert and a trustee, had talked seven major U.S. educational institutions into donating $25,000 each to the lab, including Harvard, Princeton, Yale and SUNY at Stony Brook. Harvard's representative to the board was James Watson, who had received his Nobel in 1962 and who used the prize money to buy himself a little house near the Harvard campus, where he was teaching. But he had always wanted to be the director of Cold Spring Harbor, always wanted to live in the director's fine pillared house. Soon he got the part, but he kept his teaching position at Harvard until 1972, at which point he and his wife moved to Cold Spring Harbor full-time.

His arrival coincided with good things, according to Witkowski. "One was that the war on cancer began," said Witkowski. "He had some hand in that, with Mary Lasker, to get things going." President Nixon was in power; like biologists, he seemed to think an attack on a big disease suggested worthy motives. Watson saw this war as a way to fund work on the genetics of cancer-causing viruses, as a way to pick the lock on the cancer door. Soon Watson invited Joe Sambrook to be the associate director and to start a cancer group. Sambrook won a cancer research award—a five-year grant—that gave stability to the whole organization. "We've held that grant ever since." said Witkowski. The grant was just renewed for the sixth time and had grown to $12 to $15 million over five years. Charlie Robertson's first gift to the lab was another good thing that happened during Watson's tenure: Robertson created an $8 million endowment for research. And according to Witkowski, the fourth good thing was the new technology of recombinant DNA. "It enabled a whole different approach to research . . . You could do things with mammalian cells that you couldn't do before. That was the foundation that enabled Jim to transform this place."

So when did corporate America first come calling to give money?

"Corporate America came calling quite early, 1980ish." The laboratory also spun off a company based on its cancer research around the time the U.S. Congress wrote the Bayh-Dole legislation, making it obligatory for research institutions getting federal grants to market their innovations. I pointed out to Witkowski that this was also the point at which biologists began to see science as a way to get rich, but he didn't agree with me at all. He thought it was only lately that a small minority of scientists had been able to expect wealth from their work. "The currency of science is primacy," he said.

Primacy, I thought, is often the currency of business.

What I really wanted to know was when corporate sponsors began to donate money to the laboratory, apparently in return for an inside view of leading-edge work. I'd noticed that two of the corporations that supported the Banbury Center's meetings also enjoyed the prestige of having James D. Watson on their boards. Some of the trustees on the laboratory's board were also executives with large pharmaceutical and biotech companies. This crossing over between charitable and business interests reminded me of similar relationships between corporations and politicians. Corporate donations to politicians, political parties and leadership campaigns are usually explained as having nothing to do with advancing vested interests— just an expression of corporate good citizenship and a wish to aid democracy. Such protestations aside, most observers conclude that these relationships often involve a quid pro quo: we help you now, you help us later—or, as one Canadian prime minister used to put it, ya dance with the guy that brung ya. Should a corporation give money to a charity as part of a quid pro quo, such as getting an inside look at science no one else will see until years later? The leaked information about James Thomson's stem cell work in primates gave Geron a real marketplace advantage. These connections and crossovers are perfectly legal, but they also seemed like inhibitors to free inquiry.

Not a problem for Witkowski. "Nice to have ties to these people," Witkowski said. "Actually, we should do more of it." Each sponsor gave $25,000 a year to the meetings program. "They are allowed to send eight scientists to meetings but only three to Banbury meetings and only one at a time . . . They have no say in how the money is used, topics for the meetings, and at the main lab [decisions] are made without reference to corporate sponsors."

Well, I said, there's a reason why they're giving you this money, and it's got to be because they get a line on things no one else hears about. Why else the secrecy? Why else no journalists allowed?

"Why secrets?" he repeated. "We can only have forty people here." He pointed out that the meeting room is small—and it is small, but not that small. "And the second is we want people to talk about unpublished stuff." He meant work not yet in print, but also work that had not borne fruit but that, if shared, might save others wasted time. This sort of thing is only rarely published in journals.

But why would a corporation pay for its scientists to watch a meeting if the information had no commercial value? And what about the *private* public policy meetings?

He said there had been very few of those. Once a major drug company had tried to take over all the work product of the Scripps Institute for $300 million, which had prompted a meeting to discuss corporate funding of private research institutions. And by "public

226 • ELAINE DEWAR

policy," he certainly didn't mean anything that could drive legislation. He seemed nervous that I might draw a connection between public policy discussion and political activity.

Do you have news conferences? I asked. "No news conferences," he said.

I told him I'd been told that PPL Therapeutics had held a press conference on the Banbury Center lawn, in the middle of a conference on cloning, to announce the first cloned pig. After Watson's performance the previous evening, it now occurred to me that the Banbury Center might have permitted it because it wanted to be associated with a cloning advance. "I have to say I can't recall it," he said. "We do have TV crews here. Tom Brokaw did a piece . . . The BBC came. Those take place outside the meeting itself. If we do it, we do it as a courtesy." (Later, he added that "The last thing anyone wants to be associated with is science by press conference.")

So, I asked, do people talk freely in your meetings?

He said it again: the primary currency in science is credit and, as a result, "they're always wary what they talk about . . . even with friends or colleagues in the bar. But if your deadliest competitor is in the group too, you watch your tongue. . . scientists have always kept secrets until their work is published for fear of someone else getting there first."

We fenced back and forth until we ran out of time. He offered to drop me at the Cold Spring Harbor train station so I could see the main campus. He drove fast—very fast—over roads that wound around past gracious houses with wings and dormers and pools, the kinds of places used by Hollywood directors like Frank Capra and George Cukor to signify great wealth—think Katharine Hepburn's home in *The Philadelphia Story*. Every now and again, the sun would bounce off hidden Atlantic waters and slivers of light would blind me. Finally, we came over the top of a hill and there, below, was Cold Spring Harbor. The sun blazed down on glass-smooth water. Birds walked stiff-legged across a broad sandbar. The road clung to one arm of the harbor, then rose up into a little village of clapboard houses, mostly hidden under fir trees. Witkowski pointed at a huge, rounded, orange-colored yurtlike house, set halfway down a hill, perched above the water on the opposite side of the harbor. "That's Jim's house." The house revelationary biology built.

At the station, as I thanked him and said goodbye, my eyes drifted down to the car floor. That's when I noticed what Witkowski was wearing on his English schoolboy feet: cowboy boots.

Part Two

CHAPTER ELEVEN

ROOTS

March 2002. The wind and snow and misery of a true Canadian winter howled in before dawn, two weeks before the official start of spring. The ice on the freeway was as black as the sky. Rat-a-tats of sleet stung the taxi broadside, and bitter cold seeped right through the plate glass windows in the airport. Eight hours and two movies later, at Heathrow in London, warm, moist night air soothed my frigid bones. There were flowers out there—I could smell them through the open window as the taxi beetled past the Serpentine. When the sun came up, I could see them: pink and lilac hyacinths in clay pots in front of the polished doors of Kensington and in the window boxes and on the tiny decks set out upon Peter Pan roofs. On the grounds of the Natural History Museum, a yellow-brick Victorian pile with a gilded wrought-iron fence, there were small hills of nodding daffodils, dazzling rows of acid yellow flowering shrubs and trees fuzzy with green buds, as if their limbs were swathed in the most supple, airy stockings.

I had to armor my spirit against such gentle beauty, clamp down on the Other One. Softness, gentleness had nothing to do with revelationary biologists. They had the tongues of adders. They seemed not to know or care that one of the worst fallacies in argument is ad hominem, criticism aimed at the individual instead of the idea. Almost everywhere I'd been, someone had called someone else a bad name, pointed to personal motives or immoral deeds. It wasn't just James D. Watson who called his critics "ignorant," although Watson certainly led the pack in this behavior. Worton had dissed his colleagues in Pennsylvania. Freda Miller had made remarks about *Science*'s decision making, its willingness to bend the knee to those with clout while slamming doors in the faces of the diffident. Michael Bishop and Alan Bernstein had made remarks about Michael West, who in turn had cast aspersions on a former colleague at Geron. Even Ann Kiessling had thrown rocks at critics of Advanced Cell Technology's first human clone: "First, they don't understand why it's done and, second, they're jealous," she'd told me. It was as if the start and end points of biological arguments were permanently set at the personal.

At first I'd believed John Maddox, the former editor of *Nature,* who pinpointed 1980 as the year when such behavior began. Certainly after the U.S. Congress mandated, in 1981, that institutions getting federal grants must make every effort to get innovations to the market, it made sense that American biology would change its character as it opened itself to commerce, but why would such a change take hold so quickly among scientists of different nationalities who also publish in *Nature?*

Neverthless, I had at first accepted that the marketing of purity and the vicious personal competition arrived with the privatization of biological science, and as a direct result of it. Certainly a see-saw battle between public and private interests revealed itself everywhere I looked. It was the hot theme of the story of the Human Genome Project, which had started in 1988 with taxpayers' support from the United States and Britain for academic researchers trying to sequence whole genomes. But then private interests took up sequencing too. Those spending public money, who called themselves the "public consortium" (although they were neither public nor a consortium) rarely ceased railing against these private competitors, and vice versa. In 1998 the brand-new company, Celera Inc., led by former NIH staffer Craig Venter, announced that it would sequence the human genome, and do it faster than the inefficient academic researchers who were wasting taxpayers' money. Those spending public funds had then behaved as if they were in a desperate struggle for the very soul of science, moving heaven and earth to get the job done first. Both sides claimed to have finished their first drafts of the human genome at the same time—in June 2000. People from both sides went to the White House and 10 Downing Street and made nice speeches. But then, leading figures in the consortium claimed that

Celera could never have compiled its draft if it hadn't relied on the consortium's work; some leaders of the consortium even tried to keep *Science* from publishing Celera's version. Vindictiveness between the parties now flowed like cheap champagne at a ship launch.

It made sense that in a battle for survival, all parties would claim the same ground, stop at nothing and try to pull their competition down; it made sense that since so many biologists are in business, they adopted business behavior. But just because a story makes sense and is supported by evidence doesn't make it true. Sometimes a focus on contemporary events obscures a much larger tale. Ad hominem attacks certainly had a long history. In the very first chapter of *The Double Helix*, James Watson recounted how, in 1955, years after he and Crick published their first DNA articles in *Nature*, he ran into a former colleague of Maurice Wilkins's on a hiking trip in the Alps. The man didn't stop to talk, just asked, in a most derisory way, "How's Honest Jim?"[1]

By the time I got to London, I had begun to suspect that all biology's themes—the cruelty, the culture war between materialism and vitalism, the hubris, the flat-out pursuit of fame, and this apparently new one, the use of personal attacks to shape opinion—descended from a common source; they were like the contradictions and complexities of individual character. As every mother knows, character, a basic attitude toward the struggles in life, is apparent in a child almost from birth. It was as if a tiny original population of biologists had flourished in a particular place and their descendants had spread out, taking their characteristics everywhere.

One drizzly morning, I walked to the Natural History Museum to see its display on Darwin. Over the years from 1831 to 1836, when Darwin traveled on the British naval vessel the *Beagle*, he amassed a huge collection of specimens. He journeyed to the Canary Islands, Brazil, Argentina and Tierra del Fuego, up to Chile and then across to the Galapagos. From South America, the *Beagle* struck out across the South Pacific to Australia and Tasmania and, eventually, back to England. According to his most recent biographer, Janet Browne,[2] Darwin was seasick whenever the boat was in motion, which created a certain incentive for him to make his brave and often lonely explorations on land when the *Beagle* put into port. On his ramblings, Darwin collected and studied many examples of flora and fauna unknown to British science, including ancient fossils and bones of gigantic, extinct species, as well as small birds and mollusks and barnacles. He gathered with them the intellectual tools to ask sharp questions about the origins of the fabulous variety of species he encountered.

Charles Darwin was certainly not the first to develop a theory about why nature changes. His grandfather, Erasmus Darwin, had developed his own theory of evolution many years earlier, during a time when the doctrine of materialism was much more popular among thinkers than it was in his grandson's day. By the time Charles Darwin was a young man, Erasmus Darwin was considered a radical materialist, and his views had been shoved to the fringe of science. Darwin's father, Robert Waring Darwin, was a doctor and a man well known for his practical capacities, especially for his interest in science and technology rather than adherence to religion. Charles's older brother, also called Erasmus, was a naturalist and much admired by Charles, who followed his brother to medical school in Edinburgh. There, Charles recoiled in horror from the nature of a medical education. In those days surgery, even on children, was performed without anesthetic. Human anatomy was demonstrated upon dissected corpses ripped from their graves—and sometimes upon people who had been murdered for the purpose of the sale of their bodies to anatomists. Dead bodies and surgeries were presented to theaters full of young medical students who bought tickets to lectures as if they were going to see a play. Erasmus never practiced medicine after his graduation and never married, but he set himself up in a house in London, becoming a well-known figure in scientific and literary circles. Charles, left alone in Edinburgh, avoided medical lectures and threw himself instead into his first research work. He joined the student natural history society, called the Plinian, and fell in with Robert Grant, who would later become professor of comparative anatomy and zoology at the University of London.

Grant was a transmutationist, which is what evolutionists once called themselves; he believed that species adapt to fit changing circumstances and pass their changes on. He was inspired by the work of Jean-Baptiste Lamarck and Erasmus Darwin. Charles Darwin worked with Grant on the study of the spongy life forms found slithering along the harbor bottom. Grant taught Darwin how to dissect under a microscope, how to study these organisms' developmental cycles. He also spoke to Darwin of his theory of evolution. But though Darwin had already read Lamarck and his grandfather's work, he paid Grant little heed.[3] As Browne explains it, this was because Darwin fell out with Grant over Darwin's observation that the genus *Flustra* reproduces by eggs that swim. Darwin was very proud of his discovery because it had been missed by the greats of biology. But when Darwin told Grant what he'd found, Grant, who was working on the same thing, told him to stop working on this subject and not to publish on it. Three days before Darwin read his paper on *Flustra* to the Plinian, Grant read one of his own to a much more illustrious natural history society, incorporating Darwin's work without giving Darwin credit. As Browne put it, Darwin's "first scientific discovery . . . was marred by being also his first introduction to 'the jealousy of scientific men.'"[4]

Darwin's father was not amused when Charles quit medical school. He wanted his son to have a profession and not spend his days in idleness. In his late teens and early twenties, Darwin liked to hunt more than just about anything else, and his father thought he was going to ruin his life. He announced that if Charles didn't like medicine, he should be a clergyman, a logical profession for a person interested in natural history. Gentleman-clergymen could get a reasonable living from tithes, maintain themselves at the correct social level and have plenty of time to pursue their private interests. He sent Darwin to get a BA at Christ's College, Cambridge, as the first step toward becoming an Anglican clergyman.[5] Darwin soon realized he would be unable to swear to a belief in the creed of the established Church of England, which he would have to do to be ordained. He began to dread having to face this moment of truth when he would have to declare himself. Meanwhile, he spent more and more of his time in the company of the naturalists—botanists, zoologists and geologists who lectured at Cambridge. He became a particularly ardent student of the botanist John Stevens Henslow[6] and the geologist Adam Sedgwick.

Henslow was politically active, and also well connected with important people in the Admiralty in charge of exploration. When Darwin was in his final year at Cambridge, Henslow was asked if any young gentleman there might act as a companion to Captain Robert FitzRoy of the *Beagle,* who was being sent out by the British government to explore and map the southern coast of South America. FitzRoy had decided he must have a gentleman companion on board in order to avoid going batty, as his predecessor had done, after being confined for months on the small ship with uneducated sailors. Henslow thought this was a great opportunity to collect specimens from places none of his colleagues would otherwise get to, and he recommended Darwin.

In those days, almost all professors at Oxford and Cambridge were ordained. Fellows of the various colleges were bachelors. Professors were either gentlemen with private incomes or they lived from the sale of tickets to lectures.[7] Science as a profession was in its late childhood; to have to earn an income through one's work was considered demeaning to any British gentleman, and certainly to a gentleman of science. Few scientists were employed by the government. The great universities were religious institutions, dedicated to spreading knowledge of God's great design. Science meant the elucidation of that design, although there were some who agitated to create universities that were places of free inquiry. Mainstream scientists held that God made the variety of life in one act of creation: species may die out, but they did not change. As Browne put it, "science in a sense was religion."[8]

Biologists were at the early stage of knowing life by sampling its variety. Rare specimens of animals and plants were much sought after by the few public science institutions, such as the British Museum and Kew Gardens, but they were also prized by private collectors, who would pay for them. Rare or unique specimens could be of significant commercial value. Collectors who returned with specimens from distant climes were considered experts. The public were fascinated by those who went off on adventurous explorations and returned to tell about it. Such people showed off what they'd found through lecture series and publications and made names for themselves. Books were available at reading rooms. Alfred Wallace, a man of the lower classes, educated himself by attending these public lectures and public reading rooms. He and a friend managed to get a collector's agent to back them on their own voyage of exploration. They went by ship to the Amazon and slogged through the jungle for years hoping to find unique specimens and make their fortunes.[9] Wallace became an inveterate traveler.

Many strings were pulled to ease Darwin's way to the *Beagle* by those who thought he could bring back information valuable to British science. When he was finally given permission to go, instead of Darwin's being provisioned by the navy, his father bought everything his son would need for years at sea and arranged for thousands of pounds sterling to be at Darwin's disposal, wherever and whenever he needed it. While Darwin's unpaid job was to make himself a pleasant companion to FitzRoy, his unpaid avocation was to build a collection of specimens of flora and fauna new to science. The ship's surgeon was the official naturalist assigned by the Admiralty to the *Beagle,* though; he soon quit in protest when Darwin was given help to gather samples by the ship's crew under the captain's orders, while the surgeon was not. Darwin dined in private with the captain; the ship's surgeon was not a gentleman, so he was shut out. Darwin, a private person subsidized by a shrewd private investor, his father, piggybacked on a public venture and yet retained the ownership of his valuable collections himself.[10]

Darwin armed himself with the latest reference works, including the work of Charles Lyell, a geologist who advanced the theory that the world's form was made slowly by the accumulation of small changes but that there was nothing progressive about this, nothing inevitable or planned, and that large uplift on one side of the world would be balanced by subsidence elsewhere.[11] Darwin and FitzRoy were fortunate enough to witness a major earthquake in Chile and to note that the high-water mark on the nearby dock actually went down eight feet after the quake, meaning the land had thrust up. The letters Darwin wrote home describing his various forays, along with the samples he boxed and sent off for identification, were of sufficient interest that his work had already been announced at a meeting of the British Association and publicly read out and published by the Cambridge Philosophical Society, before he returned home.[12] Charles Lyell was particularly grateful that Darwin and Captain FitzRoy had found evidence to support his theory.[13]

Lyell, president of the Geological Society, soon began to advance Darwin's career; Darwin's success reflected well on Lyell, and a man as eminent as Lyell could ease Darwin's way into the major scientific societies of the day. Soon, through Lyell, who conducted a fashionable salon in his home in London, Darwin met many people who were politically or intellectually important in London society. Later, he also became friends with J. D. Hooker, deputy director of the great botanical gardens at Kew. Hooker sent him numerous plants from Kew, which Darwin used in his own research. Hooker also often helped Darwin with identification of plant species, sharpening Darwin's arguments. Darwin was also introduced to Thomas Huxley, the zoologist and comparative anatomist. Darwin was older than both Hooker and Huxley, and better established by the time they met. Lyell's theory formed the basis of Darwin's apprehension that species emerge from gradual random changes in organisms that allow adaptation to circumstance. Together, Darwin and his friends destroyed one dogma and built a new one.

The main hall of the Natural History Museum is an immense barrel vault of light yellow brick supported by intricately carved, spiraled stone pilasters. It was built to display the remains of dinosaurs and other natural treasures from around the empire. Giant skeletons flew from the ceiling or stood as if frozen in mid-stride in the main hall. Schoolchildren in uniforms skittered here and there, their shrieks and laughter echoing everywhere. I was directed up the stairs to the permanent Darwin display. The stairs rose, then split to the right and the left. On this mezzanine level, there was a larger-than-life-size bronze of a thinnish man, tall, with a strong nose: Richard Owen, who founded the museum in 1881. Just behind the stairs there was a large marble sculpture of Thomas Huxley, seated in a great Tudor armchair, staring off into the middle distance, his back to Owen's. Huxley was a handsome, fine-featured man who had a wickedly caustic tongue. He had ridden out to do public battle on behalf of

Darwin's theory when Darwin couldn't do it for himself, most famously in a debate with Bishop Samuel Wilberforce at Oxford. Huxley and Owen quarreled repeatedly over the meaning of anatomical evidence—and over whether species were the result of design or of natural selection. After the publication of *On the Origin of Species,* Huxley became the leader of the Darwin faction in London's top scientific circles. Owen led the opposite camp.

But in the beginning, they were friends. It was Owen to whom Darwin had first sent many of his fossils for identification. Owen was the Hunterian professor of zoology in the Royal College of Surgeons, and Lyell introduced him to Darwin at his home.[14] Before Darwin was pushed into publishing his theory of natural selection, Owen was Darwin's supporter. Owen agitated for years to get the museum built, to free himself and his colleagues from the glacial pace at the British Museum,[15] where samples from official British naval expeditions were sent for curation and display. There they sat in the cellars for years until someone finally got around to them. Darwin eventually fell out with Owen over the theory of natural selection, which Owen publicly abhorred, and Owen brought this place into being against the footdragging and representations to Parliament of Huxley, Hooker, Darwin and their friends, who didn't want all the power in natural history in Britain collected by this museum to run through Owen's hands.[16] The museum opened the year Darwin died. To get to the Darwin exhibit, I had to pass by Richard Owen.

In the upper hallway, on the way to the Darwin gallery, there was a triptych that illustrated Darwin's theory. At first I thought it was works on paper, three beautiful line drawings. One suggested a nude man joined to a woman; the next showed busts of chimpanzees and gorillas and humans, implying a relationship. All three crosslinked humans and animals and other living things; apples and other fruits were displayed beneath symbols of human culture—a hand, a hammer, a cross, and knives and forks. When I got closer, I realized these were not line drawings, but friezes pushed up out of a plasticized mesh. While each body, head, hand was almost sculpted in three dimensions, they remained conjoined by this common material, the common weave. When I stood at a distance, the different figures seemed separate and distinct, but close on, man and beast and fruit and culture all emerged from the same fabric. It was a brilliant evocation of Darwin's ideas and went way beyond his first simple notions of the origin of species through natural selection: it captured the concept of the role of sex in randomizing traits and providing variety for selection to work on, and it made clear the idea of common origin, of continuous shared connection forward, backward and sideways across geography and time, across tissue, blood and egg, which erased the notion of immutable barriers between species. This was a physical representation of life not as separate organisms, but as one giant chimera that includes humankind.

Beyond, a series of glass display boxes held the pathetically crude microscopes and tweezers and glass specimen jars Darwin had used in his work, along with photographs of Darwin and his famous study at his house in Down, Kent. The juxtaposition of the sophisticated image to the simple tools bore witness to the power of something possibly immaterial—mind.

In the exhibition hall there were the usual interactive booths. At various stations, bits of Darwin's theories were set out. Here was the story about the Galapagos Islands and their thirteen different species of finches, the importance of which Darwin didn't really grasp until he got home. This presented the problem of how new species are nurtured: was it by isolation, or was it by the power of competition? In another example, two species of fish diverged from a common ancestor in Lake Victoria. The lake's levels had risen and fallen, with the result that, for certain periods, one lake became two and then two lakes became one again. Did this separation allow two species to develop, or was it some other agency? The problems with Darwin's gradualist theory were also alluded to in a display about how quickly rats adapted in England to the use of the rat poison warfarin and how quickly grasses were able to adapt to high levels of copper in soil. There was a booth with a discussion about what selection acts on—DNA and its arrangements. There were text and images about genetic modification through cloning, including pictures of Dolly and the genetically modified Polly.

I passed the glass cases again on my way out. One statement beside pictures of Darwin's house in Down made me open my notebook. The curators had written, "It was here, in 1844, that Darwin completed his famous work, *On the Origin of Species by Means of Natural Selection,* published in 1859 . . ."

This was obviously wrong or, worse, propaganda. As Janet Browne's volumes (the second would be published within months) laid out, Darwin wrote out a choppy outline of his ideas in 1842—a sketch he discussed only with a few carefully chosen friends. He wrote a letter to Hooker about it in 1844 and showed it to him. He was more than a little rattled when an anonymous author published a book later that year about how species evolve; it was called *Vestiges of the Natural History of Creation,* and was scandalously popular among general readers, but it was utterly smashed in a review by Darwin's geological mentor, Adam Sedgwick, in 1845. According to Browne, Sedgwick said the book was so bad "it could have been written by a woman. But not even a woman could stoop so low as to put all the phenomena of the universe together as 'the progression and development of a rank, unbending and degrading materialism, etc.'" Such ideas, Sedgwick thundered, were repulsive and turned every civilized value upside down; if accepted, they would destroy everything worthwhile in life. Not surprisingly, Darwin was shaken that someone else shared his ideas, but even more by the devastating venom that his mentor had unleashed and that he realized would be heaped upon his own version of this theory if he ever found the courage to publish it.[17]

In 1857, after a few more papers hinting at evolutionist ideas were published, including one by Alfred Wallace in 1855, Darwin wrote out an abstract of his evolved ideas in a letter to Asa Gray, the American botanist.[18] But his book, *On the Origin of Species by Means of Natural Selection,* which he started to work on in 1856, wasn't completed until thirteen months after Wallace's paper setting out the same theory appeared in Darwin's mail. It was only the extreme pressure of being forestalled that made Darwin marshal his evidence, drafting and redrafting, cutting and cutting again, whipping himself on, to publish in 1859. Clearly, the museum didn't want to even hint that Wallace deserved priority.

The Down house appeared to be white stucco, not beautiful, but large. The study looked small, crowded and fussy. Darwin's father had given him the money to buy the house and staff it a few years after he married his first cousin, Emma Wedgwood, in 1838. Darwin had first settled in Cambridge then moved to London, but he found the city dirty, crowded, stressful. He had suffered from what seemed to be hypochondria as a boy and young man, but he had caught something in Chile that had almost killed him. After his return to England, he suffered from bouts of illness that he described in great detail to his close friends. He experienced uncontrollable flatulence, boils and prolonged bouts of nausea and vomiting, especially when he was under pressure. These pressures had intensified in London as he made his way through the various scientific societies and salons and took on obligations to produce books and papers describing his travels and discoveries, at the same time as his wife began to give birth to their children. His own mother had died of a stomach ailment when he was eight. When his daughter Anne developed similar symptoms and died at age ten, he became seriously concerned that through this marriage to his first cousin he had transmitted inherited weakness to his children.[19]

The photo of him in the glass case was taken in his later years, when he was a world celebrity—his face was so familiar that people spoke to him in public as if they knew him, and even arrived at his door without invitation. This was the image of Darwin that lived in my imagination: a heavy man with a full, white beard, big hat, big cape, beetling eyebrows. But I could see now that there was something terrible and sad about his eyes—a man in pain, a man who was not comforted by any belief in a life beyond, who had cast out the idea that human existence fits into some larger plan. He'd built his theory of change upon the winnowing of the unfit; death was selection's agent. By the time this photograph was taken, he had already suffered the losses of many loved ones; he knew that his own suffering would increase as he aged, and that his big reward would be his end.

I fumbled for my watch. I had to get to the train—I had an appointment at Cambridge University Library to read from its vast collection of Darwin's correspondence. He'd recorded his observations minutely and carefully throughout the voyage of the *Beagle*, sending home reports in letters to his family in case he should lose his material at sea. On his return, he wrote to anyone he thought could provide him with information, sending out his pleas for help (larded with oily flattery), writing again and again if help came too slowly. According to Adam Perkins, who keeps the collection at Cambridge, 13,500 letters that Darwin sent out and received are known, of which about 9,000 are at Cambridge.[20] The library has organized and summarized this correspondence both in print and online. I had spent many hours online, wading through the summaries, organized year by year, and had sent the library a long list of letters I wanted to read. I had contacted the library because I was interested in Darwin's relationship with Francis Galton and Galton's eugenics movement, but increasingly I had become fascinated with exactly what Darwin thought when faced with Wallace's paper, with that moment of moral choice.

King's Cross Station is a nineteenth-century monster, surrounded by narrow streets jammed with unruly traffic. The train was late twentieth century, electric and soundless. It glided through myriad neighborhoods of Edwardian row houses, through industrial zones and new suburbs, finally carrying me out into the most beautiful groomed landscape. I was so happy to drink in that vital green, the tumbles of ivies and creeping brambles, the small brooks winding under tiny footbridges, the spreading oaks. The farm boundaries were marked out by clipped, ancient, leafless hedgerows. There was not a scrap of paper, not a single rusting abandoned car or tractor or barrel, as if litterers had been banished.

It was dark when I arrived, so I set out the next morning on foot to find the Cambridge University Library. I had a map and a tourist book. At first, as the road wound past a very ordinary girls' school, I was horribly disappointed. This Cambridge was like any town anywhere: waves of traffic, modern boxes and a reeking smog drifting above the pavement. But then I crossed some invisible line. Small stores on either side of Downing Street held behind their glass windows spectacular antique furniture of great age and exceptional quality. Eventually I passed a high wall with an open gate. Inside I could see a green field and, around it, glorious buildings in the Greek revival style—pillars, porticos, pediments, sculptures—an aged palace for learning. The street narrowed, and narrowed again. I turned left, looking for a street called Trumpington. All the buildings around me were suddenly very old, and the streets narrowed to the point where a horse-drawn cart would have had trouble passing. Finally, the road emerged into a kind of plaza leading to a stone bridge. Down below was the placid river Cam. Bumping into docks on each bank were rows of punts ready for hire. On my right there was a magnificent stone college. A wooden footbridge arched out from its second floor and crossed the river to the green sward on the other side. College fellows could walk over the willow-draped water directly to the mown fields, among the daffodils, forget-me-nots and dogwoods.

On the far side of the bridge, a stone wall about twelve feet high hid most of a stone house that faced a hidden quadrangle. The plaque on the house said Darwin College. It was small and new by comparison with the colleges on the other side of the river, most of which seemed to have religious names: Christ's, Emmanuel, Jesus. This is when I finally grasped how painful it must have been for Darwin to describe to his teachers here why he believed creation could be explained by natural forces. The courage it must have taken to stand up to their belief in God's omnipotence as represented everywhere in nature, and as celebrated in this ancient city by these beautiful acts of stone, made my knees shake.

I was shown to a table and handed a pencil, and I took up a volume of Darwin's letters from the trolley. On each page lay sheets of blue paper folded in two, sometimes many sheets folded together. I was thrilled by this physical connection to a man dead for over 120 years, but thrill immediately gave way to frustration. His hand was unreadable. He had a chicken scrawl, the kind of penmanship developed by someone hurrying to get his thoughts down. I turned page after page, thinking I couldn't make out a single word. But then I picked out a phrase, and then another, and then whole sentences. His voice began to roll in my head.

On May 9, 1856, Darwin wrote to his friend Hooker regarding Hooker and Lyell's pressure to get Darwin's theory into print in a journal before he was beaten to it by someone else. They had urged him on, but he dragged his heels. He was working on something large, and he refused to send a short version of his work on species change to any journal. He was afraid of the ructions his ideas would surely raise: "I am fixed against any periodical or Journals, as I positively will not expose myself to an Editor or Council allowing publication for which they might be abused. If I publish anything, it must be a regular thesis and little column giving a sketch of my views and difficulties, but it is really dreadfully unphilosophical to give a resume without exact references . . . I am in a peck of troubles and pray forgive me for troubling you."[21]

The volume I was most interested in contained letters from June and July 1858. Alfred Wallace's paper arrived in Darwin's mail in June. Wallace was someone Darwin had written to before—in fact, he wrote to Wallace after Wallace's 1855 paper appeared. Janet Browne suggests that Darwin's letter to Wallace can be read as a warning to Wallace that Darwin was already working in this area—a polite suggestion that Wallace should back off. Wallace apparently didn't read it that way. Darwin had thought Wallace was somewhere in Malaysia; he'd asked Wallace to send him samples of Asian poultry skins.[22] But now Wallace had sent him a package postmarked that February in Ternate, an island in the Dutch East Indies. When Darwin read Wallace's essay, he had to face the fact that in spite of more than twenty years of work since he got off the *Beagle*, in spite of being a published and prize-winning author (Hooker had arranged for him to get a Royal Medal for his published work on barnacles), he had been beaten by a man with none of his advantages. Wallace had written out his ideas succinctly, arguing that species arose by a process of natural selection. He set this process in the same Malthusian context of a grim competition between all living things for life and scarce resources that Darwin had adopted. Competition led to the obliteration of the weak and the triumph of the fit, whose successful adaptations eventually marked out the population carrying them as a new species. If primacy, as Witkowski had said, is the currency of science, then at this point Wallace should have been rich and Darwin broke. But the currency of science is minted by varying means—including the application of social power.

The letters I found referring to the matter showed that Darwin had quickly passed the dreadful news of Wallace's paper on to his friends Hooker and Lyell and mentioned to them that he'd sent a letter outlining his views to Asa Gray some years earlier. The first letter I turned to in this volume also showed that something else was going in Darwin's life at this time—something horrible.

June 29, 1858.

My dearest Hooker:
Be most sorry for me when you hear this. The poor boy died yesterday evening.

This letter was a cry from Darwin's heart. His youngest son, Charles, a toddler and his favorite, had passed away, probably from scarlet fever. He described how his son's face, at first contorted in suffering, returned to placidity, the "sleep of death"—"Thank God he will never suffer more in the world." But at the bottom of the letter, he returned to the business at hand—what to do about Wallace's paper, and a possible solution: "I can get my letter to Asa Gray copied."[23]

Darwin, a man who had no faith, thanked God for relieving his child's suffering through death, using the language of Christian grief. And at the same time as he was brought low, here he was advancing his own interests. The letter to Asa Gray was his proof that he had written out his ideas in advance of Wallace and might establish his intellectual priority. Darwin lived in a world drenched in the loving sensibility of convinced Christians and, at the same time, in a much rougher, self-interested one.

That same day, he wrote to Hooker again:

June 29, 1858.

My dear Hooker:
I have just had your letter and I see you want papers at once. I am quite prostrated & can do nothing, but I send

Wallace and my abstract of abstract of letter to Asa Gray, which gives most imperfectly of the means of change and does not touch on reasons for believing species do change.

I daresay all is too late. I hardly care about it—but you are too generous to sacrifice so much time & kindness . . . I send sketch of 1844 soles [sic] that you may see by your own handwriting that you did read it.—I really cannot bear to look at it. Do not waste more time. It is miserable this to care at all about priority . . .

. . . I wrote another similar but shorter accurate sketch . . . I can write no more. I sent this and servant to you,

Yours,

C. Darwin.[24]

By this point Lyell and Hooker had cooked up a scheme to read Wallace's paper, along with an abstract of the larger piece Darwin was working on, to the Linnaean Society meeting. After that, they would publish them together in the Linnaean journal. All three men were council members of the society at that time, and Hooker virtually ran the journal; Wallace was known mainly as a supplier of specimens to its members.[25] According to Browne, they wanted to move with great speed in case Wallace had also sent this paper to others and some one else published it first, without Darwin's work alongside. But they had to wait for the grieving father to recover himself. Finally, on July 5, Darwin responded to Hooker:

But in truth, it shames me that you had lost time in a nice point of priority . . . but I am quite indifferent and place myself entirely in your and Lyell's hands.

I can easily prepare an abstract of my whole work but I can hardly see how it can be made scientific for a journal without giving facts . . . Can you give me any idea how many pages of Journal?[26]

By July 13, only fourteen days after the death of his child, Hooker and Lyell had arranged matters. Now Darwin had to inform Wallace. Hooker sent Darwin a letter to pass on to Wallace explaining how the paper had been read out at the Linnaean Society and how it would be published alongside Darwin's letter to Asa Gray. But Wallace had not asked Darwin to publish his paper; he had just asked Darwin to show it to Lyell.

> My dear Hooker:
> Your letter to Wallace seems to me perfect, quite clear and most courteous. I always thought it very probable that I might be forestalled, but I fancied that I had great enough soul not to care: but I find myself mistaken . . . I had . . . quite imagined myself and had written a letter to Wallace to give up all priority to him and I certainly [could] not have changed had it not been for Lyell's and yours quite extraordinary kindness. I assure you I feel it and shall not forget it.
>   I am much more than satisfied at what took place . . . I had thought the . . . letter . . . to Asa Gray was to be appended to Wallace's paper . . . [27]

At this point in the development of his ideas, why on earth would Darwin talk to Hooker about his soul? And how could a man with a soul of any size move in on Wallace in this way at the same time as he was burying his son?

Darwin's biographer, Janet Browne, describes him spending the next several months convincing himself that his motive and morals were pure: "Restlessly he told himself that everything had been done with the highest moral intention."[28] While pointing out that if he had tried this with any competitor of the same social rank there would have been a terrible fuss, Browne suggests that Wallace was lucky: he would not have been able to defend his ideas as well as Darwin could because of his low position within Britain's class system. Darwin had access to all the top societies and was taken seriously, while Wallace didn't and wasn't. Being published alongside Darwin in the Linnaean Society's journal in August 1858 and being championed by Lyell and Hooker elevated Wallace far beyond anything he could have managed on his own.

But I didn't entirely buy her argument. Why couldn't they have written Wallace and asked his permission to print his article along with Darwin's work? Browne calls Darwin's subsequent focused determination to write his book and publish as soon as possible the demonstration of his inner steel. It looked to me like the adaptation of his moral nature to a changed circumstance in a desperate struggle to keep his theory—and his name—alive.

Darwin published *The Origin of Species* in October 1859 and waited with trepidation for the reviews. He kept track of enemies who wrote about him. Even though reviews were written anonymously in that period, it didn't take long for people to find out who had written what. He was also very methodical in the way he sent off copies of his books to anyone who might help him, picking important figures who could sway the debate in his favor, cheering on friends to review him anonymously and, of course, favorably. Darwin's friend Huxley, for example, wrote a very favorable anonymous review of Darwin's work in *The Times*. In addition to inventing the first advance on royalties as a means of payment in book publishing, which Darwin insisted on from his publisher, Darwin also seems to have invented the first book-promotion campaign. As Browne put it, "his active intervention in the post publication process was hidden but intense. Paradoxically, the intimate process of writing personal letters, one individual speaking to another, became an integral part of his public voice, an activity that could be just as shrewd and tactical—even predatory—as any polemic dreamed up by Huxley. Without moving out of his home Darwin came to dominate through letters. Promoting the finished book became the directing theme of the life to come as completely as his earlier years had been governed by constructing the theory."[29]

It was around the time of the publication of the book that Darwin and Owen parted ways. According to Browne, Owen had come to believe that Huxley and Hooker and Darwin were combining against him to stop his proposed museum. Owen had been mild about the Wallace-Darwin essays published in the Linnaean journal. Darwin went to see Owen after sending him an advance copy of his book; Owen was pleasant on the surface but made it clear that he disagreed with Darwin's thesis and with many of his facts. Owen reviewed the book anonymously in the *Edinburgh Review* in April 1860. It was a slashing pan. Browne recounts that Hooker and Huxley were visiting Darwin at Down when the journal came out. The review took on several recently published works, including a piece of Hooker's, and described Darwin's work in the most humiliating terms. One of Darwin's examples of species change (a relationship of descent between bear and whale), which he had amended after discussions with Owen, was thrown in his face: "We look in vain for any instance of hypothetical transmutation in Lamarck so gross." Owen slyly referred to his own work favorably, intimating, as Browne would put it in her second volume, "that Professor Owen had already pondered these issues and had come to wiser, altogether more philosophical conclusions." Owen suggested that Darwin's work would be forgotten in ten years.[30]

The feud was on. Arguments about facts and ideas were washed away by a flood of ad hominem attacks. Darwin wrote to Hooker on April 18, 1860, about Owen: "what a base dog he is."[31] In July, Darwin wrote to Hooker thanking him for his kindness and affection and unflinching public support: "Talk of fame, honor, pleasure, wealth, all are lost compared with affection . . . I have read lately so many hostile reviews I was beginning to think that perhaps I was . . . wrong & Owen was right . . . But now that I hear that you & Huxley will fight publicly (which I am sure I never could do) I fully believe that our cause will in the long run prevail."[32]

What an interesting turn of phrase—"our cause." Science is supposed to be objective, disinterested. The truth of propositions and theories is to be methodically demonstrated or disproved. Darwin seemed to see his theory as something more, as a great overturning, a revolution, and his friends as comrades at the barricades. He left the public defense of his work to Hooker and Huxley and stayed at home in Down, compiling a body of work. With each book published, translated, reviewed and discussed, Darwin's fame grew, and with his increasing fame, the power of "our cause" advanced. When Hooker's autonomy at Kew was threatened, Darwin and friends organized a letter-writing campaign to protest. His friends also created the X Club, a discussion group that pushed forward members' interests while shutting out their enemies. Yet no increase in power and recognition diminished the sting of Owen's attacks. Long after Darwin's works were being read in translation around the world, Darwin was still venting his rage at Owen privately. If anything, the more famous he became, the more he hated Owen. He wrote to Hooker on August 4, 1872:

> . . . I used to be ashamed of hating him so much, but now I wish carefully [to] cherish my hatred and contempt to the last day of my life . . .
> Your affectionate friend,
> Charles Darwin[33]

By 1877 Darwin had achieved such fame that the emperor of Brazil asked to come and see him. Darwin turned him down (the pressure of work being too great) and apologized in a letter to Hooker, who'd tried to arrange the visit, on June 16; in the same letter, he also congratulated Hooker on becoming a knight of the Star of India. Eighteen years had flown by since Hooker and Lyell had managed the Wallace affair so adroitly. Lyell had long since been granted a knighthood. Huxley too was well known, well respected. All of them had risen, like linked boats on a rising tide. Wallace, on the other hand, had embarrassingly become a believer in spiritualism, the channeling of the voices of dead souls through a medium to those who wished to make contact. In the 1870s, going to séances was all the rage among people of a certain class. Darwin sent his younger cousin, Francis Galton, to investigate for him. It had to be fraud, otherwise how would a materialist explain it? And it was.

Francis Galton had made his admiration for Darwin known to him by sending Darwin a copy of his own book, on hereditary genius. He replied to Darwin's thank-you letter in December 24, 1869:

> My dear Darwin,
> It be idle to speak of the delight your letter has given me for there is no one in the world whose approbation in these matters can have the same weight as yours. Neither is there

anyone whose approbation I prize more highly on purely
personal grounds, because I always think of you in the same
way as converts from barbarism think of the teacher who
first relieved them from the intolerable burden of their superstition. I used to be wretched under the weight of the old
fashioned "argument from design" of which I felt, though was
unable to prove to myself, the worthlessness. Consequently
the appearance of your "Origin of Species" formed a real crisis
in my life: your book drove away the constraint of my old
superstition as if it had [been] a nightmare and was the first
to give me freedom of thought.

I now look forward anxiously to your final opinion after
you have quite gone through the book . . .

Believe me very sincerely,
Francis Galton.[34]

It was a cruel revelation that Darwin and Wallace had brought
into the world: individuals against each other, a blender called "sex"
to ensure variety sufficient to meet any challenge, early death to sort
wheat from chaff. Darwin's observations freed his followers from the
idea of nature as the immutable creation of a kindly, Christian God,
overseen by mankind imbued with a moral purpose. But having
thrown out the notion of a plan—and by extension the Planner—
what about that moral purpose? Darwin's mail filled with the mus-
ings of readers trying to figure out how to think and act
in the light of this new lamp. Were Christian charity and love still
proper guides for human behavior, or did they actually subvert the
driving principle of nature? Did one remake morality to fit this new
description of life? Should the focus of human concern now shift
away from the value of the individual soul to the health and pleasure
of whole populations? Should an enlightened person take up the
utilitarian principle of a previous generation of materialists? Was it
one's moral duty to mankind, to all of nature, to let the weak die
while supporting the strong?

Darwin received two letters from G. A. Gaskell at the end of
1878 dealing with these questions. Immorality did not have to
triumph, Gaskell opined. The arts might play their part in the

survival of the fittest, and good, moral parents might have more successful offspring than immoral parents who paid their children little attention. So even if the Christian view of nature was wrong, Christian morals might still have a fitness value. But Gaskell also thought that the birth of only the fit should be enforced by the state. Eugenics meets politics, I thought as I made my notes: "It may not be utopian to expect that someday a medical certificate may be required to define the rectitude of adding a new member to society. The weak in body or mind may be cared for and protected so long as they conform to the social mandate not to continue their race. They may . . . love but must not have offspring." The dilemma of what the strong should do about the weak could best be solved if the unfit were never allowed to be born: "In conclusion, I submit 'The Birth of the Fittest' offers a much milder solution of the population difficulty than the 'survival of the fittest' and the 'Destruction of the Weak.'"[35]

Darwin apparently replied. Gaskell's second letter makes it clear that while Darwin's thesis that mankind is in no way elevated above the rest of nature is true, certainly some populations of mankind are much more fit and worthy of success than others: "There is certainly great danger in lessened fertility of some races—that the pressure of other races upon them might extinguish them. The lessened fertility commences . . . in the races which are stronger socially—I trust they will endure. The nations, guided by reason, could not long submit to having their standard of comfort lowered, or their means lessened, by the influx of an inferior race . . . [T]here are some points of danger . . . as shown in regards to the recent Chinese exodus and its more useful action may some day be to preserve a civilized nation against the social encroachment of an uncivilized."[36]

Francis Galton would expand on such ideas a few years later, and pass them on to Charles Davenport and Harry Laughlin at Cold Spring Harbor. And, of course, they also found their way to Germany.

Darwin and his colleagues grew up in a place and time in which biological science was supposed to describe a sanctified creation, when moral meaning was absolutely central to biology. Their description of life as connected and mutable and arising from random events eventually prevailed, but they did not pry loose biological inquiry from its moral foundation. Darwin's tree of life was firmly planted in ancient soil. Habits of faith, habits of mind, endure. A kindly deity and the incalculable value of an individual soul are not such frail concepts that they could be routed by a few books and papers, however well promoted. If one considers one's work a moral act and someone attacks it, one can feel justified in pointing out that person's iniquities. The moral imperative and venomous relations with critics are complementary; the one calls up the other. The vaulting ambition and the passion to be known were also character-istics of Darwin and his colleagues and mark their descendants still.

But somewhere along the line, biological science also acquired a more benevolent guise, a pretty and idealistic story about itself, a Red Cross sort of story. By Maddox's day, biologists were supposed to be objective, collegial and even selfless, as if to pursue this science was also to dedicate oneself to the Greek notion of *vertu,* of something done for its own sake. Above all, biology was supposed to be a public pursuit for the common good. This idea of biological science as a public endeavor clearly arose long after Darwin's bones were set in the floor of Westminster Abbey.

CHAPTER TWELVE

DESCENDANTS

I had formed an image of Frederick Sanger as a kind of gray wizard, the anti-biologist. David Baillie had described him as a modest person, and several others had said that his tremendous contribution had been overshadowed by those who blow their own horns. Nicholas Wade, a *New York Times* reporter who had recently published a book dealing with the Human Genome Project, called *Life Script,* mentions Sanger but mainly points to his two Nobel prizes.

Wade writes that Sanger "launched the field of genomics in 1977 by sequencing the genome of a small virus 5,375 units in length. But without advanced computers, automation, and a method of amplifying DNA not invented until 1985, Sanger had been unable to take his tour de force further."[1] Wade might also have said that without Sanger, who also invented the concept of sequencing long before 1977, there would have been no insight that at the very center of human life there is a mystery: communication between unlikes using different codes. Most of all, Sanger's descendants brought forward a whole new way of doing biological science.

The story I had was that Sanger had selflessly retired from his laboratory at the age of sixty-five to make room for others. Alan Coulson, his former assistant, then head of nematode genetics at the Sanger Centre, had supplied me with Sanger's address. Unlike all the other worthies I was chasing, Sanger had no computer and didn't use e-mail. I'd given up hope of a reply when a handwritten note arrived, on blue paper. Sanger explained that he had lost my letter under the Christmas mail but had found it again. He would be pleased to talk to me.

The Wellcome Trust Genome Campus, which includes the Sanger Centre for Genome Research, is a fifteen-minute taxi ride from Cambridge, outside the village of Hinxton, in the middle of an eighteenth-century estate. There was a gatehouse with a sentry, then a large, new Frank Gehry–style compound with two buildings joined by a courtyard. These structures had wavy postmodern roofs and walls of glass that revealed several floors of offices plus an atrium. From a distance, they looked like two transparent lunch buckets dropped in a field. There were white umbrellas shading tables and a great long log set on its side between the two buildings, signifying God only knew what (I was later informed that it symbolizes the DNA molecule).

The people smoking under the umbrellas were very young and friendly. So were the two security men at the information desk. "I'm here to see Dr. Frederick Sanger," I said. One of them typed the name into a computer. He looked at his colleague and said, "Not here." "Oh," said the other man, "Fred's not in the system anymore."

Fred? Two Nobels, and he's "Fred" to security?

I was shown into a small conference room on the second floor. It had a window that stretched the whole width of the wall and provided an achingly beautiful view of a large pond with a small footbridge. Swans. Willows. A wave of green hills, capped by a thin line of dark, graceful, leafless trees. On the far side of the pond, there was a red brick house with many chimneys and mullioned windows. The sky was gray watered silk.

A tiny man sidled in. In the photos I'd downloaded from the Internet, Sanger looked like a tall, thin fellow, with straight dark hair and square black glasses. This Sanger was small and gray, a shrunken Mr. Dressup. He wore a green figured wool sweater, old and well used, ordinary work pants, dusty shoes. He held out his hand diffidently. The fingers were large and flat, with dirt ground into the weathered skin and underneath the nails—gardener's hands. In fact, he looked as if he'd spent his morning bent over bulbs in a potting shed. He couldn't have been more than five foot four, closer to Bilbo than Gandalf.

He scuttled past me and took a seat across from mine. I had been to the library and had pulled out a memorial book called *Selected Papers of Frederick Sanger* that presented his major work with his commentaries.[2] The publisher had listed his honors. Sanger was elected a fellow of the Royal Society in 1954; got the Royal Medal in 1969; got the Copley Medal in 1977—honors also given to Darwin. He was a Commander of the Order of the British Empire and a Companion of Honor, and had been awarded the Order of Merit. He was awarded the Nobel Prize for chemistry in 1958 and again in 1980. There were numerous honorary degrees as well. In his preface he confessed, "I have rarely made extensive plans for my research but have been guided to each new experiment more by its intrinsic interest and technical feasibility than as the result of any long term aim. I must admit too that my choice has often depended on which experiment would be the most fun to do. The experiments reported in these papers are only a small fraction of those that my colleagues and I did; most were complete failures, inconclusive or only preliminary."[3] I had wondered if this modesty was humbug. Now I knew that it wasn't.

Sanger started slowly and hesitantly, his voice thready, as if he wasn't used to talking, let alone about himself, and found it hard. His story paralleled Darwin's, at least through his early years. He too was the younger son of a doctor, born in 1918 in Rendcomb, a village in the Cotswold Hills. Sanger's mother, like Darwin's, was from a family that made a lot of money during the industrialization of Britain, specifically from cotton. While Sanger's father pursued the normally parsimonious life of a country doctor, the family lived well from his mother's income. Like Darwin's brother Erasmus, Sanger's older brother was a big influence. The two of them were fascinated by nature's variety, hunting down newts and snakes slithering through the countryside. Like Erasmus, Sanger's brother was also an extrovert, whereas Sanger, like Darwin, was shy. Sanger's father often took his children with him on his rounds. As a child, Sanger thought he'd also be a doctor.

But there were differences, and they were profound. Darwin came from a family of liberal materialists; Sanger's family were Quakers, Christians without a dogma. His father had been a missionary in China, and Sanger was sent to a Quaker school, and then to a private school called Bryanston, very different from the rigid, cold, dirty private school Darwin endured. "It was fairly progressive," said Sanger. "It gave a lot of freedom, which worked for me. I was prepared to work. I was keen. I wanted to do well. The turning point was, after I did the full coursework for entrance to Cambridge, I had a year to spare without serious work. I had a good chemistry teacher . . . He allowed me to play in the lab, and it was very exciting—use your hands and make multicolored crystals. I enjoyed doing things with my hands."

He arrived at Cambridge in 1936. He'd seen how hard his father worked, how every day there was another problem, how the problems in a doctor's practice are not always truly solved. He didn't want to be a doctor, but he wanted to do something that could help. "I wanted a project; I thought it would be more interesting as a scientist. I was thinking of chemistry rather than doctoring, and when I got to Cambridge, I decided on science. I had to have three subjects." One on offer was biochemistry, which he'd never heard of before. "I was very excited. It endeavored to explain biology in chemical terms, as an exact science—hopefully useful in medicine. So I signed on."

At this point the genetics, biochemistry and biology disciplines were still sharply separated from each other. "In Cambridge there was a department of genetics and one for biochemistry and one for chemistry," said Sanger. "We in biochemistry had very little contact with genetics. The language was different. At that time, if I were asked what I thought genes were made of, I'd have said made of protein. I'd have said it well into the '40s, before it was established that DNA was important."

His doctorate was "just a simple chemical problem—well, not so simple," he amended. Sanger was assigned by his supervisor, Albert Neuberger, to work on the metabolism of the amino acid lysine. It was Neuberger who taught him how to do research, which is quite different from doing the experiments laid out for a student by a professor to demonstrate various already-proven theories. A researcher has to invent his own experiment, something that will answer a question in a methodical and repeatable way. "When you start, you haven't a clue. All previous experiments are planned for you. When you do research, very often it doesn't work. You've got to learn to live with it," said Sanger.

Science had grown up some since Darwin's day. By the time Sanger earned his doctorate, people could spend their professional lives working in private laboratories for industry or teaching in institutions like Cambridge. Government support for research grew with the war effort. But labs often were run as hierarchies. Many lab directors put their names on their underlings' papers. It was difficult to get a name, to be known for one's own work by peers, unless the director was generous and allowed underlings to publish on their own. "I was very lucky in that way," said Sanger. On the other hand, the age of the English gentleman of science wasn't finished yet. (English gentlemen of science were still at the Laboratory of Molecular Biology when Baillie got there in the early 1970s. He recalled a man who arrived to do his doctorate in a chauffeur-driven Rolls-Royce with his own privately paid-for technician in tow.) Sanger lived from his family money; there wasn't any public money to support him.

In 1943, the year Sanger got his doctorate, A. C. Chibnall became head of Cambridge's biochemistry department, and Sanger worked in his lab. World War II was in full howl. Sanger had not joined the service because of his beliefs. "I think I did believe in God at that stage," he said. "I think my brother and I accepted the philosophy. One of the important things was pacifists wouldn't take lives." During World War II—a war in which Britain's existence hung in the balance, a just war—pacifists were hard for others to understand. To be a conscientious objector was to walk a lonely road.

"It was unpleasant," said Sanger. "On the whole it worked out. A lot of my friends were Quakers. My first school was a Quaker school. You wondered a bit about it, but accepted that it is wrong to kill people and that's it. People in the lab were worried. A few thought we were getting out easily. Anyhow, science students didn't have to go. They could finish the course, and might be asked to do war work. Neuberger thought we should work on different potatoes, feeding the country, the nitrogen content of potatoes. It didn't excite me."

But Chibnall presented him with the problem that carried him to his first Nobel Prize: deciphering the structure of the protein insulin, a substance vital for life. When Sanger started, not much was known about proteins in general, or this one in particular. There were various theories. It was usually assumed that each protein has a unique sequence of amino acids organized in a chain, but no one knew if that was true. Chibnall and his colleagues had found, through chemical methods, that insulin has a relatively high proportion of free amino acid groups, apparently at the ends of the insulin molecule. He wanted to know what those amino acid groups were. Sanger came up with a reagent that would react with them.

This was where Sanger's genius lay, in his capacity to think and fiddle and play, and finally to come up with some method that no one had thought of before, in this case a chemical reagent that could form bonds with free amino acid groups. After breaking down the protein chains with acid "you could isolate the amino acids coupled to it. One had to fractionate them."

Fractionation means the separation of large, complex organic molecules into their smaller constituents. In that period, the standard or classical fractionation technique was by crystallization. A new method known as partition chromatography had just been developed by chemists working in wool research in Leeds. Sanger's reagent-bound amino acids, identified by color, could easily be seen in the partition columns.

Partition chromatography was improved by paper chromatography, and again by electrophoresis on paper, a method of using electrical charges to separate larger from smaller groups of molecules.

Other researchers were using X-ray crystallography, used later by Rosalind Franklin and Maurice Wilkins to produce an image of a crystallized DNA molecule. Max Perutz would use it at the Cavendish to learn the structure of the enormous hemoglobin molecule, and Linus Pauling was trying it in the United States. But Sanger wondered about its usefulness. "They thought they might get the complete structure just by the X-ray crystallography. It was a difficult technique. It was more physics . . . It was mathematical. I'd go to lectures by Perutz and not figure out what was happening. I was not good at relating mathematics to pure physical phenomena. I did physics as a subject in university. I gave it up after a year—I got completely lost."

Sanger found the two amino acid end groups of insulin—glycine and phenylalanine—with his reagent. The critical thing was that these derivatives formed by the reagent, DNP, had to be stable, but the DNP-glycine derivative was not stable, so the time of hydrolysis had to be reduced. At the end of this second reduction process, Sanger got a new product—what amounted to a sequence of two amino acids. "It was the first sequence ever determined in a protein, really," he said. "That was 1945."

He went on to create other methods of fractionation, to get other amino acid sequences out of insulin, employing various new technologies of separation. "It sounds simple," he said, "but a lot of the time it didn't work." By 1954 he had a complete description of the arrangement of the amino acids along the insulin molecule. In the process, he confirmed one of the theories about the organization of proteins: he demonstrated that "proteins are definite chemical substances possessing a unique structure in which each position in the chain is occupied by one and only one amino acid residue . . . They seem to be put together in a random order, but nevertheless a unique and most significant order."[4] Sanger first sequenced insulin taken from cattle, then he compared that sequence to insulin taken from pigs, sheep, horses and whales. He found that they were almost identical.

As a result of this work, the whole molecule could be synthesized in a laboratory. "We didn't do the synthesis," he said, "but others did much later . . . Knowing its structure helped to find out how it worked. The main thing was, it showed it was possible to work on proteins." And what he'd done in a protein, he hoped he could also do in a nucleic acid.

Sanger knew by the 1950s that nucleic acids were important. He knew that DNA was probably the chemical that conveyed genetic information, because of Heinz Fraenkel-Conrat's work with tobacco mosaic virus, which "showed a virus could get infection from nucleic acid when there was no protein there. No one believed him," said Sanger. "I think people were rather reluctant to accept [that] a substance with four simple compounds could carry a lot of information and be so important, whereas proteins, with twenty compounds, were very active biologically. People thought they were more likely to be the things that carried the genes. How wrong they were!"

None of the nucleic acids were small enough to study easily. He tried small RNAs first. "When we started on RNA, the smallest were transfer RNAs," which are about seventy-five base pairs long. The first problem was to actually get pure tRNA separated from all the other molecules shooting around in a cell. "We didn't succeed at that," said Sanger, who also referred to this frustrating period as his "dry patch." Bob Halley managed to purify the tRNA for the amino acid alanine. "He used the same protein techniques to sequence it. That was the first nucleic acid to be sequenced. I think that was 1960."

Sanger modestly failed to mention that during this period he won the Nobel Prize for chemistry, awarded in 1958 for the sequencing of insulin. He was forty years old. The Nobel wrought changes in his life. "With most people when they get them," he said, "they get a big administrative job. For me, it meant I had a steady job and I was free to carry on working." He had married in 1940 and had three children. From 1944 until 1951, he had no job but was supported by a Beit Memorial Fellowship. He was finally hired in an actual job by the Medical Research Council in 1951. When Sanger got the Nobel, his position was secure. In 1962, he joined Perutz, Crick, Brenner and their colleagues in the MRC's newly created Laboratory of Molecular Biology.

In spite of the fact that he had mostly supported himself, the idea that biological science is a public endeavor, rather than a private pleasure to satisfy private curiosity, was firmly in Sanger's mind by the time he gave his first Nobel lecture in 1958. Social historian Susan Wright has argued that this idea of biology in the service of the public good arose after World War II with the new primacy of government science funding in the United States, primarily from the NIH:

> Work sponsored by the National Institutes of Health was pursued with the general goal of understanding and controlling disease, and a major assumption underlying this support was that progress in basic research was a necessary condition for progress in medicine. A duality of purpose thus characterized

molecular biology in the United States: the research itself was aimed at basic problems in biology, but the support for it was justified in terms of solving problems in medicine. Although most molecular biologists at this point probably saw their work primarily in terms of the first purpose, the ultimate rationale was practical.[5]

In Britain, the Medical Research Council had begun to play a similar role to the NIH. It provided jobs and grants, as did Canada's MRC. Top scientists in most Western nations worked in government institutions or in public universities and competed for growing pools of public money. Peer-reviewed grants and publication in peer-reviewed journals available to anyone became the measure of a scientist's worth. The Darwinian desire for personal profit and fame was not supposed to drive scientific curiosity. There was, instead, a new contract, which shaped a new ideal: since taxpayers put up the money, scientists' results must be available for the betterment of all. Scientists must seek knowledge not for themselves, or even for its own sake, but for the greater good.

Sanger's Nobel lecture on insulin clearly articulates this view: "These studies are aimed at determining the exact chemical structure of the many proteins that go to make up living matter and hence at understanding how these proteins perform their specific functions on which the processes of life depend," Sanger wrote. "One may also hope that studies on proteins may reveal changes that take place in diseases and that our efforts may be of more practical use to humanity."[6]

The Nobel Prize was very helpful to Sanger when it came to asking for grant money. His alternatives after the war had been to teach, to work in industry or to support his own work. He didn't want to be a big businessman, as he put it. And teaching was also out. "I did do a few lectures on biochemistry and protein. I wasn't keen on it. It takes a lot of preparation. I wasn't knowledgeable on biochemistry." Sanger didn't read the journals to keep up with the whole pattern of discovery in his field, which a university lecturer must do. He was interested mainly in solving problems, and almost no one else in the world was working on his problems. "I always had somebody in the lab keen on keeping up with the literature."

After his disappointment with tRNA, Sanger moved on to riboso-mal RNA, which is about 120 base pairs long. His colleague, G. G. Brownlee, worked out methods to break this RNA down into frag-ments, then he put the sequences of the fragments together into a whole sequence. And then Sanger decided to try the same thing in DNA. "We understood it was the critical thing we are made of. There were surprisingly few people working on it."

There was no problem about getting grants, because he had the Nobel, and this work was cheap. He used the tools of chemistry, which were much less expensive than X-ray crystallography. All that was asked of him by the MRC was that he make a report of his efforts every three years. "There was not a lot of paperwork, as peo-ple complain of nowadays." And of course he was expected to pub-lish articles as he refined his techniques and made discoveries. This was the kind of support—without a direct commission, without a deadline, without an intrusive and prying patron—that artists can only dream of.

At the Laboratory of Molecular Biology, Sanger also had ami-able collaborators, usually about seven doctoral candidates and postdocs. He learned from them as they learned from him. Most of them were Americans.

Sanger had only tiny quantities of DNA to work with; Kary Mullis had not yet invented the polymerase chain reaction, a method for splitting DNA strands in two and getting them to copy themselves exponentially. He had heard of one enzyme that cuts up RNA into fragments that end with guanine, but "there was nothing like that in DNA. We messed about a lot with fractiona-tion techniques."

As he remembered it, the big breakthrough was with the devel-opment of suitable enzymatic techniques. "First one I used was from a chap called Englund. He had an enzyme. . . a polymerase. [T_4-DNA polymerase]."

He fell silent for a moment and put his hands in front of his glasses. "The memory is getting terrible," he said. "An enzyme that was a polymerase but also an exonuclease. It cuts down the chain from the 3' end.[7] If you had one nucleotide triphosphate present, say TTP, the enzyme would degrade the chain and stop at that particular residue—

and keep putting back triphosphate. If you do it long enough, all fragments will end in T."

He also needed a system that would separate these fragments by size and lay them out side by side so he could see which was which. He could then get fragments ending in each one of the four bases and, in effect, read a sequence from the way those fragments lined up. With enough redundancy of fragments, laid out end to end, he could eventually learn the total sequence of a short strand of DNA. "That was the breakthrough. The problem was to get a fractionation system that would do it exactly according to size. We spent a lot of time working on that."

Eventually, he and his colleagues hit upon using a slab of acrylamide gel, which could carry a small electric current, to separate the fragments by size. The bigger the fragment, the more slowly it would crawl along the gel in the direction of the current. Tiny fragments would move right to the top of the gel; larger ones would stall farther down. This method is still used in some labs, although contemporary sequencing machines now use a capillary system to channel fragments of different size. Initially, in order to label these fragments so he could see them on the gel slab, Sanger tagged them with radioactive phosphorus and then exposed an X-ray film. But that was time consuming. Soon it was replaced with something simpler. "A chap called Leroy Hood developed the labeling with fluorescent dyes," said Sanger. He called this system the "plus and minus" method. "We got sequences with it." The first DNA fragment sequenced was about 50 base pairs long. The first whole genome he sequenced—the first complete DNA molecule—was that of the bacteriophage virus phi-X174. It was 5,386 nucleotides long.

Sanger and his colleagues didn't just work out the virus's sequence; they also determined how it coded for the ten genes already mapped out by geneticists. Through the use of mutant forms of the virus, brought to the lab by Clyde Hutchison, a postdoc from the United States, Sanger's lab found that at least some of those sequences coded for more than one protein. Hutchison had brought a strain of the virus with an altered gene called *E*. "By comparison of sequence data on the normal and mutant DNA, it was found that the change in the mutant was in the region coding for gene *D*, demonstrating that one DNA sequence was coding for two protein sequences." This demonstration that genes might overlap had been previously dismissed as unlikely.

Sanger's name doesn't appear on that paper[8] because he didn't work on it directly and, like his own teachers, he did not get in the way of his students' achieving their own fame. He and Alan Coulson were involved mainly in perfecting a simpler sequencing method for DNA. They called it the "dideoxy method," and today it is used everywhere. "It was a way of getting fragments ending in the same nucleotide," Sanger explained. It involved putting a special chemical, dideoxy triphosphate, into a mixture of DNA to be sequenced. This chemical would be incorporated into DNA chains as they were being copied, but it would stop the chain from extending any farther once it was incorporated. All the fragments would end, for example, with the thymine nucleotide. "Once it was incorporated, you couldn't add more. If you have a dideoxy thymine and a normal thymine, where there was a normal T in the sequence, the dideoxy T would stop and you'd end with fragments all ending in T."

This was a method that combined fractionation and labeling. It made DNA fragments with known endings. Arthur Kornberg, a biochemist at Stanford, had already shown what dideoxy triphosphate could do, but when Sanger asked Kornberg for some, he had none left. Then, at a meeting, Sanger ran into a man named Klaus Geider, who had made a small quantity of this ddTTP, as it is abbreviated. Sanger found that it produced a clear pattern of fragments on his acrylamide gels. "It gave beautiful patterns," he said. But to make this method work, he had to have three other variants that would stop assembly at guanine, cytosine or adenine. It took him and his associate Alan Coulson almost a year to do it. Why? "We hadn't done nucleotide chemistry before," he said.

At first I couldn't understand what on earth he was saying: what was all this sequencing if not such chemistry? But he meant he'd spent most of his time learning about molecular structure by prising big molecules apart. Now he and Coulson had to synthesize molecules that would, when mixed with DNA and a polymerase, stop the copying of a DNA strand at the right points.

"This was synthetic chemistry," he said. "It's very different, isn't it? I contacted someone in an industry lab who said they'd make them. We waited a year. I met him at a meeting and he said, oh, we decided not to bother. So Alan and I had to get down to it."

Once they'd managed to make the three other variants of dideoxy, they decided to try to sequence human mitochondrial DNA. Meanwhile, another group in the United States, led by Walter Gilbert, had invented its own DNA sequencing method. In 1980, Sanger and Gilbert and Paul Berg shared the Nobel Prize for chemistry; Sanger and Gilbert each got a quarter of the prize for their sequencing methods. In Sanger's commentary volume, there is just a hint of irritation as he describes Gilbert's work. "The method was first described in a lecture at a Gordon Conference by Gilbert, and protocols of the techniques were circulated, but it was not published till 1977. It is interesting that although they used many of our techniques they had carefully avoided giving any reference to our work."[9]

Even with the new dideoxy method, Sanger needed one more technique to take on mitochondrial DNA. Phi-X174, the virus he had sequenced, contained single-stranded DNA, but human mitochondrial DNA is made of two strands. He decided to try the techniques of cloning, developed in California. First, fragments of double-stranded DNA were spliced into a single-stranded virus. "It was a lot of work," said Sanger. "One problem was you had to cut it up to get reasonable pieces. You had to start synthesis at a particular point to get a result. You needed a pure product to catalyze the initial reaction. We used restriction enzymes to get the large fragments. They had to be fractionated and purified. Then you have to extend those fragments with the dideoxy method. With bigger products, fractionation gets increasingly difficult. These were big pieces and little ones, and there were not terribly good methods to deal with the larger fragments." The answer was cloning by way of a circular virus. "You can split it and insert what you want. Then you infect bacteria and clone the fragments to isolate them. This is an unlimited fractionation technique. . . We didn't realize the possibility of cloning until we started to work on it." In the end, he'd get fragments of single-strand DNA starting at a known point, ending in one of the four nucleotides, running in channels up a gel, displaying a different color. The sequences could be read off just by looking at the gel.

But why did you try human mitochondrial DNA? I asked.

"It was an important and interesting molecule and John Donalson suggested it," Sanger said. "This was the next size up, about 16,000 base pairs. We just thought it was the right length."

By the time Sanger moved to sequencing whole DNA molecules, the old barriers between disciplines—so real when he had started—had begun to break down. Lots of his postdocs were geneticists who knew how to handle microorganisms, how to splice DNA into a virus and drop the virus into a bacteria and make unlimited copies of these hybrids. Some were biochemists, such as Elizabeth Blackburn, who wanted to know how to sequence the ends of chromosomes. They came to learn, but they also taught.

"You had to practice with these techniques. They were brought by people who came to the lab. Phi-X was difficult to handle. I couldn't have done it myself. John Sedat came from Robert Sinsheimer's lab, where it was developed. That's why we studied it."

Sequencing human mtDNA was a big jump up in complexity, but it also provided vital information for geneticists, and for evolutionary theorists. There was a theory when Sanger began that mtDNA descended from organisms without a nucleus—that an ancient, primitive cell had formed a symbiotic relationship with a more complex cell and "gradually lost most of its genes."[10] Sanger's group discovered something completely unexpected about human mitochondrial DNA. Like so many discoveries, this one happened almost by accident. They had begun to store the sequences of their DNA fragments on a computer when they worked on phi-X174. They had made mistakes when recording sequences by hand, and they thought a computer would keep them accurate. "Bart Barrell was the guy who collected the data. He was my technician, then a PhD working here. He used to correct the data," said Sanger. It was Barrell who noticed the overlapping genes in phi-X174. Now he also noticed that there was one sequence in mtDNA that coded for a protein, but that the same sequence was a terminator in nuclear DNA. "Thus mitochondria has a different genetic code. It was a bit of a shock," said Sanger.

Until that point, it was believed that all organisms use the same code, that a sequence means the same thing when spelled out in nuclear DNA base pairs as it does in mtDNA base pairs. More experiments and papers were done, in which this other code was worked out. Sanger's lab showed that "mitochondria are unique in many ways." The mitochondrial DNA code was primitive, it was likely older than the nuclear DNA system, and it was thoroughly different. "It suggested that mtDNA must have been—that it had a life of its own," Sanger said, "separate from the rest."

They also compared human mtDNA to that of cattle and yeast being studied in another lab and found differences and similarities.[11] So Sanger's lab also began the practice of comparisons across species. When they were finished, the group had both the entire sequence of 16,569 nucleotides of human mtDNA and a new insight into evolution: it was no longer a simple story of descent from a single ancestor, a single tree.

"This was the first human DNA sequence to be determined and can be regarded as the start of the human genome project," Sanger had explained in his commentaries. This was a terse sort of truth.[12] It was also the start of bioinformatics, the use of computers to spot patterns in DNA sequences, to predict which might be genes and their protein products—and, as I would later come to understand, a turning point in biology. "We started to store stuff in computers," Sanger said, "at the end of phi-X work. Then we got a chap from the X-ray department who knew computers. He worked out things for us; his name was Roger Staden. That would be, oh, late 1970s, wouldn't it? I don't think I used a computer for analysis. Just for storing . . . Then I think we probably had a program for translating the DNA into protein sequences. We wanted to know what it was doing."

"One wondered if the whole string is functional," he continued. "I think I'd admit we don't know, really. It is a question, the way it's built up. If you think of the evolution of the thing, it's all done by chance—you're bound to get junk . . . I never presumed to say it was junk. If you think of evolution, I don't see a method to get rid of stuff that's not used. It's easier to add than to take it away . . . I still think it's possible to have rubbish in there, still possible, I think. The whole system is more complicated than one would think necessary.

We find all sorts of proteins and other proteins that influence them. It's a wonder it works at all, really."

While they were waiting to get started on mtDNA (they needed to put in containment facilities), Sanger and Coulson also began to sequence a bacteriophage well studied by geneticists— lambda. Its DNA is about 48,000 base pairs in length, much bigger than human mtDNA. They tried their cloning procedures on lambda first. They used what is called the "whole-genome shotgun method": they chopped up the genome with enzymes over and over again, made a great many copies of the fragments by cloning, and sequenced them; then they tried to reconstruct the whole sequence by dropping out all the redundancies among the sequences. It's a little like putting together a jigsaw puzzle when many of the fragments are similar. "In this way," Sanger wrote, "the work goes very rapidly at first, but as it proceeds more and more of the results are redundant and new data are obtained more slowly. Eventually it becomes necessary to use specific methods for obtaining the missing pieces, and they are much slower than the random approach."[13]

Before Sanger retired in 1983, he and his technician Alan Coulson, and their colleagues, finished the sequence of lambda. There was already talk then of a project to sequence the entire 3 billion base-pair genome of human beings. Sanger heard that Robert Sinsheimer, chancellor of the University of California at Santa Cruz, had applied for a grant of $1 million to get it started but had been turned down. "The next thing was when Alan Coulson and John Sulston started to do the nematode on an MRC grant. And then this place was built."

He looked around him, and rubbed his thick fingers across the plain surface of the conference table, as if he was surprised to find himself in such a room, surprised that so much money had been made available to carry on his work. He had once huddled with his postdocs in a lab not much bigger than Coulson's small office down the hall from this room. But he couldn't tell me how all this money was put together, or why the Wellcome Trust had built this place and taken on the role of funding one-third of the Human Genome Project.

"I think they suddenly got a lot of money and didn't know what to do with it," he said. "Talk to Alan and John . . . I just visit occasionally. I'm a gardener now."

"What do you grow?"

"Flowers," he said.

"Don't you miss your work?" I asked. I was thinking of Watson, still in the thick of things although he was in his mid-seventies. I no longer expected Sanger to say something self-effacing; though he never put his name on papers he hadn't taken a hand in, he was not averse to swiping at a rival in print, and he was a practical man. Even Sanger was Darwin's descendant.

"I wouldn't be able to do any research now," he said simply. "The memory goes. The last year or two, there were several people working. I'd talk to them, and a week later I'd be asking them the same questions."

Nicholas Wade's *Life Script* describes the history of the Human Genome Project from an American point of view. For Wade, it began under the direction of two U.S. government agencies: the Department of Energy, first, and then the NIH. It was carried forward through grants to researchers at places like Cold Spring Harbor Laboratory, Washington University in St. Louis, and in Britain at the Laboratory for Molecular Biology, with money from the Medical Research Council and then from the Wellcome Trust. Wade doesn't spend too much of his time describing the Wellcome Trust other than to give a brief sketch and describe it as the largest medical charity in the world.

According to the Trust's website, its wealth derives from the death and will of Sir Henry Wellcome in 1936: he founded the pharmaceutical company Wellcome Foundation, later called Wellcome PLC and then various other names subsequent to various mergers. The Trust owned Wellcome's shares. When the company was still called Burroughs Wellcome, it developed AZT, the first drug effective against HIV/AIDS. In 1992, when the company's shares were at their peak value, the governors of the "Wellcome will," as this charity was called, sold a lot of shares and

272 • ELAINE DEWAR

made a fortune. Suddenly they had £200 million to spend every year.[14] By 2000, the Trust managed about £15 billion. It was re-organized under the authority of the Charity Commission for England and Wales. At first at the request of the Medical Research Council and then for its own reasons, the Wellcome Trust gave money for sequencing, but it soon displaced the MRC as Britain's primary sequencing funder.[15]

James D. Watson, the NIH's Human Genome Project's first director, had helped make the project attractive by selling it in the usual way—as a moral endeavor, as a means to learn how to cure dreadful diseases. This was important to emphasize because Walter Gilbert had estimated the cost of the project at $1 per base pair, and therefore about $3 billion.[16] As other countries supported sequencing projects of their own, the U.S. Congress also came to see this kind of science as vital to the national economic interest.[17] Watson tried hard to ensure that other countries were involved so no one could say the Americans were trying to buy control of the human patrimony. Money flowed from governments through grants to academics' labs in universities and charities that became universities, such as Watson's own Cold Spring Harbor Laboratory. American money also flowed across borders to the U.K., where it helped support a pilot project—the sequencing of the genome of the worm *C. elegans*.

But there were those who thought the people in charge were running the Human Genome Project as a closed shop, favoring their friends, shutting out others—that taxpayers' dollars for genome sequencing was money lost to more important biology.[18] One such shut-out was J. Craig Venter, then a government researcher working for the National Institutes of Health. In the early 1990s, Venter was using expressed sequence tag (EST) technology to fish for genes; he could infer what a gene fragment's sequence must have been from the sequences of RNAs found in cells. He used these RNAs as templates for complementary DNA, and sequenced the ends of these cDNAs, as they are called, to make unique tags for each gene. These EST tags could then be compared to sequences already known, in public databases.

Venter announced at a Senate briefing that the U.S. government would soon file patent applications for these ESTs, and for the genes that gave rise to them, and for the proteins they made. Venter also applied for a Human Genome Project grant to use this EST method to find human genes. He thought this would be faster and more useful than trying to sequence the whole genome. He also began to use this method to find genes of the worm.

Watson and others were appalled that the NIH might try to patent these tags and the genes they referred to. Venter's Human Genome Project proposal was rejected by the peer review committee dishing out money for the project. John Sulston, the former assistant to Sydney Brenner and the British leader of the worm pilot project, at this juncture took note of Venter as a dangerous person, a direct competitor—the kind who didn't ask for permission to enter one's own marked-out intellectual terrain.[19] The worm's genome sequence belonged to Brenner's disciples, surely, not to an interloper like Venter.

Soon all guns were blazing at Venter. Watson said that filing patents on gene pieces, whose function was unknown, was sheer lunacy. Patents are supposed to reveal something novel that can be used to do something. If everyone filed for patents, the Human Genome Project would fall apart. Watson also butted heads in public with Bernadine Healy, then the director of the NIH and his boss as director of the NIH's Human Genome Project, when Healy argued that the NIH was merely keeping its options open by filing for patents.[20]

At around this time, Healy was approached by Frederick Bourke, a businessman interested in starting his own genome sequencing venture. Bourke complained that Healy's employee, Watson, was interfering with American business interests—his— and that Watson himself had conflicts of interest. Bourke had tried to hire worm sequencers Bob Waterston and John Sulston for his own venture. When Watson had heard about it, he'd fought with Bourke and warned the MRC that Sulston, their employee, was being wooed away.[21] Watson didn't want his prize sequencers plucked by private business. His prize sequencers, on the other hand, saw Bourke's offer as a whip they could use to secure their own shaky public funding.

Healy paid heed to Bourke and investigated Watson's business relationships. The *New York Times* later named the companies

Watson was involved with as Amgen Inc., Glaxo Inc., Eli Lilly and Company, Oncogen, and Merck & Co. There was a suggestion of self-dealing, too, since some Human Genome Project grants were given to Cold Spring Harbor Laboratory, which Watson still ran.[22] No one directly accused Watson of wrongdoing, but he resigned in fury in 1992. Francis Collins, a renowned researcher and a born-again Christian,[23] eventually took over Watson's position, but not before Watson had explained to the MRC how important it was for Britain to support its own end of the sequencing project.

The MRC quickly asked the Wellcome Trust, suddenly awash in cash, to help fund the U.K. side of the worm genome venture. The Wellcome Trust saw for itself a more active role than simply supporting the MRC's worm project: it wanted direct involvement with human genome sequencing. The MRC then had to find more money for Sulston or lose its position as lead sequence funder in the U.K. It found the money, but by 1993 the Trust had bought this property at Hinxton, convinced the European Bioinformatics Institute (one of the three data centers for public deposit of sequences) to move in and placed John Sulston in command of their own genome sequencing facility. Sulston had tried to get the Trust to put up one-third of the money for the whole Human Genome Project; the Trust settled on one-sixth. By the mid-1990s, Sulston commanded hundreds of millions of pounds of charitable funds.[24]

Both the MRC and the Trust were no doubt spurred to these actions by Craig Venter, as well as by Frederick Bourke. In 1992 Venter announced the creation of a nonprofit called the Institute for Genomic Research, or TIGR, to take a private run at genome sequencing. His backer, an entrepreneur named Wallace Steinberg, started a company called Human Genome Sciences to put money into TIGR in exchange for right of first refusal on its discoveries. In other words, TIGR, a nonprofit, was funded by a business that would exploit its work. Venter also received shares in the commercial company and made many millions when Human Genome Sciences licensed the right to a first look at TIGR's findings to SmithKline Beecham for $125 million. Venter retained the freedom to publish in academic journals, but only after the company had spent six months to a year examining his findings. John Sulston later characterized Venter as follows: "I felt he wanted to have it both ways: to achieve recognition and acclaim from his peers for his scientific work, but also to accommodate the needs of his business partners for secrecy, and to enjoy the resulting profits. This apparent determination to have his cake and eat it set a pattern for what was to follow, when Craig launched a privately funded effort to sequence the entire human genome."[25]

In 1995, not very long after switching to the whole-genome shot-gun sequencing method at the suggestion of Hamilton Smith, Venter announced TIGR had successfully sequenced the whole genome of *Haemophilus influenzae*. This success was trumpeted as being a demonstration of the efficiency of the private realm over the bureau-cratic, inefficient public consortium that was still bogged down with the much larger *E. coli*. Venter had applied for a public grant for this project too; the reviewers had turned him down saying the whole genome shotgun method he proposed was too unlikely to succeed.

What Venter could do, so could others. Sulston and his colleagues grew afraid that their fellow researchers would strike deals with com-panies and the human sequence data they'd discovered would end up privately owned. They organized a conference in Bermuda in February 1996 to hammer out a deal on how human sequence data were to be treated. Funders and scientists doing sequencing attended. Sulston and Waterston, who had been working together on the worm sequence, released their raw sequences right away. They wanted to do the same thing with the human sequence data—to make them avail-able for anyone to use as they saw fit. Immediate publication would make patenting impossible. Sulston said that anyone who wanted to collaborate with the Sanger Centre would find their work posted in public. Michael Morgan of the Wellcome Trust made the same case to the other funders, and everyone at the meeting—even Craig Venter[26]—agreed that raw sequence data were to be published, prefer-ably on a daily basis, electronically, by everyone, and that this was going to be a condition for human genome funding by public entities as well as by private charity—the Wellcome Trust.

In 1997, after Steinberg died, Venter and TIGR parted company with Human Genome Sciences.[27] A few months later, Venter was approached by Michael Hunkapiller, whose company, Applied Biosystems, had developed the first automated DNA sequence machine. Only 3 percent of human DNA had by then been sequenced by the consortium. They had focused first on smaller organisms, on the theory that practice makes perfect and that they could get that $1 a base pair cost down. Hunkapiller thought his machines would reduce costs and increase speed radically. His capillary machines could separate fragments more accurately than gels and could run twenty-four hours a day with little technical support. He and Venter calculated they could do the whole 3 billion base pair human sequence by the year 2001 with a few hundred of these new machines. They could beat the public consortium by four years.

Venter and Hunkapiller, with the backing of Hunkapiller's parent company, PerkinElmer, formed Celera Inc. Their business plan included making money from selling access to databases of their discovered sequences and software for recognizing unknown genes. They said they intended to patent only a few hundred genes and that they would make their sequences available to academic researchers free—but on a quarterly basis. The company's plans were announced just before a meeting on the Human Genome Project at Cold Spring Harbor in May 1998. A favorable article appeared in the *New York Times,* by Nicholas Wade.

Venter also approached Francis Collins, Watson's successor, and Harold Varmus, Healy's successor as director of the NIH. He suggested the consortium should join forces with Celera. Venter attended the Cold Spring Harbor meeting, where he told his colleagues that Celera would take on the human genome while the public consortium should work on the mouse. Collins and Varmus were apparently willing to cooperate with this proposal, but when Sulston heard about it in the U.K., and Jim Watson, who still kept his hand in, was informed, they hated the idea. They thought that without huge amounts of duplication, Venter's shotgun method wouldn't work. Worse, Venter would not publish his material daily as the consortium was doing, but quarterly, so who could say what he might try to patent? And worst, Venter would be able to mount an argument in Congress to cut off their money. He'd told the *New York Times* that his draft human sequence would cost one-tenth of what the public consortium was spending. At the same meeting, Venter invited Gerry Rubin, an academic who headed the consortium's *Drosophila* sequencing effort, to work with Celera. Venter promised that Celera would publish raw *Drosophila* sequence data daily; Rubin agreed to join forces with him.

Under the pressure of Venter's announcement, the Wellcome Trust agreed to fund one-third of the Human Genome Project to make sure it stayed in the public domain. The Trust fired off a press release. Then Sulston flew to the meeting in Cold Spring Harbor with an executive of the Trust, Michael Morgan.[28] Morgan told the meeting that it was plain stupid to leave such vital information in private hands.

From that point forward, the two groups were like two male dogs snarling over territory. Venter argued that the shotgun method was the way to go, that he would produce an early draft, not the painstakingly complete sequence the consortium wanted to produce, and that a draft would be useful. His opponents argued that they were interested not in drafts, but in an accurate finished sequence that would stand the test of time, something the shotgun method alone would not provide. There was a battle within the consortium about whether to ramp up and meet Venter's challenge or to ignore him and continue as before. "I could see no other interpretation than Craig was aiming to gain total control of the information contained in the genome for commercial gain," Sulston later wrote.[29] Watson was so apoplectic about Venter's proposal that, according to Sulston, he'd compared it to "Hitler's invasion of Poland."

It was more like the cold war: what one side did, the other had to mimic. Soon both sides ramped up their efforts. And in the interest of speed and efficiency, smaller sequencing labs were shut out of public grants, while MIT's Whitehead Institute for Biomedical Research, under Eric Lander, put in a forest of the new sequencing machines and took on a third of the work. The rest was mainly divided between Washington University's sequencing center, Baylor College of Medicine in Houston, the Department of Energy's Walnut Creek Joint Genome Institute and the Sanger Centre.

In April 2000,[30] a year earlier than planned, Celera announced
that it had a draft of the human genome sequence; its share price ran
way up. Prime Minister Tony Blair and President Bill Clinton jointly
announced that the sequence information would be available to all
scientists; Celera's share price fell. Negotiations between the two
camps, which had started and then failed, began again. On June 26,
2000, five years ahead of the original schedule, success in sequencing the human genome was announced. American leaders of both
public and private teams, plus James D. Watson, arrived at the
White House. The British contingent, including Fred Sanger, arrived
at 10 Downing Street. Demonstrating the international, public-
private nature of the effort, President Clinton and Prime Minister
Blair made grandiose speeches claiming that both teams had
independently arrived at the first draft of the Book of Life at the
same time. The international effort in fundamental science
envisioned at the beginning had devolved to an Anglo-American
public-private race to a dead-heat finish.

The announcement was a public relations fraud: the date of
this celebration had been picked because it was a day when both
Clinton and Blair had free time.[31] The drafts of sequences strung
together by the consortium and Celera were still full of holes and
question marks. And they hadn't finished at the same time:
Celera's draft was less complete than the consortium's and came
together later. The consortium's sequence wasn't assembled by the
consortium, but by a graduate student at the University of
California at Santa Cruz, who, as Nicholas Wade has recounted,
spent so many hours frantically writing computer code he had to
ice his wrists.[32] The leaders of the consortium argued that Venter's
private team had piggybacked on their publicly available data. An
agreement that both teams should publish jointly in *Science* and
Nature collapsed. The consortium published with *Nature,* Celera
with *Science.*

In a *j'accuse* published in the spring of 2002 in the *Proceedings of
the National Academy of Sciences,* members of the consortium again
argued that Celera's assembly incorporated such significant amounts
of the consortium's work that they had not done their own independent sequence or properly tested the whole genome sequencing

method.[33] But Celera had other problems—Craig Venter announced he was leaving to rejoin his nonprofit, TIGR.

The Sanger Centre, same floor, different day. Alan Coulson had forgotten the appointment, but that didn't matter, he was happy to make time for me. I followed along as he loped to his office. Coulson was a tall, big-boned man in his middle years, with graying, flaxen hair thinned to nonexistence on top, wisping at the back, a blondish beard, blue-gray eyes. He wore a black T-shirt and black cords. He was awkward and shy. In 1967 he'd arrived at the Laboratory of Molecular Biology to answer an ad for a lab technician. He spent the next sixteen years working with Fred Sanger developing sequencing methods.

I was curious about the mtDNA sequence, which was where this whole inquiry had begun for me. Sanger's sequence had become the reference sequence that had launched a whole new approach to anthropology; it had been compared to sequences sampled from various living human populations in order to find patterns of relationship and descent. Many arguments had been constructed on the back of this sequence.

"There's a funny story about this," Coulson said. "We sequenced 90 percent from one person's placenta. There was a little we couldn't clone. The sequence for that was from mtDNA extracted from HeLa cells, an immortal cancer cell line. People use it for all kinds of things. So part of our so-called reference sequence was derived from the HeLa cell line. A couple of years ago I got a letter. I was asked if we had the original placental DNA. This researcher wanted to resequence it to see if there were errors."

Coulson was afraid they might have thrown the sample out.

"I went to the freezer and found this raindrop of frozen solution." He was laughing now, remembering his own relief. What if he'd thrown it away? What if there was none left? "I tested it on a gel and it looked intact. Sent it to the guy up north, a British guy. He sequenced it, along with the missing bit, and found one or two errors we made. He also wanted to know the racial origin of the person we got the placenta from . . . No one seems to know."

When Sanger retired in 1983, Coulson began to work with John Sulston, who had become interested in sequencing. "The cell biology division was one floor below. John had worked with Sydney Brenner on the nematode since 1969," said Coulson. ". . . Anyway, John was doing beautiful work on the cell lineage of the worm. Then he started thinking about the genome." Sulston was interested in making a good map of the whole worm genome, as geneticists had been doing with less complex organisms. "At one time, you could locate yourself on a genome knowing you were near a marker, a gene of interest. To get there by clones, you had to get there by walking."

His fingers made little upside-down walking motions, as if to show how one moved around on partially mapped circular genomes of simple organisms like viruses, where genes are arranged side by side without the intron interruptions found in larger organisms. One stepped from a known gene, or fragment of a gene, to the unknown beside it. But when this kind of mapping technique was applied to larger genomes, things got very difficult.

"John was at a seminar with Matt Scott, who was talking of a walk in the *Drosophila* genome. John thought that one could work out a parallel process and generate a map for the whole genome so that any fragment could be located." Sulston talked to Brenner and others, and then came up with a scheme to make a library of clones and, by organizing them, to make a map. One could gather up all these bits and pieces of chopped-up DNA and insert them into bacteria to make many, many clones. One could generate a kind of fingerprint identifier for each clone by using restriction enzymes to cut DNA at specific locations. One could tie clones together when fingerprints overlapped. Overlaps were found by computational analysis of digitized images. This physical map of linked clones could then be compared to known gene locations. A linked clone map could help produce an ordered gene map.

But surely there were alternative ways to lay these fragments out? I asked. How did they ever figure out which way was the right way?

"You put it together conservatively," Coulson said. "Someone corrects you eventually."

I asked how that would work.

"While doing construction, we made it available to the whole worm community. They'd say, this looks wrong."

This sharing of unfinished business between researchers working on the same organism, this openness and lack of concern about error, had allowed consensus maps to be constructed. As Coulson explained and explained, it dawned on me that this collegiality over the map of the worm was completely different from the other biological projects I'd been told about. It was the opposite of the closed, secretive, don't-even-tell-your-friends style I'd encountered in Washington, Toronto, Wisconsin, London, Montreal.

"That openness carried on to the worm genome and the human genome, on the public side," said Coulson. By "public," he meant the consortium, which included publicly supported sequence centers in locations such as France, Germany, Japan, China and Canada.

They began to make this map in 1983. "It was cheap to begin with. Sydney [Brenner] had money . . . We were salaried members of staff, and there were no expensive machines . . . We did get stuck. Here's an interesting strand. We were trying to do this with cosmids." These are bacterial clones that can carry fragments of about 40,000 base pairs of foreign DNA. They needed to have 2,500 clones of this size to cover the worm genome from one end to the other. But many bacteria were unable to copy such large fragments. Eventually, they were able only to get these pieces organized into seven hundred overlapping sets called *contigs*. "And then, apparently, no progress."

Eventually Bob Waterston came back to work in Brenner's lab again on a sabbatical. "He was in the same lab as me and John Sulston. He was very interested in the genome." Waterston worked away at this problem of getting the worm's larger fragments cloned properly in bacteria. *E. coli,* which they were using, wasn't happy about some of the worm's repeating sequences—it just wouldn't copy them. When Waterston returned to St. Louis, he heard about researchers using yeast to clone large fragments of DNA that *E. coli* couldn't handle. These were called "yeast artificial chromosomes," or YACs. Waterston tried them. "He made a YAC worm library and these clones were incorporated into the map to stitch the cosmids together. It worked."

Sulston and Coulson and Waterston ended up working together long distance. The worm community—people trained by Sydney Brenner, or trained by people who'd trained with Sydney Brenner—watched and carped as the corrected map was put together. Worm colleagues communicated on an early version of the Web called the Joint Academic Network (JANET). "We used to fax people drawings," said Coulson. "I think John sent a tape copy to Bob Waterston. He was the node of access in the U.S. I had an interactive version of a program that allowed me to make the map, and we distributed a read-only version. We displayed a printout of the whole map at Cold Spring Harbor at the biennial worm meeting there."

It was 1989. Jim Watson had already been appointed the director of the NIH's Human Genome Project. Sulston and Coulson and Waterston wanted Watson to get them the money to do the full sequence of the worm.

Watson had the clout to get them what they needed, but Watson had already drawn up a list of organisms to sequence before trying humans, and the worm wasn't there. "We think deliberately on Watson's part, to make us mad," said Coulson. "Watson's strategy worked. He likes to stir people up . . . Bob Horvitz, a prominent worm researcher at MIT, was the one who got exercised about the worm not being on Watson's list. Then I guess he set up the meeting we had with Watson—Sulston, me, Waterston. John made a famous statement: 'Just give us $100 million and we'll do it.'"

Coulson slapped his forehead. "Hang on now! What was that figure? You must read John's book . . . And Watson said, 'Things don't quite work like that in this country, John.' I can't remember that number," he moaned. (He remembered later. It was $100 million.)

He sank down in his chair, his chin in his hand. He was silent for a moment, collecting the various strands of the story that drifted and reformed in his mind. John Sulston's book was called *The Common Thread*; it described the history of the genome project from his point of view. Get a copy, Coulson kept saying. But I wanted to hear the story from him.

They got the money for the pilot project on the worm. Sulston insisted that Coulson get his doctorate, and Coulson got it for his work on the map of the worm's genome in 1994. But in spite of working so closely with Sulston, Coulson was vague about how Sulston had become the leader of the British team working on the human genome sequence—how, and why exactly, the Wellcome Trust had become such a keen funder.

"It was always mysterious how John got involved in all this," said Coulson, staring at his bookcase, as if he'd find the answer there.

The more questions I asked, the more Coulson seemed to squirm in his chair, and the more frequent became his protestations of ignorance, the passing of questions on to the invisible Sulston and his published version of events. We went down the hall to the coffee/conference room to get something to drink.

"The worm sequence was essentially finished in 1998. Well, it's still not entirely finished," Coulson said. (It's finished now.) "There's a tendency to declare some kind of point to say, this has reached a useful state, then publish and finish in private. That's what we did with the worm."

And that's what you guys did with the human genome too, isn't it? I said. It wasn't finished, it wasn't even close to finished, when the draft sequence was announced.

"There were political pressures for that exhibition," he said crisply.

"But it wasn't true, was it?" I pressed. I was curious how such senior scientists had been able to justify making a public announcement about a draft of the Book of Life when what both sides had was a far cry from a complete draft.

"Do you believe Venter or Celera finished the genome? They took the public data, a blodge of their own data, recomputed it . . . and claimed they sequenced the human genome," said Carlson.

Celera hadn't sequenced enough clones to get sufficient redundancy to actually be able to put an adequate draft assembly together, Coulson argued. He believed incorporation of the consortium's public data permitted them to construct a much better draft sequence, although the extent of the influence of the public data on their sequence is an ongoing debate.

"Their PR would lead you to believe they had completed the fly sequence by the whole genome shotgun method," said Coulson. "They are now using the traditional method to finish it on a clone-by-clone basis. Celera has a big PR force behind them. That's the point. We learned you have to compete against this crap."

That was an interesting excuse—it amounted to fighting fire with fire, or meeting PR with more PR.

Why such anger? I asked.

"Venter tried to get the public process closed down," he said. "So he could corner the market. There was a congressional hearing . . . We could have lost the funding."

I asked him why that would have been so terrible.

"I think that would have been awful—no free public access to the human genome sequence. We probably wouldn't have had one." He meant by this that Celera might have decided it made no commercial sense to complete it. "To me it's unthinkable that the human genome should be privatized."

His cheeks were red. He was irate. He was disgusted. Did I know that Venter had just resigned as president of Celera? "Venter is out," he said. And did I know that Celera had changed its focus from a genomics information company to something else? "They want to turn it into a drug discovery company," he said.

Meanwhile, the consortium (or at least various publicly funded centers), which he insisted on referring to as the "public group," was still hard at work sequencing other organisms. Two yeasts and the worm were completed; the mouse was on its way. "A whole slew of bacteria," he said. "I couldn't say how many."

But I kept circling back to the fact that both sides had willingly participated in a public event announcing they'd got the human sequence drafted in June 2000 when neither had a good one. As far as Coulson was concerned, it was all PR, and the PR was all about the share price of Celera and the U.S. commitment to biotech. Sulston, in his book, freely acknowledges that the draft was announced before they had achieved the goals they'd set themselves, and that this was nonsense: "It was not clear that the Human Genome Project had quite got to its magic 90 per cent mark by then, and Celera's data were invisible but known to be thin, so nobody was really ready to announce; but it became politically inescapable to do so . . . We were sucked into doing exactly what Celera has always done, which is to talk up the result and watch the reports come out

THE SECOND TREE • 285

saying that it's all done. Yes, we were just a bunch of phonies! But we were trapped by Washington politics."[34]

"Political pressures had to be accommodated," Coulson insisted. So, I asked, what new knowledge has been produced?

"Well," he said, " . . . It's not complete, and we don't understand it. But we will. It's a bit like getting a book from outer space . . . Philosophically, it's difficult to overhype it. I find the fact we have a linear code making us what we are . . . I still find that mind-boggling."

But we aren't just the expression of our nuclear DNA, I said.

"I'm not saying we are our genes," he said. "I find the concept that we have this linear code at all astonishing. And something we can read even more astonishing. That's why it's so important. It should be freely available to all scientists."

But I was beginning to think that the way it was elucidated was just as important. It was a project both individual and collective, both collaborative and competitive. The constant correction and cross talk, the public airing of incomplete work, discovery inside computer programs: this was a way of doing biological science that Darwin could never have imagined.

But then Darwin's propensity for secrets, his concern with ownership of his specimens, his determination to profit from his work were also foreign to Coulson. Coulson's attitudes had been formed while working with Sanger, a man who selected people to work in his lab for their congeniality as well as their skills. Coulson insisted that it is dangerous for science to rely on money from private interests. "Science has to be open or it's nothing," Coulson was saying. "It has to be a free exchange of data and ideas. If you stifle that, where does science go?"

But at the same time that he put this view forward, he also contradicted it. "You have to think," he said. "Here we have a field over the last fifteen years that has burgeoned at an incredible rate. It's hard to see how it could happen in a purely academic environment. There isn't the funding. In the San Francisco Bay area alone there are eight hundred biotech companies. How could you accumulate the funding to employ those people?" Although he is still an MRC employee, he was one of the last in the building still employed by the government. Everybody else works for the Sanger Centre.

Coulson walked with me down the long, wide halls, through a set of swinging doors, then into a long, glass-walled room filled with sequencing machines from one end to another, from one side to another. The machines came up to my chest. They were set out, row on row, each one with a clear Plexiglas top, like the cover of a record player, displaying a small pick-and-place robot arm dropping samples into a tray of capillaries. Each machine had a silver foil umbilicus snaking up to the ceiling, where heat could be carried away. There were other rooms devoted to computers and servers. The marketplace had created the wealth of the Wellcome Trust, and therefore this capital, and it was the Wellcome Trust, not the public funding agencies, that kept the consortium in the human genome race, its results available to all, even its competitors.

John Sulston, in *The Common Thread,* portrays the Wellcome Trust as the steel backbone of the public consortium. But the Trust is a private entity. Its Wellcome Genome Campus is privately owned by a limited company directed by the governors of the Trust. The limited company has the broad power to spend in the cause of research in human and animal health. From 2002 to 2005, the governors of the Trust plan to spend about £3 billion, about four times the budget of Canada's CIHR. But unlike the CIHR, the Trust's governors are not responsible to the public or its elected representatives, although they are allowed by their articles to try to influence public opinion and legislation. They are elected by each other, responsible to each other, and confined in their actions only by the law on charities and the restrictions imposed by the Trust's articles of incorporation. Governors are paid well—£57,000 a year—yet remain active in their normal spheres. Research grants and contracts can be directed to their own institutions, so long as the other governors permit it and the governor with a conflict refrains from voting. Six out of nine of the governors in 2002 were engaged in research, three out of those six in genetic or genomic or DNA research. The Trust, like so many American universities, also has an affiliated company to take research it has supported into the market. It set up Catalyst BioMedica Ltd. in 1998 to "work with researchers and their institutions to translate the useful results of Trust-funded research into practical benefits and so contribute to the Trust's mission of improving human and animal health."[35]

The difference between Venter's TIGR and the Wellcome Trust is the size of the Trust's assets. Both are nonprofits run by scientists who control privately made money. Yet John Sultson was right about the vital role played by the Trust in defending the public interest. A British private charity, not the democratically controlled public agencies, had insisted that work it supported be made public immediately so that it could not be patented. In doing so, the Trust nurtured a new way of doing science—an ongoing, public, international and permanent state of inquiry. If there was a more significant mutation in the behavior of some of Darwin's descendants, I couldn't imagine it.

CHAPTER THIRTEEN

HELLO DOLLY

The plane soared up from the little business airport at London's Docklands, and set me down at Edinburgh International, in a bowl surrounded by high hills. Wherever I looked there were rough meadows, dark stands of spruce and pine, stone fences. The hills were splashed with vast patches of a low-lying creeper, like a brown wash poured across a blue-green palette. The air was fresh with cold rain. There was no sign of Edinburgh at all.

The road curved in and out of one little village then another. Ancient houses of cut stone with steep slate roofs set their grim faces right at the roadside, as if a front garden was a wasteful affront. Farm fields surrounded new housing developments with homes built so close together you couldn't stick a newspaper in between them. Finally there was a guardhouse and a barrier.

Beyond the barrier was an industrial park, very new, with a sign on one long boxlike building that said PPL Therapeutics. Beyond that, there was a group of Quonset-style buildings arrayed around a courtyard. Roslin Institute, which works at the weirdest edge of biological science, seemed to be housed in Korean War–era barracks.

Since its early days as a government agricultural research station, Roslin Institute has grown into a public-private hybrid— now a private nonprofit that competes for government grants and that gets a considerable portion of funding through contract research. All its intellectual property and assets were passed from the old government agency to the newly formed nonprofit for nil consideration in 1996, the year Dolly was born, when the Conservative government decided to reduce its science budget by throwing academic sheep to the wolves.[1] Roslin Institute had already spun off profit-seeking companies such as PPL Therapeutics, and has gone into business with other private ventures as well. It is still committed to improving the breeding of cows and sheep and chickens for agricultural benefit, but also to something new: improving human beings through something it refers to as "regenerative medicine." Inside the main doors, a lady at the information desk called for Ian Wilmut, whose secretary led me to him.[2]

The very well known Wilmut, co-father of the cloned sheep Dolly, Polly, Megan, Morag, Cedric, Cecil, Cyril and Tuppence and their descendants, had a large office partway down a very long corridor. There was a desk and a shelf unit at one end, where a young woman sat at a computer with her back to us. Wilmut's desk was on the opposite wall, under another shelf unit. There was a table with chairs set across the middle of the room, and a large window that overlooked a courtyard garden and the opposite wing. A large rabbit (or was it a March hare?) nosed out from under a bush, then bounded to shelter.

Wilmut pushed himself away from his desk on his wheeled chair and turned to face me. Was he a direct descendant of Darwin, or more like Coulson and Sulston—Darwin mutated? Wilmut's name was first on the Dolly paper published in *Nature*.[3] Keith Campbell had done a great part of the work, Campbell's theory had produced the results, and Wilmut was the senior scientist so it was unusual that Wilmut's name was first, not last. There was also the matter of Wilmut's role in the sale to Geron of the spin-off company Roslin Bio-Med. Government funding had enabled the cloning of sheep by nuclear transfer, but exclusive rights to the nuclear transfer cloning patents in biomedical applications, worth conjectured billions,[4] had been licensed to Roslin Bio-Med, even though the patents were also owned in part by a government department and the state-owned Biotechnology Council.[5] Roslin Bio-Med was sold to Geron in exchange for 2.1 million Geron shares and the promise of £12.5 million in research funding for Roslin Institute over six years. Wilmut was doing Geron's research at Roslin and was on Geron's science advisory board, yet he was still employed by Roslin.[6] He led a complicated life.

290 • ELAINE DEWAR

Some critics had trouble figuring out why the British government had transferred such vital assets into private hands. William
Cash, a Conservative member of Parliament and of its public
accounts committee, had asked for a full-scale investigation by the
National Audit Office.[7] That investigation had dawdled on for two
years. Keith Campbell, whom I had already asked about this transaction, had refused to answer questions but instead printed off
certain stories that had appeared in the British press. He'd suggested
I ask Wilmut directly about it, since Wilmut had acquired shares of
Roslin Bio-Med. He had also suggested that I should not mention to
Wilmut that I had been to see him first. Campbell, it should be
pointed out, had by then become a member of the science advisory
board of Advanced Cell Technology, Geron's competitor.

Campbell, of course, had his own detractors; one colleague,
with significant power in British science, had described him as a bit
of a loose cannon. He hadn't rolled around and fired off explosives
when I went to see him, but there was something about him that was
certainly loose. He reminded me of Dustin Hoffman in *Midnight
Cowboy*. He was a small fellow with long, graying rasta curls and a
certain way of rolling his eyes. He had made it clear that he was
afraid to speak of certain matters openly. One of his former students
had explained that the way things work in the U.K. is that people
applying for grants must put their names on their applications, while
the people reviewing the grants do it anonymously, taking the trouble even to disguise their normal spelling quirks so as to evade recognition. The peer review system was like the book review system in
Darwin's day: anonymous persons can administer punishments to
those who get out of line. I'd been made to understand that
Campbell had been warned that his grant applications would be
looked at very carefully in future.

Wilmut was medium sized, in his middle years, with pale skin, a
square head, a red halo of thinning hair, a trimmed beard. In contrast to Campbell, he looked like a banker. He wore gold-rimmed
aviators and pinstriped pants with a white shirt and a nice tie. He sat
in his chair with his arms akimbo, his pale fingers tapping out an

impatient rhythm on his shirtsleeves or on the tabletop, almost as if he was nervous, which was unlikely—he had given countless interviews after the birth of Dolly.

Both Wilmut's parents were teachers, his father specializing in mathematics. Wilmut was modest about his early scholarly attainments. He explained that he studied biology because he was "least bad" at that. He never entertained a thought of doing research. He grew up in a city, but longed for the countryside and worked in the country on weekends, aspiring to be a farmer. He went into agricultural science at the Sutton Bonington campus of the University of Nottingham and, in his third year, won an undergraduate intern scholarship from the pig industry to spend eight weeks in a research lab. He sought out labs willing to take him that were near his girlfriend (the woman who later became his wife), who lived in Cheltenham. He got a reply from Chris Polge's reproductive physiology and biochemistry lab, a unit of the British government's Agricultural Research Council, at Cambridge. Polge was trying to find methods to safely freeze cells. "He laid the basis for cryobiology," said Wilmut. "I started there in 1966. Then I came back in 1967 for a PhD project for four years."

Wilmut's project was to freeze boar semen so that it could be unthawed later and still be viable. This earned him his doctorate in 1971 from Darwin College, Cambridge. He stayed on to do his postdoc, during which he was the first to freeze a cow embryo, then unfreeze it and implant it. This achievement brought the press running. "The first calf produced from a frozen embryo," he said. "That was my postdoc . . ." He rose up and checked the date in one of the bound volumes on the shelves. "That was 1973."

He knew little about transgenics then, although he was aware of Berg and Cohen and Boyer's recombinant work in California. Neither did he have anything to do with the MRC's Laboratory of Molecular Biology, where Sanger and Coulson and their colleagues were figuring out how to sequence the genome of a virus. Wilmut was focused on the reproduction of farm animals and how to make it more efficient. He joined Roslin Institute, then a government agency called by another name, in 1973. He had become a civil servant of science, not unlike Coulson and Sulston, although his research was to have a direct practical application. The idea was to scientifically improve herds. British breeding at that time was "very conservative."

His boss was interested in developing embryo transfer technology. Wilmut was expected to direct his own research, and he chose to investigate why so many embryos die. In all the years he worked away at this problem, there was no improvement in the statistics. "Possibly worse now," he said. "In cattle, 20 percent of embryos will not be calved. Most are lost in the first three weeks. In humans, it's 50 or 60 percent lost . . . For ten years, that was my main interest."

But then, in the early 1980s, Wilmut was forced to switch areas. The idea of using transgenes to improve breeds had entered the picture. The first transgenic mouse paper he could recall had been published in 1981. These methods were then applied to farm animals. As it was with Michael Bishop, so it was with Wilmut. Transgenic farm animals got his attention—and so did the possibility of cloning. In January 1981, Karl Illmensee and his colleague Peter Hoppe published in the journal *Cell* that they had cloned mice by nuclear transfer from late embryo cells into denucleated embryo cells called *zygotes*. These reconstructed embryos divided in culture for four days, then sixteen of them were implanted, along with normal embryos, into surrogate mice. Three cloned mice were successfully carried to term, and these cloned mice in turn bred normally. Or so the researchers claimed.

"So what happened here is that molecular biology was brought into the institution," Wilmut said. At around the same time, private companies also entered into the lives of government-employed scientists. When Wilmut was a graduate student, and even in his early years at Roslin Institute, no civil servant "could hold shares in a company, and it was unthinkable to spin out a company." Now the idea that publicly funded science should be made to pay for itself and create a nation's wealth came floating up from London. Two of the molecular biologists recruited to Roslin and affiliated with the University of Edinburgh set up a biotech company to produce human proteins in animals. This company eventually morphed into PPL Therapeutics, which merged with a U.S. pharmaceutical company based in Virginia. Some of its founders remained on Roslin's staff.

THE SECOND TREE • 293

By 1984, Wilmut, like Michael Bishop, had retrained himself. The molecular biologists who joined Roslin took the time and trouble to teach. Wilmut soon began to try to derive stem cells from sheep embryos. The idea was that if a particular transgene could be introduced into the DNA of an embryonic stem cell and one could make a stem cell line, then eventually one could reliably derive farm animals with exactly the characteristics wanted, instead of trusting to the random luck of breeding. In other words, he took his first steps on the long road to genetically altered clones.

By 1985, Illmensee and Hoppe's much-heralded work still had not been repeated. Davor Solter and James McGrath tried to follow Illmensee and Hoppe's cloning recipe exactly but could not get any live mice. After methodically going through a number of experiments to test out embryonic cells and transfer techniques, Solter and McGrath declared that simple nuclear transfer cloning in mammals was not biologically possible.[8] Nevertheless, "it stuck in people's heads that it was possible," said Wilmut.

"The first person to want to do nuclear transfer here was a student named Lawrence Smith," he said. "He is now in Montreal, at l'Université de Montreal." Smith's work at Roslin Institute produced a wealth of insight into the developmental processes of mammal embryos, specifically that the early stages of development are controlled by the cytoplasm of the egg and then, after a certain number of cell divisions, controlled by the genes in the nucleus. Neal First had argued that one could not transfer nuclei between embryos unless the two were essentially at the same point in development.[9] In retrospect, according to Wilmut, this made sense. It seemed clear that if nuclear transfer cloning was going to work, the rhythms of communication and development between egg cytoplasm and implanted nucleus had to be somehow matched.

Steen Willadsen, a veterinarian, had joined Chris Polge's lab at Cambridge after Wilmut left. In 1985, Willadsen published an article in Nature on an experiment that produced lambs by nuclear transfer between eight- to sixteen-cell embryos. There were two possible explanations for this success: either the nuclei in many of the cells of these embryos retained the capacity to make a whole lamb for a long time, or somehow nuclei could be reprogrammed to their developmental start points. "A critical paper," said Wilmut.[10] Willadsen left Cambridge and went to work for a cattle company, Granada. Later, Wilmut ran into a colleague of Willadsen's at a conference in Dublin. "He talked of cloning from blastocysts [a stage in embryonic development far beyond the eight-cell morula]. This wasn't published. I was excited. This was January 1987. I didn't get details. In March, I was to be in Australia. I arranged to fly over Canada to come back."

He met in Calgary with Willadsen and confirmed the details he'd heard. Wilmut assumed that the first explanation for Willadsen's success was correct—that the inner cell mass of sheep embryos contain stem cells that retain all of their potential to create a whole animal and that, by nuclear transfer, one could "clone from them." He returned in June, and by early October he and his colleagues had decided to try to clone sheep by nuclear transfer. His task was to identify these special embryonic stem cells. Then he would introduce the human genes for therapeutic proteins into these cells, transfer the altered nuclei and reliably produce genetically altered offspring. PPL had been trying to engineer chickens to produce eggs with such proteins. It would be much more efficient if one could produce sheep or cows with human proteins in their milk.

But Wilmut never identified stem cells in the embryos of sheep.

Keith Campbell didn't share the view that there were stem cells in the embryo that could make a whole animal. He thought that the second theory, the rewrite theory, was correct. By the time Campbell was hired at Roslin Institute, he had worked his way up through British science in a most unorthodox fashion. He had told me the story in his office at the Sutton Bonington campus of the University of Nottingham, where he is now a professor, and then over a ploughman's lunch in the village pub. Campbell, five years younger than Wilmut, came from Birmingham, where his father was a "seedsman," a horticulturist. He was a very bright boy who attended a very good boys' grammar school but left at sixteen without his A levels (necessary for university entrance), which made his mother cry. She'd imagined him being the first in his family to attain higher education; she'd imagined him becoming a doctor. He worked instead at a hospital medical lab as a technician, studying for his technical diploma at a college at the same time. The day he got it was the day he quit the job. "The thought of doing it for forty years scared the pants off me," he said. So he went to the University of London, graduated with a degree in microbiology and took off again, first for Yemen, then a series of biology jobs abroad and at home. Finally he decided he wanted a doctorate. He got a cancer research job at the Marie Curie Foundation in Surray in 1980 and spent three years working on chromosome abnormalities and DNA damage. But then the director of the institute died. A new director came along, his boss took early retirement and Campbell was stranded without a doctoral supervisor. He was offered a scholarship—to leave.

He was interested in the cell cycle, so he wrote to Paul Nurse, who handed him over to Chris Ford at the University of Sussex. There, in 1983, he started his second attempt at a doctorate, focusing on the maturation of frog oocytes (immature eggs). Campbell freed himself from cell-development dogma—once a cell differentiates, it stays that way forever—during his time in the cancer lab. He had observed that all kinds of cells are found in cancer tumors. In cancer, developmental progression becomes deranged, and fully differentiated cells become capable of becoming anything. Because he was working in frogs, he was also aware of John Gurdon's early cloning work at Oxford, and he attended a lecture given by Karl Illmensee. As he learned about the factors involved in controlling development in the cell, he kept thinking about cloning by nuclear transfer. And then, in 1991, when he was at the University of Aberdeen, he saw an ad from Roslin Institute.

Roslin Institute, according to Wilmut, had found money to try nuclear transfer in sheep. It came from the British government and from private sources. The private money was from a start-up company called Animal Biotechnology Cambridge, whose scientific director was Wilmut's former doctoral supervisor, Chris Polge.

Wilmut was acutely aware of what everyone else in the field was doing. Willadsen was making progress; Neal First in Wisconsin had had success cloning cattle from early stage embryos. "We were aware and watching each other," he said. "It was desperately important to be first."

"Why?"

"That's the way we're made. Anyone who tells you otherwise is kidding. It's always the case. There is more competition and secrecy now, but it was always the case."

The competitors had managed to clone by nuclear transfer between very early embryos, but only after many tries. Wilmut wanted to find a way to make cloning an industrial operation. "We brought in Keith Campbell to look at the cell cycle," Wilmut said.

Campbell was beginning to think that the mammalian egg could reprogram any transferred nucleus with the proper number of chromosomes back to the beginning of development. If they could just work out the timing properly, they should be able to take the nucleus of almost any cell, so long as it had two sets of chromosomes, put it in an egg and watch the nucleus express the genes associated with the first phase of development. But Campbell's ideas directly contradicted Wilmut's.

"The Central Dogma was, to do nuclear transfer, we would need a special cell, an embryonic stem cell, to get development. I didn't believe a word of that," Campbell had told me. "Frogs had been produced from intestinal epithelial cells of tadpoles. And there was the history of my work in cancer: either there were stem cells in tumors or cells were switching phenotypes. The majority of the cells in the body have a complete set of genes . . . If the majority have a copy of the genome, it's a matter of tricking embryos and DNA to cooperate to produce animals."

Campbell started at Roslin in 1992. Eventually, he got the first cloned sheep, Morag and Megan, out of a cell taken from the inner cell mass of an embryo. This cell had first differentiated in culture, and then its nucleus had de-differentiated when it was transferred into a denucleated egg, which was tricked into dividing as if fertilized.

"They were born in July of 1995," Wilmut said. "The patent was 31st of August. The priority was the patent, then to write a paper." The paper was published in February 1996.

Campbell, who explained to me that the intricacies of cell development are still mysteries within a black box, had nevertheless figured out enough about them to make the box work. He and Wilmut and their colleagues had shown proof of the principle that a differentiated cell's nucleus, inserted into a denucleated egg, could instruct an embryo to develop all the way to live birth. As far as Campbell was concerned, this was the true cloning breakthrough, not Dolly. Roslin Institute issued a press release and stories were splashed across many front pages, but there was nothing like the media tsunami that followed the announcement of the birth of Dolly. Why? Campbell had noted that the *Nature* article on Morag and Megan came out on a Friday. On the following Monday, "a man walked into a school in Dunblane, Scotland, and started shooting."

The British government gave them almost half a million pounds more. PPL Therapeutics also had some available cash. More important, PPL also had a cell line established in culture taken from an adult sheep's mammary gland that the company wanted to try to clone.

At the same time as Campbell and Wilmut worked on trying to clone from this mammary cell line, they also tried to derive live lambs from cultured embryo cells and fetal cells. Two hundred and thirty-one nuclear transfers from this cultured embryonic cell line eventually yielded four live males that were genetically identical but physically and temperamentally different. The fetal cells yielded three live clones. The mammary cell line yielded Dolly. For Campbell, the work was grinding, unrelenting. "We had the nuclear transfers three days a week for the month." They put most of the reconstructed embryos right back into receptive ewes, and only a few into culture dishes, since they were not certain they could keep embryos developing in the dish. In the first group of ewes, the embryos developed into blastocysts, which were removed and then put in another group of ewes. "There were just a few of us, and it had to be coordinated. One-half dozen people were involved plus farm staff," Campbell had explained. Dolly and the seven other lambs were born in July 1996, but no paper was published until February 1997. "The adult cell went against a central dogma. That's why the world press went lollygaga," Campbell had said.

Campbell wrote the patent applications. When they were licensed exclusively to Roslin Bio-Med, he was paid £5,000 as his inventor's share. He had already left Roslin to work for PPL because he wanted to try to clone pigs, but he then left PPL to take his academic appointment at Nottingham in November 1999, before the first cloned pigs were born, in February 2000. By November 2000, he'd become a scientific adviser to Advanced Cell Technology to help with human cloning.

Wilmut had been glancing at his watch. His next appointment was coming up, and he seemed anxious to get away.

How did Roslin Bio-Med get the Dolly patent? I asked.

"When Dolly was born, we were encouraged to license," he said. Roslin Institute had been encouraged to license the nuclear transfer patents in exchange for the promise of cash to be used for more research. It got "£6 million over three years from 3i."

3i Group is one of Britain's largest venture capital companies. It became one of the majority shareholders in Roslin Bio-Med, along with the Roslin Institute, after the company was spun out of the institute. "We started a virtual company," Wilmut said. Two Roslin staff members became directors of the newly formed Roslin Bio-Med. It had no offices or assets. "The institute owned the patents; the companies have licenses to use the technology," he said. Roslin Bio-Med got the license to the nuclear transfer cloning patent's biomedical applications.

But why was the company then sold to Geron?

"We had contacts at the high business level," Wilmut said. In fact, Roslin approached Geron after it announced Thomson's success in human ES cells. "I was a member of the board. I had a vote— I voted yes. It brought together several technologies and . . . brought us more resources."

This is exactly what the Geron press release said. The Roslin Bio-Med purchase by Geron brought together the technologies of nuclear transfer cloning, human embryonic stem cell line manufacture and the human telomerase gene. I could sense Wilmut wanted to bring this discussion to a rapid close.

So, I asked, how exactly did you end up with shares in Geron?

"I had shares in the originating company."

But this made no sense to me. Roslin Institute, which created the originating company, is a nonprofit, or a company limited by guarantee. How could he have acquired shares in a company spun out of a nonprofit? 3i Group had purchased its 42 percent ownership position in Roslin Bio-Med with a £6 million investment over three years What about him? Why was he a shareholder at all?

"Because 3i insisted on it," he said. "They say they won't handle [it] till many are at risk. So I had to buy them." By "handle it," he meant, do what venture capitalists do—find the early money to get a company going, then groom it for sale, either as a publicly traded company with its own listing or to another company, like Geron. Venture capital companies often like to have the directors and senior staff of companies they invest in at risk of losing money if things go wrong. The theory is that companies do better when those who run them rise or fall with the company's fortunes. Of course, such an economic interest in a company can erode objectivity, which is what makes people worry about the truth value of science done by business. Wilmut and his colleague John Clark purchased shares in the new, profit-seeking entity.

Had anyone from the National Audit Commission talked to him? No, said Wilmut. The press had written nonsense on the change of ownership, and he told me not to believe the nonsense in the press. He did not tell me what he paid for his shares but I learned later that Wilmut paid £12,000 and got 2.9 per cent of the company.

"So did you get rich when Geron bought you?"

"For a government employee," he said, "yes, I got rich."

By now, his face was flushed, as if this whole talk of money and personal enrichment from his endeavors was deeply embarrassing.

You're a famous man now, I said. Has it been worth it?

"Well," he said slowly—leaving me in no doubt as to his displeasure with me—"there's people like you . . . I tell my wife you can't have the invitations to nice meetings and the shares without the other, without being willing to explain."

He was angry. He also wanted to make clear that he wasn't in science to get rich.

"I'd say I'm a social democrat," Wilmut said. "I have anxieties about free-market capitalism . . . There is a sense this is a big social experiment for British science. It finds favor in the Department of Trade and Industry." He realized that privatization provided money to do science, but he also knew that it could remain an opportunity for him only so long as Geron's desires coincided with the interests of the scientists at Roslin Institute. His whole working life had now changed. "I read manuscripts—I look to see if anything is patentable. When I did my PhD, I never heard patents mentioned. But British science was lousy at commercializing."

Geron and PPL Therapeutics are both publicly traded companies. They are also competitors. Wilmut, still a Roslin employee, also serves as a consultant on Geron's scientific advisory board. He is considered an insider of Geron's when it comes to trading his shares. The patents arising from his work on Polly have been exclusively licensed to PPL. He has to be careful that he does not disclose Geron secrets to PPL, and vice versa. PPL's share value has ridden up with its various successes in animal cloning, but also down. Geron's shares have risen and fallen in value too. And Geron had recently changed its focus. Where once it had been willing to finance work at Roslin in animals, now "it's all on human cells," Wilmut said. "Now they don't support animal work anymore."

This brought us to the difficulty of aligning interest in basic inquiry with profit, and profit with truth. According to Wilmut, everything he has to say on animal cloning has a spin related to human reproductive cloning. Wilmut had become one of the leading voices denouncing the very idea of human reproductive cloning, warning that animal clones are not healthy.[12] Keith Campbell, like Michael Bishop and Michael West, found that view extreme. Who was Wilmut speaking for when he made these claims—Roslin Institute, Geron or himself?

"Jaenisch and Wilmut were over the top in *Science* where they said there are no normal clones born," Campbell had said. So far as Campbell knew, it is rare that commercially bred sheep live to the ripe old age of six, as Dolly had done. So what did Dolly's arthritis actually say about the dangers of cloning? "Dolly is only one sheep," Campbell had said, "and not kept under normal conditions for sheep. To compare to others is outrageous." Cloning was a triumph for British science and it shouldn't be bad-mouthed, Campbell had declared.

But who was Campbell speaking for, Advanced Cell Technology or himself? As he'd shown me out of his office, he had done a funny little dance at the top of the stairs, waggling his bum as if he was trying to protect his precious parts from a searing flame—his uncomfortable position so long as the matter of the nuclear transfer cloning patent was being adjudicated. Geron had spent a great deal to acquire the license to the patents. Advanced Cell was betting they didn't apply to human beings. But why would that matter to him? "Well," he'd said, "if they lose the patent case, they'll be after who wrote the patent, won't they?"

"Are any clones healthy?" I asked Wilmut. I explained to him that Michael Bishop said most of Infigen's are.

"I've asked that question. The answer is we don't know. Animals haven't lived long enough. And here I worry about companies—they put a rosy glow on things. Let me suggest you ask where Gene is," he said. His fingers thrummed aggressively on his arms, which were crossed over his chest.

Gene was Michael Bishop's first successfully cloned bull. The Infigen conference room I sat in is named after Gene. Bishop had told me that Gene was doing well, but he hadn't said where he was exactly. As it happened, at the moment Wilmut suggested I ask that question, Gene was in the Wells Fargo Farm in Minnesota, a pioneer-style petting zoo.

"Ask where Gene is," Wilmut said again. "You get a clear answer to that."

"Well, do you know where he is?" I asked.

"I have the answer," he said, "from someone close . . . If you'd asked me is Dolly healthy six months ago, I'd have said yes."

So, later, I did ask after Gene's health. The zoo returned him to ABS Global in the fall of 2002, in excellent health for a bull of his weight and age. He was put down by the veterinarian on December 10, 2002, because he had "outlived his scientific usefulness." An autopsy was performed; there was evidence of arthritis in one fore-limb, which the vet considered quite normal since arthritis often appears in bulls around the age of five.

I assumed that Wilmut's objections to human reproductive cloning would disappear if it turned out that animal clones can be safely brought into the world. No, Wilmut said, he would object even if it was as safe as riding a bicycle. Why? Though he had pointed out that four male sheep cloned from the same embryonic cell line were different in looks and temperament, he now said he was afraid of unfair expectations that would be placed on children born from cloning. People would expect them to be exactly like the person from whose nucleus they arose.

John Bracken, a veterinary technician, had brought a little red car to Roslin's front door to drive me to where Dolly and the other cloned sheep are kept. We arrived at a low, concrete blockhouse and then walked up a small hill to the sheep house.

Concrete block walls, a peaked, corrugated roof. The dampness that gathers inside on a wet day, filtered light. Inside one of two pens, lying down on the concrete floor by a big bale of straw, there was a white-faced sheep, a puffball of dirty, cream-colored wool. She looked like one of those old *New Yorker* cartoons, a furball with stick legs. Bits of straw and dirt erupted from the wool. There was a heavy stench of lanolin. Dolly, said Bracken, pointing to her.

She watched me with sweet eyes from under her thick white eyelashes as I moved close to the barrier holding her in. Dolly had had six lambs naturally, Bracken was saying, a single, triplets and twins. The two full-grown sheep in the pen with her were her nat-urally born lambs. In the pen beside them were Morag and the genetically modified Polly and Molly. They chewed placidly and in unison.

These were the products of a chimerical science: the fruits of wonderfully free inquiry fused to science-for-profit, Darwin's progeny hitched to Sanger's.

When Dolly saw Bracken, she pushed herself upright, leaning on her front knees and heaving herself forward like an ungainly cleaning woman rising up from a wet, scrubbed floor. There was nothing to show that there was anything wrong with her—that there was arthritis in the joint of a hind leg—except that she was slow. She seemed to expect a treat from him.

I reached out to pat her woolly side. My hand bounced off. She nibbled at my fingers. A phrase I'd picked up in childhood, from my father, kept running through my head: first do no harm. I thought, then, that it was the opening phrase from the Hippocratic oath. In fact, as I would later learn, it is merely the essence of Hippocratic teaching.

Bracken was explaining all the hoops to be jumped through and licenses to be got by anyone wishing to experiment on animals in Britain. He is a person licensed to do certain kinds of experiments. I'm not sure how this came up, but it was as if he'd read my unease. I was staring at Dolly and she was staring back at me. She may be stupid, I was thinking, but there's a being in there, a soul of some sort. She was aware of herself and of me and everyone else around her. Does it do harm to make living beings by putting two cells together in a dish, or by plucking nuclei from here and putting them there, without knowing the outcome for them or their descendants? If it is right to do this to other animals, can one at the same time maintain that it is wrong to do this in human beings?

This was the third time I had found myself staring into a research animal's eyes, wondering about the line so neatly drawn between doing anything imaginable to them and refraining from doing the same to us.

Dolly looked as if she was waiting for me to do something, and she wasn't sure whether she'd like it.

There is no treatment for arthritis, Bracken was saying, it's a degenerative disease and when it gets too bad, Dolly will be "euthanized." My father also has this degenerative disease, and he puts up with the many miseries it engenders. A cousin has it. I will probably get it. If I get it as badly as my father, would anyone suggest that I be euthanized? Why don't we say the word "kill" when that is what we mean?

I liked Bracken; he seemed a gentle person. Nevertheless, I was upset. With the "sacrifice" of the mice, I'd felt queasy. In Wisconsin

I thought there was something obscene about the way the pigs were kept in permanent confinement, with nothing to do but chew on metal chains, while waiting for the needle and the knife. The more I learned from these biologists, the less able I was to see barriers between species, between families, even between orders and kingdoms. Several copies of the gene for a human protein, Factor IX, and the genetic promoter for a milk protein in sheep had been put together and introduced into the nucleus that was transferred to the egg that became Polly. A human protein could be extracted from her milk.[13] To sequence the human genome, bacteria and yeast cells became libraries for human DNA. So how could anyone draw lines between humans and all the rest? These divisions are arbitrary at best, leftovers of the dogma from before Darwin, when biologists strove to recognize each distinct member of God's miraculous creation. And if, as the Buddhists and McClintock say, we are all one, then doesn't the adjuration "first do no harm" apply to all? How could one know whether one was doing harm to the animal or plant life altered, the microbial life engineered, until that life sickened and expired?

Bracken was telling me about how he had applied to be transferred here eight years ago, just in time to get involved in the births of the cloned sheep. He had spent twenty-four hours straight in the pen with the ewes due to lamb. Keith Campbell had worked right alongside him. When PPL was trying to clone pigs, he'd scanned the gilts with ultrasound from midnight till six in the morning, checking every two hours to pick out the exact moment when oocytes moved from their follicles, so they could know exactly how old oocytes were when they were retrieved. When the first cloned pigs were farrowed, Bracken had sat for four or five days in the pen with the pigs to make sure the sow didn't roll over on her brand-new cloned piglets.

He agreed with Campbell that the cloned sheep were just fine; even Dolly was perfectly fine considering she had never been kept in conditions normal for sheep, and she had lived long past the norm. Morag was older than Dolly and showed no signs of any untoward conditions.

I couldn't stand Dolly's watchful eyes anymore. Bracken and I walked back to the car.

"I never in my wildest dreams thought I'd be involved in something so . . . that could . . ." He looked at me, helpless to express himself.

"Change the course of evolution?" I supplied.

A great smile rolled across his face. "Yes," he said. "So, if I do nothing else, I can say I was part of opening this amazing door."

But in the intimate space of the little car, he talked in a different language, phrases and ideas left over from a time before science enjoyed such intimate relations with life. I'd done the same myself only a few days earlier, when talking with Alan Coulson in his car. We were talking about belief in God. I had to say that I have no real religious beliefs, although I certainly indulge in religious practices. And I hadn't yet written off the idea of a larger power. I was still troubled by our mtDNA cooperating with our nuclear DNA though they speak different tongues; by the way cells from two different animals learn to coexist in one; by the way a nucleus ripped from one cell will operate normally in a cell taken from someone else; by the way an egg can apparently rewrite a nuclear program; by the idea of life as a program; by the unwritten something, hovering like a Platonic ideal, insubstantial and irreducible, that brings two different programs into synchrony. Isn't there something that smacks of a higher power in all of that?

I had also told Coulson that I am all layered up with superstitions. I throw salt over my shoulder when I spill it. I avoid cracks in sidewalks, think twice about crossing the paths of black cats. The Other One insists upon it. It's absurd, ridiculous, but I do these things, and I'm not alone in this. In fact, I think this layering up of ideas that don't fit together, that contradict each other yet survive and hang over from one cultural shift to the next, is like the information content of DNA—so loaded with sequences from other times and circumstances, from random doublings, from viral invasions, that it stretches out to the crack of doom.

"It was fate that brought me here; you could say that it landed me here at just the right time," said Bracken. "Oh, don't write that down."

CHAPTER FOURTEEN

FRESH FROM THE NEW GARDEN

M y cousin picked me up at Ben Gurion International
Airport very early on a Saturday morning. The terminal
was empty, except for the few hardy souls from my plane.
The British flight crew had gotten off at Larnaca, Cyprus, and
Israelis had taken their places because the British didn't want to fly
into Israel anymore. The home of three of the world's great religions
was in a bloody war called the Intifada. Palestinian kids threw rocks;
Palestinian men had taken hostages in the Church of the Nativity
and had shot up Orthodox Jewish illegal settlements. Then came the
suicide bombers. Israeli tanks had rolled back into territories
returned under the Oslo Accords. Checkpoints had multiplied.
Soldiers conducted house-to-house terrorist hunts, breaking through
the walls between people's homes to avoid being shot at in the street.
The terrorists' spiritual leaders made snide remarks about how good
Muslims are not afraid to die while the miserable Jews are obsessed
with life.

"She deserves a medal," my cousin said to her neighbor, as she
introduced me. The neighbor had just announced that she had it on
good authority that the United States would launch its war against
Iraq in exactly six weeks—on May 1, 2002. (The authority was not as
good as she supposed.)

Why does she deserve a medal? the neighbor asked.

"For coming here," my cousin said.

In fact, I had plotted and schemed to get to Israel, because it was
home to the creators of the Garden of Eden story; I was thinking of
the clever use I could make of biblical allusions. But I had reporter's
reasons, too. Israel, a Jewish religious state, is among the world's most

free-wheeling jurisdictions when it comes to human cloning. Human embryonic stem cell research is permitted, and the ban on human reproductive cloning is temporary and to be reviewed. Michael West had feared with good reason that the Israelis would beat him to nuclear transfer cloning of human embryos: several Israeli researchers had been first to differentiate several types of human cells from embryonic stem cell lines and to set them to work. They had been helped along by a bioethicist, Michel Revel, a professor at the Weizmann Institute of Science, who formed the ethical guidelines for their efforts. I had found a website that described Revel as the 1999 winner of the Israel Prize in medicine, the molecular biologist who identified the gene for human interferon, a founder of one of Israel's first biotech companies, a religious Jew, a bioethicist representing Israel on a UNESCO committee and chairman of the Bioethics Advisory Committee of the Israel Academy of Sciences and Humanities.[1] Now that, I thought, was efficiency—scientist, businessman and ethicist all in one. Revel saw nothing wrong in principle with human reproductive cloning. He had said so in print.[2]

But it wasn't just the allusions and the creations of science and the efficiency that pulled me to Israel. I was thinking about the inevitability of my own death; I was thinking of myself as a material being on the downslope to the end. I had had recent losses from age and disease—a favorite aunt, a favorite first cousin, two childhood friends. My uncle was on the brink, as was a neighbor who'd been a friend for twenty-three years. So I was trying to thicken my fleeting life, reduce it like a wine sauce into a denser blend of business and pleasure and surprise. The pleasure would arise from seeing my family, being close in the flesh instead of just on the phone, telling stories, sharing gossip, a meal.

My cousin and I drove up to Jerusalem. Well, she drove. I listened to the radio with my heart in my mouth and watched the landscape scroll past my window. It was blissfully warm. In spite of morning rain, which released the fragile scents of jasmine and lemon, the hills were dry. There were waves of wildflowers, whole fields of acid green grass and bobbing yellow daisies, followed by dark walls of pine and cedar erupting from outcrops of ancient rock. It was all so alive and so compelling.

But something bad had happened, or had been about to happen, on the train line to Acre that morning; the trains had been stopped and emptied, schools in the north closed. The radio barked out the news that in Kfar Saba, a gunman had sprayed a crowd with two magazines of bullets from his handgun until someone had shot him down like a rabid dog. A young woman had been critically injured. We turned the radio off while we ate lunch at a childhood friend's house. She and her husband, a more distant cousin of ours, live in the country near Rehovot, Israel's high-tech center. There was a view from their swimming pool of an old cemetery, on the hillside opposite their house. The headstones were set tight together, yet they also leaned haphazardly, like an open paper fan. "That's where I'm going to be buried," my cousin said. Since we are the same age, since we have been best friends our whole lives, I was disturbed by her growing interest in where she would one day lie.

It was late afternoon when we turned the radio back on in the car. The announcer said the young woman shot in Kfar Saba had died. Before we got very high into the hills below Jerusalem, my cousin had made enough cell-phone calls to find a friend who knew her. The dead woman was a musician, a sweet and talented girl. She had played at a friend's son's bar mitzvah only a month before. She could have been, from this description, one of my nieces, one of my children. Now she was gone.

Immortality suddenly had its charms.

We pulled into the parking lot below the Hadassah Medical Center's Ein Kerem campus on the outskirts of Jerusalem just as the sun was going down. The sky was a pellucid indigo. The pines smelled absolutely delicious. I slowly picked my way up the stony path leading to the entrance, trying to take the time to savor everything, particularly the sharp clarity of the air. White lights winked on every hilltop as far as I could see. We emerged from the pines into a courtyard set between the hospital and its research buildings. My mother had reminded me to look at the famous Chagall chapel. She and her friends had worked for years to raise money for this hospital and its research efforts, holding bazaars, selling greeting cards. There the chapel was, a small building made of Jerusalem stone with a barrel-shaped roof, with large windows of stained glass. The light was on inside and color poured forth—rich reds, soothing blues, stabbing yellows—from Marc Chagall's ethereal figures. Animal and human, they drifted upward like stories from a campfire.

I had an appointment with Benjamin Reubinoff, who, along with Alan Trounson and Martin Pera, had come within a whisker of beating James Thomson to the first human embryonic stem cell line in 1998. Since then, Reubinoff had been learning how to turn these cells into neurons. Reubinoff's office was in the Mother and Child Center of the Goldyne Savad Institute of Gene Therapy, a new building with a big glass atrium. My cousin rode up on the glassed elevator with me, in case I got lost.

Reubinoff's office was tiny. There was space for his desk, a chair, a computer, a visitor. Along the edge of the desk he had arranged a group of learned papers, each one set precisely at the desk's edge, each one overlapping its neighbors by exactly the same span, like a cardsharp's slide shuffle. "Very, very brilliant," one of my brilliant friends, an Israeli pediatric urological surgeon, had said of him. Very, very neat too, I thought, but my friend, a neat freak in his own right, wouldn't have noticed that. Reubinoff looked so young; his skin was like a smooth olive, and his glasses gave him an owlish, boyish look. In fact, he was forty-three.

He is a physician, born and educated in Israel. He studied medicine at the Medical School of Hebrew University and took his residency at Hadassah in obstetrics and gynecology, specializing in IVF. After he finished his residency and worked at Hadassah and taught at Hebrew U. for several years, his sabbatical rolled around. In 1996, he began to look "for a place and subject." He found Alan Trounson, of Monash University's Institute of Reproduction and Development in Melbourne, Australia, already one of the leading figures in the quest to establish cell lines from human embryos.

Trounson had been working at this—and failing—for some time. He collaborated with Ariff Bongso at the National University of Singapore, the city-state which has identified biotechnology as an engine for surefire economic growth.[3] Both Bongso and Trounson were frustrated that they had found no method to keep embryonic cells alive and dividing in culture without differentiating. Reubinoff thought there was great promise in these cells, and he wanted broader horizons than research on infertility. "So I came to Australia in January 1998 and started to work on this project," he said.

Experimental work on human embryos is illegal in the Australian state of Victoria where Monash University is located, but there was no problem working on stem cells. In Singapore they were able to get ethical approval to experiment on human embryos surplus from IVF. So they split the task between the two jurisdictions. Martin Pera, a developmental biologist from Oxford, had been experimenting on human embryonic carcinoma cell lines, very similar to human embryonic stem cells except that they usually have too many chromosomes.

Reubinoff did the initial derivation process in Singapore. "After we established the cells, I took them to work on in Australia," he said. He carried the cells in a small test tube snuggled in his shirt pocket next to his body on the flight. Then he spent months trying to keep them multiplying without differentiating. A one-year sabbatical became two, and he got a PhD when he succeeded,[4] but not before James Thomson's success in Madison, Wisconsin. Thomson also worked with Israelis: Michal Amit and Joseph Itskovitz-Eldor, Amit's former teacher and head of Rambam Hospital's IVF department in Haifa. Itskovitz-Eldor had sent Thomson a dozen surplus frozen embryos donated by his Israeli patients, and four of the five cell lines Thomson made came from Israeli embryos. Itskovitz-Eldor carried cells from all five lines back to Israel before Thomson's article appeared in Science in November 1998.[5]

"We had them going before he [Thomson] published in Science," said Reubinoff. "So we took it further on to emphasize points not covered . . ." Reubinoff and his colleagues tweaked some of their stem cells into becoming different precursors for adult cells in the dish. "We call it 'somatic cell differentiation in vitro,'" said Reubinoff. They published in May 2000 in Nature Biotechnology and showed that human embryonic stem cells could make progenitor nerve or muscle cells.

So, I asked, how did you do it?

He explained that he borrowed from the recipe for how mouse embryonic stem cells are cultured. To support proliferation and to prevent the cells from differentiating in culture, "you need a feeder layer of cells. These are mouse embryonic fibroblasts that support mouse embryonic stem cells." To make these feeder layers, one must "mince a midgestation embryo," Reubinoff said. "You take out the organs and the head first, so it's mainly muscle and bone and connective tissue left. You mince it, use enzymatic digestion, plate it and grow cells, and you mainly get these fibroblasts. You plate them and put on the embryonic stem cells from mouse or from human origin."

There had been criticism of Bush's decision to fund study of human embryonic stem cells only from lines created before August 2001 because all of these early lines had been started on layers of mouse cells. Some thought it might not be safe to use tissues derived from them for human therapies, because they might be tainted by the mouse cells. I assumed Reubinoff and his colleagues had since found something better to grow human cell lines on, something safely chemical.

So, I asked, what do you grow these cell lines on now?

On minced-up human embryos, he replied. He explained, later, that they still use the mouse fibroblasts, but also human fibroblasts taken from a human embryo or foreskin. The embryonic fibroblasts come from a single human embryo.

I cringed. "Isn't there an ethical issue in that?"

"You can take it from abortions. In the human you can use earlier embryos, from the first trimester," he said. While this is ethically challenging, it can be done in several countries with the approval of the appropriate ethics committee—in his case, from the Hadassah Helsinki committee. He was trying to differentiate these cells so that one day he could make mature tissues to transplant into those in need.

But what about the problem of rejection, I asked. If these human embryonic stem cells did not come from my own eggs, for example, wouldn't I just reject the tissue made from them? Michael West and Jose Cibelli had pointed out many times that nuclear transfer cloning was necessary in order to provide patients with cells directed by nuclear clones of their own genomes.

There are strategies for dealing with immune rejection, Reubinoff said. They could take advantage of hematolymphoid chimerism, for example. One could create blood-generating stem cells from the cell line and then transplant these cells to the person in need. "There is some evidence you might be able to induce tolerance to these cells in the patient if you do transplant the blood cells derived from them first," he said. "It may be that embryonic stem cells will induce tolerance to their own progeny. So there are many ways to overcome this issue. Nuclear transfer is just one of them."

Well, I thought, there goes Advanced Cell Technology's business plan.

Reubinoff had just published a paper in *Nature Biotechnology* on human embryonic stem cells that he developed into three types of neurons.[6] He turned his computer screen round to show me. First, he and his colleagues had made a highly enriched preparation of human neural progenitor cells from a cell line. Then, he said, these were transplanted to the brain ventricles "of newborn mice."

I wasn't sure I'd heard him correctly, but I had. He had put human embryonic stem cells, tweaked to become neural progenitor cells, into the brains of live baby mice. He had first labeled these cells with a marker that can be incorporated into DNA. "You culture the cells with the marker and it incorporates into the host cell's DNA so you can identify the cells that have it." The mice were killed at certain developmental intervals and their brains and bodies examined to see where the human cells went and whether they developed.

These human-derived cells had developed in these mouse brains into three types of neurons: astrocytes, oligodendrites and mature neurons. They had migrated along established tracks and differentiated according to the hosts' brains' signals. He found some human neurons in the mice olfactory bulbs, the area responsible for processing smells, the same area from which Freda Miller had first taken adult stem cells to derive neuronal cells. His transplanted cells had continued to differentiate and perform, but, perhaps more important, they hadn't made tumors.

So you made mouse-human chimeras, I said.

"It is not mouse-human chimeras, but transplantation of human cells into a specific location within fully developed newborn mice," he said. "The important finding is that there was a conversation between the recipient and the transplanted cells. If you think that you want to transplant such cells in the future in Parkinson's patients, you want them to talk to the cells . . . and function there. These results are a clue this may be possible."

Okay, I said, I know this works between human beings, but how can this possibly work between humans and mice? Weren't you surprised?

He was not—he had expected it. It isn't just arrangements of DNA that are conserved across evolution, it is also gene functions, and the signals that cells send and receive.

The very idea of human nerve cells integrating into the brain of a living mouse made my neck prickle with dread. This reaction I was having clearly marked the limit of what I was willing to tolerate—I had looked at a dead body ravaged by cancer with much more equanimity. Would it have bothered me so much if the experiment had been run the other way, if a few mouse embryonic stem cells had been inserted and integrated into the brain of a child? I didn't think so. Yet putting barely differentiated human cells into a developing mouse seemed plainly outrageous. I could not help but think that a mouse had been given something of the human capacity for thought, which was silly, wasn't it? How could a young mouse, even with transplants of precursor human neurons, ever develop the capabilities of a human brain? The structures and layers of cells that communicate with each other to make our minds would have to be replicated too, wouldn't they? On the other hand, if human and mouse cells are so similar that they can communicate about development, and if human cells can move smartly along in accord with the signals of the mouse, the structure of mouse and human brains cannot be that dissimilar. Would the genes relating to human brain development operate fully in the mouse but produce a mouse brain's features? Would mouse stem cells put into a human brain integrate just as easily? Would the mouse cells make the structures of a human brain in the right circumstances? What shapes an organ if not the genetic program?

These questions made my hands sweat. Somehow they exposed the meaning of Darwin's insights into the oneness of living things in a way that nothing else had done for me.

Reubinoff was cautious about the meaning of his results. He had shown that transplanted embryonic stem cells can interact with the cells of the mouse and continue to develop, but he couldn't be certain that the same would hold true if these cells were transplanted into the brain of a person with Parkinson's disease. To be useful, these cells had to be able to correct the functional deficit caused by the disease.

This was the core question about this research that many critics pointed to. Why teeter on ethical razors if these cells, when transplanted, might not do the job? There had been very few papers showing that mouse embryonic stem cells restore lost function when transplanted into mice with disease. According to Reubinoff, there had been one in *Nature* that showed healing of Parkinsonism in rats. "There is an indication [that] they can induce or generate functional cells," he said. But he agreed that it was also possible that adult stem cells could do the same thing. "Efforts should go on with both adult and embryonic stem cells," he said.

Money to do the work was an issue. He handed me an article that had just appeared in *Science* the previous week on him and his Israeli colleagues.[7] There are four different research groups in Israel that collaborate on embryonic stem cells under one Israeli government grant. Their leaders include Reubinoff, Itskovitz-Eldor, Karl Skorecki and Nissim Benvinisty, who made the first stable genetic alteration of a human embryonic stem cell. (He inserted a gene for a green fluorescent protein that shuts off when the stem cell differentiates.) Such cooperation seemed strange to the writer at *Science* but familiar to me; it reminded me of the Canadian Stem Cell Network. These groups can cooperate because they have divided areas of expertise. One works on blood cells, another on neurons, another on liver, another on cells that make pancreas. The story did not mention that Reubinoff was also involved in another collaboration; a company headquartered in Singapore to exploit the work of

Reubinoff and his colleagues. Alan Colman, formerly of PPL Therapeutics, had been hired as its chief scientific officer.

This company, ES Cell International, was put together as a consortium of institutions; it includes Hadassah, Monash University and the National University of Singapore. The Hubrecht Laboratory, of the Netherlands, had just joined as well. The company will "commercialize the intellectual property of the consortium." Reubinoff and his colleagues were members of the company's scientific advisory board. ES Cell was set up when Reubinoff was in Australia, with investment from the government of Singapore (through Life Sciences Investments) and private investors in Australia (ES Cell Australia, Pty. Ltd.). Reubinoff handed me a slim corporate brochure. Among other things, it stated that the company had seed capital of $17 million in Singapore dollars and intended to have products to sell by 2005.

I was surprised Reubinoff had taken a hand in starting a company to sell the products of his labors. Reubinoff seemed to be completely absorbed in being a doctor, a teacher and a researcher, and utterly disinterested in profit. Why a company? I asked.

"The issue that motivated the scientists who derived these cell lines is to try to propagate the research of the area of embryonic stem cells for the benefit of mankind," he said primly, apparently offended by the idea that he might try to profit from his research. The problem, he explained, is that they were all dependent on grants to do their work, and the granting agencies at first wouldn't help them. "At an early stage, we had cell lines, but we couldn't raise funds in Australia because of the ethical issues in this area. We thought if we need to propagate the area we require funding and it's a problem to get it from academic agencies, so we need commercial backup." James Thomson, in the United States, had solved this problem by getting money from Geron. "All the scientists share this view," said Reubinoff. "We want to collaborate with commercial entities to support the development of the area. We are not working for a firm. We are free to do research we are interested in."

The grant situation had changed when President Bush made his announcement. Since August 2001, U.S. granting agencies had asked Reubinoff, as they had asked Freda Miller, to put in applications.

"How do you see the human embryo, anyway?" I asked as I gathered up my notes. "Are you religious? Do you see it as a repository of the soul, or is it a collection of cells?"

"I am not religious," he replied. "I would say I see the embryo as a collection of cells. My perspective is this is just a group of cells. Although there is potential, I didn't have the feeling this is a human being."

Lior Gepstein and his colleagues were the first in the world to turn human embryonic stem cells into cells that act like heart-muscle cells.[8] Gepstein was also the soul of kindness. He had a major experiment going, mapping the interior of the heart of a pig, when my cousin and I called him for directions from the car. We were lost in the welter of narrow one-way streets that girdle old Haifa. He set us straight, and even arranged for us to park in the completely jammed lot between Rambam Hospital's campus, where he works, and the Israel Defense Forces base next door.

Gepstein's lab is in the Bruce Rappaport Faculty of Medicine, which is part of the Technion-Israel Institute of Technology, a major Israeli research center affiliated with Rambam Hospital. "Rambam" is the English transliteration of the Hebrew acronym for Rabbi Moses ben Maimon, sometimes known as Maimonides, a philosopher and physician who fled his native Spain to North Africa as Christians began to gain back territory conquered by Moslems. He eventually became Saladin's doctor. The Rappaport Institute is a high, concrete tower of 1970s vintage. In the front, beside the wide steps, a waterfall of bright red bougainvillea poured over a low stone retaining wall. Inside, past a very large, empty lobby, we found an empty cafeteria, where we waited for Gepstein. My cousin had told him what we look like. "We'll be the only old people," she'd said. This turned out to be true. When he shambled in, dressed in hospital greens, he looked like a graduate student in his mid-twenties; he was actually a ripe old thirty-four. He was a little heavy, with sloped shoulders, gold-rimmed glasses, very pale skin, a sharp nose, a wide and constant smile. His eyes shunted here and there at great speed, no doubt in time with what was going on behind them.

He took me to a tiny, shabby office with a window on the corridor. There were no pigs in sight, but he hadn't stopped worrying about them. I had noticed his name on the Technion's website when I was searching for a phone number for Itskovitz-Eldor, whom everyone seemed to refer to simply as Itskovitz or Jossi. Itskovitz was out of town and unavailable, and the Technion PR person had recommended Gepstein instead. Gepstein is cardiologist with a PhD in physiology. He spent eight years working on two of the most common causes of heart disorders (cardiac rhythm disorders), and heart failure, one of medicine's biggest problems. "We know why it happens, and the consequences, but there is no cure to date," he said. "It is responsible for more hospitalizations than all cancers combined. About two million Americans are suffering from heart failure, the inability of the heart to pump enough blood . . . This leads to reduced strength because the heart doesn't eject enough blood forward; fluids stay in the lung . . . The person feels short of breath, they may develop pulmonary edema with the lungs filling with fluid that prevents proper oxygen diffusion. Patients have reduced ability to work . . . In advanced cases, 50 percent of these patients last less than two years."

The only option for severely ill patients is a heart transplant. Those with less severe problems are treated with drugs to flush the lungs and reduce congestion. Other drugs help the heart contract better; still others reduce blood pressure. These drugs slow the advance of the disease, but they don't cure it. When an area of the heart is denied sufficient blood long enough, the muscle "undergoes necrosis." In other words, the cells die. Muscle is replaced by scar tissue.

"Adult heart cells cannot regenerate," Gepstein said. "They don't divide. . . " Or this was the dogma. There were too few hearts to transplant, and those who aren't quite sick enough to warrant an organ transplant have had to make do with a much-reduced quality of life. There had been various experimental attempts to transplant cells instead of whole organs, but the problem is lack of sources for human cardiac tissue. In most animal studies, the cells used were either not heart muscle cells, or were heart cells taken from fetuses. "For ethical and practical reasons, this is not good," Gepstein said. "So we tried skeletal muscles. A small population of cells in skeletal muscles, the satellite cells, do divide. Investigators isolated them, propagated them in vitro—up to 100 million myoblasts [muscle cells] to implant in hearts."

But there were difficulties using these cells too. Heart cells need special kinds of connectors, called *gap junctions,* "through which electrical activity is propagated in the heart. That's how they get synchronous contraction of hearts. Skeletal muscles have no gap junctions, so they can't contract as heart muscle."

The evidence from animal studies had shown some improvement in heart function after the injection of skeletal myoblasts, but "it would be ideal if we had heart tissue," said Gepstein.

One day he bumped into Joseph Itskovitz-Eldor, who was looking for Israeli collaborators to see if something could be made of his embryonic stem cell lines. He asked Gepstein if they could be used to treat heart disease.

"I didn't know people were working on this. I thought it was crazy to transplant cells," said Gepstein. "But I read and saw it was an opportunity, the magnitude of the problem and the interesting questions. Now 60 percent of my lab works on this."

And what were these interesting questions?

"Why do cells become this kind of cell and not another during development? This is the most interesting question in biology. And after they become heart cells, how do they start to beat?" he asked. And how do developing cells know which part of the heart to move to? Is there some kind of program, or is it all random—just a matter of a cell being in a certain place at a certain time getting corralled into a particular job? None of these questions had even occurred to him when he was in medical school.

When he talked to Itskovitz-Eldor, Gepstein already had his own lab, with his own projects on the electrophysiology of the heart. He got three of his graduate students to look at whether or not these stem cells could be useful. And no, he had no ethical qualms about doing experiments on cells derived from embryos. He knew that they were derived from blastocysts left over from IVF treatment, and that the patients had donated them.

The first thing he and his group did was to try to derive heart cells from a cell line. A number of research papers have since been published by several groups demonstrating that these cells can make all two hundred tissue types in the human body. "What's interesting is that ten out of the first twelve papers are from Israel," he said.

He and his students grew the cells in a colony on a line of feeder cells, which allows them to multiply in an undifferentiated state. Then they removed them from the plate and grew them in another dish in suspension. The cells formed clumps called *embryoid bodies*. After ten days they put some of these clumps in a flat plate. And lo and behold, two to twenty-two days later, about 8 percent of those cells began to beat spontaneously. "They beat for weeks," Gepstein said.

His computer showed a digital display of a dish with a layer of cells. Sure enough, they pulsated. Beat.

His students ran these beating cells through a battery of tests. They looked at their structure under an electron microscope and saw that they looked like fetal heart cells at a very early stage. "We can follow it through time, and it changes," he said. "It doesn't reach the adult type, but it matures to near an adult cell. You can see how the structure forms. It starts to beat early in the embryo, at four weeks. It has to work against a load of resistance. Does this shape the development of the heart? No answer to that yet. We know that without working against a load, it doesn't reach a full phenotype. We did immunostaining to look for cardiac-specific proteins. We found it was positive for all types of heart proteins—all we tried—and negative for skeletal muscles."

Then they looked at the function of the cells. They gave them adrenaline to see if they beat faster. They did. They checked the way these cells handled calcium (which plays a role in the penetration of molecules into cells). "If it looks and acts like it, then it is a human duck," Gepstein said, mincing his metaphors.

"The next question we asked is whether we only get individual unrelated heart cells in a dish, or whether they'll function together as one tissue. To test that, we used an electrophysiological recording system of sixty electrodes, each 100 micrometers apart, in a grid." They lifted out the cells that were contracting, or beating, and put them on these electrodes and recorded their electrical activity. The activity at each electrode was analyzed and the information was color-coded. Maps were generated, which showed the propagation of electrical activity from one cell to another. The source of the activity was a pacemake area generated in these cells.

"We know [that] in the heart there is a pacemaker that drives this. It starts spontaneously and spreads to the rest of the heart. In our generated tissue we see a similar phenomenon with the electrical activity starting from this area and then spreading. We can measure this for weeks. It always starts in the same area."

After going through all these tests, he realized what he had. The beauty of these almost-heart cells beating in a dish is not only that scientists might be able to transplant them as healthy cells into diseased hearts, but also that they might be used to safely test heart drugs on human cells. "This is a great way to see what the drugs are doing. We usually go from animal to human clinical trials. But to specify what does the drug do to a human cell—this may help," said Gepstein. "If it does terrible things to the electrical activity in vitro, it won't go to a clinical trial."

Having established that these cells are just like human heart muscle cells, they tested them to see if they would integrate with other cardiac tissue. They took a rat heart, minced it, grew its cells in culture. Gepstein showed me an image of the resulting rat cells, which had been stained. In one quadrant, the rat cells beat slowly.

Then came another one of those Dr. Frankenstein experiments that these researchers seemed to arrive at, inevitably. "We added human tissue, put it in beside the other to see if they would integrate and beat together. We saw clear integration."

And so did I. The next image showed the two kinds of cells, red for rat and green for human, beating as one.

"How do they do it?" he asked me. It was a rhetorical question, of course, because I was watching his screen with my mouth indecorously open. So he answered himself. "They form gap junctions between the two tissues, and electrical activity can propagate from one to another."

But of course, this didn't answer the real question. That they beat in concert is interesting; that junctions formed between two different species' tissues was much more strange and alarming. Gepstein's computer was livid with red and green cells, the green on one side, the red on the other. A slow beat started in the red and then, gradually, it moved to all the cells in the dish.

"Again, it starts in the red, the rat, and activates the rest of the tissue—it propagates as one unit," he said.

Gepstein and his colleagues had also tried this experiment in the opposite direction, propagating the beat in human almost-heart cells, watching it spread to the rat cells beside them. They put a yellow dye in the rat tissue, and as the two groups of cells integrated, they saw the dye spread, in twenty-four hours, to the human cells. "It propagates through the gap junctions," Gepstein explained.

I was almost as unnerved by this two-way communication between rat heart cell and human as by Reubinoff's human cells integrating into a living mouse's brain. These almost-heart cells were deeply troubling. First, they beat all by themselves, without being organized into an organ. In fact, one cell, for reasons unknown, always kick-starts the process. These cells somehow self-select for this particular behavior. But in addition, and possibly more peculiarly, two different kinds of cells, from two different species well separated by evolution and which normally have very different heart rhythms, also beat together as one.

I asked what in hell this meant.

"This means the basic proteins are very conserved in evolution," he said, smiling, as if this was an explanation. But it simply raised more questions. They are conserved, and yet they obviously differ from one kind of animal to the other. They are not identical, and yet they can cooperate, and either can lead the beat.

"There is almost no change in the amino acids," he continued. "Nature doesn't allow many mutations, and the gap junction is a very important intercellular structure."

Gepstein was a doctor: he was focused on how he could make use of this phenomenon more than on why this phenomenon exists. "But what if we inject in the heart?" he continued. "They have to survive in a hostile environment. Cells already died there—there's been an infarct."

His mentor during the preparation of his PhD thesis, Shlomo Ben-Haim, had invented a method to map the area of a heart killed by an infarct, developing a catheter with a magnetic sensor. When the catheter moves through the artery to the heart, its progress can be monitored by following the magnetic tip, which can also be used to record electrical activity. The normal use for this device is to find rogue pacemaker areas in a heart that are beating to their own time instead of following the pacemaker cell, which causes certain arrhythmias.

"We zap it with a radio frequency current which can heat and destroy the rogue cells," said Gepstein. "These patients are cured. This was developed here at the Faculty of Medicine, and this is a state-of-the-art technique for treating arrhythmia." Gepstein thinks this machine can be adapted to directing these almost-heart cells to the area of an infarct. "So you want to inject these cells only in the scarred area," he said. "You can see the scar area because there's no electrical activity."

He had already done preliminary experiments on pigs, doing test burns in the area of an infarct. When they kill the pig and examine the heart to see if they have actually burned the right place, they find it in the area of the dead heart tissue.

"In the future we hope that following a heart attack, let's say two weeks later, we can transplant our cells together with a procedure that opens the occluded artery with a balloon or stent. We can then use the mapping system to identify the dead area, and inject the cells in that area," said Gepstein.

I was delighted by this cleverness.

But there were still problems, he said. He needed to have pure colonies of these cells, and lots of them. While he could grow a few thousand almost-heart cells in a dish, to do proper cell transplants he would need to have millions. "We need to increase the yield of heart tissue, to push them to be heart cells by using a combination of different growth factors. There is no single growth factor that can by itself push an undifferentiated cell to become a heart cell. . . . heart development is such a complex process."

If they did manage to figure out which growth factors do which job and figure out the signaling system that operates between cells, they would still face the problem of rejection. But he wasn't worried about that, not really—his colleague Karl Skorecki was grappling with that. One point he did want to make, though. "We are using for three years something generated by a single embryo that was to be destroyed," Gepstein said. "The public should get the right information—this should be discussed."

Karl Skorecki was obviously a man of greater parts than Lior Gepstein. In fact, he is the director of the Rappaport Family Institute for Medical Research, which also supports Gepstein's work.[9] He was delayed somewhere, so my cousin and I were asked to sit inside his office to wait for him. It was a large room, well furnished, with a wonderful view. The sun was hot and there was a brisk sea breeze that kicked up a fine haze of salt and beach sand. Two walls of windows revealed a compelling arc of the Mediterranean all the way up to the border with Lebanon. Down below, young soldiers in full battle dress were bobbing in the water in a flotilla of inflatables, practicing assaults from the sea.

Skorecki came in with a colleague from the University of Washington. He was dressed in cords and vest, with a keepah on his head. I had first noticed his name on a paper coauthored by Mike Hammer of the University of Arizona. Hammer's lab uses Y chromosome markers to trace human population movements and descent from a male ancestor in the same way mtDNA markers are used to trace descent from a female.[10] Their paper described markers on the Y chromosome that are particular to Cohens, the putative descendants of the ancient Jewish priest class, founded by Moses' brother Aaron. These men, according to the Bible, were set the task of caring for the Holiest of Holies: the ark of the covenant. Descendants of these original Cohens still take precedence over others in some Jewish religious ceremonies. They are also prohibited from washing the dead and from marrying divorcées. Skorecki wanted to know if their Y chromosomes had markings indicating a true line of descent from some original small group of Jewish males. He found they did. It was odd to see this materialist science used to test the claims of a patriarchal class system ordained by God. It was not the kind of work one would expect from someone trying to coax human embryonic stem cells to develop into pancreatic cells, Skorecki's current task.

Skorecki, a Canadian born and educated in Toronto, came across stem cells very early in his career. In his second summer at University of Toronto Medical School, he worked on stem cells in E. A. McCullough and Jim Till's laboratory. Skorecki got his MD in

324 • ELAINE DEWAR

1977 and went to Boston for postdoctoral work in nephrology and internal medicine; his special interest was why human kidneys fail. This eventually brought him back to stem cells. He came to Israel on a sabbatical, to the Weizmann Institute, at the start of the Gulf War in 1991. He wanted to learn about gene expression and the promoters that regulate genes. He and his wife decided then to move to Israel permanently, which took him four years to organize. By 1996 he had managed to scare up a $2.2 million donation from the Canadian Technion Society, some of which goes to his work. Shortly after that, he had a talk with Joseph Itskovitz-Eldor.

Itskovitz had organized a small meeting at the Weizmann Institute to coincide with the award of the Wolf Prize, a major Israeli honor in arts and sciences, to Neal First, to celebrate his success with bovine cloning. First and various others involved in stem cells and cloning, including James Thomson, were invited to attend the meeting. Itskovitz-Eldor knew that Thomson had made progress on getting embryonic stem cells from primates, and he made an agreement with Thomson to provide him with donated human embryos. He also knew that Skorecki was working on telomeres and gene expression and how DNA replication is modulated by kidney disease; Skorecki was parsing the human version of the intricate dance between signals and genes that Don Riddle was searching out in the dauer behavior of the worm. Itskovitz-Eldor told Skorecki about Thomson and asked Skorecki if he would be interested in trying to work with human embryonic stem cells.

When Itskovitz-Eldor came back from the United States with the first human embryonic stem cell lines, Skorecki joined the collaborating groups and offered to focus "on insulin-producing cells and the whole issue of telomerase."

He was interested in beta cells, as insulin-producing cells are called. "A leading cause of kidney failure is a result of complications of diabetes . . . The beta cell is responsible for sensing blood sugar and releasing insulin in a regulated way," he said. He wanted to derive these beta cells from human embryonic stem cells taken from one of Itskovitz's lines, and eventually to use them for therapy.

"We first said, let's see if we can find insulin-producing cells." His group easily detected gene expression markers for insulin in the general population of cells derived from a stem cell line. This work was done by Suheir Assady and was published in the journal *Diabetes*.[11] Skorecki was surprised at how easy it was, because beta cells are a very small fraction of the trillions of cells in a human body, and they have not been so easy to derive from mouse embryonic stem cells. On the other hand, he said, "why shouldn't they be there? Now the idea is to enrich them, to push the cells to beta cell fate, then purify them."

In this task, he had the same problem as Gepstein. He could probably get a pure population of a few thousand insulin-producing beta cells in the dish, but to cure a diabetic he needs 200 to 300 million such cells.

So he was working on a way to use genetic means to trap only the cells he wants, following the method of Loren Field, at Indiana University, and others who were trying the same thing in mice. The insulin promoter is turned on in beta cells. To find these cells, one can stick in a marker and hook it to that promoter. One could use a fluorescence marker, for example, or antibiotic resistance.

"If we succeed, we'll have to contend with patents," he said.

What about rejection?

"I think about rejection a lot," he said. "It's one of the least problematic areas." He thinks there is evidence for greater tolerance of tissue transplants than organ transplants as a result of chimeric mechanisms. The second solution is to create banks of cell lines and use one that is compatible. An expert immunologist at Harvard, Terry Strom, had given him the opinion that only fifty to eighty different cell lines would be able to produce cells compatible with 95 percent of the world's population. Skorecki thought it might be possible as well to instill tolerance-inducing molecules into cells, something impossible to do with organs. Like Reubinoff, he didn't think nuclear transfer would be necessary to forestall rejection. In fact, he thought there were still questions about it: "The question is whether my forty-nine-year-old nucleus will be reprogrammed, rewritten—it doesn't need to be made young again."

He hadn't mentioned adult stem cells. Surely stem cells from one's own body would be the best solution of all? But he didn't think it had been shown that adult stem cells would be able to proliferate enough to be of more use than embryonic stem cells.

"But haven't you seen Freda Miller's work?" I asked.

He hadn't, which I thought was strange. After all, it hadn't been in some obscure journal—it had been published in *Nature Cell Biology*.

Do you guys avoid reading stuff that might get in the way of using these embryo lines? I asked.

He tilted his head to the side to consider this. "I think we are so preoccupied, and there could be some bias—maybe a prior notion makes you put it on the bottom of the pile."

"You're religious, right?" I asked. "Do you have any ethical qualms about working with human embryos?"

Yes, he replied, he is observant. And when this first came up, he thought about it a lot. "I talked to rabbinic authorities. I wrote a few pieces about the ethical and religious dimensions." He had come to the conclusion that there is no problem with this work in Jewish religious law; his colleague Michel Revel had laid this all out very clearly. He was certainly aware that other religions went in other directions. "Scientists have to be responsible and not belittle concerns nonscientists might express. But"—he wagged a finger at me to emphasize his point—"not be paralyzed by them."

So, I said. Do you see the embryo as a human being or as a collection of cells?

"I see it as a collection of cells," he said, without a moment's hesitation. "That's how Jewish halachic law sees it. There's a progression, an increase in sanctity, from implantation. Until it's in the uterus, there's no such progression. It's more than a piece of paper, but way far from the realm of the sanctity of human life."

Getting an appointment to see Michel Revel had been, to be mild, difficult. He had ignored my e-mails. My phone messages were never returned either. I'd enlisted my cousin, who caught him in his office by telephone one day. He yelled at her that he had no intention of making time for the likes of me. Then I had a little talk with a public relations person at the Weizmann Institute. Finally, Revel had e-mailed to say that because of the "security situation" in Israel, other important talks had been canceled. He would be able to see me after all.

I poked my head into his lab at the Weizmann Institute's Arnold R. Meyer Institute of Biological Sciences to see if he was inside, but there was no one there. The place was in remarkable disarray. Books were strewn, lab glass was every which way. I had just about given up waiting when he marched through the building's glass doors, half an hour late. He wore one of those navy blue European student hats on his head, the kind with a visor so small it's barely there. He was a short, tubby man, with white hair and a goatee and a certain officiousness, as if he had modeled himself on a first-class French bureaucrat but had since gone to seed among bumpkins. He led the way to his office, which if anything was messier than the lab: bookshelves lined the walls, books and papers rioted for space and a whiff of mildew arose from leaning towers of paper on the floor.

Michel Revel was born in 1938 in Strasbourg. He obtained a doctorate in medicine and another in biology at the University of Strasbourg by the time he was twenty-five. One of his supervisors was François Gros, one of the first generation of molecular geneticists. But Revel was also a religious scholar; he had studied with his uncle in one of the better-known yeshivas at Montreux. He specialized in kabbalah—mystical writings about the hidden meanings of sacred texts—as well as in the ethics of medicine, which he also teaches. In 1968, after the Six Days War, he moved to Israel permanently with his wife and four children. He set up the Weizmann's Department of Virology, and then its Department of Molecular Genetics. He wore so many hats, he was practically a hat tree.

It had fallen to him and the Bioethics Advisory Committee to figure out what Israeli scientists should do about embryonic stem cells and human cloning. Of course, his views were known when he was appointed committee chairman. In his 1998 article in *The Scientist,* published before James Thomson and Benjamin Reubinoff both successfully derived human embryonic stem cell lines, let alone perfected nucleur transfer in humans, he had argued that cloning research must be allowed to continue within agreed-upon guidelines "because it entails benefits that should not be scrapped outright because of perceived risks." In effect, he presented this work as the lesser of two evils. He raised the specter of the alternative: the use of abnormal fetuses "with no identifiable head and chest" to provide organs for grafts. He raised the possibility that such fetuses, considered human beings by no one, might be generated on purpose. Cloning could forestall this: "The unavailability of organs is becoming a major death factor that will worsen unless scientists successfully culture fully formed organs. Today's organ traffic could appear to many a more dangerous peril than producing one's own cloned embryo for auto graft. Ethics should weigh the rights of the individual to benefit from scientific advancement against the risks and strictly regulate or sometimes delay rather than condemn. Demonization of science and upfront banning of potential new technologies is not a dignified human answer."[12] Revel's bioethics committee reported its findings, which were pretty much in line with his earlier statement, in August 2001, around the same time George Bush announced his decision on funding research on human embryonic stem cells. Among other things, the Israeli committee recommended the creation of embryos by nuclear transfer into oocytes for research purposes, as well as the use of surplus IVF embryos to create stem cell lines. "It should be recalled," the report argued, "that the purpose of bioethics is not to ban upfront scientific advances, particularly in the field of medicine, but to define the limits of the socially desirable and ethically permissible."[13]

I barely got my first question out before it was lecture, lecture, all the way. Revel was a man with a nicely ordered mind, with plantations of ideas connected by precise roads of reasoning. He would start down a path, then digress down a side path, which would generate more side paths, and then he would swing back, quite neatly, to the main line.

He and his committee had based their ethical arguments on the Jewish and Islamic agreement that life begins about forty days after fertilization. So my first question was, when does Jewish law say the soul enters into the living body? This is the critical problem for those against cloning and stem cell research: they argue that it is not right to use an embryo because ensoulment happens at conception and sacred human life continues until flesh is parted from immortal soul at death.

Revel pushed back in the chair and looked over my shoulder as he talked, as if he was addressing an imaginary class. "The Jewish tradition doesn't say when the soul enters in," he said. "Muslims say it occurs in forty days, or three times forty days. Most [religious authorities] we talked to stick to the first forty days. Jewish tradition doesn't speak of it at all."

Revel said some tracts in the Talmud, a vast compendium of Jewish civil and religious law and commentary, describe a forty-day embryo as being like water, without form. He could not say whether Jewish scholars in ancient times actually studied embryos at early stages, although he was certain that some observations were made because many genetic diseases were well described in the Talmud. In his view this number—forty—was symbolic, recapitulating the number of days Moses spent on Mount Sinai and the number of years Jews wandered in the Sinai desert after the escape from Egypt.

"A period of formation [that is] symbolic," he said. "Again, it's not saying the soul, but the embryo is formed. Kabbalah says the soul is only at birth."

Aha. A mystical *Roe v. Wade*.

In 1998 Israel passed a law temporarily banning human reproductive cloning and germ-line genetic alterations, to be reviewed in 2003. "It is a balanced law," Revel said. "Not like many other countries, where cloning is considered against human dignity." France and Germany, for example, had tried to organize an international declaration to that effect. "It's not the way we look at it," he said. "We look at any technology as part of the duty of man to mend creation. If it can be proven to have application, and it's safe, the question of dignity does not change the situation . . . Twenty years ago, the in vitro fertilization technology was branded as being against human dignity. I remember a priest of the Pontifical Academy saying man is born in an act of love, not in a test tube. We all see millions of parents have children through IVF. I don't think it destroys human dignity."

But the new law did not mention stem cell technology or therapeutic cloning.

"After Thomson and Itskovitz's work was published, we realized we needed a position," said Revel. "The first one we worked out was the report on stem cells."

Viewed in the light of Revel's report and suggested guidelines, Itskovitz's initial acquisition of embryos for stem cell research from patients had not been handled as one might have wished. Itskovitz works with infertile couples at Rambam Hospital, and the embryos used by Thomson were donated by some couples being treated in Itskovitz's department. Itskovitz had applied to his local hospital's Helsinki Committee, which rules on whether planned experiments on human beings meet ethical requirements, and he got permission to go ahead. But Revel's committee had recommended that such donations should be asked for by someone independent of the IVF physician, someone patients would not fear to displease by saying no. "If he did today what he did then, there would be a violation of the regulations," Revel said.

The Israeli Ministry of Health had established a body to examine and approve proposals for genetics and stem cell research. Physicians who want to do embryonic stem cell research cannot ask their own patients to donate surplus eggs or embryos.

As for reproductive and therapeutic cloning, a problem for Revel is that to do the research to make it succeed will require the donations of hundreds of eggs, and there is already such a shortage of eggs in Israel that it had led to a scandal. The head of a department of gynecology had been caught taking too many eggs from women treated in the IVF clinic in his hospital and then selling the extra eggs to women in his private clinic. Revel's committee had made it clear that embryos are not to be made into commodities for sale. The committee took no position on the sale of immature eggs, or oocytes which is already regulated in Israel. "New regulations will allow women to donate oocytes," he said. "They will be paid for the service."

The Canadian Standing Committee on Health had been adamant that buying and selling of egg and sperm are just as wrong as the buying and selling of embryos, I said. They don't even want anyone reimbursed for expenses.

"It will not be a transaction of paying for oocytes, but for reimbursing for time and pain and trouble," he replied. "The life has been disturbed." He thought oocytes needed for nuclear transfer experiments could possibly come as donations. "But another possibility is to use nonhuman oocytes."

Oh Lord, I thought, spare me. This mixing of the human and the animal made my skin crawl. Yet the more I thought about it, the more it seemed to me that my disgust was irrational, just another facet of the superstitious layers wrapped around the Other One like leaves on a cabbage. What Reubinoff and Gepstein's work showed is that cells of some mammalian species are virtually interchangeable with ours. Genes might be arranged in a different order on a different number of chromosomes, but the basic proteins are the same and their behavior is too. If I was willing to eat a cow, why wasn't I willing to consider using a cow's egg in this way?

Revel said he could accept it because what matters is the attempt to cure the sick. He had no problem making a stem cell derived from the transfer of a human nucleus into cow's egg to help a diabetic. "That it has gone through the cytoplasm of cows doesn't make a difference," he said. "But it doesn't look nice in the eyes of the public."

Revel's committee's recommendations were accepted by the scientific community. The Knesset had not held hearings, although its Science and Technology Committee had the right to inquire. Little of the argument pouring forth from the U.S. Congress or the Canadian and British parliaments had been heard in Israel. "We do not consider cloning [to be] against human dignity," Revel said.

But who constituted this "we"? The list of his ten committee members appeared to include only Jews and Moslems, no Christians, at least none I could spot by name. I found this miracle of consensus hard to believe, especially in Israel, where disagreement proliferates like Russian mushrooms in springtime. Surely some of the rabbis in Israel, not to mention the various Christian and Muslim religious authorities, had raised concerns about these manipulations of human nature? Ultraorthodox rabbis forbid the completion of a house excavation if a fruit tree might be harmed or if human remains might be exposed. Revel himself had said, in his article in *The Scientist*, that religious couples consider the use of donated sperm or egg adulterous.

"The rabbis would prefer reproductive cloning to anonymous gamete donation," he said.

"Why?"

"They hate that there is the danger of unknown donors," he said. They were afraid of inadvertent incest, of a daughter being impregnated by her father's sperm anonymously donated to a sperm bank,

332 • ELAINE DEWAR

or that a brother and sister, conceived through IVF but raised by separate families, might unwittingly marry each other.

"Sweden has decided for that reason to have the name of the
donor available. They can look it up when they get married," said
Revel. He knew that legislators in other jurisdictions fear that people will stop making donations if children of IVF could grow up, consult the donor lists and present themselves at donors' doorsteps.
"The rabbis say reproductive cloning is a better technique. At least
you know who is the father . . . I'm not saying everybody—some
don't accept IVF. But those that do will accept cloning. It's very different when I go to UNESCO. They look at me . . ."

He fell silent and shook his head. He seemed sensitive about
how Israel's policies are seen on the international stage. It was not
that he was ashamed of the Jewish religious view of these matters,
but he was of an age and background to have intimate knowledge of
the worst forms of anti-Semitism.

"They condemned cloning," Revel said. "The Universal
Declaration on the Human Genome includes a sentence that
human reproductive cloning shall not be permitted. This was added
afterwards on the requirements of the German government, who
threatened to leave UNESCO if this was not included. This was in
1997. It was a purely political move . . . We could only take exception to this. Israel, Canada, the U.S. made exceptions to this issue.
We said it's not fair to say this technology is against human dignity.
There is a lot of politics. The German position in bioethics is
extreme. They don't allow embryonic stem cell derivation." He had
received many visitors from Germany, including the head of the
official opposition. Germans had had a serious debate on embryonic stem cells. "Their position is they can't allow this type of
research," he said. "They are so afraid of what the Nazi physicians
did. They fear everything. They are the main motor in the UN to
ban cloning. The French are alongside . . . The French prime minister is in a position to allow therapeutic cloning, but President
Chirac calls it the worst crime against humanity . . . It's all politics;
they are not interested in science."

It was not that Revel had no sympathy with those who, for religious reasons, could not accept the idea of embryonic stem cell

research or any kind of cloning. "It's fair to recognize that for Christian believers the problem of the embryo is serious," he said. "If you are a person from fertilization, then it can be considered as murder to end the life of an embryo . . . If you consider it a crime, you can't commit crime to save the life of another, to take a heart or kidney, it's a crime. You could say in the Jewish view it's not a crime because the embryo does not have human status. The real halachic point is whether or not it is implanted."

The Jewish position is logical, just as the Catholic position is logical; both follow from their premises. But in Revel's view, the Catholic premise is also "extreme." "I heard from one professor who says every embryo [produced] in IVF has to be implanted. That's wrong in medicine." Only 30 percent of embryos are actually healthy enough to be implanted, he said. "I can understand their position, but it leads to decisions that seem to me irrational," he said. And Britain, a predominantly Christian society, accepts the use of embryos to make stem cells.

Well, most of the British, I thought, but not Tory MP William Cash. The same man who raised the alarm about the privatization of the Dolly patent had insisted years before that Parliament examine the moral problems of the creation of embryos.

One of Revel's British colleagues on the UNESCO committee had taken things too far in the other direction for Revel's tastes. Alexander McCall Smith had suggested that UNESCO declare that the embryo is a collection of cells like any other tissue. This was not far from Gepstein's and Reubinoff's views. But Revel could not agree: the embryo is something more than that. "It has the potential to be a person . . . We all are the issue of this collection of cells— but it's not enough to be a human being," he said.

Oh yes, I said, necessary but not sufficient—this was the first really important distinction I had learned studying reasoning back in the palmy days of my youth. The embryo is a collection of cells that is necessary, but not sufficient by itself, to do the job of making a living person. It must be placed in a receptive uterus and carried to term to make a human being. But surely the same will soon be said about human oocytes? An oocyte is necessary, but not sufficient, to make an embryo; in fact, in the right conditions, oocytes can become embryos without fertilization. In the case of Dolly, an unfertilized oocyte with a transplanted nucleus became a living sheep. It is also conceivable that one will eventually be able to say the same about certain skin cells. What if further study of Freda Miller's stem cells derived from dermis shows that if they are placed in the right growth factors, they can be transformed into the equivalent of an oocyte? What if eggs, embryos and skin cells can all be described as cells that have the potential to become human beings? It seemed to me that the insights that these new experiments have provoked into the utter plasticity of life make many previously reasonable arguments obsolete.

So, I asked, how did you deal with making an embryo for the purpose of doing experiments?

"Maybe we are Jesuitical," he said, "but we said to produce an embryo a priori from sperm and an oocyte shall not be allowed. It makes the embryo a means and not an end. It's a means to make an embryonic stem cell. It is not morally permissible. In IVF the purpose is reproduction—but with leftovers to be donated. In therapeutic cloning, we allow it to be produced for research."

This left me scratching my head. For the life of me, I couldn't see any difference. In the manufacture of embryonic stem cells, one fertilizes an oocyte with sperm, grows the resulting embryo in culture, then destroys it for its stem cells. In the manufacture of a cloned embryo, one would take out an oocyte's nucleus, put in another, stimulate this reconstructed oocyte to divide as an embryo and generate a cell line. One method makes an embryo; the other would make an oocyte behave like an embryo. It seemed to me that both turn oocytes into means. Another little mantra I learned in school is this: a difference, to be a difference, must make a difference. So what difference was there here?

"You are using nuclear transfer and these cloned embryos can't be used for reproduction," he answered.

Well, sure they can, I said. If you can make the cloned embryo divide, you can implant it.

"I think it will take a long time to be sure that technique will work to make a cloned baby."

He hadn't answered my objection.

"I'm sure we should not prevent this research," he said. "It's the only way to overcome the technical difficulties."

That was not an ethical answer either. Wait, I said, this isn't about technical difficulties, this is about principle, isn't it? The difference between means and ends? Ends cannot justify means?

"So we allow to produce a cloned embryo a priori for research. Some don't understand," he said. "You can't do it with sperm, but you can do it with nuclear transfer. I used my prerogative as chairman to introduce it. People said it's not logical."

I agreed with that. It certainly was not logical.

"My only argument is, it gives the message the embryo is the means. We don't want to give this message. I agree this position is weak. I know how it looks abroad. I don't think we want to be seen to produce them commercially."

His concern seemed to be that if Israel allowed its scientists to use donated eggs and donated sperm to make embryos for research, they would end up creating a market and selling these embryos to the world. He didn't want Israel to turn this into business. He shot me a certain look, which could only be described as the we're-both-Jews-together-and-we-know-non-Jews-think-ill-of-us look. He explained that similar issues were raised by collecting ethnic groups' DNA. Private companies own these samples and mine them for particular genetic characteristics that might be exploited commercially. He thought there should be a statutory authority to regulate this work. "This issue won't go away. It involves individuals and ethnic groups that can be stigmatized."

His brown eyes fixed on mine as if to say, we both know about stigma. "These companies want to collect and sell to giant pharmaceutical companies. Will they be used to stigmatize the Jews?"

I have been getting looks like this all my life. I don't think pointed looks substitute for good argument, but, on the other hand, what has good argument got to do with human history?

"Why would you jump to the conclusion that this will be bad for Jews?" I asked.

"Here we have a study showing a genetic problem of Ashkenazi Jews, let's say showing they are mentally inferior. Anything can be said! We are accused of killing children to make matzohs!" Suddenly, he was shouting.

"Oh, come on," I said. "That's so old it creaks."

In fact, the last time I'd heard this outrageous accusation about Jews was also the first time. My high school mathematics teacher, a nice man, had invited me and several of his other Jewish students to come to his Anglican Sunday school class just before Easter. He didn't explain this to us, but it soon became clear that he wanted to demonstrate to these children that Jews are just like everybody else. We had dressed up—I still remember the dress I wore and the nylon with a run behind the knee. The first question a little girl asked me was, are Jews allowed to wear stockings? I had stood up and pulled it away from my leg to show her, and the run became a huge hole that spread right around to the front. The next question, from a sweet-faced boy, was, is it true that Jews steal Christian babies and use their blood to make something for the Passover? I had laughed out loud; I'd thought it was a weird kid joke, but my teacher had almost died of mortification. This is known in Jewish communities as the "blood libel."

No, no, Revel said, this isn't so old. In fact, this craziness had just appeared in a column the previous week in a Saudi Arabian newspaper. The Israeli newspapers were full of this story. "I didn't take it seriously," he said, "but if Ashkenazi Jewish women in New York are asked to bring a BRCA1 test [a test for a gene thought to be specific for breast cancer] before being employed . . ."

He'd heard of this happening. Some genetic diseases don't show up until the middle years, and insurance companies and employers would not want to be saddled with people who have them. As ethnic groups are studied for genetic variety, many more such genes will become known and many insurance companies will make these demands. "BRCA1 is just one case," he said. "If you test on a genome-wide scale, every gene possible, they will find differences between this and that ethnic group. Most differences will be within ethnic groups, but those between ethnic groups will be taken as a sign of inferiority or superiority. I think bioethics is here to deal with these problems. It should say how to do it, not don't do it."

Bioethics seemed to be quite distinct from medical ethics. Medical ethics are about what doctors and nurses should and should not do to patients. Bioethics seemed to enable scientists to do things that make people nervous.

How did you get involved in this? I asked.

He said he was trained as a physician and in science. The company he started in 1979, Interpharm, is now doing $400 million a year in sales and had just won FDA approval for a product of his—he'd used recombinant technology to produce interferon, which is used to treat multiple sclerosis. "I've contributed to biotech in Israel," he said. "I'm not trained as a bioethicist, but I'm a Jew, studying all my life. I'm not a rabbi, but I am a learned man. I have a weekly study group where we study Jewish philosophy." He was also appointed to the UNESCO International Bioethics Committee as the representative of Israel in 1993.

I could see why Revel had been chosen—he embodied all the modern conflicts: scientific reductionism versus belief in the interactive creative capacities of culture, the random facts of materialism versus belief in a God-ordered universe, the particular interest versus the public.

"When I started in 1993, there were thirty-six members on the UNESCO committee. Not all were scientists—there were lawyers, philosophers, clergymen . . . The aim of the Bioethics Committee is to say the limit of what is permissible . . . I think it's a good definition. We're not saying what is good, but what is permissible."

I had found on the Web an impassioned argument Revel had made for the importance of this sort of ethics, particularly for those studying humankind through the sequencing of the human genome. Scientific disciplines, he argued, had "sketched a complex image of biological *Homo sapiens,* whereas paleohistory, structural analysis of myths and psychoanalysis have sketched a complex image of the human mind and soul. Therefore, it must be made clear that sequencing human DNA is an extreme reductionist approach to the human species, a species characterized by intellectual creativity and interpersonal communication of knowledge and emotions . . . The human genome project will reveal the genetic program, through which joined human gametes develop into a human being, but it would be an intellectual disaster to equate Humanity with a DNA sequence."[4]

"Science is basically utilitarian," he was saying. "Go back to Darwin's paradigm: the fittest should survive. But if you follow that logic, then we should not care for the sick. And not make surgery in the mother to close the hole in the baby's heart. Why change nature? This was Nietzsche's approach, that Christianity is a crime against nature. The law of nature is the poor are not good enough to be wealthy and the sick have a defect. Well, science cannot decide what is good and bad. You need other sets of values to define morals . . . It's not by accident that man developed religion to have a value to base their ideas of good and bad. I think we can't leave decisions in the hands of scientists. We need community with philosophers, with lawyers to say if we are not infringing a law. The philosopher Jürgen Habermas has said the ethical decisions we see, living today, are not won by argument but by consensus."

Or, as Revel had put it in his article, the best we can hope for "is an Ethic of Discussion, in which consensus is not obtained by force, be it force of logical reasoning or of empirical evidence."[5]

Revel gave me a rare and gentle smile. I thought the hope for any consensus about the start point of the human life and soul was ridiculously faint. And I was not too certain about the virtues of consensus either. The Nazis developed a remarkable consensus about the Aryan place within human variety. This enabled individual Nazi physicians to do terrible things to the physically and mentally impaired, to homosexuals, to Roma and, of course, to Jews.

It is certainly true that a consensus had enabled this research in Israel. A structure of beliefs about the moral status of the embryo and the duty of the physician was shared by two of Israel's major religious communities long before this work was thought of. But this Israeli consensus was not enough for Revel: he was concerned about the contrary beliefs of other communities. In the global context, where agreement does not exist, his forceful arguments gave courage to his colleagues—scientists and businessmen and even some religious men like him, up to their elbows in revelationary biology.

Distinctions between activated eggs and embryos, and the sanctity of the same, shrank to the size of a pinhead in the glare of an Israeli noon. We had lunch, my cousin and I, on the patio of a restaurant in Haifa's German quarter, at the bottom of Mount Carmel, just down the road from the bottom of the amazing Baha'i peace garden. The garden is terraced, with phalanxes of red snapdragons, purple pansies, gray creepers, set out in arcs and curves under waving palms. Seen from below, broad rivers of violent color seem to run down the mountain face like swarms of lemmings rushing for the sea.

We were practically the only visitors at these gardens, or at any of the restaurant patios set out on both sides of the street below. A few weeks earlier, a bomb had blown up a Haifa restaurant, just like this one, owned by an Israeli Arab, just like the man who set out our drinks. The bomb had snuffed out many lives, tearing flesh from bone and murdering the hopes of the maimed survivors. And on this particular day, what with the talks between Israeli and U.S. officials about the repression of the Palestinians by Israel, and talks between Prime Minister Sharon and Vice President Cheney about the looming war with Iraq, there had been three more attempts to splatter limbs and brains and blood around the Israeli streets. A group was caught trying to smuggle bombs into Israel from the West Bank. A suicide bomber was foiled by some miracle in Jerusalem. Katyusha rockets had flown in from Lebanon, exploding on the border farms of northern Israel. Who could worry about embryos in petri dishes in a circumstance like this?

We sat resolute, my cousin and I, tearing into pita and hummus and salad, to the great delight of the restaurant owner. A photographer from the local paper asked us to pose—this would be proof the restaurants were safe, if two old women (one a foreigner, a Canadian!) sat and ate in broad daylight.

CHAPTER FIFTEEN

IS IT SAFE?

It was one of those days in San Francisco that cannot be equaled. There was a sweet wind off the glittery bay, where sea lions hung out to dry on rocky outcrops. The sky was a cerulean helmet (or maybe a bell jar). The street people lay on benches, on the chipped sidewalks, on the gritty pavement, their faces upturned to the brilliant sun. On the BART train to Berkeley, I read the newspaper. Francis Fukuyama had a new book out. His previous effort had declared the end of History, which he seemed to define as wars between states. With the end of the cold war and the rise of the one great superpower, state-to-state war was supposed to be over. People in San Francisco probably hadn't heard. There had just been a huge demonstration against the repression of the Palestinian nation by the great Satan Israel, the U.S. war against Afghanistan, the coming war with Iraq. Fukuyama's new book, according to the review in front of me, now declared the inevitable end of humanity as we know it as a result of the coming triumph of biologists over nature. This belief in prowess, this suggestion that we have nature in a box and can make it dance, almost made me laugh. I had learned enough to know that this is both true and false, ridiculous and ominous, like the question of safety in *Marathon Man,* a Dustin Hoffman film. Hoffman is caught in the toils of a Nazi conspiracy to escape justice. One of the worst of the Nazis is hiding in plain sight in New York, waiting for certain instructions. When Hoffman stumbles across him, the Nazi thinks Hoffman knows more than he does, and he tortures Hoffman with a dental drill. While Hoffman writhes and screams, the Nazi asks, over and over, "Is it safe?" Hoffman can't answer the question, though his life depends upon it.

The man I was going to see, Ignacio Chapela, and his graduate student David Quist had indirectly raised this question about genetic modification in the previous November's issue of *Nature*. The same question could be asked about any of these biological technologies: Is it safe to place a gene from one life-form into another, inject stem cells into a human brain, put a human nucleus into a cow's egg and use the result? If all life is interconnected, how will these things change us, and everything around us? The way Chapela and Quist's article was received also raised the question of whether we can trust the answer we get from those who do this work. And where better than the Bay area to ask about that?

San Francisco, the chimeric heart of new biological science, is driven by two conflicting ideals: the post-Darwinian notion that inquiry must be free to arrive at truth and the neoconservative theory that a free market is the true source of innovation. The Bay area is where genetic engineering evolved and biotech exploded. What happens to scientific inquiry for its own sake when business invades the academy?

Chapela and Quist did a molecular study of apparently wild, or genetically unaltered, Mexican corn growing in the backwoods of Oaxaca. The corn they sampled grew many miles away from any genetically modified crops. They said they found three different kinds of engineered gene sequences in the corn they tested and that these foreign genes had broken up and scattered across the corn's genome. This suggestion—that engineered genes might not be stable and that foreign genes had spread to corn in a backwoods area that was supposed to remain untouched—landed in the world of biotech like a stink bomb, not least because Chapela and Quist were at the University of California at Berkeley, ground zero for the conflict of science ideals.

Environmental groups were quick to say I told you so. But *Nature* had just published letters challenging Chapela and Quist, written by their Berkeley colleagues and former colleagues.[1] In an editorial, *Nature* also took the bizarre step of washing its hands of the article it had published only four months earlier. *Nature* said it had given Chapela and Quist time to answer the critics, who claimed that the initial publication was so flawed it shouldn't have been published, but they had been unable to satisfy the editors: "*Nature* has concluded that the evidence available is not sufficient to justify the publication of the original paper." In a reply to their critics, Chapela and Quist acknowledged some error and the possibility of different interpretations of data, but they stood by their conclusions and contended that their findings were also supported by those of the Mexican government.[2] The GE Food Alert Campaign Center, an anti–genetically modified food advocacy group, asserted on its website that *Nature* had caved to pressure. Chapela told the *Washington Post* that his work had been targeted because he led a campaign to stop a science-for-money deal between Berkeley and the chemical/pharmaceutical giant Novartis. Chapela had repeated this to me and added that Quist, his student, was being libeled as an ecoterrorist.

I had asked Chapela how he thought cauliflower mosaic virus genes, whose pieces he said he found in Oaxaca and which is used to carry foreign genes into corn, could have broken up.

Read about *mariner*, he'd said, cryptically. Look up *Sleeping Beauty*.

I found papers on *mariner* and *Sleeping Beauty* soon enough. These were names for a family of genes, and a particular gene that makes a protein, a transposase, that helps genes jump around. Barbara McClintock first noticed movable genes, or transposons, in corn in the 1940s; she called them "controlling elements."[3] Some of these corn transposons were silent for many generations, but then, having moved, they expressed themselves, or suppressed other genes. These movable elements had effects similar to mutations, producing different colors or kernel shapes. McClintock's observations that such transposons controlled development had at first been doubted, but then transposons were discovered by molecular biologists in all sorts of simple organisms' genomes. By the 1990s, transposons had become genetic tools—used as mutagens, as reporters for genes of interest, or for changing chromosome structure. In 1994 transposable elements were found in animal genomes, and in 1996, in the human genome. Different kinds had by then been identified and given names like "integrons," "shufflons," "mobile introns" and "retrons." These elements seemed to be a

way to ensure variety, and seemed to predate sex. "There is now substantial evidence," wrote David Sherratt in a preface to a book on this
subject, "that the mobile elements of all types have crossed species
boundaries, have infected new populations, and may become extinct
. . . Much of the time they hide within genomes where they can exert
subtle as well as dramatic effects on cell and organism biology."[4]

In 1997, an ancient, inactive gene to help other genes jump was
identified in fish. The original gene had called up a transposase that
cut out and then pasted genes into new places. The original gene
sequence of a gene family called *mariner* was calculated and synthesized and dubbed *Sleeping Beauty*. It was then used to catalyze
the cut-and-paste of genes in fish, then in mouse cells, and finally in
human cells. Those who revived it proposed that it could be used to
move all kinds of genes around genomes.[5] But to be predictably useful, once inserted, such genes would have to stay put and express
when they were meant to. What with shufflons, integrons, mobile
introns and retrons—not to mention cauliflower mosaic virus genes
splitting up in corn—how could one assume that any inserted
gene would keep doing the right thing?

The BART train let me off in the center of Berkeley. I followed a horde
of students who climbed a wooded hill to the university campus. We
all emerged into a glorious garden set out around Federal-style buildings. There were mounds of planted annuals, but also wild areas
where riots of vine and flowering shrubs reached for the water in
deep gullies. A brook sang its way through a stand of trees; students
leaned on a wooden footbridge to watch it race over smooth stones.
A gigantic eucalyptus danced at the center of a small plaza and, with
every movement of branch and leaf, showered the air with fresh,
medicinal perfume. Wherever I turned, there was sparkle and dapple
and fragrance. Up the hill was Sproul Hall, where in 1964 Mario
Savio denounced the university for bending over backward for business interests, launching the Berkeley Free Speech Movement. I was
looking for Hilgard Hall, but I found myself staring instead at the Life
Sciences Building. Each corner below its square roof was guarded by
winged human-animal chimeras.

Ignacio Chapela, a fungi specialist, had a small office in a department called Environmental Science, Policy, and Management. He was on the phone when I arrived, so as I waited, I got a chance to study him—a heart-shaped face, Elizabeth Taylor eyebrows set above tilted hazel eyes, thick black hair shot through with gray. When he hung up, he wanted to know what I was writing about. I explained as best I could. His eyes lit up when told him I was fascinated by chimerism and the integration of different life-forms operating on different instructions. As a microbiologist by training, he said, he sees life as massively interconnecting. He wanted me to understand that there is only a thin membrane dividing one individual's DNA from another's. Then he invoked the name of Richard Dawkins, who had argued that genes operate almost as if they are organisms, maneuvering for their own continuance. "DNA has a biology of its own that plays on the stage of the cell," he said. "It organizes the system, but there are very important processes below that level, most of the time undetected. This DNA is playing by its own rules. This is what I see as the conceptual background of why I dare step into the hornet's nest . . . There is also politics . . . my critique of what we are doing with genetics . . ."

He was a member of a working group of the National Academy of Sciences reviewing the scientific basis for the nonregulation of transgenic crops released into the marketplace. The United States, Chapela said, doesn't regulate transgenic plants, it deregulates them. Once a plant has been sufficiently tested and its product shown to contain no traces of known toxins, there is no more scrutiny. In his view, deregulation of genetically modified organisms has been rushed for reasons of politics and economics, and it's time for science to look at modified plants and see what has happened to them over time. This was the context for his *Nature* article on corn, and for his later forays into the newspapers. "I am guilty of being a general biologist," he said. He is concerned about the interactions of life, particularly plants with soil, worms and fungi. "Our own dean here . . . goes around telling California farmers we are already looking at the time when soil will not be needed for agriculture. I say, what? Being a biologist in this context is seriously political."

Of course, when Chapela began university, in 1979, at the National Autonomous University of Mexico, he had never imagined that science and politics had anything to do with each other. He fell in love with bugs, with their methods of communication and development. And that brought him to the world of fungi, incredible organisms with a dual nature, both microbe and macrobe. He soon realized he could work with fungi on the cellular level and yet ask questions about multicellular organisms. "They are close to animals," he said. "A good model for humans."

He did his doctorate in Wales, where he studied "a whole universe of fungi called endophytes." Since they live in and with plants, endophytes taught him to doubt Darwin's concept of a Malthusian struggle going on everywhere in nature. They infiltrate plants but do not struggle for dominance. Yet the idea that you could have fungi in a healthy plant was then considered anathema.

"Trees are chimeric organisms," he said, "made up of tree tissues and fungal tissues. The dogma then was the fungus was a disease." He began to see that there is a dynamic interaction between the tree and the fungus, and that disease is just one possible result—he found himself interested in the view of nature as symbiosis not struggle. He went on to Cornell to do a postdoc in plant pathology, and then to Switzerland to work for Sandoz, which would eventually become Novartis. There he worked with a pioneer in the study of endophytes. Although he had doubts about working for a corporation and was interested in the politics of the environment, for the first year he liked working in Basel. But in his second year, he was asked to work on a project on biologically derived insecticides.

"I thought it was the greening of the company, replacements for nasty chemicals sprayed all over the world," he said bleakly. Sandoz had purchased rights to the work of a British academic who had labored for ten years to isolate a toxin produced by a bacterium, a pseudomonad. It kills insects, but not the plants the bacterium lives on. He was told that Sandoz had isolated the effective compound and they could use it simply as a chemical spray. At first Chapela saw this as greening the industry from within. He asked if they could talk about what they were doing, got a wall of silence and realized there were questions he could not ask at Sandoz.

Chapela returned to Cornell, to the lab of Thomas Eisner, an entomologist who was then working on a biodiversity agreement between Merck and Costa Rica that would be lauded at the 1992 Rio Earth Summit; the agreement was to provide Merck with rights to survey Costa Rican rain forests for life-forms of interest in exchange for money, giving Costa Ricans an alternative way to get value from their forests to cutting them down. Chapela was still at Sandoz when this agreement was announced, and he saw plenty of flaws in it. "I wrote to Thom and said can I contribute to the brilliant idea. He got MacArthur [Foundation] money for me."

From that point, Chapela immersed himself in the environmental politics around the Convention on Biological Biodiversity signed at Rio. His interest was how drug discovery from natural products could be used to promote conservation of diversity. Eventually he became a consultant working for organizations such as the World Resources Institute, and even for the World Bank, until Berkeley called, offering him a position in its new environmental policy department within its College of Natural Resources. He arrived in 1996. "This is one of the few places in the world I can do what I want," he said. His lab houses interdisciplinary graduate students, undergraduates and postdocs who work at ecophysiology, biochemistry and ecological biogeography. There are also graduate students interested in policy on energy, genetic resources, access, ownership and stewardship. Berkeley is partly a land-grant college, built on land contributed by the federal government to the state for a public university. Chapela thought he'd landed in academic Eden.

"And then Novartis came here," he said.

And did what? I asked.

He waved a copy of the March 2000 *Atlantic Monthly* with a cover story called "The Kept University." The story, by Eyal Press and Jennifer Washburn, opened with a description of a 1999 Berkeley faculty meeting to review a survey Chapela drafted regarding Novartis's offer to fund one-third of the plant research conducted by professors in the Department of Plant and Microbial Biology of Chapela's college. Some professors, along with the dean, Gordon Rausser, had invited Novartis in, and Novartis had offered $25 million over five years in exchange for exclusive first access to anything arising from the work of those who signed on to the agreement.[6] Chapela had been elected to represent the faculty on the College of Natural Resources executive committee. One week after his election as chairman, he was informed by Dean Rausser that a deal would be signed in one week between Novartis and Berkeley; Novartis would get two out of five seats on the committee that would decide which science projects got Novartis's money.

Chapela knew very well what kind of mess this could lead to. He had been at Sandoz when it had tried to make a deal to buy rights to the intellectual product of the Scripps Institute. This was the offer that had so alarmed scientists that Cold Spring Harbor's Banbury Center had held a symposium about it. "They ended up in the U.S. Congress," said Chapela. "It was bad for Sandoz."

Novartis wanted the right of first refusal to exploit patentable or proprietorial discoveries made by individual professors who signed on to the agreement—professors whose salaries are supported by the public purse. The university would own the patents, but the company to be formed between the university and Novartis would have a say in the patenting process. About $10 million of the total would be used by the university; $3 million a year would be available for the committee to grant to department members on a competitive basis. Participating scientists were also to get access to Novartis's private genomic database.[7]

"All I could say was, Gordon, have you thought what the faculty will say when they read this?" said Chapela.

And that was the beginning of a ding-dong battle that lasted for two years. No one was allowed to see the proposed contract unless they signed a confidentiality agreement. Chapela said many of the faculty were "outraged . . . They said the deal was secret . . . We pulled out the details slowly through pressure."

His job was to get the faculty's views. He did this on the phone, in the halls and, finally, through a survey. People told him there would be "serious consequences for you." Chapela's survey showed that many in the college opposed the deal because of the impact they thought it would have on academic freedom within a public university. But four days after the survey was taken, according to Chapela, the agreement was signed anyway. As Chapela told *Atlantic*

Monthly, "I'm not opposed to individual professors serving as consultants to industry. If something goes wrong, it's their reputation that's at stake. But this is different. This deal institutionalizes the university's relationship with one company, whose interest is profit. Our role should be to serve the public good."

Not long after the cover story appeared, the California State Senate Committee on Natural Resources and Wildlife and its Select Committee on Higher Education held a day of joint hearings, despite the fact the agreement had been signed eighteen months earlier. The chairman was Tom Hayden, former husband of Jane Fonda, lifelong environmental and social activist. The Senate hearing room in Sacramento was packed with students. Get a copy of the Senate hearing record, Chapela said. There I would find the contributions of Matthew Metz, who had signed one of the letters in *Nature* refuting Chapela's corn article. Others who signed the *Nature* letters had been connected either to the Department of Plant and Microbial Biology or to Novartis. *Nature* had played a role in the Novartis controversy. The editor had come to the Berkeley campus and spent half a day "with me and others and wrote an editorial."

It was all a circle, according to Chapela. He had raised a ruckus about the Novartis deal, and now those in favor of Novartis had raised a ruckus about his corn article. This wasn't just disagreements over science, this was also the politics of science—Darwin's descendants squabbling in the dirt.

But it had started over whether it is safe to assume that genetically modified corn is the substantial equivalent of any other kind of corn. "We review the assumptions, particularly the substantial equivalence," Chapela explained. This is the doctrine developed in the late 1970s and early 1980s by the FDA and other regulatory agencies to deal with genetic changes to plants and animals for human use in the United States. Products made through genetic manipulation are considered substantially equivalent to naturally made products if they can't be distinguished from each other by a broad range of tests. The underlying theory is that one should examine the product and compare it to the ordinary, unaltered version, not examine the process that makes it.[8] "If you get a cosmetic from a tobacco plant, it's the same as one from the avocado,"

Chapela said, rolling his eyes. The U.S. Environmental Protection Agency, he explained, had a similar rule, which goes by the acronym GRAS, which stands for "generally regarded as safe." "That means that we all agree it's safe until it's proven differently. So the EPA do not look. The FDA—it's the familiarity principle. We are familiar with these things."

Well, spell out your problem for me, I said. Why is a gene put in corn not the same as any other gene in corn?

His answer boiled down to this—context makes a difference. "My position is that when we put it in corn, it is subject to all the forces of genomic selection . . . that pertain to the corn context," he said, "but it evolved in a bacterial context."

But bits and pieces of genes move regularly in nature, I said. So what's the problem if we do it in a lab?

"The answer I have, which is not media-ready," he said, "is that the scales are important in this." While he recognized that genes jump in genomes and move between species, such events happen rarely. But when corn or any other organism is modified and then grown on an industrial scale, we make a very large impact over a short span of time. Evolution happens over 100,000-year or one-million-year timescales. This industrial alteration of genomes is happening every month, every week. Not only that, he said, the vectors used, the biological tools that have been developed or borrowed to move things in and out of organisms, are by definition those bits that move easily.

"We take those that are promiscuous—that is why we select them. That's why we call them vectors and promoters. They are not just any piece of DNA or any rate of exchange. It's massively increased. The scale, the incredible distances breached . . ."

Well, I said, what's a bigger evolutionary distance breached than virus to man? And that happens all the time.

"There are boundaries," he said. "There are mechanisms to reinforce connectivity within and discourage connectivity across boundaries."

Was he saying that the *Bt* gene put into corn could some day enter the human genome?

"I know what is in corn today will come to my genome eventually, that *Bt* will come in many generations down the line," he said.

"Well, how?" I asked. This had been troubling me ever since I saw that a gene from a jellyfish had changed the color of the descendants of a genetically modified mouse. A gene taken from one kind of complex, multicelled creature was put into another and went to work. Viruses penetrate barriers, take up permanent residence within human genomes and change human cells. We have *E. coli* in our guts that permit us to digest food, a vital function, but their genomes don't merge with ours. If enzymes can digest the skin and muscle of a rat to produce a colony of cells that will interact with human cells, why can't all kinds of foreign genes enter the human system—through digestion, for example?

"Do you mean that by eating genetically modified corn, I might take that bit of a gene into my stomach, and it would find its way into my reproductive system and my eggs would carry it?" I asked him.

He didn't want to go that far. "There is a connectivity," he was saying. "But going to the extreme of total connectivity is totally wrong."

So how had this worked with corn? The Mexican corn that Chapela sampled in Oaxaca had probably been hybridized with genetically modified corn distributed as food aid, the kernels used as seed by poor farmers. Chapela wasn't really interested in how foreign genes found their way to corn in an area of Mexico where they aren't supposed to be—everyone knew this kind of transmission of modified plant genes into the wild environment would happen, eventually. In the end, so far as Chapela was concerned, it didn't matter how it got there. What mattered was that this foreign DNA was also moving through the corn genome; it had broken up in pieces.

The critiques of his article focused on the techniques he used to find and identify these foreign gene fragments in corn. He had pooled many grains of corn and used a method called iPCR—inverted polymerase chain reaction—to explore the context in which gene pieces were found. This is a fairly new method of amplifying DNA sequences and is prone to error. He said he found a sequence that bound to the probe for the cauliflower mosaic virus, which is often used as a vector to carry gene sequences into corn. The critics argued that Chapela hadn't found any such thing—he had merely amplified contaminants by mistake.

He had compared the sequences he found to known gene

sequences engineered into corn through the use of a computer program called BLAST (which we'll see again in the next chapter). The critics said the sequences he found actually were more similar to known repeat sequences in corn, not to the cauliflower mosaic virus. Chapela argued that these data were open to interpretation. "We call them this, they say that," he said. In two out of the eight sequences he and his student published, he agreed that they had made errors. In the other six, he was adamant that he and Quist showed where the transgenes ended and where the corn genome continued in the sequences studied. The critics argued that unless he had the whole sequence of a foreign vector, what he had didn't matter. "It goes back to the conceptual start point," Chapela said. For his critics, the transgene can be identified only if it's complete, and if you find a piece, it's just an artifact of a poor process. "They say if two are wrong, therefore six are wrong, the lab method sucks and we question your motivation . . . And your student has been accused as a crop trasher, a known bioterrorist."

He sat back in his chair. In fact, none of the *Nature* letters had said a word about motivation or crop trashing. His defense was in fact an attack on the motives of his critics, not on their words.

What is it your student is supposed to have done? I asked.

Chapela explained that some of his colleagues who had signed on to the Novartis deal at Berkeley made and grew transgenic crops in test plots, and that some of these crops had been destroyed by unknown vandals. Quist was accused of being the vandal. The evidence, Chapela said, was that after Quist was questioned by the police, the trashing stopped.

Well, did he do it? I asked.

"Of course he hasn't," Chapela said. "He was accused. The police could not find evidence—the police confronted him. I am thinking of using in court the e-mails sent to the world, the character assassination." (Quist also denied it. He was in Mexico when the vandalism occurred.)

Chapela pulled from his desk photocopies of the letters published in *Nature,* pointing to the names of the signatories. "In the two letters printed in *Nature*—all are connected with the Berkeley-Novartis deal."

There were six names on one letter. On the second letter there were two. Later, when I compared these names to the list of people who had signed onto the Novartis-Berkeley agreement, I found that only two who signed the first letter had also signed the contract. Neither Johannes Fütterer nor Matthew Metz, who signed the second letter, had signed the Novartis agreement.[9]

Yet, according to Chapela, all "are behind the wall of Novartis." Fütterer, who is now in Switzerland, was formerly the postdoc of Wilhelm Gruissem, who brought the Novartis deal to Berkeley. Fütterer had returned to an institute involved with Novartis. Metz had played a role in the debate.

Nature requires its correspondents and authors to declare their conflicts of interests. No conflicts were declared beneath the first letter, although two of the signatories were also signatories of the Novartis-Berkeley agreement. Beneath Metz and Fütterer's letter to *Nature,* a competing financial interest was noted.[10]

Chapela explained that his application for tenure would come up shortly. His colleagues were very supportive in public. But critics of the *Nature* article had told the *Washington Post* that Chapela's work is "mysticism masquerading as science." He worried that his critics "are fronting for bigger names—Watson, Paul Berg, [Daniel] Koshland—that caliber."

Was Chapela simply trying to deflect appropriate criticism with accusations, or were his critics people with reasons to grind their axes on him? Chapela asked me to read the record of the California senate's joint committee hearing in May 2000 concerning the Novartis deal. Peter Rosset, a man with an impressive string of credentials and director of a public interest group called the Institute for Food and Development Policy, or Food First, commented on the risks of genetically modified organisms and the risk of truth's being twisted by business interests. Rosset pointed out that the Swiss Federal Research Institute had shown that "the *Bt* molecule incorporated in the crops is a shorter chain molecule than the molecule originally found in the bacteria used as a natural insecticide. Industry has not been forthcoming with this information. But in fact, while the original insecticide was selective and did not harm natural enemies and other kinds of insects and mammals and other organisms, this shorter molecule proves to be toxic to a much broader spectrum of organisms, including natural enemies that regulate natural insect populations. And no human health testing has been mandated on this different molecule." Rosset asserted that regulatory agencies had been "asleep at the wheel and that the sedative that they've received is called money and politics." He worried that even if regulatory agencies were properly funded to evaluate risks appropriately, it would be hard to get them staffed. So where would science properly skeptical of business claims come from in the future? "Our publicly funded research centers and educational centers are becoming joined at the hip, one after another, with the very industries that will pose the principal regulatory challenge for the remainder of this new century," Rosset said. "With that in mind, the situation at the College of Natural Resources—I will not mention the word Novartis, except this one time—is alarming."[11]

Rosset had a point. It is well established that business relations tend to reduce scientific skepticism. Major medical journals have been forced to demand that reviewers declare all their conflicts of interest because such conflicts are endemic in biomedical research and had often been hidden. A study led by Canadian scholar Joel Lexchin, published in the British Medical Journal, showed that drug testing funded by the pharmaceutical industry is four times more likely to favor the sponsor's product than is publicly funded research.[12] If grants to test a drug induce bias, how about a joint venture like the Berkeley-Novartis deal?

When Chapela's critic, Matthew Metz, testified, he was still a graduate student in the Department of Plant and Microbial Biology at Berkeley; he has since moved to the University of Washington in Seattle. Metz said several things of interest. The first was that Novartis's side of the deal had not been fully honored: the researchers did not get easy access to Novartis's intellectual property, as promised, nor did the money help graduate students other than to replace funds that had been withdrawn by the National Science Foundation. The credibility of those in the department had been questioned, and the agreement had been the target of "a misrepresentative survey." In addition, "we have had research vandalized" and "been painted as 'bourgeois labs' by rather Marxist attacks on distribution of research funding." But there had been positive effects, according to Metz. "Unrestricted research funds are enabling adventurous experiments that would never gain support under the cautious inhibitions of public granting agencies."

The word "adventurous" made me sit up in my chair. Metz had asserted that private money takes risks that public funders won't, a charge often made by Craig Venter in another context.

> SENATOR PEACE: Can you give me an example of a "cautious inhibition"?
> MR. METZ: For example, a colleague of mine, working in the lab that studies corn and various other grasses, had a grant that was refused because it was too risky. I believe it was submitted to NIH. And now they're working on it, using Novartis funds.
> SENATOR PEACE: What was the risk?
> MR. METZ: The risk involved the uncertainty of the hypotheses they were investigating. In other words, the certainty that the experiment would succeed in finding something. . . [13]

This was the same argument Benjamin Reubinoff in Israel had made for getting involved with a company. What if Metz had meant physical risk? Biotechnology critics routinely warn about companies taking undue risks, to protect the bottom line.

Metz went on to argue that universities had no choice but to get private money since public authorities had been withdrawing from supporting public universities such as the University of California. He was right about that. Gordon Rausser, the dean of the college, testified that the State of California funds only 34 percent of Berkeley's budget, leaving the university to find the other 66 percent elsewhere.[14] In his written testimony, Metz summed up: "To conclude, genetic engineers are not practitioners of black magic, nor are academics with private collaborators practitioners of treason. There are many legitimate concerns regarding both. But rhetoric and opposition can become uncompromisingly apocalyptic, shamelessly misleading, full of conjecture, and entrenched in dogma."[15]

Did that comment about rhetoric also apply to those who complained to *Nature* about the work of Chapela and Quist?

Stanford is a private university near Palo Alto, a place with a business-is-good sensibility. I had come down from San Francisco on the train. The closer the train got to Palo Alto—the gap junction between private money and biological science—the greater the number of plumbing pipes, signifying bathrooms, signifying wealth, piercing the roof lines of the houses. The Stanford bus dropped me at the plaza behind the oval, a grass playing field in front of Stanford's Mission-style administration buildings, which proclaim California's Spanish roots. That was it for charm. Most of the other buildings had great corporations' names hung over their front doors. A SWAT team covered in body armor rushed around hunting a rabble of protestors rumored to be on their way from San Francisco.

Paul Berg, who shared the Nobel Prize in chemistry with Frederick Sanger and Walter Gilbert, worked away on his word processor as our allotted time ticked by. He is Cahill Professor of Cancer Research and Biochemistry, emeritus, at Stanford's School of Medicine. His office is in the office of the director of the Beekman Center, a new research building behind the hospital. Berg, brought up in Brooklyn, came to Stanford in 1959, along with Arthur Kornberg and most of their colleagues in Washington University's microbiology department. Stanford didn't have a biochemistry department, so it hired them all to create one.[16] Yet Berg didn't quite fit with Stanford's business-is-good atmosphere. For one thing, he had a row of good paintings and one marvelous drawing on one wall. For another, he had played a vital role in shaping public discourse about public control of revelationary biology: after his invention of a method to recombine different species' DNA, Berg and a small group of colleagues brought the issues into focus through the famous conference at Asilomar. They were the first to be assailed by the question "Is it safe?" about recombinant technology.

Berg was small, freckled, his hair pepper and salt, parted high up on the side, giving him the look of an aging Archie. He was also vigorous, his handshake firm, chin strong, eyes focused. There was a sharpness to him, a kind of flinty sparkle as he told the story of what led to the Asilomar conference.

In 1972, Berg had sent his student Janet Mertz to Cold Spring Harbor to learn certain techniques for experiments they intended to do. They were already making hybrids between a known tumor virus called SV40 and the phage lambda. They planned to put the new hybrid DNA into E. coli to see if the SV40 genes would be expressed in the bacteria. Berg knew a lot about SV40—he'd worked with tumor viruses for a year at the Salk Institute in order to learn how to handle it. When Mertz explained to her Cold Spring Harbor class-mates what she aimed to do, Robert Pollack, the instructor who taught the cell culture course, didn't like it one bit.

"He called me from Cold Spring Harbor and expressed his deep concern, accusing me of deadly intent to spread human cancer," said Berg. "I knew him. We went back and forth. I explained why I thought that was unlikely. In the end, I decided not to do Janet's experiments in bacteria."

Pollack had reason to be alarmed. The tale of the SV40 virus and its cancer-causing propensity was a chilling reminder of what hap-pens when no one asks if it's safe until it's too late. As Michael Rogers recounts in his 1977 book Biohazard,

SV40 was first isolated from the kidneys of rhesus monkeys in 1960—a batch of kidneys that had been destined for use in the production of the new poliomyelitis vaccine. The virus had escaped notice for years previously as it caused no apparent disease in the monkeys themselves. Injected into newborn hamsters, however, it caused tumor growth: and it did the same, moreover, to human cells in test tubes.

SV40 attracted instant attention because by the time it was discovered, vast amounts of polio vaccine had already been pro-duced in monkey kidney cultures rife with SV40, which in some cases survived to inhabit the vaccine itself. Thus, from ten to thirty million American children in the years between 1955 and 1961 received, along with their polio shot, a dose of live SV40 virus.

By now, considerable discreet medical surveillance has been directed at the known SV40 recipients—and thus far there has been no evidence of any mass onset of malignancy which, considering the numbers involved, could have made the thalidomide tragedy look insignificant in the annals of self-inflicted human suffering. Some recent studies, however, have suggested the presence of SV40 in association with some human neurological diseases that involve progressive central nervous system degeneration—a class of disease not included in the original epidemiological surveys of SV40 infectees.[18]

In fact, contaminated vaccine was used until 1963, two years after the tumor-causing nature of SV40 was first discovered. The "medical surveillance" Rogers refers to was certainly discreet. No one identified and followed actual recipients. Epidemiological surveys were done instead. The possibility of vaccine contamination was certainly not widely known, at least among Canadian professionals. My father, who struggled against outbreaks of polio as a general practitioner in Saskatchewan, never heard about it. According to Berg, however, in the Soviet Union, where excellent records were kept and it was possible to follow individuals, no clear evidence of higher incidence of cancers emerged.

It was also known by 1969 that SV40 had successfully recombined on its own, in a lab, with adenoviruses—the kind that cause the common cold.[18] Yet Berg never believed that work he or his students wanted to do with viruses could lead to problems in their health or in the health of the population at large. He wasn't alone; other biochemists at that time routinely did experiments with noxious chemicals, later shown to be carcinogens, without gloves or special hoods to vent away fumes.

"It was a bit of a shock to be accused of irresponsible action," Berg said. "Robert Pollack and I organized a conference to examine what we did know of viruses being potentially dangerous. The result was a series of recommendations on containment." These included the use of laminar flow hoods and gloves to prevent organisms from escaping. After that first meeting of January 22, 1973, the general feeling was that any recombinant DNA experiments "oughta be put on hold." This was no hardship on his colleagues. Berg's recombination method was so laborious that if his group didn't do it, it was unlikely that anybody would. But that changed with the discovery that when certain enzymes cut DNA at specific nucleotides, the severed ends of the DNA will spontaneously rejoin and the hybrids could be cloned in bacteria. "That made the whole field of genetic engineering of molecules wide open," said Berg. It became "trivially simple" to make and propagate recombined DNA in living organisms. Anyone could do this—and anyone would.

At the request of the National Academy of Sciences, Berg organized the one-day meeting at MIT in 1974 to discuss the implications. Those invited included James Watson, David Baltimore, Norton Zinder, and an observer from the National Science Foundation. "Watson had just been made director of Cold Spring Harbor and was promoting work on tumor viruses," Berg said. "David Baltimore and Zinder were not in the field, but smart. I got people who were smart and critical that I trusted and for whom doing this research would be likely." A day of conversation led to the suggestion that "we should call attention to the fact that there might be danger, and call people to put on hold experiments that were in theory risky, but not stop research completely." Their letter to the president of the National Academy of Sciences was published in *Science, Nature* and the *Proceedings of the National Academy of Sciences.*[19] Stanley Cohen and Herbert Boyer, whose experiments had helped open this door, as well as Ronald David and David Hogness, asked to sign the letter too, though they were not at the meeting.

This is generally referred to as the "moratorium letter," but in Berg's view, no blanket moratorium was asked for, but rather a deferral of two main types of experiments: those that would confer antibiotic resistance on organisms that didn't have it already and those that would link tumor- or cancer-causing viral DNA to replicating plasmids. The letter asked the NIH to create an advisory committee to oversee experiments to explore the risks and to create guidelines to prevent the spread of these new, manufactured, self-copying molecules in the environment. The letter also called for an international meeting of scientists to be held quickly to review "progress" and discuss potential hazards. The letter said flat out that such experiments, while important, might "result in the creation of novel types of infectious DNA elements whose biological properties cannot be completely predicted in advance."[20] In other words, the letter made it plain that no one could answer the question "Is it safe?"

"The media picked it up and immediately made it a big story," said Berg. Berg thought the media paid attention because of the heightened suspicion of government and science and authority that arose with the Vietnam War. Scientists had created awful weapons such as the hydrogen bomb and then napalm, whose searing results were available to be seen on the evening news. And what about those Nazi doctors who had caused such suffering in the name of science, and those Japanese scientists who did unspeakable things to prisoners of war to test their theories? "The feeling among those who signed the letter was there was suspicion of scientists. All of us felt a sense of responsibility to do no harm."

This adjuration encapsulates the Hippocratic School's respect for the complexity of nature and its humility before it, its belief that doctors should refrain from intervening unnecessarily, that a doctor's interference may cause more harm than good. Although these ideas were more than two thousand years old, there were no other competing ethics to rely on so far as Berg knew.

The international conference at Asilomar was organized under the auspices of the National Academy of Sciences and supported by the NIH and the National Science Foundation. Berg, Maxine Singer, David Baltimore, Sydney Brenner and Richard Roblin, a professor of molecular genetics at Harvard Medical School, formed the organizing committee.

Why Brenner? I asked.

"He was a giant in the field," said Berg. "After our letter was published, there was a British commission chaired by Lord Ashby. Brenner was the most influential biologist on the commission. The Medical Research Council prohibited all recombinant DNA experiments—all of it . . . They recommended that the British moratorium be lifted but that research be under stringent controls. He was important to have on our committee."

In other words, Brenner agreed with Berg and Singer that freedom to inquire could still prevail under reasonable guidelines. But many others didn't want to be scrutinized by government or even to regulate themselves. When the conference opened at the private conference center near Monterey in February 1975, the meeting included the presentation of new science and was followed by discussions of risks, much of which devolved into "raucous shouting." Although Watson had agreed at the meeting at MIT that self-regulation and containment were necessary, he had changed his mind. Others might have come thinking some rules were necessary, but not for their own experiments, just for the other guy's.

"Watson was saying this is the stupidest mistake ever made, signing the letter," said Berg. "You had guys drawing circles around their area, as no-risk, and pointing at others. We had a whole disagreement on how to go forward. Would we recommend strict guidelines . . . or say nothing? The lawyers said we could not leave without saying something. . . We sat all night, the organizing committee, to make a consensus statement. We admitted we don't know risks, but we think the National Institutes of Health needs to make guidelines."

The report summarizing the committee's recommendation that the NIH create guidelines, published as a letter in *Science*, was also like an instruction manual for people entering the field. Berg and his colleagues knew that this new area would draw in all sorts of biochemists who had never worked with live microorganisms before. Without guidelines, they might do dangerous things such as throw live bacteria down the sink or suck up samples with a pipette, running the risk of spreading infectious organisms around in the water supply or swallowing them. It was Brenner who proposed at Asilomar that containment be matched to risk and who suggested different levels of risk: experiments one could do in the kitchen sink; experiments that need a certain level of containment; experiments that need a profound level of containment, including special facilities behind airlocks; and experiments so dangerous they should not be done at all, such as recombining DNA in listed dangerous pathogens such as anthrax or plague. These recommendations appeared in the letter, which also suggested that perhaps time and experience would show that much less containment was actually needed.[21]

The foxes stayed in charge of their henhouse, Erwin Chargaff charged in *Science*, when the NIH began creating guidelines for funding and for a committee to oversee this work. Who were the people in charge at NIH, Chargaff wrote, if not scientists with an interest in recombinant experiments? True, replied Singer and Berg, but who else had the expertise to provide oversight?[22] While no law was passed to regulate this science, if one wanted NIH funding, the guidelines had to be followed. Institutions had to buy expensive equipment and properly train their workers. Those planning danger-ous experiments had to put them first in front of the NIH's Recombinant DNA Advisory Committee. Eventually, the experi-ments that at first were required to be done in airlock containment were safely moved to the lab bench. Eventually, too, researchers decided the DNA molecule isn't dangerous, it's the infectious organ-ism one puts that DNA into. Some organisms that were feared turned out to be fine. Mice infected with bacteria carrying genes from cancer-causing viruses didn't get cancer. "SV40 turned out to be harmless," Berg said.

How do we know that? I asked.

Well, he explained, tumors could be examined for the genes of the SV40 virus. Berg thought there had been a few studies, mostly on brain cancers that had been found to be related to SV40.

In fact, there had been a recent study by the Institute of Medicine that did not take as sanguine a view as Berg's about SV40. It claimed that previous epidemiological work claiming a benign impact of SV40 in polio vaccine had been so flawed that it was impossible to say whether or not SV40 had led to an increase in human cancers. The report noted that both children and adults had been given vaccines that could well have been contaminated with SV40, although it also found evidence that some human beings who had never got the polio vaccine had SV40 in their systems. The biological science demonstrated that there was a moderate chance that there had been some increase in cancers attributable to the SV40 virus in polio vaccines. The institute recommended that more tissue studies and properly crafted epidemiological studies be done.[23]

While the guidelines applied only to researchers getting NIH grants, other countries adopted rules very much like them, and so did industry. Why? It might have seemed like a lot of money to James Watson to outfit Cold Spring Harbor labs with containment systems, but "to Merck it's a drop in the bucket," Berg explained. It would have been much more expensive to defend itself if a company released dangerous organisms into the world. "Why run the risk? They'd have been sued to the wall. It was cheaper to conform, and they had much tighter control over their scientists."

Yet the stem cell and cloning research debate had followed a different trajectory. Instead of everyone following an NIH standard, the NIH had withheld funds and therefore regulatory oversight, so researchers had looked for corporate backers and easy jurisdictions—and they found them.

Did you get into the biotech business? I asked Berg.

"I stayed clear of all the early developments," Berg said. "I thought it would cheapen my credibility. After the guidelines were made public, there was huge agitation against this work at Harvard and other university campuses. I went through years of testifying to city councils, universities, trustee meetings, in Congress. Congress was this close to prohibiting recombinant DNA research." He made a circle with his thumb and first finger to indicate how close "close" was.

Berg stayed clear of business interests until 1981. Then a friend, a distinguished entrepreneur, prevailed on him. "I was persuaded to set up an academic pure research biotech," he said. He did this with his colleagues Arthur Kornberg and Charles Yanofsky. "We helped recruit our former students to work in it. NAX still exists—it was bought one and a half years later by Schering-Plough."

"Did you get rich?" I asked.

Berg's smile vanished. Perhaps he was aggravated by my bluntness.

"I became moderately well-off," he said. "I had no qualms."

I asked him if he'd followed the fuss about Quist and Chapela's article in *Nature*. All he knew about it, he said, was that people were disputing the original claim. He did not feel able to comment.

Look, I said, Chapela raises important questions, even if there are flaws in the paper. There's the problem of taking things from one genome and putting them in another, which might induce unexpected results. There's the question of the evolutionary impact of these changes introduced on an industrial scale. Is that safe?

"I've no evolutionary concerns," Berg said sharply. "Evolutionary biologists have argued that there are considerable indications that intermixing of genes among widely different species has been going on throughout evolutionary times." Then he shifted in his chair, as if he was not happy making absolute statements. "That same issue was debated during the evolution of the guidelines and was dismissed as so much hand-waving, but maybe I shouldn't be categorical. My knee-jerk reaction is, it's another one of these what-ifs. No matter what evidence you cite, another what-if arises . . . At a gathering to reflect on Asilomar there was a lot of discussion on genetically modified foods. All the papers came down that there is no evidence of allergy or killing butterflies." This discussion was burdened by an irrational fear of reshaping life, which he doesn't think is amenable to science. "I see it as a cultural thing, where people react about killing God's creatures." While he'd been born into a Jewish family, had had a bar mitzvah and attended other people's life cycle events in synagogues, he had no belief in God or a divine power. In fact, if anything, he seemed antireligious. He had no need to rely on a higher power for his views of life and the world we live in, he said. Besides, religions have all too often been the cause of war.

The qualms Berg did have were about patenting. He learned during his earliest training that knowledge is to be shared, so he has never patented anything. Cohen and Boyer's fundamental patent on recombination started the long march to the U.S. Supreme Court ruling of 1980 that said that engineered life-forms can be patented.[25] "And I opposed it around here," he said. "I am maybe the only one." Two universities were involved, Stanford and the University of California at San Francisco. "They laughed . . . Do you know how much Stanford got? Maybe $150 million.

This was where, in Berg's view, great danger lay: in the transformation of the university itself into big business through the commercialization of their discoveries. Universities' business interests were beginning to inhibit the free flow of information and materials that makes science possible. "I think it will change the character of science," he said. "People make choices of what to pursue and may turn toward a money-generating outcome rather than seeking more fundamental discoveries. I can see the message students get as people patent left and right."

Berg had tried to lobby his colleagues to get around this maw of commerce. Researchers regularly borrow study tools from each other—mutated organisms, like Brenner's worms, or useful chemicals, like Don Riddle's intercalating agent or Sanger's four variants of dideoxy. With each university bound to commercialize anything of value, university technology-licensing offices treat these items as commercially exploitable. Each transfer requires a material transfer agreement, consuming months of time and acres of paper and fleets of lawyers. Berg had drafted his own anti-agreement so he and his colleagues could pass things back and forth as freely and quickly as they once had. He handed me a copy:

> Material Transfer Agreements (MTAs) have become increasingly onerous barriers to the free and open exchange of scientific information and materials. Seeking to return to the era where collegiality and sharing take precedence over commercial consideration, the institutions identified below have adopted the following policy. Subject to obligations we have to existing research sponsors and licensees, MTAs will no longer be used for the exchange of biological materials between our faculties and their academic and institutional colleagues elsewhere, and, where assured that such materials will not be used in commercial processes, with scientists in the commercial sector as well . . .

He'd sent it around to his colleagues at the leading institutions. Although they were mostly in favor, these colleagues took a different view. "None would sign it," he said.

CHAPTER SIXTEEN

COMPUTING LIFE

I'm still surprised that I had the wit to travel down to Santa Cruz, California. It was far out of my way and I didn't really expect any revelations. What could a little school perched on the western edge of North America have to offer? Yet I got on the train in San Francisco, following the scent of a story. There was something about the accounts of the first assembly of the human genome sequence that cried out to be explored. Who figured out which areas of the 3-billion-base-pair-long human sequence are genes, and by what method? Who was able to tell a president and a prime minister that there are only 30,000 genes in the human genome instead of the 100,000 everyone had so confidently predicted for years?

Nicholas Wade, in his book *Life Script*, devotes several pages to the sequence assembly. I thought it must have been a terrible problem, like trying to put thousands of clear glass fragments together in the right shape after you've smashed the vase. Wade quotes Craig Venter, who said the consortium had been in disarray until James Kent, a graduate student of David Haussler, a professor at the University of California at Santa Cruz, wrote an assembly program that was finished at the last minute. "We were truly amazed," Venter said, "because we predicted, based on their raw data, that it would be nonassemblable. So what Haussler did was, he came in and saved them. Haussler put it all together."[1] Yet John Sulston, in *The Common Thread*, only briefly refers to this problem of tying all the bits of human DNA fragments together in a seamless, assembled draft that bore some relationship to reality. "We just put together what we did have, and wrapped it up in a nice way, and said it was done," he'd written.[2] So which was it? Had David Haussler and James Kent saved the day, or was the draft assembly a trivial thing, slung together by unnamed elves between elevenses and teatime?[3]

When I went to Santa Cruz to sort this out, I had no idea that Haussler and Kent were part of revelationary biology's latest mutation: that they and their colleagues are turning biology into a theoretical science based on computation.

I took Amtrak to San Jose and changed to a bus at the crumbling pink stucco station. The bus carried me over forested hills that grew into mountains. Each hairpin curve in the road gave a more vertiginous view down treacherous ravines gentled by fern and flower and vine. The forest was deciduous for a while, but as we climbed, evergreens crowded in. Then, as the bus plunged from the mountaincrest down to the Pacific, the canopy reared so high and the sunlight dwindled down so low that it was as if the bus morphed into a submarine cruising lake bottom. The far side was a whole other world from the Bay area. By the time we emerged from the forest where the town of Santa Cruz crouches at land's end, it was blazing hot.

My stomach cramped, again and again. Which meal, carrying which bacteria, was to blame? I had spent one night out in San Francisco with friends, and another with a niece and her pals. The niece and her pals were twenty-somethings and thirty-somethings who had found their way by various paths to the city. I had listened to them tell their stories, envied their pouchless eyes, their taut skin, their so-white flashing teeth. Most of all, I had been fascinated by the various ways in which my niece expressed rearranged bits of my aging self and our shared relations. It was not just the turn of eye and ear, it was the temperament—the curiosity, the sharp impatience, the ragged flag of buried rage. But she was also utterly unique and surprising, and it was these unexpected qualities I took pleasure in. Sameness, I found myself thinking, is dreadful; sex is what makes surprise possible. The fear of sameness is what makes people fear cloning. That night I made a fool of myself lecturing them on the vital importance of sexual attraction.

I was pretty sure the cramp came from the evening out with old friends. He was a lawyer, she an Episcopalian priest. I'd met him in New York when I was twenty. He was then a law student at Columbia University, which was in the grip of a student uprising, part of a wave of outrage against authority that rolled across the world in the spring of 1968 like a pandemic. Now he was walking the back nine of his life, but still feisty. His wife had become a priest after spending the early years of their marriage trying to read nearly everything of literary value written in English. He'd warned me that a close friend of hers had just died and so, over the Indian food we'd shared, I'd said I was sorry. She'd astonished me by saying, "He who guides my steps led me to her." The idea was that she'd met this woman at just the right time, and that God had been their matchmaker. By the time the woman was diagnosed with cancer, their friendship was firm and she was able to help her friend accept her death—which, she explained, is not to be confused with material cessation, but is a passage of the soul to another stage. Ms. Rational thought this was unlikely in every way, but it went down with the Other like a good champagne. I had eaten the unfamiliar Indian food with gusto; the cramps and worse had set in a few hours later.

The bus put me down in a round plaza in the center of town. I'd been told Santa Cruz is quaint. It seemed to have no taxis, if that was what was meant by quaint. There were many tattoo parlors and teenagers with too many piercings. I shouldered my bag and briefcase and set off in search of my hotel, something I'd found on the Internet, described as an elegant beachfront inn on Main Street. I decided, as I walked, that there was something about this town that didn't agree with me. It wasn't just the bad combination of sticky humidity and blazing sun or the fact that every other store seemed to feature tie-dye. It was not even the significant number of strange people talking to invisible friends. There was a stench blowing in off the water, as if all the leftover hippies of my youth had heaved up here to die.

The inn, when I found it, did overlook a huge, flat beach. There was a round casino too, painted bright orange, like some overgrown fruit left behind after a monstrous picnic. The hotel was a faux-deco down-on-its-luck motel. The tiles in the bathroom were so retro they were heading back to chic. It was as if I'd rolled down to the end of the earth and fallen through a time warp.

I walked to Jim Kent's place. At first the road was full of cars well past their prime. Then the neighborhood's character shifted from the weedy, rundown haunts of recent immigrants to reclaimed cottages with teensy precious gardens and heritage-colored doors. Kent's house was painted a teal blue-green with lavender trim and had a front porch with a suspended swing. A very large gray stuffed animal stood guard duty. In the tiny wildflower garden, a miniature bridge crossed over a Japanese stream of pebbles.

Kent's wife, Heidi, sat me down in the kitchen and called her husband in from the garage out back, a garage with French doors, modified into a kind of studio. "So let me guess," she said, with more than a hint of sarcasm. "You're working on the Human Genome Project, right? Come to see the savior of the public program?"

Kent shambled in. He moved like a man who takes as little exercise as possible. He was short, square shouldered, thick bodied, with a black, bushy beard shot through with wires of gray. He had on a green T-shirt, a pair of chinos, sandals. A green wool beret sat precisely on the middle of his head. He lifted it high and plunked it back again in exactly the same place; he had dark hazel eyes, black lashes, and a way of looking not at me but at his own hands or out the window above the kitchen table where the afternoon light poured in. The kitchen had been smartly renovated, and Kent's redone garage was almost pretty. There seemed to be computers everywhere. None of this was what I'd expected. I'd had in my head an image of a reed-thin, struggling grad student, living in housing hell, not this middle-aged man with a wife, two kids and taste.

Heidi sat between us and listened; whether this was to make him more comfortable with a stranger or because she enjoyed hearing his stories, I couldn't tell. He was born in Hawaii in 1960; there were a slightly older sister and a younger brother, a father who spent twenty years serving in the armed forces as a psychiatrist. There were years in Texas, then San Francisco, where he went to the Cathedral School for Boys.

"Are you religious?" I asked. Heidi snickered and rolled her very blue eyes.

"Not really," he said, with a little smirk of his own. "In my mom's opinion, it was one of the two best schools in San Francisco." It was run by Grace Cathedral, which was headed by Bishop Pike, "quite a character," according to Kent. "He was a mystic of some sort. I don't know. Not all that religious, not straight and narrow, and neither was my mom."

Mom was a painter. Kent was the kind of child who read encyclopedias and was sent to school early. He went to Berkeley for one quarter, when he was sixteen, but somehow a communication glitch prevented his father from knowing he should pay the fees. So then Kent came down to Santa Cruz, the alternative school of the University of California system, and got a job as a janitor in a Chinese restaurant. He had signed up to study molecular biology, art and honors calculus at Berkeley but decided at Santa Cruz to do psychobiology. For that he needed biology, and he wanted physics, and he was good at math, and there were the art and the writing courses. "I wanted to be well educated—not associated with a career," he said. "This was the '70s—especially here, people were trying to get away from defining themselves by what they did . . . People worked out of genuine interest and no incentive to be competitive, and we could help each other."

Oh good God, I thought—learning for its own sake, sharing, no competition. It was half a lifetime ago that I had thought like this myself, and lived it, too.

And then?

"I had a, well—then we, well—then I left to study for the summer and then more Tibetan." This came out of him very, very slowly.

"Tibetan," I said. "You mean the language? Why Tibetan?"

Because, he explained, he was studying Tibetan Buddhism under Lama Yeshe.

Well, well, I thought. A biologist with an interest in the soul.

He thought he could better learn the Tibetan language at a Tibetan monastery in Switzerland, so he went there. While there he also learned French. The monastery was just outside a village not far from Lausanne. "It took a while to figure out that the Dalai Lama was basically the pope," he said.

He seemed slightly embarrassed by his former interest in such matters. He had once thought Buddhism was something other than a religion but had then come to see that its followers organize themselves like any church—in order to purvey a dogma.

"But were you interested in the spiritual aspects of Buddhism?"

"Spiritual?" he said. "I think it was another model of the mind. I was really never a joiner. The irony of a room full of people chanting in unison, in a tongue they didn't understand, 'We will think for ourselves,' was never lost on me."

How the mind is formed was of great interest to him, but he was not convinced science could deal with it either. He had personal reasons for this interest. His full name is William James Kent, after William James the American psychologist and philosopher. His mother was a manic-depressive. "There are certain aspects of human experience that science doesn't touch," said Kent. It doesn't deal with aesthetics and religious experience and emotions such as falling in love, all of which are necessary to understand human beings. While the idea of a God was "pleasantly missing" from Buddhism, in the end this could not make up for very uninteresting Tibetan science. So he came back to school at Santa Cruz, studied biology and math, got his MA in math and began to combine his interest in art with mathematics. "I was in computer graphic programming for ten years."

He went into the personal computer software-creation business in the early 1980s, when home computing and the games markets were taking off. He lived in San Francisco and in Sausalito. He worked for a company for two years, and when it changed direction, he bought the rights to develop further the software he'd developed for them. He produced a number of successful software packages, the last one of which was a desktop animator series, which he copyrighted in 1989. "Put me in the black permanently," he said. He made so much money on this desktop animator software that he thought he'd never have to work again, at least if he moved somewhere cheap, like Oklahoma.

"But I like working," he said. He continued as a freelance software writer and was soon developing programs for Microsoft. But it was expensive, and eventually, the complexity became unbearable.

With each new generation of its software, Microsoft would send out a new development kit to freelancers. "First one CD. Then the next year twelve. I was appalled—the human genome fits on one CD that doesn't change every three months."

So in 1997 Kent returned to Santa Cruz for graduate work in biology. His first two years he studied chemistry, immunology and yeast genetics. "At the end of it we'd cloned a gene, pasted it in to see where it landed in the genome, which had been sequenced by then. In the next quarter I went to the worm lab, just after the sequence of *C. elegans* was finished."

Kent reentered the world of biological science at a moment of intense change. When he left school, many biologists hoarded their results until ready to publish. When he returned, the Sanger Centre had just forced all those institutions sequencing the human genome for the consortium to post all sequences publicly and immediately. The genome sequencing side of biology was all about sharing.

The man in charge of the worm lab, Al Zahler, soon got to know about Kent's background with computers. They were all having problems with the way genome sequence information was organized by the Sanger Centre and Washington University. They were able to get access to the sequences of the worm genome from the National Center for Biotechnology Information over the World Wide Web. They could hunt for genes and other items of interest, but the program used to display sequences was awkward to work with and way too complicated, while the server that directs Internet traffic at Santa Cruz was way too slow. Kent set to work to fix things.

Kent was interested in the regulation of genes—what turns them on, what turns them off. "Not all of them are on at one time; they are orchestrated. This is one of the mysteries," he said. He was also looking at how a single human gene sequence can call up different proteins according to the order in which its exons are transcribed and put together by transfer RNAs. In multicelled organisms, the exon portions of the gene, which code for proteins, are embedded within long sequences of other bases, introns, which don't. When the gene is active, these intron sequences are snipped out of the RNA's message by a little genetic machine called a "spliceosome." The result is a much shorter version of the original RNA transcript, which includes both exons and introns. It is this short version that guides the ribosome to make the right protein. But there is more than one way to splice together those exons. "In humans there is a middle-stage RNA, where the gene is spliced in different ways," Kent said. "It can behave in different ways, depending on how it's spliced." Just as a protein can fold in different ways, creating different domains which have different functions, so one human gene can produce several differently spliced transcripts which will produce a different order of amino acids, and thus a different protein. Instead of one gene, one protein, it was one gene and who knows how many different versions of the message. "This process appears to be interactive."

He drew a diagram to illustrate. He made two sets of four squares, each square symbolizing an exon in the same gene, and sketched linkages in different places to show that each exon could be tied into a series of exons in several different ways. Something causes the gene to be spliced in different ways; this interactive process regulates the gene's operation.

To study alternate splicing, Kent needed a useful way to compare RNAs to gene exons, so he wrote a program to show all the possible RNA alignments a gene sequence could create. He also made what he called "utilities," additions to the program to make it more easily usable, including a visual component that makes it easier to imagine different splicing possibilities. He also found a way to compress images so that the department's computer system could handle more information much faster. His fellow students and people on the faculty began to use his software instead of the software being developed by the university's own bioinformatics department, whose leaders soon came to know him.

Bioinformatics describes the use of computers to search through mounds of biological information, especially sequences, for patterns. Kent's lab supervisor, Zahler, worked closely with those who ran the Santa Cruz bioinformatics department: David Haussler, Kevin Karplus and Richard Hughey. And Haussler liked the way Kent solved these problems.

Haussler encouraged Kent to study comparative genomics—looking at sequences in one animal and comparing them to another. Kent also took Haussler's graduate course in bioinformatics in the fall of 1999. This was when the Human Genome Project moved into Kent's very focused line of sight. Around Christmastime, Eric Lander, of the Whitehead Institute at MIT, called Haussler and asked him to help the consortium find genes in the human genome sequence.

Lander, who also came late to the genome project, had rapidly become a very important player. He is a mathematician and had taken charge of the problem of gathering meaning from the human sequence. It wasn't enough to simply put the giant string of 3 billion base pairs of DNA together in the right order; they also needed a way to figure out which areas of this sequence are previously unknown genes, what proteins or RNAs they make and how they are regulated. Several computer programs had already been invented to search through raw sequences and make clever guesses about which portions could be genes, but these programs needed large sequences to be of use, and the Human Genome Project had chopped up human DNA into a large number of very small clones.

The cloned libraries of human DNA sequence were mostly about 10,000 base pairs long. But the average human gene is much bigger—about 50,000 base pairs long. It would be extremely difficult to find genes unless these small fragments were strung together in much bigger segments in reasonably accurate order. Somebody had to find a way to do that.

Kent heard that David Haussler was very worried about this problem of assembling the clones, and he knew the consortium was in a hurry because Celera might assemble their sequence first, find the genes and file for patents. A group at the Sanger Centre and another at the National Center for Biotechnology Information were already trying to create computer programs that could assemble these cloned sequences into larger units from which genes could be predicted, but so far, no cigar.

But how could they have come so far, I asked, and not have thought through this problem of assembly?

"The plan, the big plan, was well thrashed out and would work—first to test the sequence methods on larger and larger model organisms, first bacteria, then yeast, then the worm, and then to do 10 percent of the human genome," said Kent. First this 10 percent would be mapped, then the rest would be chopped up at random so many times that each area of the genome would have eight or ten different clones to cover it, and eventually they would end up with enough redundancy to put the entire sequence together with great accuracy. At the end of this process, sequences about 1,000 base pairs in length would still have to be done by hand to tie the whole assembly together into one long string of A's, G's, T's and C's. "That was the plan, to get them all into one piece. It would still be a deal to piece together," said Kent.

He took a piece of paper and laid out what he called a "tiling path," where overlaps in base pairs at the beginnings and ends of sequences were laid beside each other, aligned where their sequences were identical. But, of course, many would not align, and one would need a great many of the tiles laid out on a very large plane to get all of the sequence after stripping away redundancies. The consortium's original plan was to have an accurate, finished sequence by 2003, but Celera had a great many of the new sequencing machines, and its work was going very well—or so it said. It claimed it would be done by 2001. So the consortium decided to change course. Instead of taking their time to do a very accurate sequence, they decided to do a rough draft. Doing a draft assembly meant someone had to write a program that could tie together 30,000 small clones in the right order, which could then be computer-analyzed for genes. "They needed a product to compete with Celera's," said Kent.

When David Haussler first mentioned to Kent that there were problems with getting the genome assembled, Kent didn't get excited. Although he was very concerned that Celera might be able to lock up a genome sequence that had taken three billion years to accumulate and that in Kent's view ought to be freely available to all, he also knew that three groups were working on it. Besides, he had his own problems working out a program to align worm RNA to worm DNA. He wasn't satisfied with any of the alignment programs available. He created a cluster of three computers, in his garage. In effect, he made a very powerful computer out of three fairly ordinary ones. "I could do all the worm RNA against worm DNA in four days with all my computers in the garage hooked up together." He and Zahler were also working on a program that compared the entire genome of C. *elegans* to the genome of its close relation C. *briggsae*.[4]

Kent wanted to do the same with known human gene sequences: he wanted to line up all the known human RNA sequences to all the known human DNA sequences and see what kind of alternate splicing arrangements there might be. But to try to align all that data was a much bigger computer problem. Kent calculated that using the same program, with the same number of computers to deal with thirty times as much RNA and thirty times as much DNA as in the worm, it would take him twelve years working out of his garage. "I'm not that patient," he said. He needed a much bigger cluster of computers.

Meanwhile, Haussler had decided to try to assemble the human genome clones with a cluster of one hundred computers working together. This kind of supercluster, which allows for the division of the large computation job into many small jobs among many computers, was ideal for Kent's project too. "We got one hundred computers, borrowed from the various computer labs at UCSC. We hijacked their upgrades for three months. They were okay with that. Managing one hundred computers needed a lot of work—controlling them from the same place, breaking up the jobs to run them and keep them all busy." Computers are fairly reliable; they may make a mistake or have a glitch once every two or three weeks. But with one hundred computers working together, glitches occurred in the cluster twice a day, so Kent had to write a program that would not be disturbed by such faults. By March and April, he and Haussler's other students and employees had the one hundred–computer cluster working. Then he had to break away to study for his doctoral oral exams. He finished his orals by May and wondered how the human genome assembly program was coming. Celera had announced in April that theirs was almost finished. When Kent had a look at what the Santa Cruz group had achieved, "it didn't look like it would work. We had 30,000 clones to put together."

Not only was Haussler's group unable to get its program to make an assembly, the other major centers working in the consortium were failing too. "No one had succeeded," said Kent. "It turns out it is hard." He was staring out the window again. Within a few feet of his house, the new leaves on a small sapling glowed like emeralds in the afternoon light. "The human genome is more like a poem or a song than prose," he said.

There are many repeating sequences in the human genome, particularly in the central areas of the chromosomes called *centrosomes*, but also along the telomeres—areas collectively called the *heterochromatin*. Kent used the example of trying to put together in the right order all the words on a page of a newspaper after cutting the page up into pieces or thin strips. One could put it back together into a whole sheet by overlapping or aligning all the similar sequences of letters, because a newspaper is written in simple, direct prose. But songs and poems have many repeats in their choruses—recreating a song like "Mary Had a Little Lamb" wouldn't be simple. Kent ran through the various ways in which the lines of the song could be repeated. If you didn't know the song and you just saw the letters laid out, how would you know whether to sing "little lamb" once, twice or three times? How could one know how many times human sequences should be repeated to reproduce the sequence accurately?

Celera hoped to solve this problem by using a vast number of random fragments. While many in the consortium didn't think Celera would be able to assemble an accurate human sequence from these bits, Kent believed Celera could because it had already succeeded with this method when they assembled the sequence of the *Drosophila* genome. Kent became extremely worried that Celera would get the job done first and file for a slew of patents. He also calculated that if he worked on this assembly problem for a short stretch of time—and succeeded—it might save the scientists who would use the draft sequence much more. "It looked as though if I worked two months, it could save 10,000 scientists four months. It seemed clear I had to do it."

Did anyone call you and ask you to do it? I asked.

"This was self-called," he said. "I didn't want to undermine others' work, in particular David's. I thought it was delicate to say

'I don't think this thing of yours will work.' I worked privately for four days before I told David. I knew I could do better than no assembly at all—the alternative for the near future."

Kent got his assembly program running only two or three weeks after he started. He used every piece of public data he could, including expressed sequence tags, messenger RNAs, known gene sequences, the map Bob Waterston made in the earlier days of the project and the large number of clones from the consortium.

Other groups had worked for many months and gotten nowhere, yet Kent solved the problem in a matter of weeks. How? He explained that he had long since developed a pattern of work in which he finds the minimum structure that works in the shortest possible time. He had built a structure of logic, a scaffold, which allowed him to arrange the fragments he had and to correlate them with the map and the other forms of information in public databases, cross-checking his arrangements and building bridges between the sequences. Once this initial scaffold was in place and the pieces were ordered upon it, he would keep adding to it and subtracting from it over the next two years as more clones were sequenced by the various sequence centers. But he had enough to make the first rough assembly in May 2000.

According to Nicholas Wade, Celera managed its own first assembly only on June 25, one day before Bill Clinton and Tony Blair held their congratulatory parties. But Kent actually had his program, which he called GigAssembler, working well a month earlier than that. He finished a second, more polished assembly on June 22.[5]

"How long would it have taken the consortium to make a draft assembly without your program?" I asked.

"Four or five months," he said. He knew this because the National Center for Biotechnology Information continued with its own assembly program development long after he'd finished. "By October they had one that worked."

So, I said, did you get a grant, or did you get paid for this?

He shook his head. No, he didn't get paid and there was no grant. "Deliberately on my part, not," he said. If he'd taken money from the university, he wouldn't have retained all rights to his assembly program. Similarly, he had taken no money from anyone for his alignment program, which he called BLAT. Because he didn't take money, he owned them both.

Wonderful, I thought. Yet another twist. The assembly of the sequence was a crucial problem for the consortium, which spent about $300 million a year; the man who made the assembly program got nothing. Yet William James Kent, graduate student and businessman, who did this work in an altruistic manner, didn't do it for completely altruistic reasons. This wasn't a triumph of free inquiry over the marketplace. Kent had honed his skills in the marketplace; he was used to working at top speed and with great efficiency because in private software development, time is money and the market never sleeps. He also realized that there is great value in his assembly program, so, by refraining from taking money from the university, he kept it. Darwin would have understood this well: Kent had taken the risk, now he could reap the reward—if he wanted to. Kent was thinking he might hand GigAssembler over to the university. He hoped the program would continue to be made freely available to academics and nonprofits. He didn't need to make money from it—he already was making money from licensing his BLAT alignment program. "I got a good business selling BLAT to bioinformatics companies, big pharmas," he said. "It's five hundred times faster than any existing published program. You can do human against human in less than a day in a cluster of one hundred computers."

"So, did you work so hard on the assembly program you had to ice your wrists?"

"Yeah," he said. But he worked a fourteen-hour day, not twenty-four, and he had a lot of help, from his wife, from Haussler and from Bob Waterston, whose map he used. Waterston and Haussler tested out his assembly as he was putting it together. In all, it was one very intense month.

"Were you invited to the White House?"

He looked embarrassed again—not for himself, but for others. No, he had not been invited to the White House to celebrate the achievement of the draft human sequence. He thought it was perhaps because so few people knew about him outside of his small circle. He had communicated in the early phase through David Haussler. "It took a while for the community to get to know me in my own right," he said. "David is so good at sharing credit."

The phone had rung several times, and twice he had put off calls from Haussler. His younger daughter had asked for help drawing a sumo wrestler, and, in the same way that he had written the assembly program, with the most minimal chain of logic that would be operational, he had sketched her one out of a series of stacked ovals. But he kept glancing over his shoulder at the open French door to his garage, as if someone inside were calling. He was impatient for me to leave.

"So what's a gene?" I asked.

He laughed. He gave me the biology-textbook answer first. "A gene is an abstract unit of inheritable information that can be seen to be transferred parent to child," he said. "Molecular biology, it gets more complicated. Loosely, it's something that makes protein, although there is also noncoding RNA that gets transcribed sometimes by the next gene over." And, of course, none of these definitions take into account the alternative splicing of the same sequence. "There is no gene-finding program that takes into account alternative splicing," Kent said. "We need to get experimental findings . . . My role is to use computers to organize this."[6]

Kent had just described a new relationship between biologists and computer theorists, like the one that has existed for three hundred years between mathematicians and experimental physicists.[7] But biological discovery has rarely been driven by theory before.

He stood up, extended a limp hand and walked in his stiff-legged way out the door to the garage.

Stories are like ideas: both change in the mouth of every speaker, in the mind of every reader. The gene as idea, as story, seemed to shift and shimmer like a lake surface in a breeze. First it was an unknown something, a gemmule floating in the blood, that somehow determined an organism's characteristics and behaviors. Then it was a specific unit conveying information. Then it was specific array of nucleotides, like beads on a string, calling up a protein.

Then it included operons, which turn things on, and transposons, which move genes around or suppress them. Now it could embody many messages, depending on how they were snipped and how they were linked, which in turn depended on the needs of the cell. The gene, as idea, as story, had taken on such layered complexity, such interactivity, that it was well beyond the orchestral into an almost Zoroastrian metaphysic, as if molecules could think and make choices but these choices are restricted and pressed upon by circumstance. The accumulating complexity of this idea was mirrored by the story of the Human Genome Project, first described as a public and altruistic effort but which had turned into something much less and much more. The story was structured like a good melodrama—a race, bitter enmity, crucial action and, now, rich surprise. The good guys had been saved in the nick of time from the evils of monopoly by a formerly Tibetan Buddhist entrepreneur who valued sharing while holding on to his own intellectual property, who seemed to be a fusion of the two species descended from Darwin: scientist-artists and scientist-businesspeople. He and his colleagues were changing biology's character, too—into something communal.

The entryway to the University of California at Santa Cruz is distinguished by a fence and by a wide-open field of waist-high grass, traversed by walking paths. I could see no buildings for a long time, just open space, but then, abruptly, the road plunged down into a forest primeval. I could smell fern and mold and rot. Redwoods soared so high their pinnacles could not be imagined, and the campus buildings hunkered down in their deep shade. David Haussler's third-floor office was in an engineering building made of raw concrete, with wraparound balconies serving as corridors. His desk sat under an open window with a view of a diminishing series of vast trunks and feathery branches—like the vista from the ewoks' tree house in *Star Wars*.

Haussler had arrived on a bike. A tall man, broad shouldered and slim, he was dressed in a Hawaiian shirt and shorts and sneakers. His hair was the color of wet sand, his small eyes bright sparks beneath eyebrows that met over his nose, which described the arc of an unfurling fiddlehead—not a pretty man, but strong. He also seemed to have a quality seldom found among academics: emotional balance. There was a sense of ease, even joy, in his description of the ideas he plays with, and none of the competitiveness, envy and bitterness I had previously encountered. He praised others constantly.

David Haussler is a Howard Hughes Medical Investigator, meaning he has been hired by the $12 billion Howard Hughes Medical Institute at a large salary, with a large research budget, for seven years, to think about computation and genomics. He doesn't have to teach if he doesn't want to, but he does. There are only a few hundred Howard Hughes investigators in the United States. In the beginning, those appointed were in medical schools. Haussler is among the first computer engineers hired by the institute.

And this was why I would later be so grateful he had agreed to see me. It wasn't just that he had played an important role in assembling the human sequence, it was that he described to me how mathematics and computer logic came, so recently, to be so vital to revelationary biology. I was perplexed by this business of computer programs predicting unknown genes. How could that be done when the definition of the gene keeps changing?

Haussler explained that he first used mathematics to solve a problem in biology when he was twenty. His older brother, Mark, Regents Professor in the Department of Biochemistry at the University of Arizona, already had his own lab when Haussler enrolled in math at Connecticut College. In 1973 Haussler went out to Arizona to work in his brother's lab for the summer. His job was to go downstairs to the basement every third day, cut the heads off chickens deprived of vitamin D, homogenize their intestines, then do an assay. These little murders produced a lot of data, which then had to be arranged and analyzed. At a lab meeting, the question of how to lay out the results was discussed. "I said we need to transform the data and then use a linear regression analysis. They all looked at me. They said, 'Okay, you do that.'" This project resulted in his first publication credit—in *Science*.

He went on to a doctorate at the University of Colorado at Boulder in computer science because he was fascinated by formal logic—"abstract reasoning for its own sake," as he defined it. His adviser was Andrzej Ehrenfeucht, "an astounding genius." Ehrenfeucht was from Poland, a logician and a mathematician who also taught linguistics and computer science. "He is omnivorous in his appetite for ideas," said Haussler. "He didn't have time to write up all his insights . . . He was like a firehose of ideas."

"My favorite Ehrenfeucht story is this. He lives off campus about two miles. One night, at about 2 a.m., he got a call from a security guard on campus. He said he was doing his rounds, saw Andrzej's office door was ajar, he looked in, the room was completely ransacked. Andrzej didn't drive, so he gets someone to drive him. The desks are piled so high, the papers slough off six inches high on the floor, but there's a narrow path to the chair. Andrzej looks carefully and says, no, this is exactly the way I left it."

Until the early 1980s, when Haussler joined the group around Ehrenfeucht, very few mathematical tools had been developed to analyze problems in biology. "The main work to be done was understanding complexity that doesn't reduce to a few equations," he said. "The way a cell works, and the way a genome of 3 billion bases of DNA in all of these cells works, is a product of accident and structural events over billions of years. First DNA has to be documented at the qualitative level to tease out all the players. Then you start looking at interaction. Increasingly, there are lots of neat math things to say of DNA, but a quantitative theory is not possible without algorithms. With a computer, you can make a hugely complicated number of contingencies in the algorithm—in vast excess of a human using a pencil and paper. For example, we build a mathematical model of DNA with a hidden Markov model. It's a particular model that has in it a set of states that represent the different functions that pieces of DNA can have. States represent exons, the bases at splice junctions at the end of exons where the intron is spliced out, and many other functional features. For each state there is a description of the probability of seeing each of the four bases—explaining the probability of each occurring in a gene. When the computer looks with this model at 100,000 bases of DNA, it tries to interpret it to find out if a gene is there . . . The purpose is to analyze DNA by the algorithm and make predictions so you can say a gene is here—link it to the protein, see whether it is looking like, for example, a histone or a kinase. You can push on to make a prediction of function. But a gene-finding hidden Markov model has literally tens of thousands of parameters, each with a specific meaning. For a human being to take such a model and by hand apply it to 3 billion bases, it would be ludicrous. Computers are essential.

"We still find it darn hard to find the genes . . . There's no way we have found so far to reliably predict genes in human DNA. We hope by comparing to other genomes we will enhance reliability, but we're a long way from that."

Then this computer gene-finding is just inference? I asked.

"Everything in science is based on inference," he said. "It's hard to say when it crosses to fact . . . All the absolutes of science eventually get broken. When tempted to say something is absolute, I bite my tongue. Nature seems to have explored every possibility; it seems as if, if there's a crazy way of doing something, it found it. That plays into this complexity."

Haussler's fellow graduate students included Eugene Myers, who would go on to create the assembly program used by Celera, and Gary Stormo, now at Washington University. Haussler became interested in applying mathematical models to DNA when Stormo gave a talk about DNA sequences in 1981 at one of Ehrenfeucht's weekly research meetings. At that point, Sanger and his colleagues and other groups had managed to identify the sequences of only a few small genomes. Stormo was trying to find a pattern to predict a good ribosome-binding site in sequences of *E. coli* DNA. He had wondered if maybe someone, sometime, had come up with a computer algorithm that could recognize this pattern. After a literature search, he found one called a *perceptron*.

Perceptron was a computer model created in the 1960s, in the early days of trying to create an artificial form of intelligence.[8] Various perceptron programs were devised to teach a computer to distinguish one thing from another. The perceptron Stormo found could teach a computer to distinguish the gender of people from photographs showing only facial features. The algorithm had long since fallen into disuse when Stormo tried it, and it would become useful again years later to help model the complex way human neurons divide tasks in neural networks. "Gary dug it up; he was a real leader," said Haussler. "That was the first instance where I saw a chance to apply math models to DNA sequences."

As a graduate student, Eugene Myers developed an algorithm to predict secondary structures inherent in RNAs, which can bind to themselves. "They have a much richer topology than DNA molecules," Haussler said. "Their interactions can be nested. Could a computer predict the secondary structures they could form?" Yes, it could. Later, Myers became the key author of a simpler but more flexible program, called BLAST, "the sequence algorithm that transformed bioinformatics," said Haussler. This algorithm allows one to relate DNA sequences to RNA sequences and proteins. "You could put in a sequence and it would search all DNA, RNA, proteins for a match. It gave back a list. The BLAST paper is the most-cited in bioinformatics, I think."

Haussler got his doctorate in 1983 and stayed on as a postdoc with Ehrenfeucht for three years while teaching at the University of Denver. He spent most of that time working on statistical models, without any regard to how they might be applied. In 1986 he took a job at Santa Cruz, where he has been ever since. By then, he was interested in how statistical models "could explain scientific data on a broad scale." He was interested in using computer algorithms to mimic how neural networks adapt and shift constantly to organize information in the brain—in other words, how we learn.

A postdoc named Ander Krogh came to work with him in 1991. "We were interested in speech recognition, to design an algorithm that could recognize words." Haussler and his students had been using a hidden Markov model, which had proved to be very useful. (Mathematician Stephen Wolfram writes that this type of model was built around the presumption of a hidden layer of neurons, and had first been applied to making models of neural networks in the early 1980s.[9]) Haussler told Krogh that he thought these hidden Markov models might also identify patterns in proteins. He thought they could teach a computer to recognize a member of a protein family by the sequence of its amino acids.

"For weeks we designed an application for hidden Markov

models to see if they could distinguish globin proteins from others." They set the algorithm a problem. Given a particular protein sequence, is it a globin? The BLAST program could not recognize as globins many proteins that are in fact members of the globin family. "We built a hidden Markov model and proved it could recognize them all," said Haussler.

They presented their first results in April 1993, at a neural networks and computing conference in Snowbird, Utah, an annual meeting of the top thinkers in the field. By this point, hidden Markov models had proven to be useful in several biological applications. "We said, look, it should also work on DNA. They loved it."[10] Later, he and Krogh also applied a related model to finding genes in *E. coli* DNA.

By 1994 Haussler had decided to devote himself entirely to bioinformatics. He spent the next seven years developing these hidden Markov models. He was not alone. Computerization of biology took off. The sequencing centers were using computers to make databases of the sequences of clones available over the Internet. These hidden Markov models, with their thousands of parameters, their chains of logic built on statistical probabilities, gave researchers the chance to make predictions about what a particular DNA or RNA sequence might mean. If the model said a sequence looked like a gene that would make a certain protein, a researcher could go into a lab and check. For the first time, genes could be discovered without the messy business of making myriad mutations in the laboratory. Biology had been driven by experimental findings that required explanation. Now comparison and prediction could drive experiment.

"This is a definite paradigm shift from where molecular biology worked before," said Haussler. "It is a different way of thinking." As time went by and computer capacity increased, the capacity for sequence comparison kept scaling up. The increase in computing capacity achieved by running cheap computers in parallel led to an explosion of predictive capacity. Haussler had increased the size of his computer clusters from three to one hundred to one thousand computers, which are capable of "analyzing a whole genome in one computation every day."

Yet he didn't get involved in the Human Genome Project until the day Eric Lander called and asked for his help. Lander told him that while there had been some negotiations about cooperation between Celera and the consortium, he thought the consortium had to be prepared to go it alone and do a full analysis of the sequenced genome, "including finding the genes."

But why you? I asked.

"My student David Kulp, who is now a vice president at Affymetrix, developed a hidden Markov model called Genie to find genes, and applied it to *Drosophila* . . . He said, 'Your group developed hidden Markov models to find genes—what about analyzing the human genome? But it's in pieces, it's not done.'"

Haussler thought the problem of finding and analyzing human genes—tiny exons in a vast sea of noncoding introns—would be almost impossible by computer. It had been terribly hard with *Drosophila,* a much smaller genome of much less complexity. No money was offered, and there was no time to write up a grant application. Nevertheless, he signed on.

"The first thing I did was invite Jim Kent to lunch. I can say with 20–20 hindsight that was a great decision."

"Why did you pick him?" I asked.

"He was unmistakably brilliant," he said. "He was in my class; we did research projects together. He had worked with his adviser, Al Zahler, at an astounding pace . . . The fact of ten years in industry weighed heavy in my mind. That level of maturity, knowing software development would be important."

Kent explained about his need to prepare for his PhD orals. Haussler said fine, and then called on David Kulp to do gene prediction. Kulp was focusing hard on his start-up company, Affymetrix, but joined in nonetheless. Haussler hired a math student and a former doctoral student and set them to work with Kent trying to identify gene sequences from the various messenger RNAs, expressed sequence tags and so on, that had been piling up in public databases for years. This was only going to get them sequences and genes already known. "To find new genes, the sequence had to be assembled," said Haussler. "I got worried: there was no clear plan to assemble all this data—it wasn't a part of the plan till the last minute."

In February 2000, he talked the dean of engineering and the chancellor of Santa Cruz into kicking in $200,000 for one hundred Dell Pentium III computers, which he set up in a cluster to try a sequence assembly.

"The problem of assembly is so complicated there are only two ways to solve it," said Haussler. "One is to put together a team of experts to work together and fit their solutions together. That approach in software takes a minimum of a year—we had a few months. The only other way is that one person sits down and pulls it all together. In a sense, it had to play that way." One of his group tried an assembly program that went partway but wouldn't bring all the chunks of data together.

"Jim had developed key foundational data structures to handle the messenger RNAs and expressed sequence tags. But then he took six weeks off to do his orals," Haussler said. There were phone calls back and forth between the various groups trying their own programs, but it wasn't going anywhere. "Jim came back and asked, what is the status of the assembly? I said it's looking grim." When Kent said he had a method that might work, Haussler said Godspeed. "I knew him to be an extraordinary person, but this was so far beyond the realm . . . It is an amazing task. He wrote thousands of lines of code."

Haussler explained that in the business world, there is an understanding of what the typical programmer can be expected to do in a day, and it was well known that on troubled projects, as more people are added, much more time is consumed. "I think Jim's productivity was well over ten times what is considered reasonable or possible. It is off the charts that an individual could produce that amount of code—to do that complex a problem is absolutely staggering. He tested it himself . . . I'd go to his garage, we'd have whiteboard discussions . . ."

The result was a miracle, because it worked. "Anything could have gone wrong. In most cases the dean would have said, 'Write a grant.' Or the configuration wouldn't work, or we just wouldn't have someone so far off the charts in ability—it was an amazing chain of unbelievable things. It's a beautiful, creative act when you create that complicated a program—it has hundreds of modules with complicated interactions with other modules. Bill Gates once said his computer operating system software is more complicated than a Boeing 727. You wouldn't think of an individual building a 727, but some individuals can do this in their heads."

On May 24, they ran Kent's assembly program. "Everybody was free to look at it," Haussler said—and to suggest modifications. "Then it was modified and ran again on June 15. It finished four days before the White House announcement." Haussler spent the weekend responding at all hours of the day and night to questions from Francis Collins, the director of the project for the NIH. Collins wanted to have great statistics to flash in front of the world's press at the White House, including the number of contiguous stretches of DNA in the assembly of the human genome. With the sequence assembled, consortium scientists, in particular the Ensembl group at the European Bioinformatics Institute, were able to apply gene-prediction programs and come to the conclusion that there are fewer than 30,000 human genes, one-third the number expected. The Celera program predicted even fewer. Of course, the gene-prediction programs didn't take into account that some genes are spliced in multiple ways.

The University of California at Santa Cruz quickly put Kent's version of the human genome sequence assembly on its own website; it gets 50,000 hits a day. There are many links to the other sequence centers' websites—the NCBI, the Sanger Centre, the Whitehead Institute and so on. Each of these centers keeps adding data to the assembly, which means continual new versions of the assembly, as well as various programs for finding genes, retrotransposons, promoters, predicting messenger RNAs, proteins and so on, plus comparisons of every other genome sequenced, all of which are also revised constantly. Normally, scientific publication involves peers looking for error and asking for corrections until a final and accurate paper is ready to be published; because of this review process, the publication can be relied on. As Haussler talked about these Internet connections, the rapid changes in all these assemblies, I suddenly saw it—like a crack of light beneath a door opening onto a new world: these assemblies on all these interlinked websites are a whole new way of communicating science.

"None of this is like a normal scientific publication, is it," I asked Haussler. There's no peer review of the sequences before they are posted in the databases, and no real peer review of the sequence assembly itself. There are multiple versions of a truth. "Is this even science?"

"It is true it wasn't science done in the normal fashion," he said, slowly. The way the sequence is presented now, on the Santa Cruz website, over half of it "is coming from labs willing to share."

"But are they doing this sharing before peer-reviewed publication?"

"This was the amazing thing of the Human Genome Project," he said. "All of it was available publicly during the project. Only in February 2001 was the paper published. This is much more open than almost any other scientific project has been. It's been amazingly open. But it wasn't perfect. It did not go through a full scientific review at every stage, or get vetted, but how do you do that with 3 billion bases—it'll be years to sort out—but it'll be so valuable you cannot withhold it. We put it on the Internet on July 7, 2000."

His eyes seemed to mist over when he thought about that day. "I should say it was more exciting to me personally than the White House," he said. "I mean, it's very exciting to go to the White House. Francis Collins gave an impassioned speech, and Craig Venter did too. But for me, I wasn't speaking. I was very proud our team had made a fundamental contribution. There was a lot of posturing. The scientists needed the data. We said we're different from Celera 'cause we'll show you the data. We made good on that on July 7, free, no strings attached, on the Internet."

He wanted to know if I had checked out the assembly on the Santa Cruz website. I confessed I had tried, but my computer was agonizingly slow and I had given up.

He was already turning to his own computer on the desk behind him to demonstrate for me. There was no way he was going to let me leave without showing this to me, even though his time was short, we'd had no lunch and, when we set up this appointment, I'd asked if I could take him out. He appeared to have forgotten the concept of lunch.

"The Net is an amazing thing," he was saying. "The fact the human genome was on the Internet captivated the imagination of enormous numbers of people. This university charts the activity of outgoing traffic over the Internet. We looked at the chart of July 7th . . ."

Just like that, on his screen, up jumped a bar graph in glowing chalkboard green. The page recorded use of the various websites on on the UCSC campus on each day over several weeks in 2000. It showed one giant spike on July 7. "One half-trillion bytes of information were broadcast in a twenty-four-hour period. It broke all the university records. That is the power of the Internet."

So, I said, does this sequence on the Internet constitute the actual publication of the human genome?

"How does a result that big get shared with the community?" he asked in return. "The Internet is up to it," he answered. "To me that was a transcendent experience. It was the most poignant experience in my scientific life—I saw that this ocean of A's C's, T's and G's that is our genome, and we put it on the Internet. It was beyond words."

The screen on his monitor announced the working draft of the human genome and the mouse genome. The mouse sequence draft assembly was being posted in the same way as the human and could now be compared to the human. The latest version had been assembled the previous December. He explained how the page works, how a scientist looking for something starts the process by typing in the number of a chromosome of interest. The sequence assembly marks and indexes anything defined as a gene on that chromosome. This sequence can then be related, through the use of programs like BLAST, to proteins, RNAs and similar sequences found in other organisms. He urged me to pick a gene I was familiar with, or even just a chromosome, so he could start the process of a search. I asked to see the gene for Duchenne muscular dystrophy, the huge gene so brilliantly tracked and cloned by Ron Worton and his colleagues in Toronto in the 1980s.

He typed it in, and the screen changed. Long lines of letters stretched horizontally across the page. Above the line of letters were sections marked off by red brackets and blue brackets; blue brackets indicated known or predicted genes, red ones showed exons within the gene. Other marks showed known single nucleotide polymorphisms, or SNPs—places along the sequence where experimenters have found one nucleotide difference between one person and the next. "Maybe that makes a difference. It could change the amino acid in a protein. It's all the difference in the world in cystic fibrosis—one amino acid change," Haussler said. The Duchenne muscular dystrophy gene, which is on the X chromosome, is so big it could not be displayed across one page at the ordinary font size. Haussler wanted to pan to the left and right of it so I could also see what is known or predicted of genes on either side. The human genome is so vast and so much of it remains unknown, he said, that he was sure we could find evidence for a gene that no one else had ever seriously looked at until this very moment. All I had to do was randomly point at a particular exon and he would zoom in and see what we could find. This is what this website, and the others it is linked to, is great for, he said. He could spend hours trolling at random and find things no one else knew existed.

The scientific literature, he explained, is completely unable to keep pace with the production of data. In the old days, one would read the journals to keep up. The journals would be about a year behind what was actually going on in labs around the world—one could search the literature most diligently and miss vital things that were already known somewhere but not yet published. Waiting for journal publication, "we would miss this single little SNP submitted to the public database a month ago," he said, pointing at a small area on the screen. There were a lot of areas marked as SNPs. In fact, as Kent had told me, it is believed there is a variation between individuals every thousand base pairs on average, meaning each person will typically have 3 million single base-pair differences in their DNA sequence from any other person's.

"We have gene predictions here," he announced. "If we sit here another hour. . ."

He was leaning close to his computer monitor, working the mouse deftly to show more and more of the Duchenne muscular dystrophy gene on a single page by shrinking the font size several orders of magnitude. Within the framework of the large gene, smaller blue-bracketed areas indicated many smaller genes. These were in some cases already known; in others, the result of predictions made by the algorithm. I don't know why, but I found my heart beating faster. Haussler was excited about this, about using this tool, with its gene-prediction algorithm, to find things no one else knew about, no one else had put together.

He clicked and clicked. The type shrank and shrank again, but he still couldn't display the outer edges of the muscular dystrophy gene on one page. "Jim designed this software," he said. "The beauty of it is it's so responsive. I go to meetings and people tell me, this browser is hands down the best. We just zoomed out tenfold and tenfold again. That is very hard to do efficiently. We're still not seeing the whole gene—it's still running off the ends of the page. Now we're at 5 million bases on chromosome X. There it is."

The letters had become extremely small, but we could now see the whole gene, which was marked with a blue-bracket closing at each end. "It is 2.2 million bases, peppered with tick marks of exons in an ocean of introns," he said. The tick marks he referred to were the red exon brackets, which at this resolution looked very, very tiny in comparison to the very long line of letters and the blue bracket enclosing them all. "You know the lifeblood of many scientists is in those ticks," he said. "All that effort, the lifetimes of effort for this gene—all at your fingertips, in an instant, to explore."

I wondered if Ron Worton turned on his computer at the end of the day and came to this website, just to look at his handiwork and see what others were adding to it. I told him Worton's story, the years of struggle, the public praise snatched away so unfairly.

If I pointed to any given section marked as an exon, Haussler said, he could compare its sequence to other animals also sequenced—the worm, the fly, the fish, the mouse. Everybody posted their findings as fast as they could. The fact is, this posting of new material without waiting for scientific publication was so common that even companies were being forced to do it.

"We're talking about a new way of bringing work to life," he said. "It grows within the fabric of science as it evolves."

The old practices of secrecy, of waiting for patent applications—the usual strategies of corporations seeking monopoly advantage—were being swept aside by the accumulating power of this instrument and the impatience of researchers to know more. And yet, I thought, without the scrutiny of the old kind of science publishing, in which claims are examined by peers first, this Internet form of publication would soon be riddled with error. "How do you get rid of mistakes?" I asked.

"We show the latest version," he said. "They evolve." All previous versions of the assembly were available in archives, and each center had its own methods of dealing with this old information. He could flip back and forth immediately from one center's website to another's to see who had something his website might be missing. "I can connect to NCBI and view the same part of the genome."

To demonstrate, he clicked a few times on his mouse and we entered the NCBI web page at its version of this part of the sequence. Instead of the letters being laid out horizontally, NCBI's ran up and down the page. The gene predictions aren't identical either, although they are substantially similar.

"I think this is the beauty of the Net," Haussler said. "Some things we don't have, but look at that, so much is at your fingertips." Someone tracking a gene of interest could look at and compare what was known about it by clicking back and forth between various websites. This was a form of publication in which the possibility of an absolute, unvarying truth was not even considered. Why would it be? Constant variation is the truth in biology.

The font size was small enough to display the whole Duchenne gene. Haussler pointed at the many smaller blue brackets riding within this larger frame. These indicated smaller genes nested within the big one. Several of these smaller genes matched those predicted in the mouse. He pointed at blue-bracketed areas on either side. "Look at all these other gene predictions—the mouse matches. Pick one," he said. "We may be the first ones to look at it with this data mapped on it."

I said "Eenie, meenie, mynie, mo," and pointed to a red tick that signified an exon. Haussler clicked twice on the area, making it bigger. Blue brackets appeared. A gene is predicted here, he said, by two predictor programs. A connection was shown between this area and another one, much farther to the right, suggesting some kind of extension of the gene into that other area. The connected region was also bracketed in blue.

"That is a known gene in blue, but the smaller gene is embedded in it. Two different programs predict it. I make the conjecture that it is a separate gene."

He clicked again and the print on the screen grew by another order of magnitude. He was homing in on a group of 7,000 bases within the larger group we were examining. "Ensembl, a prediction program, predicts a two-exon gene here," he said. He clicked several times more to check if a similar sequence was found in the mouse, and he found one. "Bingo," he said. "I bet we have a real gene."

"What does it do?" I asked.

"Right," he said. "Always the next question." He clicked back and forth between websites. He checked to see if anyone had a gene probe for this sequence. No. He checked to see if it matched any known gene-expression data. No. He checked to see if anyone had spotted any single-nucleotide changes in this sequence. No. Did it correspond to any known areas of repetitive DNA? No. Were there any known retrotransposons within the sequence? No. How about some kind of promoter upstream that might control it? No. Such singularity is unusual. "This is an interesting gene prediction," he said. He checked protein-predictor programs. He tried Ensembl. "The predicted protein sequence is described," he said. "It has similarities to known protein sequences." Then he clicked and clicked again to try Jim Kent's BLAST-like alignment tool. He pasted the 7,000-base sequence into the correct area on the screen and clicked on an icon marked BLAT.

To do this kind of cross-comparison by searching journal articles on paper would have taken a graduate student weeks, even longer. Haussler had done it in ten minutes. It wasn't necessarily an exhaustive search—some company or lab somewhere might have noticed this sequence already, discovered its function and moved to patent it without publishing anywhere—but that same problem would afflict anyone trying to search this sequence through a trail of published journals. If Haussler was right and old habits of publication were being overthrown in favor of inserting one's findings immediately into this emerging organism, this Internet communication system was as important an innovation as Gutenberg's movable type, which had made modern science possible. Science is a conversation about evidence found and meaning extracted; the speed of these conversations was absolutely breathtaking.

Haussler returned to his comparison of the protein this sequence was predicted to make and other known proteins. "We are hitting what may be a novel immunoglobin, with a high similarity to a number of these proteins. . . My guess is, we found an undocumented one. It might have been known to someone else, but there's a chance it's completely novel."

I wasn't sure how to read him. He had slowed down as he talked me through the stages of his search. His tone had gone from almost giddy to almost somber, as if he was as surprised as I was that one gesture from my finger had actually pointed at what he'd said it could—something no one had studied before.

"Are you sure it's a new gene?" I asked.

"It's a pretty good guess," he said. "I'd put it at ten-to-one odds."

I felt as if my head had been blasted open. All those fusty concepts about how science is done, about the ownership of discoveries, about peer review, about control of error, had just blown away like last year's leaf fall. What Haussler and Kent and their colleagues had made— were continually remaking—was a whole new vehicle for driving the world's biological experiments. Science is not just a method, but also a structure of ideas that changes over time. This was no rigid, old-fashioned scaffold, put together board by board; this was organic, like a language—a moving, changing, breathing Leviathan of learning; a growing tree of interconnected, patched-together, constantly changing observation. Ungainly, like a newborn living thing. Evolving, like a story.

My heart was pounding. This organism they'd all brought forth was the real discovery. This was as important as Sanger's methods, as Berg's recombination, as Brenner's parsable worm, as Watson and Crick's model of DNA. I was thrilled. And then, having been thrilled, I wanted to be sure others recognized the value. But I remembered that Jim Kent hadn't even been invited to the party at the White House, and euphoria gave way to anger: how could all those strutting politicians, with their cheap phrases about the Book of Life, have missed this?

"Why didn't Jim Kent get invited to the White House?" I asked.

"I barely got invited to the White House at the last minute myself," Haussler said.

"But look what he's made, what you've made," I said.

"People hadn't realized the contribution. It didn't fit into the whole scheme—it was all rushed," he said. "The way it was originally planned was, all the centers would just keep cranking through until they completely finished the chromosome they were responsible for. A race to do the first draft was not how it was set up. It was a while before people realized the actual assembly of the working draft was a critical thing. The data analysis is fundamental."

He clicked away at the mouse. I sat quiet for a moment, fuming about how seldom it is that important changes are understood as they are made. Then I thought, maybe he was too young to understand how vital it is to make people see the wonder of this, the beauty of it. Revolutions need protectors in their early years; they need supporters with power, or they will find themselves cut down by competing interests.

"How old are you?" I asked.

"I'm forty-nine in October," he said. "Certainly this is the biggest deal so far. Clearly, um, in my scientific career, putting the genome on the Net was the highlight. It was crazy. There was no special celebration till after." He had expected on the first day that biological scientists, particularly geneticists, would be the first to go to the Santa Cruz web page to view the human sequence.

"Naive me," he said. "I would've thought when we posted it the first people who would go to it would be like your Worton, who did Duchenne—they would go and see what was there. If I'd worked on a gene for many years, I'd be the first person there." But geneticists aren't cybergeeks. "They have no flasher up on their screens with what's new on the Web. The cybergeeks first noted the fact that the genome was on the Net . . . They were searching for mystical messages—they were counting how many times *Gattaca* appears."

"*Gattaca?*"

"It's a serious movie about the future, where kids' genomes are engineered and if they're not right, they're not part of society. It's a total dark vision, a modern-day *1984*."

Francis Collins had mentioned this Andrew Niccol film in his speech at the White House.

"The one thing I am most concerned about regarding the social impact of my work is society's ability to assimilate knowledge of human differences and to choose different genes for our offspring," said Haussler. "Now you can choose through artificial insemination—you can choose to have a child without Duchenne muscular dystrophy by selection of fertilized eggs; you can choose one that doesn't exhibit the harmful trait. You couldn't design a human being that way . . . But . . . fifty years from now?"

He was talking about enhancement, remaking the human animal. He had this look on his face that said, I am a person who thinks, I am a person with concerns—you can trust a well-meaning person like me to manage such a future. In the context of the beautiful learning tree he and his colleagues have made, which cries out to be used, to be played with, which will predict all kinds of genes worth testing and altering and moving, this expression of concern about how it will be used struck me as beyond trite. Who among his colleagues, with an instrument such as this at their fingertips, would be able to resist the allure of making human beings anew?

"It is scary to think we may some day be able to change the genetic content to get an effect. To think the state will control this is terrifying," he continued.

There was something annoyingly self-congratulatory about his concern, and naive too. How could he fail to distinguish between one kind of state and another, between a democratic and open state like the ones we live in and the dictatorship of *1984*? How could he not see that a democratic state, for all its flaws, must be a better regulator of such experiments than the heartless, remorseless, faceless and irresponsible market? His learning tree is open for everybody to see—that was the wonder of it. The marketplace would soon find a way to use it—that was the downside.

"Do you have children?" I asked.

He nodded. Yes, he did, two.

"You know how this will work," I said. "You know that if parents are offered the chance to buy the permanent improvement of their children, to make them faster, taller, smarter, prettier, to live forever, they will do it, no matter what it costs."

"The market is also frightening in many ways," he said. "Our knowledge of the effect of different versions of a gene is primitive. You have to be cautious if you have the technology to change something but not the knowledge of what it will do. That is the crux of my fear."

A little late for that, I thought.

Haussler's secretary entered the room with force. She had been trying to get his attention for some time. Someone from the Howard Hughes Medical Institute had been cooling his heels while we played with the computer. Still, Haussler didn't want to let me go. He insisted on introducing me to the two young men in charge of the computer cluster that serves the sequence assembly and its gateway to the Internet. He wanted me to go downstairs and actually look at the Dells. In fact, he insisted on it. The number of computers had increased from the original one hundred to one thousand. With new genomes being sequenced all the time, all of which are being constantly updated, amended and compared to each other, the demand for computers to manage the data is growing at an exponential rate. They would soon install a cluster of ten thousand machines.

The two young men were just about what one would expect—a little nerdy, a lot funny. Down we went to the main floor, then through a door that led to a room with computers stacked floor to ceiling on row upon row of metal shelving. The room was like a library in which all the book covers look exactly the same. A gridded floor hid all the interconnection cables. All of the computers were managed from one workstation set out by itself in an otherwise empty corner. The workstation looked like a lectern in a pulpit. The Dells looked like Dells. The whole room was white. It took about thirty seconds to see what there was to see—a room filled up with machines. I thanked them and turned to leave.

"Oh, you can't go yet," said one.

"Not yet," insisted his partner. "You haven't seen the heart."

"What heart?"

One of them crooked his finger. I followed him down a corridor between two rows of computers stacked floor to ceiling. In the corner farthest from the door, like altars to some data god, stood two wildly painted, hot-from-the-arcade video game stations.

CHAPTER SEVENTEEN

THE SOVEREIGN STATE OF BIOLOGY

I came home from California in quite a state, switching, like an alternating current, between euphoria and fear. I had inhaled the heady fumes of discovery and then turned on my teacher for his failure to be as expert in human politics as I thought I was. His fear of government was so American. Okay, so eugenics laws had been written by democratically elected politicians in Canada, too, according to the science of the day—that was then, when democracies were young, when politicians still believed in the wisdom of scientists and when politics were played fiercely in backrooms. I could see that Americans were having trouble making rules about revelationary biology, and that was where my fear came in. But things were so much better in Canada. It had been some months since I had attended the hearings of the Standing Committee on Health, and distance makes one forgetful, so I was almost smug about the Canadian political process: the committee had created opportunities for conflicting views to be heard; a bill would soon come forward.

But while I was away, political hell had broken loose. Alan Bernstein and Francoise Baylis had held a press conference to announce the CIHR's guidelines for publicly funded human stem cell research. Bernstein told reporters that as soon as he appointed an oversight group to enforce these rules, grant money to study human stem cells of all sorts would flow to Canadian researchers.[1]

At this point, there was still no legislation before the House of Commons. The father of the legislative proposal known as C-13, Allan Rock, had been handed a new portfolio, and his successor as minister of health, Anne McLellan, had been on the job only a few months. The Standing Committee on Health had suggested significant amendments to Rock's draft the previous December: the committee wanted all commerce relating to egg, sperm and embryo driven from the temples of human fertility by the scourge of law; it wanted a ban on nuclear transfer cloning to get stem cells; and it wanted researchers to be forced to demonstrate that each use of human embryos was essential to answer an important question. The new minister still had not resolved the differences between the government's views and the committee's when Bernstein made his announcement. Several MPs, including members of the governing party, exploded at Bernstein's slap in the face to the democratic process, at these foxes stealing the henhouse in broad daylight.

In mid-April 2002, Bernstein came before the committee to explain, and he promised that no CIHR funds would flow to this research until April 2003. The foxes appeared to be groveling. But when I called the CIHR for an appointment to see Bernstein, a public relations person said she thought Bernstein had done very well. She didn't sound in the least chagrined. "We Q and A'd him with questions for a day," the woman said. "There was only one question asked we hadn't asked him already." She sent me the transcript of his committee appearance.

At first, I thought her admission of this preparation was a measure of her naiveté. Then I read the transcripts: they were a measure of something more serious. The CIHR president seemed to fear no repercussions for his actions, as if it was well known that one could ignore parliamentarians and do as one pleased. His tone was not in the least apologetic. Even his opening statement suggested he was on the inside track—that he and his colleagues had the government by the tail, not the other way around—and that mere members of Parliament were entirely out in the cold. Bernstein reminded them that they already knew about his committee's activities in drafting guidelines and his intent to fund this science. He reminded them too that Janet Rossant, his committee chair, and Francoise Baylis, a bioethicist on the committee, had also helped write Rock's draft law. He explained that the CIHR's guidelines dovetailed with the draft legislation considered by the committee. He said the CIHR had merely moved into a legislative void—there was no law against any of this research in Canada, only a 1998 policy produced by another committee of scientists.

The members didn't like it. What followed was one of the more brutal cross-examinations I've read, but brutality is often a sign of weakness. It wasn't just the opposition parties who whacked Bernstein. Paul Szabo, Liberal MP for Mississauga South, not even a committee member, came to the hearing (any member of the House can attend a committee hearing and ask what he or she likes) and worked Bernstein over with verbal truncheons. Chairwoman Bonnie Brown, Liberal MP for Oakville, had lectured Bernstein at length.

Rob Merrifield, Canadian Alliance MP for Yellowhead, wanted to know if he'd gone to the minister for approval. Bernstein said he'd told the minister he was going to do this but didn't ask for her approval. He is supposed to operate at arm's length.

Was the scientific community frustrated by the slow pace of legislative change, Merrifield wanted to know.

No, not frustrated, said Bernstein.

MERRIFIELD: I'm absolutely appalled that you feel you can make a decision like this, the most important ethical decision that we have to make, as legislators—you pull that away from the legislation and you make that a decision at the eleventh hour, prior to legislation coming forward and through the House . . . You say that the guidelines and the moneys will not flow until next year. But do you realize that Genome Canada has adopted your guidelines and they have started funding for that research last week?

Genome Canada is a nonprofit with corporate and institutional members and a large dose of government funding, like the Stem Cell Network. Bernstein explained that his guidelines were important for precisely this reason. But the MPs weren't having any of it—they could not abide his assertion that the CIHR's guidelines were the same as a draft law they wanted rewritten.

MERRIFIELD: . . . . It's either the scientists have taken this under control, into their own hands, or the Minister is using you as a way to get this into the public arena through the back door.

A Bloc Québécois member asked if it was true that certain researchers had already gotten grants to work on human embryos—before the guidelines were issued. Bernstein answered that three researchers had grants to study human embryonic stem cells derived from lines made in the United States. These researchers would have to come before his new oversight committee to get further funds. Was there shock among the members over this answer? It was not apparent from the transcripts, but *I* was certainly surprised.

Bonnie Brown asked how many Canadians the CIHR had consulted in the preparation of its guidelines. Bernstein said eighty-nine individuals had responded, and perhaps twenty-seven groups. Brown pointed out that these groups appeared to be what she called "the sicknesses," charities organized to support research into such diseases as diabetes or Alzheimer's. She argued that they had been sold by scientists on the notion of quick cures that would emerge from this research, cures that she doubted would be forthcoming. She and her committee had, by contrast, spent many long hours face to face with a great many Canadians, including him. She wondered if it had ever occurred to him that perhaps he owed the committee the courtesy of informing them in advance of his plan. Specifically, she wanted to know why he hadn't sought the committee's advice on its area of expertise—how things might work best politically.

Bernstein answered that politics should be avoided by the CIHR, which exists to provide scientific advice "and where our best judgments are as to where the science is going and what the potential is. Otherwise we'd become, if you will, second-rate politicians and I think that is not of value to you . . ."

BROWN: . . . In your presentation, you continually said, our guidelines fit with the draft legislation. Every time you said that we all bristled . . . This committee was not comfortable with the draft legislation. So you can see . . . why basing your second set of guidelines on the draft legislation sends chills up our spines.

Paul Szabo reminded Bernstein that Francoise Baylis had told the committee that human embryos are human beings. He hinted that the CIHR was withholding money from adult stem cell research in favor of human embryonic stem cell research, that CIHR had a bias. Bernstein answered that though he had not mentioned this to the committee before, he now wanted to say he is a world-recognized expert on adult stem cells. This was almost an admission of interest— he was creating guidelines that would shape his own work—but nobody asked about that. He also said it was fortunate that politicians had to deal with ethical issues, not him.

Try as they might, the members were unable to get Bernstein to acknowledge any error other than that it would have been courteous to warn the chair of his intentions. Bernstein's performance reminded me of Agent Smith in *The Matrix*, who is impervious to all means of destruction; he simply flows back into working order after any assault. Would Bernstein pledge to withhold funding for experiments until after legislation passed? His answer was a straightforward no.[2]

It was a blustery cold Friday afternoon in May when I went to see Bernstein at the Samuel Lunenfeld Research Institute in Toronto. Why the Lunenfeld? Because Bernstein has an office there, down the hall from the lab that still has his name on the door. He worked in this office every Friday, apparently doing research, or at least guiding the research of postdocs. At first, the obvious problem with this —the president of the CIHR still involved in research whose funding he could theoretically influence—didn't register with me. I was too focused on the problem of how a democratic, multicultural state with many conflicting ethical views and interests can control vital science that is moving forward at warp speed.

Bernstein's Lunenfeld office was in a cul de sac near a drink machine. This time Bernstein was not dressed like a down-at-the-heels movie producer but wore suit pants, a white shirt with rolled sleeves, a tie. I asked how he'd felt about the committee's examination.

"It was like Question Period in the House, but I'm not a politician so I can't answer back," he said, making it sound as if he'd been unfairly set upon by bullies. "It really highlighted for me that politicians will take advantage of a situation to further their own agendas."

And these agendas would be?

"Frustrated backbenchers," he said, "pro-life, suspicious of science per se . . ." He thought he had been unfairly caught in the flow of various narratives that had little to do with the problem at hand.

Where had he learned such confidence? I knew he had studied biophysics at the University of Toronto under E. A. McCullough's partner, Jim Till, and so was introduced to stem cells very early on. He did a two-year postdoc at the Imperial Cancer Research Fund laboratories in the U.K. and became an expert on cancer-inducing viruses and on retroviruses. But that wasn't enough to explain this contempt for his political masters, displayed so openly to a journalist. I asked how he had become the first president of the CIHR.

He told me he first came to the Lunenfeld as assistant director in 1985 (the same year he published a paper, with John Dick, describing the introduction of a gene into primitive stem cells that could rebuild the hematopoietic system of a mouse)[3] and became the director nine years later. By 1998, discussion had begun within the medical community and in the government over how to reorganize the old Medical Research Council, at that time the primary funder of academic biological/medical science in Canada. The government's idea was to combine traditional biological researchers with social science and humanities experts working on health issues into a series of research institutes, like the NIH in the United States. Bernstein and his colleagues watched "with trepidation" as a bill was drafted in 1999. He was not an insider to the creation of the CIHR, exactly. He'd been on the MRC's advisory committees, but he'd never been a member of the MRC's governing council—the equivalent of a board.

Bernstein was appointed to a steering committee to help design the new CIHR. He wrote, along with another researcher, a five-page document on how a Canadian Institutes of Health Research could work. His notion was that instead of dividing the new organization into separate research areas, each getting an equal share of the total dollars, only two-thirds of the funds should be dispensed to researchers through competition, while the other third should be used by the CIHR for strategic investments in science. The president and the governing council would decide how to use this money to position Canadians at the leading edge.

The government then put out a call for suggested names for the president of this new organization, in the fall of 1999. "Biomedical guys were worried as hell," he said. "The tent was made bigger, but the budget was not." The biomedical guys were worried that with social science and humanities researchers competing for a share of the same pot, their own grants would shrink. The social science researchers were equally concerned. "They viewed this as the MRC on steroids," he said. In this atmosphere of mutual suspicion, a committee was struck to find a president. He was vague on who the committee was and to whom it reported.

"I got on the phone to people to put names forward," said Bernstein. "I know many phoned me to take the job. I was not out to get this job—I felt conflicted. I was very happy here. The Lunenfeld is a great place. I took over in 1994, the year the MRC budget started to shrink. Our budget doubled while others' dropped by 50 percent."

Aha, I thought. That's where the confidence came from. Bernstein is a man who brings in the money, just like James D. Watson. As government cuts to Canadian science escalated with federal government efforts to control the deficit, Bernstein found money for his team from the Howard Hughes Medical Institute's International Research Scholars program, the MRC Distinguished Scientist program and NIH grants. But, more important, he convinced local developers to make large donations and got Bristol-Myers Squibb, the pharmaceutical corporation, to put up $2 million a year. He helped talk the Province of Ontario and the municipal governments of Toronto into positioning Toronto as a biotech center. He got very proficient with buzz phrases like the "new economy" and the "economy of lifelong learning" that were sweeping through government corridors like a big wind. "To me it was a real learning experience. I got to learn how to talk to politicians and line up their votes . . . I got politicized."

Yet he wasn't sure if he wanted to become the CIHR's first president. "The biomed guys wanted a biomed guy in the chair of the CIHR," he said. "I was having a ball here. I had remarried, had two stepkids. Why jump off a cliff in the dark?"

He went to Montreal for a meeting, where he was taken out for dinner by two eminent medical researchers with a great deal of clout. "It clicked I was being interviewed for the job." Then, in February 2000, he attended a big legacy dinner for the MRC at the Chateau Laurier in Ottawa. He was late. He ran into a man who said, "You're the man. They've chosen you, yesterday at two o'clock." A lot of other people at the dinner told him the same thing. Henry Friesen, then head of the MRC, now the chairman of Genome Canada ("the Moses of the CIHR," according to Bernstein, implying that the CIHR is the promised land), took him out to dinner and asked him to let his name go forward.

The president of the CIHR is a Governor in Council appointment, made by the federal cabinet. So where were the political actors in this story?

There followed a dance, midway between a negotiation and a vetting, between Bernstein, federal bureaucrats, such as Deputy Minister of Health David Dodge, and the prime minister's political aides. "The negotiation for me was money and travel and who is on the governing council. I said, how do I get to have my board? I was told it was Order in Council." He had no idea what that meant. "I'd never heard of the Privy Council before. So I said I'd like to have some say. I was allowed to put names forward." He did get some of his nominees on his governing council, but not all.

So, you had interests, investments, associations you had to divest? I asked.

"Yeah," he said. "I was on scientific advisory boards of small biotechs in the U.S. and here in Canada, and on a scientific advisory board of a vencap on Bay Street. I had to divest. So this was a big— at every level a big change, every possible level. I was consulting with companies, on the boards of three institutions: the [National] Cancer Institute at the NIH, Roswell Park Cancer Institute in Buffalo, the advisory committee on science of the Alberta Heritage Foundation. The advisory ones I kept—I found it too painful to quit."

He did not tell the Lunenfeld that he was thinking of leaving while he was negotiating with the government. But there were rumors flying, especially after he made a "stealth visit" at the end of May 2000 to meet the staff of the about-to-be-defunct MRC in Ottawa. Finally, as the legislation enabling CIHR was proclaimed, he had to tell people. He gave the Lunenfeld's board no notice. "Very painful, traumatic. No notice. I felt very badly about it. It would have compromised the CIHR if I took longer to come, and compromised this place."

He was in the job by June 2000.[4] The first thing he did was to get the Institutes set up. In October he appointed the stem cell guidelines committee, which issued draft guidelines publicly in March 2001, months before Rock's draft bill was brought to the Standing Committee on Health. The whole point of the creation of the CIHR, Bernstein said, was that it become "a proactive agency creating events." He said he had had no idea Health Canada was also working on legislation to regulate assisted reproduction. "We put two observers on our committee from Health Canada. We wanted to have good linkage to Health. Every Institute has a representative from Health Canada—that was David Dodge's vision." It was these Health Canada observers, he said, who suggested that the government's planned bill on IVF should also include something on stem cells. "So that's what happened. So the minister did phone me. He told me he would introduce draft legislation."

I had a story in my clipping file on this subject, published in January 2001, four months after Bernstein formed his committee, that suggested that the government had long been working on a bill regarding stem cells. This story appeared in the *Globe and Mail* just after the British House of Lords voted to approve the manufacture of human embryos for stem cell research, under the regulatory control of its already-existing IVF licensing agency. The story said that Health Canada was considering allowing nuclear transfer cloning of human embryos for the creation of organs for transplant. This was an odd thing for Health Canada to be considering, since no one had successfully done even the first part of this work (and wouldn't for another three years). The story said that this proposal was included in a document outlining possible legislation then being studied by Health Canada. Janet Rossant was quoted only in her capacity as a knowledgeable researcher; no mention was made of the CIHR's stem cell guidelines committee or of her role as its chair.[5] Reading between the lines, it was obvious that legislation had been in the works for a while. Somebody was floating a trial balloon.

But the way Bernstein now described events, the inclusion of stem cells and cloning in legislation happened a "couple of months before Rock introduced it in the House." But this meant, surely, that the CIHR had been quietly planning to set up its guidelines and fund this research before any regulatory agency was established.

Where did the regulatory agency idea come from? I asked.

Rock came to a dinner thrown in Bernstein's honor by the Lunenfeld, he said. Rock sat with Bernstein's colleague Rossant, who chaired the stem cell guidelines committee. They talked about the experience in the U.K. with a regulatory agency. There was a discussion then on a Canadian agency based on the U.K. experience. "It's quite possible the agency came out of that black-tie dinner talk with Janet," said Bernstein.

The draft bill appeared very promptly thereafter, presented to the Standing Committee on Health in May 2001. "I met with the Standing Committee on Health in the fall of 2001 . . . They invited me to come on Yom Kippur."

He gave me The Look. It was the same as the look I got from Michel Revel and signaled that this conversation would now touch upon anti-Semitism. It said, everyone should know Jews don't do official engagements on Yom Kippur, the holiest day of the Jewish religious year. It said, everyone should know I'm Jewish. It said, what do you think of that for insensitive?

Did you think it was done on purpose? I asked.

"I didn't take offense," he said, but then immediately he added, "it crossed my mind. I went on Halloween instead. It was a very friendly interaction with them. Preston Manning [former leader of the former Canadian Alliance Party] asked me a lot of things; he proceeded to teach me about stem cells." He was being sarcastic now, rolling his eyes at the temerity of politicians who had the gall to think there was anything they could teach him about his own field.

The committee seemed concerned about the morals of using human embryonic stem cells and wondered about using adult stem cells instead. Why use the former if the latter could do the trick? "Papers were starting to come out," said Bernstein, without mentioning Freda Miller's name. "I said—and said weeks ago—these are issues but we need more research before the dust has settled. They heard witnesses after me who said adult stem cells—wonderful." In fact, in his April appearance he'd tried again to shut down the adult-stem-cells-are-as-good-as-embryonic-stem-cells argument. He had mentioned a few papers that he said showed that the hoopla over adult stem cells was overblown.[6]

When the committee's report came out in December 2001, Bernstein was surprised that it was "rather negative," particularly the demand that researchers demonstrate that they could answer a vital question only through the use of embryonic stem cells before getting a license to proceed. Yet he acknowledged that this reluctant-bride tone, as he called it, was not so different from the responses the CIHR had already got to its proposed guidelines in the late fall of 2001. "Many were negative in tone," he said. "Part of it was orchestrated by pro-life groups. We knew that would happen."

And was there a discussion, I asked, a political discussion, about announcing the start of funding in March?

No, there wasn't. "I thought, do we go or no, because of this issue of the tone of the health committee report," said Bernstein. "On the one hand, we had a process of high integrity. Do I pull the plug and say don't go public till a new minister I haven't yet met brings in legislation?"

Now, that was neatly done, I thought. In the midst of his rhetorical questions, he had dropped in the assertion that the new minister had taken months to sit down with him.

"I think it would have signified to the community of scientists total confusion," he answered himself. "There was nothing stopping a researcher from going ahead . . . We'd come to the standing committee. I was very explicit on our process. The guidelines were consistent with the draft legislation . . . If we aborted this process, I could see someone saying, if there's no legislation, I'm gonna go ahead. The minister asked for tea with me at the end of January. I told her about the council approval of the guidelines. I wanted CIHR to be open and transparent—no secrets. We're not an old boys' club like MRC—you scratch my back, I'll scratch yours. What do I do now that council approves the guidelines?" he asked, with yet another rhetorical flourish.

Well, I said, how about calling Bonnie Brown?

"No, it did not occur to me," he said.

I waited.

"Mea culpa," he said. And shrugged.

"But surely you could see that they'd be furious if you didn't warn them? Surely you knew enough about politics to know that?"

"My own political antennae were not working. I thought I'd done that when I appeared in front of them in October. I'm not saying I shouldn't as a courtesy tell Bonnie, [but] I think there is a danger in politicizing what we do, a danger in me contaminating the CIHR process. If you go to a politician, you risk the chance of them saying, don't. Then what do you do? There are many examples in history of politicians having their own agendas."

The more he talked like this, the more I wondered what planet he was living on. In the first place, he was going to spend taxpayers' money on research that would require, under the planned legislation, permission from a regulatory agency. Failure to get it would earn a researcher up to five years in jail.[7] Without the legislation, there could be no such authority and no permission. And what did he mean by the word "contaminate"? When he put Health Canada officials into all the CIHR institutes and put two on this stem cell guidelines committee, he opened the door wide to science formed by the government of the day. Didn't he realize a politician is entitled to have an agenda, which is the raison d'être of the political life? He seemed to think he was responsible mainly to his peers.

Don't you see that Parliament is the boss here? I asked.

"Somehow we have to set policy at arm's length," he said. "If Parliament passes a law, we're bound by it. Should I tell Bonnie Brown? Yes. Am I politically savvy? Probably I am. The scientific culture is first through the door. It didn't occur to me to call her."

There, I thought, he's admitted he had stolen a march deliberately, because that's what scientists are trained to do.

His thinking, which he reiterated in one rhetorical question after another, was this: the CIHR is supposed to get out there and make things happen, not sit on its hands like the MRC had done. The community of scientists would charge ahead without scruples if he didn't provide an ethical framework.

But Paul Szabo had already answered him on this point. All you had to do was withhold the money, Szabo had said to him on April 17.

They think you did this in your own interest, I said, not the public interest.

"If they think I'm not giving objective advice, they should fire me," he said.

It was quite clear he didn't think he would be fired any time soon.

It was only after I left Bernstein's office that it hit me: not only was he fearless about giving offense to members of Parliament, but he seemed to care little about being seen to have an appearance of a conflict of interest. He is a stem cell researcher who intends to spend public funds on his own field, and he was still working in this area while he ran the CIHR. The government of Canada, for all its many flaws, has rules about this that apply to people appointed to public office. As president of the CIHR since June 2000, Bernstein is an appointee of the Governor in Council, which makes him a public office holder[8] who must comply with the Conflict of Interest and Post-employment Code.[9] Public office holders "have an obligation to perform their official duties and arrange their private affairs in a manner that will bear the closest public scrutiny, an obligation that is not fully discharged by simply acting within the law." Public office holders are not to become embroiled in conflicts of interest, either real or apparent, and if they do, these conflicts are to be resolved in the public interest. They are not to ask for or accept "transfers of economic benefit," other than incidental gifts or other benefits "of nominal value unless the transfer is pursuant to an enforceable contract or property right of the public office holder . . . etc."[10]

Bernstein was required to disclose to the ethics counselor—and the public—certain assets and activities. While outside activities are allowed "where it is not inconsistent with their official duties and responsibilities and does not call into question their capacity to perform their official duties and responsibilities objectively," all such assets and outside activities, whether resigned or continued, must be declared on the Public Registry. Public office holders "shall not" engage in the practice of a profession, actively manage or operate a business or commercial activity, retain or accept directorships or offices in a financial or commercial corporation, hold office in a union or professional association or serve as a paid consultant. While a public office holder can accept certain gifts over $200 that would not impede his or her objectivity, all such gifts are to be publicly declared.

Not only had Bernstein continued his association with the Lunenfeld but, as he had told me, he had also failed to resign from some advisory board positions. I spent the next few days going over Bernstein's c.v., checking it against the government of Canada's Public Registry. His c.v. declared him to be not only the president of the CIHR, but also "senior scientist" at the Samuel Lunenfeld Research Institute and a professor at the University of Toronto, with four different appointments. A call to the phone number given for his office at the Lunenfeld was answered by an assistant as "Dr. Bernstein's office." Even if Bernstein was not actively working for the Lunenfeld, his use of an office in an institution whose researchers rely in part on grants from the CIHR could give rise to an appearance of bias. Use of an office in a high-rent portion of downtown Toronto could also be considered a declarable gift with a value greater than $200. Carrying on research comes perilously close to continuing a profession. His four professorial appointments should surely be declared on the registry, given the large number of researchers at the University of Toronto competing for CIHR funds.

Bernstein's declaration showed that he is still a member of the Scientific Advisory Board of the Juvenile Diabetes Research Foundation International, a charity in New York. He declared he had resigned from the Van Andel Research Institute, but his c.v. said, and the Van Andel Institute agreed, that he was still on their board. He declared he had resigned from the Advisory Committee on Research, Alberta Cancer Board, but his c.v. said he was still a member, and the Cancer Board confirmed it. His c.v. said he was a member of the Ontario Science Centre Board of Trustees; his declaration did not refer to it. His c.v. said he was a member of the Committee of Excellence, Weizmann Institute of Science, Toronto;

his declaration did not refer to it. He declared no gifts, nor his continuing relationship with the Samuel Lunenfeld Research Institute and the University of Toronto. His declaration was dated May 30, 2001, almost a year after he took office, almost a year before our meeting. I found these deficiencies when I checked the record in 2002. When I checked again in the spring of 2003, they still had not been corrected.[11]

Bernstein's c.v. also listed the various committees of the CIHR he chairs. First on the list—the ethics committee.

Failure to disclose interests to the public seemed to be catchy; it had also spread to bioethicists involved with the stem cell guidelines committee, the Stem Cell Network, and Genome Canada—people shaping the rules, getting public grants to inquire into the ethics of this research, or both. By the time the minister of health introduced her revised assisted human reproduction bill into the Canadian House of Commons on May 9, 2002, without a reluctant-bride clause, some bioethicist members of Bernstein's stem cell guidelines committee had already gone into battle for the hearts and minds of the public and of swayable MPs, without mentioning their interest in the result.

Timothy Caulfield, Canada Research Chair in Health Law and Policy at the University of Alberta, who also served on the CIHR stem cell guidelines committee, published (with two members of the University of Toronto's Joint Centre for Bioethics and University of Montreal law professor Bartha Knoppers) an article in the *Globe and Mail* only seven days before McLellan's bill landed in the House. The article attacked what it called the U.S.-style politicization of science by MPs critical of the government's draft bill and of the CIHR's decision to fund research in advance of its passage. Caulfield notably failed to disclose that he had helped draft the very stem cell guidelines he and his coauthors lauded.[12]

Caulfield's fellow authors also failed to disclose their professional interests in the matter. Bartha Knoppers, a former president of the Canadian Bioethics Society, who helped write the HUGO Declaration on the Human Genome and has written extensively on ethnic-group

DNA banks, was one of the theme research leaders for the Stem Cell Network and was on the board of, and has a grant from, Genome Canada. On the website for her project, HumGen—a database covering the worldwide development of ethical, social, legal and social aspects of human genetics—logos of sponsors displayed include the government of Canada's departments of justice and industry, the government of Quebec, the CIHR, Genome Canada and the pharmaceutical companies Roche and GlaxoSmithKline. But the line identifying her in the *Globe and Mail* described her just as a Canada Research Chair in Law and Medicine, mentioning none of the above.

The other two authors on Caulfield's article were Peter A. Singer and Abdallah Daar, both of the University of Toronto's Joint Centre for Bioethics. Singer is the center's director and holds a chair funded by Sun Life Assurance. The center's website referred to bioethical disagreement over academics' taking money from corporate interests. It noted that some argue that an academic institute should spurn money from companies "to legitimize its roles as watchdog of the ethics of scientific and medical research." The center asserted that it takes the alternate view: "Bioethics will have greater impact in improving research ethics and patient care by working with and disseminating knowledge among all stakeholders." In the interests of transparency—also known as "full disclosure"—the center's website showed the sources of its funds. A pie chart demonstrated that the largest portion of its money came from governments and academic institutions, that the money from such corporations as Sun Life, which had given $1 million over five years and had recently renewed its commitment, was nothing much in the skein of the center's strings of support. Most came in the form of long-term awards—totalling the astonishing sum of $16,990,560 to Peter A. Singer. These were primarily from Genome Canada, already funding stem cell research using CIHR guidelines, and the Ontario Research and Development Challenge Fund.[13] Singer and Daar did not refer to these partners in their defense of stem cell research guidelines in the *Globe and Mail*.

Francoise Baylis also published an opinion piece in support of the CIHR's stem cell guidelines, which appeared in the *Hill Times*, an independent Ottawa weekly devoted to public policy, and was soon posted on the Stem Cell Network's website. It was clearly aimed at influencing opinion among MPs and the public. Baylis also failed to identify herself as a person who helped write these guidelines, though the paper did find the space to identify her as a member of the CIHR's governing council (which approved them) and a bioethicist and philosopher at Dalhousie University.[14]

Baylis had also been president of the Canadian Bioethics Society. She gave a speech at the 2001 annual meeting in which, according to one of her colleagues, "she said money was driving everything. We should be ashamed." Baylis quoted bioethics critic Carl Elliott: "Bioethics boards look like watchdogs but they are used like show dogs." She pointed out that the health research agenda is driven by corporations that spend $3.3 billion on research and development in Canada alone; that CIHR has industry partners, as does Genome Canada; that the flow of money in the form of research contracts and donations is so large that it impinges on academic freedom. She pointed out that Geron's ethics advisory board had signed off on guidelines for human embryonic stem cell research only six days before James Thomson's paper was published—about three years after his research was first funded by Geron. In short, she described her fellow bioethicists, to their faces, as conflicted pawns. She got a standing ovation.[15]

Nobody, I soon came to understand, has a bigger conflict of interest with regard to revelationary biology than the federal government itself. I finally figured this out in the cavernous Metro Toronto Convention Centre, a bunker dug into a hill with a stunning view of Toronto's former railway lands. It was full up with 15,000 delegates to the annual BIO Conference of 2002, the biggest biotech trade fair anywhere. I rode up escalators, walked across a gangplank. There were sayings of famous scientists on white posters along both sides of this walkway. In a glossy handout, Pasteur was quoted: "Science knows no country, because knowledge belongs to humanity and is the torch which illuminates the world."

Down escalators. Then down more escalators. The last time I was in this convention center, it was to report on an alternative-medicine fair, and the first display I encountered then was a woman lying on a table with a candle burning in her ear. There were no such odalisques this day, just a lot of police officers checking badges. The organizers must have been thinking some terrorist might wander in off the street. Perhaps they feared the ragtag environmentalists, holding an anti-BIO meeting half a mile north, might kidnap one of the politicians—from the United States, Britain, Germany or Canada—moving through these halls. These governments' representatives were here to tell the biotech industry how much it is appreciated, to cheer on genetic engineering, stem cell research, tissue engineering as the wave of the future, and to remind researchers and corporate executives that there is taxpayers' money to be mined from flesh and blood and bone.

At one session, I had already heard Patrick Moore, a founder of Greenpeace, now a convert to revelationary biology, give a rousing, over-the-top, save-the-planet-with-biotech speech. Did you know that organic farming takes too much room, so it can never feed the planet's hungry, whereas genetically modified crops are so productive per acre they will help us save our forests? Allan Rock, Minister of Industry, had been moved to say that Canada is the second-largest biotech power in the world, with the second-largest number of biotech companies anywhere, researchers second to none, and a great place to come and do business. The science and technology minister from the U.K. made the same claim for Britain. A legislator from Germany said the same about Germany, as did various American governors from states like Kentucky and Michigan.

The BIO Convention was arranged by an organization of the same name—BIO, the Biotechnology Industry Organization—which is the main U.S. biotech lobby, based in Washington, DC. No surprise then that the convention staff, and most of the speakers, came from south of the border, and that the speakers kept forgetting they were in Canada, where a different form of government prevails. This also explained why no official in the press room had ever heard of Canada's minister of health or had a clue where she was speaking. They had to send out for a Canadian who was able to point the way.

While I waited for her to appear, I considered Alan Bernstein and his bioethical supporters. Undeclared interests and public praise for one's own work dated back to Richard Owen's pan of Darwin. Do these things matter? Does such behavior actually damage anything of value? I was thinking that I should let it go. I sympathized entirely with Bernstein's decision to fund this research—I would have done the same if I'd been in his shoes and had the nerve. He was being honest about where he thinks science is going, should go. He was paid to exercise his educated judgment. And yet, the Other kept whispering, you can't turn a blind eye to the flouting of rules, to the way he and his colleagues appeared to thumb their noses at elected representatives, to these attempts to push the public their way without telling the public why this might matter to their own careers. Even people of the finest character should be watched, particularly when they are in charge of so much public money, as Bernstein is, which translates into so much power. And, yes, a failure to disclose interests does damage something vital—public trust.

The room where the minister was to speak was half empty. Alan Bernstein moved into a chair in the second row. His head was up and his shoulders rolled forward; he fairly bristled with attention, showing by body language his sincere interest. Soon Anne McLellan moved up the center aisle, nicely dressed in a ministerial suit. She was also wearing extremely sexy black patent high heels with sharply pointed toes. I felt a pang of shoe envy. She was introduced by the president of the pharmaceutical company, Aventis Pasteur, the sponsor of this event.

McLellan's breathy, high voice soared out through the speakers. She said how nice it was that she and her colleague, Allan Rock, were both here, because it pointed out how important the world of biotech is and will be. "We want to hear from you how we can put in place what you need," she said. Platitudes piled on top of each other like ill-made waffles—how vital all this new science will be to this century, but that it's also risky; how the government's goals are safety and protection of Canadians and the environment. "Knowledge is borderless," she said, and then proceeded to extol Canadian knowledge. "In the private sector, Canada is home to the second-largest number of bio companies in the world, and the high-est per capita. Canadian Institutes of Health Research has earned praise around the world . . ."

As she mentioned the CIHR, Bernstein sat up even straighter, especially when she said the CIHR president was sitting among us. "I was minister of health no more than a few hours when he came and said he needed [the budget] raised to $1 billion," she said.

Finally she got to the main role of government in biological science. "The most important role is vigorous investment in health research. Today, I am pleased to announce the CIHR will fund fifty-one research projects worth $88 million." New generations of biology's leaders would be trained at many centers, but the locations of three projects were named: McGill University, the BC Cancer Research Centre and the Samuel Lunenfeld Research Institute in Toronto.

"I'd like to thank Dr. Alan Bernstein for his work in making this project, which is so much about the future of your industry, possible," said McLellan.[16]

I had been quietly hanging on to my faith that if legislators put their minds to it, democracies could direct the course of revelationary biology, or at least slow it down long enough for the rest of us to catch up. But governments, and certainly politicians with power, seemed to focus instead on jobs, economic growth—the wealth they hoped biologists would generate. By the time the main convention doors opened and BIO's trade booths were revealed, I had seen a lot of politicians shimmying in front of future prospects. I could hear the faint strains of an RCMP pipe band receding in front of me, but the place was so huge, so jammed with row upon row of booths, I couldn't actually see it no matter how fast I ran to catch up to the music. I moved up and down the cardboard streets, from one to the next, from country to province to state to city displays. They demon-strated that many governments had chained their futures to life commerce, that many governments were determined to use it to advance the high cause of high salaries. This marketplace, which had been set up overnight, would be taken down in a few days. According to the BIO staff, in that short time, millions of dollars in

deals would be done. No government would control this; no regulatory agency would even slow such a juggernaut. No funding agency,
and certainly not the CIHR, had been asked to: the CIHR's job
description was to speed things up.

My faith in the power of the state relative to this science also
took a battering at a convention luncheon held in a space so big that
thousands of us could sit at round tables and watch child-sized
speakers strut their stuff. Carl Feldbaum, president of BIO, told the
room, as we ate dessert, that the time had come for BIO to develop a
foreign policy.

I thought he was kidding. Sovereign states have foreign policies.
Since when does a gush of science need a foreign policy? But I sat
there, in the darkened room, and thought about it. In 150 years, revelationary biology had evolved from the mind of one man in his
English country garden, corresponding and trading specimens, to
this entity or power that countries must deal with and woo, an
entity that has, as has been said of other great powers, no feelings,
only interests. It has both a physical and a metaphysical existence.
It is an expanding set of ideas about the nature of life unfolding in
the minds of the indoctrinated, and it is also embodied as the products of their labors and propagated through their learned hands. It
had changed—is changing—the way we think of ourselves, the way
we behave toward each other, the way we replicate, and it will one
day help some of us live forever. It was more like a religion than a
country, a religion without a god. But some religions do function as
states or are so entwined with them they might as well be states
themselves, and so perhaps Feldbaum was right.

If governments have huge vested interests in pumping the
growth of revelationary biology, and if this science was turning itself
into the equivalent of a state, who was left to hold it to account but
those elected representatives of the people unencumbered by high
office, the very backbenchers Bernstein disdained?

The July day I drove out to Oakville to see Bonnie Brown, chair of the
Standing Committee on Health, Pope John Paul II was in town. On
every street corner twenty-something tourists huddled together, red

rucksacks slung across their shoulders signifying their status as pilgrims. The pope had called world youth to renew their Catholic commitment and had picked Toronto, Canada's biotech capital, as the right place. Their faith involved swallowing the Church's dogma about the nature of Nature and the human place therein, including that human life begins at fertilization; that embryos are human beings deserving of protection; that all human beings, regardless of their mental competence or disease status, are equally entitled to love and support; that science must never treat human beings as if they are means to an end. The Catholic youth were extremely clean, even though many had spent the night sleeping rough waiting for the pope to appear, which he eventually did, dropping down from the heavens in a helicopter. His face was a drooling mask from Parkinson's disease, his body bent and twisted with his suffering. He moved by sheer force of will or spirit. Even in his person, he was a complete repudiation of revelationary biology.

Brown's Oakville constituency office was right down by the edge of Lake Ontario, where a group of yachts bobbed at a marina. Her constituents were obviously a wealthy bunch, although I had been unable to confirm who most of her backers are from her campaign return. Most of her contributions came from unnamed sources, people who had contributed under $200 whose names she is not required to disclose.[17] I still liked her enough after our brief conversation in Ottawa that I kept trying to forget something she had said on the phone when I was setting up this appointment. She had been talking about the groups that had appeared before the standing committee. She mentioned that the witnesses who were most opposed to restrictions on fertility clinics seemed to be Jewish.

"What is it with the Jews?" she'd asked.

She sat at her desk, wearing a simple black linen sheath, with a small pad of Post-it notes in front of her. I asked her about the assisted reproduction legislation thrown into her committee's lap. One clause permitted the manufacture of human embryos for the purpose of improving or teaching assisted reproduction procedures. The vague wording would allow a wide range of experiments on embryos—which was weird considering that making human embryos to extract stem cells for research was banned by the same section. Why was it right to create and destroy an embryo in the cause of teaching or improving IVF but wrong in the manufacture of stem cells for research?[18] Brown said she hadn't actually read the bill, though she'd had it at the committee since June. She hadn't had time to deal with it at all over the summer, she explained. She asked to see my copy, which she read. She said the clause surprised her. If the opposition didn't raise it, as she was sure they would, she'd have a look at this herself. She didn't remember any discussion on this point at all. (When I got home later and checked the government's first draft and her committee's report on it, I found a similar clause. How could the members have failed to notice it?)

She read on into the bill and came across a clause referring to suspending embryo development, which is done by freezing. It had been in the first draft too.

"How can you suspend the development of the embryo?" she asked. "Are they freezing it?"

How could she not know the answer to that question? I wondered.

She began to talk about the number of people who get involved in the making of an embryo in an IVF clinic. As many as five people could became its parents: the desperate infertile couple, a sperm donor, an egg donor, a surrogate. She spoke of the committee members' being harassed by people who wanted to pay surrogates, pay for sperm, pay for eggs, pay for lawyers—anything to get a baby without adoption. She thought their blindness to the danger of reducing human life to commerce incredible. She thought there was something wrong with infertile couples who never even considered trying to adopt a Canadian baby, who would go to Romania to find a child there before they'd adopt at home. "It's like a status symbol," she said. "Dr. Carolyn Bennett is always saying, 'What are you going to do to help these infertile couples?' She attacked me . . ."

I was surprised at her harshness toward a Liberal colleague. In her own way, she was as fearless as Alan Bernstein. "Doctors are intervenors," she said. "They think problems should be fixed. Science seeks to solve problems."

Well, yes, I thought. I tried to deflect her attention back to the government's bill. I pointed out what I thought were ambiguities in the wording that seemed to allow the making of human-animal chimeras under license.

"Why do Jews want this?" she asked.

Okay, I thought. There is no way I can ignore this.

It's the moral idea at the core of Judaism, I explained, in the most neutral tone I could summon. Jews are ordered by God to repair the world, to make the world a better place.

"The Buddhists say nature is perfect," she said. "The way nature operates is God's will. The spectrum was Jewish to Buddhist . . . Muslim opinion, I felt it was going to be fairly restrictive. It wasn't at all."

She bent her head again over the bill and her Post-it notes. I was flummoxed. Did I hear her question about the Jews correctly or not? Was there an overtone of ill will toward Jews as a people or not? Or was I simply oversensitive, reading something here that wasn't here at all? We went over the wording regarding chimerism again. "The things we understood the least was when someone talked of chimeras," Brown said. "They'd talk about exons; they were into all this scientific . . . language nobody got. That's why the committee homed in on this human family stuff. The science was way beyond us. They love that exclusionary language—they have all the power."

The "they" she referred to were scientists and doctors. She went on about the power of the science lobby with regard to genetically modified foods. "They come up with things that cause 15 million political problems. To me, it's akin to the pharmaceutical companies who do research but want to own the intellectual property. It's a license to print money. Anything good they will patent and make money off, and certain people in the world won't get access to it. I don't believe in intellectual property things," she said. She couldn't understand how one could patent a gene or some other facet of a life-form. "It's one thing to patent a machine in the machine age. Sometimes it's just a process. Maybe that's why I don't like it—it's part of nature, it's part of life-forms on the planet."

Brown talked about scientists with a mixture of fear and suspicion that was the precise obverse of the contempt for politicians expressed by so many scientists. How could she say she didn't believe in intellectual property? A capitalist economy depends upon it. One might question its application to living things or gene sequences, but surely not the whole idea of protection? If her fellow committee members were so overwhelmed by the science and the language of science, why hadn't they got people in to help them?

No resources was what it boiled down to. Oh, they had a parliamentary staff. They had a full-time lawyer, a researcher with a doctorate from Oxford (not in the sciences), whose job it was to put the issues in front of them through witnesses, and another with a background in physical sciences, but basically, her committee had to rely upon the skills of its members to think and to ask questions.

This was why Paul Szabo had taken it upon himself to self-publish a book called *The Ethics and Science of Stem Cells* after the committee issued its report. He tried to set out the basics of modern biological science, with a close eye to research using human embryonic stem cells and various types of cloning. Szabo has a bachelor of science degree and an MBA and is a chartered accountant. He is also a Catholic who rarely misses a Sunday service. On the inside cover of his book, he had mischievously quoted Alan Bernstein declaring that the role of scientists is to be canaries in the coal mine pointing attention to dangerous issues, not making decisions on the ethical limits of research. Szabo also put a notice on his book that he called the "uncopyright," encouraging everyone to freely copy his work and pass it around.[19] Soon he would fight this bill in the House, making common cause with the like-minded in the Alliance caucus and members of the New Democratic Party who wanted certain guarantees about the regulatory agency to come.

"I used to think plots were hatched," said Brown with a sigh over the strange process of making law. "No one has time . . . When legislation comes out the other end and the public is not in an uproar, it's wonderful."

She needed a break, and we went outside. Rain had come and gone. We sat on a park bench with a view of the boats, enveloped in humidity. I asked if one was hers. She laughed and said she had only a dinky little rowboat up in Muskoka. I watched and listened as she told me about herself. I was outwardly calm, yet also beside myself. How could any democratic society ever make good rules about this onslaught of biology if the public's representatives didn't inform themselves about it? But the more she talked, the more I liked her. I liked her intuition. I liked her sharp nose for interests. I liked her instinct to raise up the underdog.

Brown was a former teacher and track coach, a mother of four girls whose second husband was a social worker. She worked her way up in municipal and regional politics, serving on all kinds of boards concerning health care. She had been brought up a Catholic in a mixed marriage; dad was a British conservative Anglican, mom a Canadian liberal Catholic. Brown was educated by nuns at the Ursuline College and Our Lady of Sorrows Separate School. "I had a lot of nun teachers," she said. "The thing you learn best is humility. It's not about you . . . or your family or your career. It's about service to others. About repairing the world."

Right, I said, that's what I was telling you about Judaism. Same idea.

As a working teacher, she never had money—she was always scrambling to meet her mortgage payments and keep the car on the road. When she took her kids out for car rides, she'd say, when they went by a PetroCanada station, you're not poor, you're a Canadian; you own that. Or, when they spotted a plane, she'd say, you own that too, that's Air Canada. "Public ownership saved me from the wrath of my children."

When Prime Minister Jean Chrétien appointed Bonnie Brown chair of the Standing Committee on Health, she had no technical background that would prepare her for Bill C-13. Like every other MP, she'd been given her own five-volume set of the report of the Baird Royal Commission on Reproductive Technologies when she was first elected. The books sat on her shelf unread until her appointment. When the proposed bill came, the committee spent some time hearing from the minister and his deputy, and then studying it with the aid of researchers. They spent a couple of weeks on embryonic stem cells. "At the end of study, before the final report, we called in people for an overview of where they thought we were going and where they thought it was important to switch gears . . . We thought

the main issues would be cloning and embryonic stem cell research."
But no one argued that human cloning should be permitted, and
there seemed to be a disconnection between the scientific effort and
society, though society pays for the science. People who came forward wanted to talk about IVF. The committee heard complaints
from witnesses about letting surrogates be reimbursed for expenses.
Then the representatives of IVF clinics came forward.

"The 'industry,'" she said, forming small quotation marks with
her two hands around the word, "is centered in Toronto. They said
there's a lot of room for economic growth with this industry. It sent
shivers up our spines. There was no talk about protection of the
children who resulted from this."

The committee thought that people who use these clinics had
little protection but that those who got the least protection of all
were the products of IVF, the babies. "We thought, if we are government, it's our responsibility to protect the weakest—these babies
should be the focus. Not the parents, the scientists, the lawyers who
will make a million, but these babies, who will be Canadian citizens." Although there had been differences of opinion between the
members over various issues, "most of us came together when we all
wanted to throw up when advocates of surrogacy came up." One
advocate who introduces surrogates to prospective parents said she
ran her service out of pure altruism, but it turned out she charged
big fees to register. A woman who had served on the Baird Royal
Commission argued that IVF is fundamentally different from adoption, in which many facets of society come together to protect an
orphaned child; in IVF, on the other hand, the child is created to
serve the adults' needs. This was an argument those in favor of surrogacy didn't want to hear. "They talk about the right to procreate. I
remember growing up, some of my parents' friends had no children.
I'd ask and my mother would say, 'Sometimes, when God doesn't
send a child to some people, God works in mysterious ways.'"

Do you believe in God? I asked.

"I think so," she said. "I think there's too much monkeying
around with things. Too much pride and arrogance and not enough
humility."

When the committee's report was published, Brown attended

public meetings and she heard from scientists. Scientists were particularly concerned about the committee's proposed rules forbidding payment of expenses for donors and surrogates, and about the proposal to end anonymity so that the children of IVF could look up their biological parents' names. The scientists were afraid these measures would stop the donations and therefore deny them the leftover embryos they need to make stem cells. She also heard from operators of sperm banks, who pay young men $65 per donation; she'd learned that each donation was split at least four to six ways, each portion sold for $350. The committee thought this was just plain greed. They were concerned that the children of these donors had no way to learn of any health problems donors experienced in later life that might someday also affect the child.

So the committee focused on this anonymity as a kind of crime, as the terminal point of personal responsibility. "Someone said, 'I try to teach my children every act I choose to do has implications.' What if I raise a boy, teach him not to exploit women, to be the best he can be as a husband and father in his future life. I'll talk to him of self-discipline . . . Now, at university, someone pays him to solve a problem, $65 for a sperm donation. How does that teach him that his ability to procreate is a gift? It's just becoming a commercial transaction. I wonder if anything is sacred anymore," she said, "if the creation of life isn't."

I didn't know whether to laugh or scream. This sensitivity she expressed, the committee's appreciation of the need for personal responsibility and of their own duty to protect the weak from the strong, was exactly what most voters hope for in their elected representatives. The extension of protection to those who need it is the prime role of all good government. And yet these same people had averted their eyes from the revolution they were asked to grapple with, which will affect countless unborn children eventually. The words were too hard, the scientists' concepts too slippery to confront. Instead the committee had focused on the parts of the bill that dealt with things they already understood.

This left the important questions, the ethical questions too vital to leave to conflicted scientists and even conflicted bioethicists, essentially unexamined. Is the human embryo a human being, a potential human being or none of the above? Does it deserve protection or doesn't it? Would a manufactured human clone be similar to a twin made by nature, and, if so, what's wrong with making stem cells and babies this way? Should we mix the cells and nuclei and genes of humans with the cells and nuclei and genes of other animals? Do we have the right to alter nature in every way we can, or are there limits? And if there are limits, what are they? In a multicultural society, who can answer these questions for the community as a whole if not the representatives of the community? How could they answer if they failed to learn?

"That is exactly the point of modern biology," I said to Brown. It has replaced the idea that the creation of human life is sacred with the idea that life is wonderfully malleable. There are people who still believe in this sanctity, just as there are people who still believe the world is flat or that spilled salt can bring disaster that may be averted by throwing more salt over the left shoulder. Biology teaches that each life is a unique event that can never be precisely repeated, but that all life is interconnected, with a shared history. Life's building blocks can be manipulated because they operate according to basic rules from which complexity emerges. Once those rules are understood, they can be applied in new ways. Unique is not the same as sacred.

"The creation of life is not sacred anymore," I said. "Now what?"

"To a lot of Canadians, it is," she said.

CHAPTER EIGHTEEN

A NOBLE PRIZE

L ate on a fall morning, I walked with my dog in the private park
at the end of my street. I had spent a year of my shrinking life
peering up the skirts of revelationary biology. I was still
beguiled by chimeras and the unwritten patterns that organize living
cells. I was bereft of former certainties, except for one: these revelationary biologists had turned the manipulation of human life—of all
life—into a worldwide industry that is basically beyond any society's
control. A brisk wind blew the reek of damp mold up from the pond;
the sky was a fantastic cobalt. Such wild, abandoned color had
erupted among the oaks and the beech swaying over my head that it
made me want to sing out in praise of perfect nature. Mind you, an
unnatural silence prevailed among the high branches of the old
trees. The advance of West Nile virus through Ontario had cut a
wide swath through the crows.

It was the season of awards, including biologists' favorites, the
Gairdner and the Nobel. The Gairdner got its own award when the
laureates of the 2002 Nobel Prize in medicine were announced. The
winners were the first students of the worm *C. elegans*—Sydney
Brenner, Robert Horvitz and Sir John Sulston.[1] They won for learning how genes program the worm's cells to make organs and how
cells are programmed to die. The Gairdner Foundation's public relations people were thrilled because the Gairdner jury had picked for
their awards the men most responsible for the Human Genome
Project and so their roster included a new Nobel laureate. The
Gairdner winners were Craig Venter, Bob Waterston, Philip Green,
Jean Weissenbach, Eric Lander, Michael S. Waterman, Maynard
Olson and—tada!—Sir John Sulston. In addition, Sydney Brenner,

who had already won the Gairdner twice, was scheduled to appear as the Gairdner Foundation's guest of honor. Once again, the Gairdner's PR people could say the Gairdner jury had sniffed the winds of science politics and picked right.

The PR people were more than pleased to let me witness the various events, even though I represented no television network, newspaper chain or cable channel. They even tried to set me up with interviews. Of course, I asked to speak with James D. Watson, who had been recognized with a special award of merit along with Francis Collins and Sydney Brenner. Of course, that request was ignored.

I'm not sure why I decided to attend the whole two days of Gairdner lectures. Maybe I thought that just this once, life would shape itself like an elegantly rounded story instead of as it usually is, lumpy and disfigured by loose ends. Surely two days of lectures from so many worthy men (all the honorees and most of the other invited speakers were men) would release me from the continuing argument between Ms. Rational and the Other One over the value or danger of revelationary biology. Yes, dear reader, I was still a prisoner of that rough dialectic between reason and superstition. I was still afraid to plunge irrevocably into the cold river of materialism and leave behind the hearth warmed by vital spirits, though Ms. Rational did think the Other One was down for the count. I could see that if it's okay for nature to make endless changes by methods so various as retrotransposons, sex, chimeras, hybrids, symbiosis and mistakes in copying, it is probably okay for human beings to do so too. The Other's reply to this was pure ad hominem. She kept warning about arrogant, powerful, heedless men bent on changing everything about what we are and that which supports us. I thought, they'll be on their best behavior at the Gairdner event. They'll display their humanity, their humility, their wisdom, and I will put such silly fears aside.

I got to the Royal Ontario Museum early for the first evening's event. It was supposed to be a cocktail party and then a dinner in the main hall. I intended to position myself where James D. Watson would likely walk right by me: I was determined to see him up close, in the flesh, whether he liked it or not. James D. Watson was now a person well known to me. I had read him, listened to him, heard all kinds of stories about him. I knew what his house looked like, what he is paid, even how much money he then owed a public company whose board he sat on—half a million dollars.[2] I had many opinions about him and his work; he was arrogant, he was brilliant, he was conflicted, but he had engineered two great changes in science—first by describing the shape of the DNA molecule with Crick, then by forcing the Human Genome Project into the world. Surely he would see me as he went by, realize that I knew all about him and talk to me.

Soon after I picked my spot, the family friend whose husband had refused me Watson's private number ambled in. I had already got the number from certain filings but had decided not to use it. Who wants to be hung up on by James D. Watson? I waved at her, and she came right over. She was wearing a white cap, a pantsuit and a thin shawl of anxiety.

"Darling, whatever are you doing here?" she asked, giving me only one cheek to kiss. She looked distinctly unhappy to see me.

"Oh, you know journalists," I said. "We worm in everywhere." She scuttled away and positioned herself by the entry to the main lobby. I'd considered that spot but had decided this one was better— I had a view of all possible entries; she only got one. I surmised that she was keeping her eye peeled for Watson too.

There were a significant number of television cameras and ink-stained wretches of the daily press. Canapés and glasses of wine floated by. The foundation had set up a very charming little stage, with the flags of France, Britain, the United States and Canada to honor the countries of the prize giver and the recipients, once again demonstrating the cross-border nature of fundamental science. There were whiteboards, set on easels, that named the sponsors: Genome Canada, the law firm of Borden Ladner Gervais, the Hospital for Sick Children, the Ontario Genomics Institute. A wall of white curtains masked the dining room from view, although I peeked behind and saw round tables, white cloths, gilt chairs and flowered centerpieces, the standard setup for a wedding or bar mitzvah. Soon there was a crush of people in business suits, flushed with the pleasure of being part of an important event. I tried not to lean too hard against a glass case holding a dinosaur skull, while at the same time defending my position.

Most of the Gairdner Prize laureates arrived in a group. I could distinguish them from the rest of the crowd because each wore a yellow rose in his lapel. I recognized Sulston by his white hair and beard. He was a surprisingly small, feisty man, a bantam rooster who reminded me of the late Tommy Douglas. And there was Sydney Brenner. He was almost dwarfish, with a very large, square head and big shoulders—the bust of a six-footer set upon tiny, childlike legs. The others I didn't know by sight. There was no sign of James D. Watson, the star of stars.

Speeches began. I shifted from foot to foot as a Gairdner grand-daughter explained that her grandfather had made money in business but wanted to do something to relieve human suffering. John Dirks, the foundation's executive director, spoke of the award's forty-three-year history, how it is unique because, in return for the prize money ($30,000 Canadian), it asks the laureates to give lectures to the public across the country. He thanked Genome Canada's chairman, Henry Friesen, for its support. He thanked the other sponsors. He read a letter from the premier of Ontario. My feet were starting to ache.

Now someone else was on the stage, saying Canada would soon be a world leader in genomics because the governments of Canada, Ontario, Quebec and British Columbia were all pouring taxpayers' money into it. The purpose of all this, of course, was "to make the world a better place." Dirks came back to the stage to tell us that the lectures would start the next morning at the Faculty of Medicine at the University of Toronto. Each laureate would have his chance to enthrall us. He drew our attention to each, and the family member who'd accompanied him. As he named them, there were waves of applause from the audience, and arm waves back from the laureates. Finally, he got to the Award of Merit winners, Francis Collins and James Watson.

"Dr. James Watson and his wife, Elizabeth," he said. "They're not here yet. We're expecting them momentarily. We hoped that James Watson would speak for all." Dirks peered out into the crowd, hoping the doors would fly open and Watson would stand revealed. "He was on a late flight. Hillary Clinton was at Cold Spring Harbor today, and she's demanding." He pointed out that the last time a Gairdner winner also won the Nobel in the same year was "when Francis Crick was here to get an Award of Merit in 1962."

As the crowd clapped again, the main doors opened and bodies began to move apart, as if flesh had liquefied and become the Red Sea and here was Moses.

"Dr. Watson is here," someone shouted.

"James Watson," said Dirks, "you are just on time. Come up and say a few words."

At first I couldn't see him. Then I caught a glimpse of my old friend's white cap bobbing and I knew where to look—and there he was, bursting out of the crowd in a whoosh of displaced air right beside me. He turned on his heel, just for a moment, and looked straight at me. As if he knew me. As if he was waiting for me to offer him a word of praise, or perhaps an interesting question. He struck a pose that was somehow both diffident and aggressive, feet set in ballet first position, hand on hip, elbow cocked. He was in the open space below the podium; he surveyed the scene. A little bit of froth clung to the corner of his open mouth, and his white hair was awry. I could hear the sound of mouth breathing, see the age spots on his head and hands. I opened my own mouth to speak, but what could I say? It was the cold blue eyes, the appraising eyes, that stuck with me long after he turned away and, remarkably spry for a person of his years, jumped up onto the stage.

His speech was a series of self-deprecations. He slurped and snorted at his own jokes. "Everyone on the list," he said, "totally deserves it. I wish others could be on it. This was truly a collective effort . . . My only role was not listening to people younger than me who said it's too soon to do it. We older ones wanted to get it done sooner. You can take total pleasure in what everyone accomplished. You rewarded science at its best—these are role models for the young . . . Yeah, it's the most important book ever written . . ."

And so on.

I went the next morning to the Macleod Auditorium, which is attached to the front of the U of T's Faculty of Medicine like a boil on a cheek. There was a large sweep of theater seats on the main floor, which sloped down to a stage, and there was a balcony. I sat and stared at the row of microphones, waiting for the show to begin. At two minutes to nine, the seats were still filling slowly with people in their middle years. Watson arrived and moved right down to the front row, the better to hear his colleagues. Alan Bernstein came in and sat there too. So did Ron Worton. Janet Rossant and Bartha Knoppers would be there the next day. Two men behind me were talking loudly about a person who works in Alan Bernstein's lab who had a problem with a grant.

One after another a long line of introductory speakers reiterated the importance of the Gairdner Awards, the importance of the prize winners, the importance of the science they had created in the greatest imaginable teamwork effort ever.

One of the men behind me laughed out loud at that. "Teamwork?" he said. "Hmmmmm."

We were introduced to Maynard Olson, who helped figure out how to clone long sequences of human DNA into bacterial artificial chromosomes, or BACs. He spoke in swiftly tumbling words about the progression of his ideas from his first attempts to sequence, using the methods of Gilbert and Sanger, to his map of the yeast genome, published after six years of painstaking work. He described how he wrote FORTRAN code to find order in the sequences, how he learned to make longer and longer clones—basically, the progress of his life in the lab.

It was affecting to hear him confess his confusion before the overwhelming nature of his task when he began, how he gained confidence and learned from others as he progressed. He stressed collegiality and its signal importance. But the two men behind me were snorting to each other over that, too, and they were right to be skeptical. Nicholas Wade had described Maynard Olson's testimony to a congressional committee, in which Olson asserted that Craig Venter's whole genome shotgun sequencing strategy would "encounter catastrophic problems" when it came time to make an assembly.[3] Even after both teams had arrived at their assemblies and Venter's had been published in *Science*, Olson went after Venter again, telling Wade, "Venter in June 1998 claimed that this was not a quick and dirty approach, that it would produce a sequence that met or exceeded best current standards. That claim was absurd at the time and remains absurd."[4] Not exactly collegial talk.

Phil Green was at the podium. He explained that he had started as a mathematician. When he decided to enter biology, the hardest thing for him to come to terms with was the fact that biology is not elegant; life is in fact a kluge that keeps on keeping on. "Have to live with it," he said. His talk was essentially about the value of various computer programs he had written. What he liked about the way these programs or algorithms were evolving is that they would soon guide experimental work. "Increasingly," he said, "biology will be a more quantitative and theoretical science, using these gene predictions." In other words, he'd been working very hard to make biology less like life and more like math.

Michael Waterman, another mathematician, dared to say that the history of sequencing did not start with the chemistry of Sanger and Gilbert, but rather with "computation and data storage." He wanted it known that this work was the invention of a woman who, in 1966, started collecting every known protein sequence. This culminated in 1982 in the creation of GenBank, which records in its databases all publicly known biologically important molecular sequences. The number of sequences of DNA known in 1982 was in the millions—so few they could still be published in book form. By 1986, known DNA sequences filled an eight-volume set, which was the last time anyone tried to publish them on paper. Computational biologists like Waterman would be even more necessary in the future than they had become already.

Sir John Sulston was first to speak after lunch and he was there early, as was I. I was bored. Nothing I'd heard in the morning had enlightened, let alone enlivened. There was no sign of wisdom, and only the occasional flash of humility.

Sulston told his story, beginning with "Sydney's worm." He described how they first made mutants to dissect the animal's developmental processes, to show the action of its genes. Then they wanted to know where these genes were and thought it would be easier to relate them to each other with a map. This abstract gene map eventually became a map of physical sequences. From there, Sulston segued into a lecture on collaboration, how he had worked with his colleague on the other side of the Atlantic, Bob Waterston. "Without that," he said, "you have nothing."

This propelled him to an anti-business rant. He showed pictures of people he worked with at the Sanger Centre, told of how they drove themselves to work faster and faster because they were concerned about the competition from Celera. If Celera got the human sequence first, "it would be proprietary if not released." He insisted that when information is treated as part of a public common instead of being patented and monopolized, then progress is made quickly and disparities between rich and poor can be evened out. He raised the contrary example, which proved his rule, of two genes associated with breast cancer—genes that had been patented. "Do they have a right to surround a gene with a ring fence? The monopoly is contrary to ethics . . . Think what it means if the human genome is in a proprietary database."

"I believe," he wrapped up, "that it's neglect of the huge inequalities in the world that indirectly feeds terrorism."

There were no new ideas here at all, although the politics were interesting. This failure to explain meaning was what Sydney Brenner had criticized when he reviewed Sulston's book, *The Common Thread*, in *Nature*: "Sequencing large genomes has nothing to do with any intellectual endeavor," wrote Brenner of his former assistant's efforts. "The creative work was done earlier by Fred Sanger and by others who improved the technology. The rest is about two things: money and management . . . What I found interesting in this account is that Sulston doesn't tell us anything about the genomes he has sequenced. What did he find there that excited him? What did he learn about genes, about life, about evolution, about worlds to come? It is the play *Hamlet* without Hamlet."[5]

For Bob Waterston, the human genome was old news. The sequencing centers had moved on to the mouse, the primary study animal for biology. Its genome had turned out to be a little smaller than the human, but so far, although they had only 75 percent of its genome in clones and only a draft assembly, they had established that in the approximately 75 million years since humans and mice split off from a common ancestor, the main changes between man and mouse seemed to be deletions of certain human sequences by the mouse. "There are fewer than fifty genes in the mouse without human homologs," Waterston said.

Did that make the mouse a genetically refined human?

Interestingly enough, the rate of mutation in genes responsible for reproduction and sense of smell is twice as fast in the mouse as in man, he said. Nevertheless, the order of genes was pretty much the same. He and his colleagues were making algorithms to reflect these similarities and differences, and these could be graphed, and one could see where natural selection was currently working and where it was not. All of this would be improved by similar comparisons between man and chimp, and dog, and rat. "Chicken we'll do after we finish chimp, and soon . . . whole species of genomes, and we will move to this goal of real knowledge of what functional elements are."

He reiterated that none of this information would be patented.

Why would natural selection be working twice as hard on these two vital areas of the mouse genome as in man? What could direct such a speed-up in mutations, or, alternatively, a human slowdown? And how did this fit with evolutionary theory, which suggests that vital functions cannot mutate much at all? Didn't these genomics experts always argue that vital genes are conserved across evolution along with their protein products? Why would such vital genes as those dealing with smell and reproduction be changing rapidly? If this isn't a random process, what directs it? Was anybody going to talk about that?

Craig Venter, the anti-colleague, was introduced. His current interest, we were told, is creating an organism from scratch—a microbe to do a specific task to clean up the environment. Venter described his history using whole genome shotgun sequencing, beginning with *Haemophilus influenzae* and *Drosophila*. He made it clear that in his view, private science has led the way, not the cooperative so lauded by his fellow laureates. His group finished a project in one year that the consortium chipped away at for ten. "The first sequences done were from private institutes," he said, forgetting Sanger. After sequencing the fly, his group went on to the human. "It took nine months to make a draft. We announced the status in June 2000. When we published, we put it on the Celera website; it was available to scientists freely." If he and his team made an error, it was in not having sufficient faith in their own techniques. Their speed had increased dramatically. The first sequence of the first genome of a species of yeast "took one thousand scientists ten years. It takes six hours now. Nine months for the human, then six months for the mouse." He'd found that mouse genes are in the same order as human genes, although they are arrayed on different chromosomes. He too was amazed at the similarity between mouse and human. "We found fourteen genes on chromosome 16 that don't have a human counterpart. We were relieved," said Venter. So far, he said, there seemed to be almost no differences between humans and chimps.

He switched focus quickly to money and management. Costs were coming down very fast. The consortium spent $3 billion; his company spent $300 million to sequence the human genome. "The new generation of sequencing centers will get to the $1,000 genome. Within ten years, we will drive the new field of genomic medicine. Your genetic profile will be on CD-ROM when a baby leaves the hospital. Predictions will be based on the genetic code."

He went on making predictions for a while.

Enough, I thought. Tell us what it means that humans and mice have genomes so similar, yet we are large, they are small; we have built many civilizations, they are study subjects. Tell us why such genetic identity as exists between man and chimpanzee can result in such very large differences in behavior, length of life, intelligence, social structure. Tell us how we get all this amazing variety out of a fabric of interactive threads, broken up in a few different ways. Tell us how selection acts on one part of a macromolecule but not all of it. Tell us something that matters.

———

Eric Lander was introduced as a man of ideas who tied Harvard Business School methods to genetics. Unlike his nerdy colleagues, who climbed the stairs to the low stage, Eric Lander leaped up onto the podium. He said he wanted to focus on the future, but he immediately retreated to the past—his personal history making maps to track multigenetic traits, then the history of the last century. In the first quarter of the twentieth century, he said, biologists observed as information was conveyed across generations, with no idea what carried it. During the next twenty-five years, the information carrier was described at the molecular level. The next twenty-five years was spent figuring out how DNA codes information and how it's read, and developing tools to read it. "The last quarter is reading first a single gene, then sets, then organisms, then the human sequence in the closing weeks of the century. Not bad for a century," he said.

He went down the same triumphalist path as Venter, giving us the numbers of species sequenced, how their genomes compare to the human. "About 5 percent of the human genome is under selection," he said. "If a sequence is extremely well conserved, we can say it's no accident."

How come? I wanted to shout. He also said that vital mouse genes are under selection, meaning changing rapidly, when the mouse assembly is only 75 percent complete. And how could he say that only 5 percent of the human is under selection when the genome assembly was still in draft form too?

A graduate student of his, said Lander, had lined up the sequences of four different species of yeast and compared them. He looked at the regions between genes, trying to spot regulatory sequences. "It turns out," Lander said, "that there's a lot to learn. There are 6,100 genes when yeast is sequenced." But, he said, when his graduate student lined up these four yeast genomes, he found that 500 of the so-called genes previously identified "are not genes. Thirty new ones were found; a bunch were shortened. There were 5,600 genes, in fact. When we line up intergenic regions, there's a lot of conservation of regions."

Well, tell us why would that be, I muttered to myself. Why are noncoding regions conserved too? Are you saying they too perform vital functions that must be conserved? So how does that relate to big changes going on in vital functions of the mouse? Maybe this theory that if it's conserved it must be vital needs to be evaluated a little more critically.

"If we look for best conserved, they correspond to regulatory regions that are known," he said. In other words, regulation is as important as the core of the gene itself. This sort of work has to be done in humans, he said. "The variation we walk with today we had in Africa." Common variants found in this audience, he said, will play a role in the risk of disease. "Why not enumerate it all and correlate with diseases?" The consortium had assembled a collection of 3.5 million single-nucleotide variations to see if there is any "structure in this variation or not. There's a fair amount of structure."

This was exactly the kind of work Michel Revel was so worried about.

I went home with a numb hand. The organizers had cleverly scheduled Watson and Brenner for the second day, or I would have thrown in the towel. How could these people fail to raise the important questions? Watson's speech was to be about ethics. That would be interesting.

A large number of students were already seated when I got to the Macleod Auditorium at 8:40 a.m. Ron Worton was in the front row near Alan Bernstein when Watson arrived, ten minutes early. The young people were apparently medical students who'd been encouraged by their dean to skip classes and come here instead. We endured a long line of warm-up speakers until Ron Worton introduced Watson, whom he referred to as Jim. "Jim has thirty-two honorary degrees, and many other awards, like the Lasker, the Benjamin Franklin Medal. He's an honorary Knight of the British Empire. I give you Jim Watson." Worton passed the microphone to Watson, who bobbled it, lost control of it and watched helplessly as it slid off the podium. The president of the University of Toronto ran to pick it up.

Watson began his address with ethnicity—his own. His grandfather on his mother's side, a tailor from Glasgow, married an Irish woman. On his father's side, he is a Lowland Scot. Both families settled in New Jersey. "I was raised in the '20s, when genes were important," he said. "They were not important after the war because the Germans thought them important."

His complaint about the Germans always being wrong wound through his speech like a hungry child's whine, because the theme of his talk was not ethics, but the misunderstood virtues of genetics, by which he also meant eugenics, just like my old guidance counselor.

When recombinant DNA research was first possible, Watson said, the protestors, who came to public meetings where its safety was debated, thought "we should not create a perfect race. I don't agree with that."

Okay, I thought. You've got my attention.

"I've always known the Irish need improvement," he said. "I always wanted the perfect girl. Why not?"

Early on in his career, it became clear to Watson that there was latent and visible opposition "to genetics." When E. O. Wilson wrote a book on sociobiology that implied that human behavior has genetic components, "water was poured on his head at an AAAS [American Academy for the Advancement of Sciences] meeting." Watson was disgusted by that, as well as by the behavior of those who wanted to halt recombinant DNA experiments. "Recombinant DNA turned me away from the left," said Watson. "I thought they were crooks in the way they dealt with genetics."

By this point, I was squirming in my seat. What did he think he was doing? He kept using the word "genetics" when he should have said "eugenics," and he obviously knew the difference, so he deliberately wanted to confound one with the other. Wilson had written of complex behavioral traits having a genetic explanation. Of course, he was derided. Where was his evidence connecting known gene to known behavior? Eugenicists who were also geneticists had gone down this road before—without evidence—and very bad things had happened in the world as a result.

When one is young, Watson said, one wants to differ from one's parents; when one becomes a parent, one sees that nurture doesn't work. As an adult he had spent a fair bit of time trying to deal with the history of genetics, particularly because he is the president of Cold Spring Harbor Laboratory. "It was the center of the American eugenics movement," he said. "It was not a benign movement." That is why when the Human Genome Project started, at the very first press conference, he had said right off that ethics had to be dealt with, that 3 percent of the budget of the project would be spent on ethics to demonstrate that those doing genetics are "concerned with consequences." He appointed as the first chair of the Ethical, Legal and Social Implications section of the project Nancy Wexler, whose mother died of Huntington's disease. "I thought ethics should not be left in the hands of male doctors, but with people who suffered . . . a collection of wise men and women."

Oh my God, I thought. What did Wexler's personal suffering from genetic disease have to do with thinking through the ethical, social and legal issues raised by revelationary biology? Watson had just admitted he chose a leader for this project who would be predisposed to making genetic changes to avoid the consequences of a genetic disease, a person with a profound interest in wiping out a disease gene. But if he wanted his colleagues to be seen as concerned about the consequences of eugenics, wouldn't it have been better to appoint someone whose parents died in Auschwitz?

Disease, he wanted us to know, does not treat people equally. Minorities have been stigmatized because of diseases they are prone to. He complained that the law has not evolved fast enough to deal with the knowledge of the links between genes and disease. Francis Collins, his successor at the NIH, had thought from the start that a bill should be passed to keep genetic information private, but no such law had been forthcoming. People were afraid to take genetic tests without such a law, fearful they would lose their insurance if they turned out to be genetically predisposed to breast cancer. He didn't think a law to prevent genetic discrimination would pass until "we have enough women in Congress."

He might as well have said until hell freezes over.

And soon there would be other issues to deal with, too, such as the idea of selection—by which he seemed to mean enhancement. Soon, he said, it would be possible to correlate various genes with various capacities and predispositions and talents. Then one could select for these characteristics. Sprinters, for example, could be selected for. Perhaps certain aspects of human character are genetic. For example, we could genetically define a cold-fish personality, and that could be selected against. "We know who they are," he said. "We can't change [them] into a warm person. We all know that. We should accept reality."

How could he say these things? I wondered. Why did he not even hint at the marvelous, myriad intricacies of human DNA—of the genes nesting within genes, of alternative splicing, of different protein domains, all of which the assembly of the human genome sequence made it possible to study? How could he suggest that we can easily wipe out a disease gene if other genes and regulatory areas are embedded within it and attached to it? What about the fact that no one even knows how to define a gene anymore? It was almost as if he still believed in the one gene–one protein theory of his youth. I looked around me, trying to read the minds of the young people sitting in the audience. Did they understand what he was saying? No one was shifting in their chairs or seemed the least uncomfortable.

The insufficiency of one enzyme could make all the difference, Watson continued. He had often asked himself, what gene could predict performance in science? Answering such questions was important for ethics and for law. What if he had been born with a gene that made him impulsive? Without a good home life, he could have ended up in constant trouble with the law. Perhaps the study of such genetic predispositions for certain behaviors at the molecular level might produce "a pill to calm down those who should be calmed down."

"It would be wonderful to convert a shy person to an extrovert by knowing why he's shy. In twenty years a mass of information will be coming out of genomics centers. So we're going to create—as Craig Venter said, you'll pay money and find out about yourself and take corrective action. We all know we won't be able to cure all diseases and that some forms of inequality we can't handle. We have to decide whether to do the diagnoses before birth, if a child has a reasonable chance of surviving, and leaving to women [the decision of] whether [or not] to raise a child best seen as a case of genetic injustice. So I think this is gonna be our main issue: should we try and prevent them? I believe women should have the right to decide to take a genetic test."

His colleagues had become so absorbed in the task of doing genomics, he said, they had not held forth on what we should do with the information from such tests. But he knew what to do with it. "I think we should use it. We all know nature can be cruel, and we shouldn't be accepting. I believe we should try to do something about this inequality we all know exists. When I was in the genome project, the papers would say, 'You just want to make pretty babies.' So why not? The idea is we shouldn't enhance, as if something is wrong with that . . . I want to show you the beginning of the movie *Gattaca*. It looks into the future, how some see how genetics will go."

There was a screen behind him on which the movie was supposed to be projected. We all waited. People scurried about; there were whispers. Finally, it was announced that there was a technical problem and the film clip could not be shown, so Watson began to describe *Gattaca* instead. He said first that the movie haunted him. It was set in the future, in a world obsessed by genetic perfection, run by a genetically perfected ruling class that controls an unaltered underclass—the invalids. "A scary world," Watson said.

The hero was conceived in the natural way, which by this point in the future has become abnormal. He dreams of besting his genetically enhanced brother. He tries to get out of this world—to go as an astronaut on a voyage of interplanetary discovery to Titan. He finds a way to masquerade as one of the valid astronauts, who has been paralyzed in an accident but does not inform his superiors because he doesn't wish to lose his status.

"Is this the way human society would move if we could put genes in germ cells?" Watson asked. "Or would we transform the weak to the mighty?"

I had made a point of seeing *Gattaca* after Haussler told me that the first people to search the assembly on the Santa Cruz website had counted instances of that sequence. The *Gattaca* story is much more interesting than Watson's rendering suggests. The hero goes daily to the astronaut's house to get samples of the astronaut's hair and urine. These allow him to get past the DNA detectors, which guard every institutional entranceway, and to go to work in the astronaut's place. A murder mystery, arising out of the corruption of this less-than-perfect world, unfolds. The detective trying to solve this mystery is the hero's nemesis and brother. In spite of all odds, the hero bests his brother and achieves his dream, demonstrating that will can triumph over anything, even a world in which hierarchy is built entirely upon genes.

I had by now irritated my neighbors with my muted snorts and groans—little commentaries on Watson's speech, which stood revealed as a marshaling of arguments built on unstated assumptions to lead the unwary to false conclusions. The first unstated assumption was that identifiable human genes determine character, behavior, intelligence and talent. The second unstated assumption was that character, behavior, intelligence and talent genes will be amenable to manipulation. Watson supported these assumptions by reference to the *Gattaca* story, a work of the imagination, but since he did not tell the audience the ending, in effect he rewrote it. The end of the film shows that the genetic hierarchy of the world of *Gattaca* is built on false assumptions about the deterministic nature of genes and that genes cannot alone predict complex human capacity and behavior. Watson had turned the movie's meaning upside down.

"Today," Watson said, "public voices stress the selfish side of human nature. I see it differently. Human beings are strongly programmed to cheer their fellows to succeed . . . Underfed, unloved human beings have diminished capacity . . . How soon we understand why some learn faster than others no one knows. Much further is the possibility that germ-line gene therapy can turn slow learners to fast. If we can give mice a better genome, why not humans?" Here he was no doubt referring to American researchers who have altered the germ line of certain mice to make their muscles a great deal bigger than the norm. "Imagine a world in which lung cancer no longer kills . . . Gene therapy can protect children from lung cancer. Would we be so angry at tobacco companies if smoking did not cause lung cancer and heart disease? . . . Recombinant DNA could make humans resistant to AIDS . . ." To my surprise, he did not add to his list of good things the prospect of immortality.

He turned then to excoriate eugenics' enemies. He said they were the religious, who had organized themselves in the United States into a political force. And then there were those darned

Germans, who, as usual, had everything wrong. In Germany when he talked about genetics in this way, he had been treated as a leper, as a proponent of evil ideas. "I'm distressed when people say enhancement is bad; it just provides ammunition to people who don't like genetics . . . The real short-term issue is how we handle genetic data today. I say, be visible proponents of protecting children from genetic injustice. Run the risk of being called eugenicists . . . It's always best that momentous decisions be made by those who bear the consequences . . . The idea that women decide got a very cold reception in Germany.

"The ethic of the genome project was why we should not leave the future of the human race to God," said Watson. "The Germans say we are following the logic of the Nazis . . . I act upon truths coming from observation and experiment . . . In Germany a professional theologian called my ethics the ethics of horror. No German scientist saw fit to rise to my defense. The secular see human beings as a result of millions of years of evolutionary change; the religious fundamentalists see evolution as only a theory . . . In turning away from religion, the truth is I don't see myself as a perpetrator of cold genetic science. Love is what makes human life thrive . . . Why not, far in the future and after germ enhancement, a world of more love and less violence? . . . Love will be the main driving force of human life. We should not despair for those who follow us."

Oh, he infuriated me. My throat did choke up with his sentimental notion of a genetically enhanced human capacity to love, but, at the same time, Ms. Rational raged at the shameless way he'd appealed to my gender as the trustworthy arbiters of humanity's genetic future and then reshaped his opponents' main argument. His opponents' concern is that a manipulated future will enhance already existing inequality, that genetic might will soon equal right. His rebuttal was that since inequality is already the case, genetics will raise up those below. His logic was fine—if genes make inequality, then manipulated genes might be able to make us all actually equal. But he knew very well that genes are not all there is to the complexity that is a human organism. Who knew it better? His failure to describe the other realities that make us what we are—the interactions between genes and cells, eggs and sperm and uterus, mothers and fathers and children, never mind political economies, wars, the capacities of markets, culture, ideas—was deeply cynical. And what about that awful rhetorical trick—the Germans are always wrong, so if they say I'm wrong, I must be right!

In other forums Watson had claimed that the problem of the old eugenics was that it was enforced by the state. Now he advised these students not to be afraid of being associated with the old eugenics, which implied that a state role in his equalizing, uplifting eugenics would be inevitable. What power other than the state could enforce equal opportunity through genetic enhancement? But he couldn't have it both ways—he couldn't assert the human right to genetic privacy and private genetic decision making and, at the same time, the right to genetic equality. He must have known that no democratic state will commit itself to eugenics again, not after Germany's fling with it. He must have known that if the state stands aside, the market will move in; market-driven eugenics could certainly change human beings, but not by making them equal. Of course he knew—he'd called on his business colleagues to make this happen only a few years earlier. He must have been familiar with Hobbes's views on where freedom lies after the community has given its authority into the hands of the sovereign—it lies in the silence of Leviathan, Hobbes said, where the sovereign makes no laws. Watson's own federal government had tried to make law in this area and, so far, had come up with nothing. The Canadian federal government was now moving so slowly on Bill C-13 that there was no prospect whatsoever that it would be passed over the course of the next year (and, of course, it wasn't). The government was afraid it didn't have the votes to get it through the House; the scientists were pushing for amendments. Meanwhile, in the silence of Leviathan, revelationary biology moved forward.

The real ethical questions raised by Watson were about power and choice: who has the power, who makes the choice. But he didn't answer them. He was doing something else—a sales job. He was selling these young healers on the power for good of this revelationary biology the way the religious sell their god, the way parties sell leaders, the way thinkers sell their ideology, the way Charles Darwin sold his theory of the origin of species by means of natural selection.

I walked out into the foyer at the break. A young student was talking to a man beside her. But wait, she said, I don't get it. Wasn't DNA discovered in the 1860s? Why do they say he discovered DNA?

Alan Bernstein stood at the podium to introduce Sydney Brenner. He talked with great warmth about Brenner, who he said had long been one of his heroes. When he had heard Brenner had won the Nobel, he'd asked, what for? He didn't mean that Brenner was undeserving, but rather that he'd done so many worthy things that he couldn't guess which one the Nobel committee had fixed on. He told a bit of Brenner's story—how he was born into a poor family who "escaped from persecution in Lithuania to South Africa." He described Brenner as an avid reader by the age of four, interested in chemistry at eight, enrolled in medical school at fourteen. This drew gasps from the medical students in the audience, most of whom were their mid-twenties. (Brenner's official Nobel biography shows he was awarded an MSc at twenty.)[6]

"Then, through luck, and skill, he arrived in England in 1952, within months of Watson and Crick's discovery that changed the world. He rushed to Cambridge with Leslie Orgel to meet Crick and Watson and see the model." From that point until 1966, said Bernstein, Brenner was a star among the stars of a new discipline. Brenner worked with Crick on the idea that DNA is a code written in triplet form and that RNA carries its message. Eventually Brenner went looking for the right organism for genetic research and settled on *C. elegans*. In addition to his many achievements, Bernstein added, Brenner is also a man of wit and humor, "the Groucho Marx of science."

The Groucho Marxist Sydney Brenner approached the podium. He had on a turtleneck and a jacket, a pair of chinos. His ears were way too big and stuck out from the heavy square of his face like ailerons. His great head and his heavy brows reminded me of someone. As he began his speech, I kept wondering, who is it? Who?

He spoke very slowly, in a deep voice, a plummy accent. He said he had thought about titling his lecture "The World Ahead," because he was not going to talk about the past. He began with a ponderous joke about his latest idea, which he thought would be useful for the launch of a small biotech. He called it a "cell map." "Of course, I shall be looking for investors," he said.

But then he became serious. The world of biology, he said, is overwhelmed by all sorts of data, small bits and pieces of this and that, a very small signal of meaning drowned by a great deal of noise. "It is important to think very hard about what we will do with this data and form the architecture into which we can imbed it. How to turn data into knowledge is the crisis of the next century. I believe the character of science is, least is best. To turn the data to knowledge, we need to gain the power of predicting the unknown . . . That means we need the power of calculating and computing, how to compute organisms from their genomes."

While it is true that one could build a machine that could extract the linear sequence of DNA for a rat, he said, so what? One would never be able to do the same thing for the myriad protein interactions going on at any given time in a given cell. All those who thought they would learn something of value by trying to trace out each and every protein interaction were doomed to failure. He was also less than thrilled with the notion that one could know one's future by knowing one's genes.

And then he brought forward a strange metaphor. We all live in cities, he said. The composition of the population of the city "disassembles" every morning and then goes on about its business, reassembling in banks and schools, "where they perform functions, then disassembling every evening. You can't understand the function of the city without understanding this architecture of assembly and disassembly," he said.

He compared the genome of the cell to the white pages of the telephone book, listing all the individuals in the city. The white pages are different from the yellow pages, which list only some of the places of assembly. "It would be great to know the yellow pages, but it doesn't describe the architecture," he said. "First we have to choose the level of biological systems to focus on. So forget the genome completely, okay?"

There was once an anarchist group in British Columbia who called themselves Groucho Marxists. Perhaps Brenner knew of them? Saying forget about the genome, to these men who'd slaved for up to fifteen years on it, was anarchist's work.

Years ago, he said, a student had come to him and asked him what he'd have to do to make a breakthrough in the study of the nervous system. He told the student he was too late: the breakthrough that mattered had already been made. "So what is the breakthrough now?" he asked. "I think the cell is the unit we need to think about. What's inside cells, how they communicate with other cells—that's why this project of mine is called a cell map. How do we structure this? We have to think that in each cell perhaps 20,000 genes are active. That's a lot of stuff. If each makes a protein, and we can wire everything to everything else, we won't understand it at all. Does *E. coli* understand itself? I think biological systems are incredibly stupid . . . It has to be simplified so the cells can use the system. They have to evolve. They can't go and redesign the cell. You either die or go forward with accretion or modification."

The cell is the place in which the protein products of genes come into being, do their job and are torn apart—the place of assembly and disassembly. Proteins are manufactured in cells by molecular assemblies, the ribosomes—"gadgets," Brenner called them—which are made out of 120 different proteins. And there is another complex that starts the assembly of a protein that is itself made of twenty-six different proteins. "We see this so-called protein-protein interaction has a structure," he said. These structures are like engines, and these engines can be studied. He proposed that everyone should pay attention to these gadgets within cells, the places where things come together and are torn apart.

"Problem: each cell has different patterns. The most important thing . . . is the problem of substantiation."

He used this word, "substantiation," in a tricky way. He seemed to mean that a gene is something that can exist in different forms according to different contexts; in each context a different embodiment, or substantiation, results. He noted that many genes have more than one promoter, and that things get added to a gene that make that gene address different parts of the cell. Splicing—the different ways the same sequence can be read—creates a whole other order of complexity. But splicing is not variable, said Brenner; it is precisely defined and has a function. Yet all the different substrates in the cell could provide thousands of different substantiations of each gene. The task ahead is to define them.

The point of focusing on substantiations and cellular gadgets is to reduce complexity by an order of magnitude, he explained. "Not 20,000 gene products, but two thousand gadgets. The cell has a topography. Mitochondria—some gadgets work there. It has a plasma membrane, a nucleus. You can get down another order of magnitude to ten regions, a few hundred things."

He wanted to simplify because the cell is brainless and yet makes complicated decisions based on signals from within and without. It wouldn't be complex mathematics that these little gadgets use to decide what to do, he said, it would be arithmetic. "Computational biology will be about that arithmetic, not differential equations . . . Everyone wants a controller, or a monitor program inside the cell, and thinks of driving it. It's not like that."

I sat up straighter in my chair now. Sydney Brenner was talking to his colleagues, but it was as if he had reached down and scooped up my doubts, questions and dissatisfactions with the image of life portrayed by revelationary biology and laid them out on the podium in front of him. He was speaking of my desire to know the ghost in the machine, to capture the writer of the program that organizes two unrelated cells and makes them develop together. He was telling his colleagues how they could answer the question that most pressed on me: if genes direct the behavior of cells and the development of an organism, how are chimeras possible? He was telling his colleagues that there is no ghost in the machine, no superprogram waiting to be uncovered that guides all the rest. Complexity arises from extreme simplicity, simple rules that cells must live by, and what that means is that no one has to write the show.

"The components communicate with each other," he said, "in a . . . pathway. There's no such thing as a path, but there are signals." The tendency had been to try to think about all these components as part of a dense meshwork, like the map tourists use to go somewhere in the impossibly convoluted Tokyo subway system, which is simple compared to the to-and-fro in a yeast cell. This language carries an imbedded idea of a destinations, or end points. "The language we have to use is the language of devices, little wiring diagrams with inputs and outputs."

But inventing cell maps involved a new way of thinking about the cells we study. One could no longer define a cell by the organ in which it is found, by its final fate. There are over one thousand different kinds of cells in the brain, he reminded them, and different kinds of cells appear in kidneys. I felt a kind of relief as he said this—it fit so well with the work on adult stem cells. It explained why Freda Miller's adult stem cells seemed to adjust their shape and capacity to their physical circumstance, why stem cells could not be picked out from the crowd of cells around them when Wilmut searched for them in sheep. "To do it in the detail necessary to do it properly, we must validate for genes every substantiation in every cell in our bodies," he said. "In fact, it's necessary to try and define cells in terms of gene expression.

"Find out what genes are turned on, then put it into a map," he said. Such a map would rid people of the difficulty of communicating about these interactions with words. "We need to define a new pictorial language to communicate," he said. "We have to teach this to others, to students and ordinary people."

This task of studying gadgets—focusing on substantiation, on the interactions inherent in different contexts—ought to start now, he said. We might not know how to do this yet, but it is important to begin to think about it long before we have the capacity to get it done. When Sanger was working on the sequence of lambda, Brenner said, people were already aiming at sequencing the human genome. And he had suggestions for where to start: biologists could borrow from geographers, who use many different methods to convey information about the formations of the earth, everything from aerial photography to contour diagrams. Using visual means, he said, we can start to map the cell.

"I was asked at the beginning of the worm project, what will I leave behind," said Brenner. "I said one thousand genes. Gilbert said, forget it, we'll do it all. I said, like someone said to Columbus, wait five hundred years, you can get a cheap flight. Perhaps in five hundred years we'll come to understand the human body."

And he sat down.

Brenner was humble; he was wise. His idea seemed small at first, and simple, but it was like a plain box with artfully hidden drawers. It opened a new way of thinking about something people had been studying in the same way for more than one hundred years. It stripped from genetics its core idea that genes are the god of the cell. He assumed instead that the cell is defined by interaction to which there is no beginning or end, and so we must study where these interactions take place, not worry about destinations. Darwin struggled for years to express a similar idea about species, to explain that the way they change is a random process, that there is no guiding hand behind their creation, but even his descendants weren't really able to let go of the idea of a guide, a purpose, an end point. They simply relocated the ghost in the machine from heaven to the gene. That Watson still spoke of perfection was proof that even he still clung to the ancient belief in some sort of telos, or end to which life is aimed. Now Brenner was saying, here is how we get rid of what holds us back; here is where we must start to clean our house, to sweep out the metaphors that cloud what we can see.

There was not an ounce of sentimentality in Brenner. He wasn't selling me on a better world, on cures for all the world's terrible diseases. He just pointed out how we might discern the simple machine beneath variety, an idea familiar to those wily Greeks who took such pleasure in portraying chimeras and spinning tales of how to slay them.

I sat through the remainder of the talks. I took notes. I even went to the evening lecture, which was open to the public, for a fee. A few other speakers, like Paul Nurse, like Tony Pawson of the Lunenfeld, amplified Brenner's suggestions. It was as if he'd thrown a ball but only a few were equipped to catch and carry it. At the evening lecture, Eric Lander put on a headset and pranced up and down like a motivational speaker. Bartha Knoppers, who gave a talk in the afternoon on ethics to an empty hall, explained at night why patents are a good thing. John Sulston went on about the common thread. Even Brenner himself lapsed from clarity into something unintelligible.

But I didn't care about that. I'd heard him in the afternoon. He'd made me forget my rage at Watson and even my fear for the future. He'd reminded me that revelationary biology is more than just the sum of its themes. At its best, it is a courageous leap into the frightening dark, with no god or angel to catch the fall. That's when I figured out who Brenner reminds me of—a clean-shaven Charles Darwin.

LOOSE ENDS AND ACKNOWLEDGMENTS

A book is not like life: it should have a tidy ending, smooth and shapely and satisfying. I hoped to gather up the loose ends, tie them neatly and deliver. Just before Christmas, in 2002, I thought the God of Books had come to my aid. The Raelian bishop, Brigitte Boisselier, held a press conference to announce that the Raelians had made science history. They were the first, they had done it—they had cloned a human being, a girl, who had just been born to an American couple in a location Boisselier refused to give. The name of the girl, she said, was Eve.

Oh, how she strutted in front of the assembled press corps, smiling like a woman once spurned who now has her cold revenge. Yes, there would be tests and proof; it had all been arranged. And several more cloned babies were on the way!

But then things unravelled. The parents refused to appear before the press. The person Boisselier named who would lead the DNA testing that would prove the truth of her claims was revealed as someone who had tried to peddle an exclusive on the Raelians and their cloning to various TV networks—perhaps not the independent skeptic Boisselier had promised. And then a man in Florida launched a lawsuit against the unnamed parents for endangering their child through the use of untested technology.

Phhhhttt, just like that, the story of the first human clone, Eve, disappeared in disarray.

Well, never mind, I thought, something else will turn up. How about the legislation moving through the Canadian House of Commons, Bill C-13? I waited and watched. The government sent certain reliable MPs to attend the meetings of the Standing Committee on Health to ensure the bill came to the House without any unexpected changes. Eventually, the bill did emerge from Committee. The debate was so hot, so heavy, that the government appeared reluctant to put it to the test, as if they weren't certain they could muster enough votes. The legislative travail of this bill was made more complex by the problem of the old prime minister retiring. After much maneuvering, a vote was held and the bill passed—but the House rose, ending the legislative session.

Would the bill be brought forward again at the turn of the new year, 2004? No one seemed to know. When the government read out its throne speech to start the next session, the bill wasn't mentioned.

Meanwhile, there had been endings of another sort. My uncle passed away after a long struggle. So did my neighbour and friend. Roslin Institute had also announced sad news: in February, 2003, they put Dolly down because her arthritis was too much misery for her to bear. Soon she was stuffed and placed on display in the British Museum. At least she was euthanized for a reason: Gene had been killed because he was no longer of use to science.

I began to wonder what was happening at Infigen. No one answered the phone when I called. When I looked at their website I was surprised to see that Michael Bishop's name was no longer there. Bishop had moved on to a human health research/clinic in another town. Why? Business troubles at Infigen, although from Infigen's website one would never know it. There was not even an archived press release announcing Bishop's departure.

Which made me wonder about Michael West. I checked Advanced Cell Technology's website too. Jose Cibelli was no longer listed there, and neither was the company Cyagra, which used to make clones for a fee. It had been sold. Cibelli, I soon learned, had taken an academic appointment in the department of animal science-physiology at Michigan State in East Lansing. As for Michael West, the January 2004 edition of *Wired* magazine carried a cover story called "The Making of a Human Clone," which described the experiments going on at Advanced Cell.[1] Once again, Michael West had allowed a journalist to get up close and personal as his associate Robert Lanza and a South Korean colleague tried to get nuclear transferred eggs to divide more than six times. Once again, the journalist breathlessly reported that at least one egg divided, on this occasion eighteen times, but it was all such delicate work, the journalist wasn't actually allowed to see it. Michael West made it known that he needed more funds to carry on.

It was a group of South Koreans, mainly at the college of veterinary medicine of Seoul National University (although the last author, Shin Yong Moon, is at the University's College of Medicine) who actually got the job done that West had set out to do. In February, 2004, they published in *Science* a report of their success making the first stem cell line cloned after nuclear transfer of a cumulous cell into a human egg. The first author, the grunt, was Woo Suk Hwang. Jose Cibelli had acted as a consultant, so his name appeared too. While the South Koreans were unable to rule out completely the possibility that the egg had become an embryo without instructions from the transplanted cumulous cell, all the tests suggested otherwise. They made a point in their paper of saying that the 242 oocytes they used, and the cumulous cells cling-ing to them, came from sixteen volunteers who were not paid for their donations. They succeeded in getting a cell line from only one enucleated, nuclear transferred egg. The line had gone through more than seventy passages before they published. Their various tests showed that these cells, when grown in suspension, made all three layers of cells found in normal embryos, and, when inserted into mice testes, made teratomas (little tumours) which, after several weeks, displayed smooth muscle, bone, cartilage and connective tissue cells. The methods they used were ones that had been proven to be effective in bovine rather than porcine cloning, although they'd added a few wrinkles of their own. Instead of suck-ing out the nucleus when preparing the oocyte, they squeezed. They reckoned that their egg reconstruction methods were as efficient as those doing SCNT in cattle. Mind you, there was no way, short of inserting one of these cells into a uterus, to make certain nuclear transfer had actually worked. The paper referred to the ethical constraints that preclude doing such an experiment in a woman but "complementary investigations in nonhuman primates might provide additional confirmatory information."[2]

Meanwhile, just as this announcement was made, the Canadian Senate, the chamber of sober second thought, was handed the challenge of C-13. Committee hearings began in February, 2004 on the bill, restyled as C-6 in the Senate. Among other things, the bill bans nuclear transfer therapeutic cloning experiments. The legislation whizzed through Committee without amendments. It became law a month later.

Canadian legislators had finally poured the salt, but the bird called revelationary biology had already flown away.

Acknowledgements usually come at the very end of a book, after the notes and the index. But the people who helped on this project are really part of the story and I owe them a great deal. Stephen Dewar, my husband, kept pointing out items in the papers and feeding my interest until it took off on its own, so he was there from the beginning, and he stuck with me until the end. He read each draft, offered his comments and criticisms, and when I was done, organized the text in packages to send out to the various scholars for their response and corrections. It took a lot of time and care to get this job done and I am more appreciative than I can say. My daughter Anna Dewar spent many hours in the library pulling out books and articles for me, as did her friend and ours, Timothy Gully. My daughter Danielle Dewar helped with phone calls and messages. Each one of the scientists, bioethicists, politicians, archivists, and others referred to below, several of whose names do not appear in the text, invited me into their labs or spoke with me on the phone, making a gift of their limited time. They offered explanations of their work, their personal stories, their opinions and their insights. Several were not featured and hence were not asked to help with the fact-checking process. A few of those who were featured did not participate in the fact-checking either. Some didn't have time. Some were annoyed with what I'd written. But most did respond. They saved me from many embarrassing errors, and their patience and care and dedication to getting things right is greatly appreciated.

And so I thank: Mick Bhatia, Ronald Worton, Andras Nagy, Janet Rossant, Freda Miller, Michael Bishop, Michael West, Carol Greider,

Calvin Harley, Woodring Wright, Elizabeth Blackburn, Don Riddle, David Baillie, Miller Quarles, Michel Revel, Dianne Irving, Gordon Tenor, Peter Lansdorp, Carolyn Astell, Dan Butler, Dave Wolyn, Dr. M., Ms. G., Dr. E. the pediatrician, Ronald Green, Anne Kiessling, Bartha Knoppers, Bonnie Brown M.P., Paul Szabo M.P., Jan Witkowski, Marilyn Monk, Anne McLaren, Adam Perkins, William Cash M.P., Frederick Sanger, Ian Wilmut, Keith Campbell, Benjamin Reubinoff, Lior Gepstein, Karl Skorecki, Ignacio Chapela, David Quist, Jim Kent, Alan Bernstein, E.A. McCullough, Trudo Lemmens, Peter Singer, Craig Venter.

I must single out David Baillie, whose patience and kindness is greatly appreciated. Similarily, I am so grateful to Frederick Sanger, who was so gracious in describing his life and work, and so disappointed by my failure to capture the early part as well as the latter. Paul Berg was equally painstaking and helpful in his reading of the text and spent time with me he did not have. I must also thank Alan Coulson, whose care in making certain he did not overstate or understate was truly remarkable. David Haussler corrected and amplified the text, but also took the trouble to repeat his steps in the discovery of a possibly previously unknown human gene so as to be certain I had rendered them correctly.

Despite all their efforts to set me on the path to truth, I have no doubt that many errors still remain. Responsibility for error is, of course, mine.

This book would not have emerged without the care and guidance of my editor, Anne Collins. We have worked together for years very happily: this time, as she worked with me, she also carried a large and important burden, never complaining, for which she will never get the credit she deserves. I am more than usually grateful for all her efforts on my behalf. Her assistant, Craig Pyette, fielded phone calls and huge electronic files with great calm and élan. The copy editor, Stephanie Fysh, took a large text full of science language and made the style consistent throughout. Maureen O'Donnell helped me navigate the Gairdner Awards. This book could not have been completed without a non-fiction writing grant from the Canada Council for the Arts. Their much-appreciated support bought me the time I needed to finish the manuscript.

This book is about living life, which is still fleeting (in spite of some biologists' best efforts), as much as it is about those who explore life's nature. I was offered a great deal of hospitality and the pleasure of good company throughout, particularly from my family and close friends, who are scattered in many places around the world. My cousin Sandra Broudy put me up in her splendid home in Vancouver. My cousin Debbie Landa did the same for me in San Francisco. Carol and Jay Luther wined and dined me well. Jessica Ruvinsky told me many stories about what it's like nowadays to be a doctoral candidate in botany in a top U.S. university. In Israel, my cousin Patricia Puterman not only fed and housed me, she also acted as chauffeur, researcher and travel guide. Her husband, Micha, put up with our meandering progress across the Israeli highway system, and her children, Shani, Adi and Moolie, entertained me royally. Elizabeth Appel walked me through King's College Chapel in Cambridge on a glorious spring morning. Tess and Alan Gully introduced me to the best curry house in London. Finally, I acknowledge with gratitude the support of my parents, Petty and Sam Landa, who encouraged their children to follow curiosity no matter where it leads. Thanks for all those stem-cell clippings, Mom.

NOTES

INTRODUCTION

1. Garland E. Allen, "The Eugenics Record Office at Cold Spring Harbor, 1910–1940: An Essay in Institutional History," *Osiris* 2nd ser. 2 (1986): 225–64; Charles Davenport, *Heredity in Relation to Eugenics* (London: Williams and Norgate, 1912), 233.
2. See *Muir v. Alberta,* 32 ALTA.L.R. 3rd, 95. This case involved a suit brought against Alberta by a former inmate at a provincial training school for mental defectives who was sterilized, or rendered incapable of conceiving children, with the approval of the provincial Eugenics Board. Evidence in the case, which had to do with actions forty years earlier and was about whether an expert witness would be allowed to give testimony as to the behavior of the Board at the trial of the issue, showed that three thousand such sterilizations had been approved by the Eugenics Board in Alberta. See also *Eve, by her guardian ad litem, Milton B. Fitzpatrick, Official Trustee v. Mrs. E., Respondent and Canadian Mental Health Association, Consumer Advisory Committee of the Canadian Association for the Mentally Retarded, the Public Trustee of Manitoba and Attorney General of Canada, Interveners,* Supreme Court Reports 1986, 2, 388.
3. Marilyn Monk, "Of Microbes, Mice and Man," *International Journal of Developmental Biology* 45 (2001): 497–508.
4. C. E. Novitski, "Another Look at Some of Mendel's Results," *Journal of Heredity* 86 (1995): 62–66.
5. Janet Browne, *Charles Darwin The Power of Place* (New York: Knopf), 11–35; Robin Marantz Henig, *A Monk and Two Peas: The Story of Gregor Mendel and the Discovery of Genetics* (London: Phoenix, 2000), 108–10.
6. Horace Freeland Judson, *The Eighth Day of Creation: Makers of the Revolution in Biology* (Plainview, NY: Cold Spring Harbor Laboratory Press, 1996), 620.
7. John Sulston and Georgina Ferry, *The Common Thread: A Story of Science, Politics, Ethics and the Human Genome* (London: Bantam, 2002).
8. Ian Wilmut, Keith Campbell, and Colin Tudge, *The Second Creation: Dolly and the Age of Biological Control* (New York: Farrar, Straus and Giroux, 2000), 253.
9. Gretchen Vogel, "In the Mideast, Pushing Back the Stem Cell Frontier," *Science* 295 (2002): 1818–20.
10. Geron Corporation, "Geron Announces Issuance of Key Telomere and Telomerase Patents," March 4, 1998, http://www.geron.com/pr_030498a.html.

11. Carol W. Greider, "Telomeres and Senescence: The History, the Experiment, the Future," *Current Biology* 8 (1998): R178–81.

12. Robert F. Service, "'Fountain of Youth' Lifts Biotech Stock," *Science* 279, (January 23, 1998): 472.

13. S. J. Jones et al., "Changes in Gene Expression Associated with Developmental Arrest and Longevity in *Caenorhabditis elegans*," *Genome Research* 11, (2001): 1323–24; Carolyn Abraham, "Worming our way to a longer life," *Globe and Mail*, July 13, 2001, A1.

CHAPTER ONE: Biologists Seize the Day

1. Browne, *Charles Darwin: The Power of Place,* 101.

2. Janet Browne, *Charles Darwin: Voyaging* (New York: Knopf, 1995), 385–89.

3. Ibid., 502; Browne, *The Power of Place,* 36–37.

4. Henig, *A Monk and Two Peas,* 140.

5. Galton's letters to Darwin may be read in the Cambridge University Library.

6. Francis Galton, *Inquiries into Human Faculty and Its Development* (London: Macmillan, 1883).

7. Ibid., 307.

8. Henig, *A Monk and Two Peas,* 139.

9. Ibid., 140.

10. Ibid., 155.

11. Ibid., 173–87.

12. Davenport, *Heredity in Relation to Eugenics,* 10–11.

13. Allen, "The Eugenics Record Office at Cold Spring Harbor."

14. Charles Davenport, "The Family History Book," Eugenics Record Office, *Bulletin* no. 7 (Cold Spring Harbor, NY: 1912); Harry B. Laughlin, *Report No. 1* (Cold Spring Harbor, NY: Eugenics Record Office, 1913), 9, 25.

15. Davenport, "Family History Book," iv.

16. Ibid., 4.

17. As quoted by the Supreme Court of Canada, re. *Eve,* 1986, SCR, 388, p. 207.

18. Barry L. Kiefer, "Recombinant DNA: Controversies and Potentials," in *Recombinant DNA Research and the Human Prospect: A Sesquicentennial Symposium of Wesleyan University,* edited by Earl Hanson (Washington, DC: American Chemical Society, 1983), 29.

19. Gregory Stock and John Campbell, eds., *Engineering the Human Germline: An Exploration of the Science and Ethics of Altering the Genes We Pass to Our Children* (New York: Oxford University Press, 2000), 77–79.

20. Bernardino Fantini, "The 'Stazione Zoologica Anton Dohrn' and the History of Embryology," *International Journal of Developmental Biology* 44 (2000): 523–35.

21. James D. Watson, *Genes, Girls, and Gamow* (New York: Knopf, 2001), 249.

22. Erwin Schrödinger, *What Is Life? With Mind and Matter and Autobiographical Sketches* (Cambridge: Cambridge University Press, 2000).

23. Health Canada, Preamble, *Proposals for Legislation Governing Assisted Human Reproduction* (Ottawa: Health Canada, May 2001).

24. Ibid.

25. Ibid., sec. 9 (3) a.

26. Ronald Worton, curriculum vitae.

27. Rick Weiss and Deborah Nelson, "FDA lists violations by gene therapy director at U-Penn," *Washington Post,* March 4, 2000, A4.

28. Deborah Nelson and Rick Weiss, "Gene test deaths not reported promptly," *Washington Post,* January 31, 2000, A1.

29. Jose Cibelli, Robert P. Lanza, and Michael D. West, with Carol Ezzell, "The First Human Cloned Embryo," *Scientific American,* November 24, 2001, http://www.sciam.com/article.cfm?articleID=0008B8F9=ACb2=1C75=9B81809EC588EF21.

30. Joannie Fischer, "The First Clone," *U.S. News & World Report.* December 3, 2001, 50.

31. Jose Cibelli et al., "Somatic Cell Nuclear Transfer in Humans: Pronuclear and Early Embryonic Development," *e-biomed: The Journal of Regenerative Medicine* 2 (2001): 25.

CHAPTER TWO: Clones, Anyone?

1. Jose Cibelli, Testimony to the Standing Committee on Health, House of Commons, Canada, October 18, 2001.

2. Kim Honey, "The man who aims to cheat death," *Globe and Mail,* November 27, 2001, A1.

3. Gary Stix, "What Clones?" *Scientific American* 286 (February 2002): 18–19.

4. Ibid.

5. Jay Ingram, "The story of the big clone story," *Toronto Star,* January 27, 2002, F8.

6. Honey, "The man who aims to cheat death."

7. Timothy Caulfield, Testimony to the Standing Committee on Health, House of Commons, Canada October 18, 2001. See also Timothy Caulfield, "I smell a cloned rat," *Globe and Mail,* January 4, 2003, A17.

8. U.S. House of Representatives, Committee on Energy and Commerce, Subcommittee on Oversight and Investigations, March 28, 2001.

9. BBC News, "Doctors defiant on cloning," March 9, 2001, http://news.bbc.co.uk/hi/english/sci/tech/1209716.stm.

10. Panayiotis Zavos, Testimony to panel of the National Academies on Scientific and Medical Aspects of Human Reproductive Cloning, August 7, 2001.

11. Tim Adams, "The clone arranger," *The Observer,* December 2, 2001, http://www.observer.co.uk/2001review/story/0,1590,624237,00.html.

12. Paul Koring, "Human cloning dangerous, unethical, experts say," *Globe and Mail,* March 29, 2001, A10.

13. Margaret Talbot, "A Desire to Duplicate," *New York Times Magazine* (February 4, 2001), 40.

14. Marissa Nelson, "Raelian sect told to stop research on cloning," *Globe and Mail,* June 30, 2001, A11.

15. Talbot, "A Desire to Duplicate."

16. Francis Crick, *Life Itself: Its Origin and Nature* (New York: Simon and Schuster, 1981), 15–16.

17. Rael, "Human Cloning Will Make Terrorist Attacks Inefficient and Will Allow the Judgements of the Perpetrators," press release, September 14, 2001, http://www.rael.org/pres.

18. Perhaps this idea of emptying one's memory into a computer for storage and later retrieval is not so ridiculous as it sounded; see Carolyn Abraham, "Chip may one day fix your failing memory," *Globe and Mail,* March 14, 2003, A1.

CHAPTER THREE: How Many Ways Can You Make a Mouse?

1. James D. Watson, *The Double Helix: A Personal Account of the Discovery of the Structure of DNA* (New York: Scribner, 1998), 197.
2. "Clue to chemistry of heredity found," *New York Times*, June 13, 1954, L17; F. C. H. Crick, "The Structure of Hereditary Material," *Scientific American* 191 (October 1954); Curt Stern, "The Biology of the Negro," *Scientific American* 191 (October 1954).
3. Watson, *Double Helix* and *Genes, Girls, and Gamow.*
4. Watson, *Genes, Girls, and Gamow*; Nathaniel C. Comfort, *The Tangled Field: Barbara McClintock's Search for the Patterns of Genetic Control* (Cambridge: Harvard University Press, 2001), 18.
5. Watson, *Double Helix*, 48–50.
6. Ibid., 225.
7. For a history of the discovery of imprinting, see Monk, "Of Microbes, Mice and Man."
8. For a description of newer work about the interactive molecular complexes that bind to DNA and suppress gene expression, see Jim Kling, "The Complexity of Gene Silencing," *The Scientist* 16, no. 4 (February 18, 2002): 31.
9. Andras Nagy et al., "Derivation of Completely Cell Culture–Derived Mice from Early-Passage Embryonic Stem Cells," *Proceedings of the National Academy of Sciences* 90 (1993): 8424–28.
10. Wilmut, Campbell, and Tudge, *The Second Creation*, 237.
11. Helen Branswell, "Chick study shows how faces could form," *Toronto Star*, December 20, 2001, A21.
12. Dawkins, along with Crick and Orgel, suggested that some genes move around on chromosomes, without creating changes in the organism, just to perpetuate themselves, almost as if discrete sequences of base pairs have a sense of intention and identity. See W. Ford Doolittle and Carmen Sapienza, "Selfish Genes, the Phenotype Paradigm and Genome Evolution," *Nature* 284 (1980): 601.
13. Andras Nagy and Janet Rossant, "Chimaeras and Mosaics for Dissecting Complex Mutant Phenotypes," *International Journal of Developmental Biology* 45 (2001): 577–82.
14. Comfort, *The Tangled Field*, 2, 207.
15. Ibid., 268.
16. Francis Galton, *Hereditary Genius: An Inquiry into Its Laws and Consequences* (London: Macmillan, 1869), 368–69.
17. Comfort, *The Tangled Field*, 260.
18. Monk, "Of Microbes, Mice and Man."

CHAPTER FOUR: Unraveling Mortality

1. Table of Contents, *Molecular Biology of the Cell*, 12, no. 12 (2001).
2. A brochure put out by the Cooperative Human Tissue Network explains that there is a fee for handling, preparing and shipping costs.
3. A year later, it was announced that Reeves had recovered some feeling in his toes and fingers and was beginning to breathe on his own. None of these miracles were attributable to stem cell therapy, but to extremely rigorous and persistent

physical therapy. See Jim Suhr, "Paralyzed Superman star can move toes, fingers," *Toronto Star,* September 11, 2002, A3.

4. Greider, "Telomeres and Senescence."
5. Carol Greider and Elizabeth H. Blackburn, "Identification of a Specific Telomere Terminal Transferase Activity in Tetrahymena Extracts," *Cell* 43, part 1 (December 1985): 405–13.
6. Greider, "Telomeres and Senescence."
7. Ibid.
8. Ibid.
9. Calvin Harley, A. Bruce Futcher, and Carol Greider, "Telomeres Shorten during Ageing of Human Fibroblasts," *Nature* 345 (1990): 458–60.
10. Christopher M. Counter et al., "Telomere Shortening Associated with Chromosome Instability Is Arrested in Immortal Cells Which Express Telomerase Activity," *The EMBO Journal* 11, (1992): 1921–29.
11. Greider, "Telomerase and Senescence."
12. Andrea G. Bodnar et al., "Extension of Life-Span by Introduction of Telomerase into Normal Human Cells," *Science* 270 (1998): 349–52.

CHAPTER FIVE: A Little Skin Magic

1. Jean Toma et al., "Isolation of Multipotent Adult Stem Cells from the Dermis of Mammalian Skin," *Nature Cell Biology* 3, no. 9 (2001): 778–83.
2. Ibid.
3. Comfort, *The Tangled Field,* 260.
4. John Maddox, Preface, *The Eighth Day of Creation: Makers of the Revolution in Biology,* by Horace Freeland Judson, exp. ed. (Plainview, NY: Cold Spring Laboratory Press, 1996), xi.
5. Freda Miller, curriculum vitae.

CHAPTER SIX: Flies, Worms and Running-Dog Geneticists

1. Watson, *Genes, Girls, and Gamow,* 74.
2. Jones et al., "Changes in Gene Expression."
3. David Leonard Baillie, home page, http://crick.mbb.sfu.ca/home.html.
4. For a recapitulation of what was then known, see F. H. C. Crick, Leslie Barnett, S. Brenner, and R. J. Watts-Tobin, "General Nature of the Genetic Code for Proteins," *Nature* 192 (1961): 1227–32.
5. Michel Morange, *A History of Molecular Biology,* trans. Matthew Cobb (Cambridge: Harvard University Press, 2000), 142–43; Judson, *The Eighth Day of Creation,* 341.
6. Crick et al., "General Nature of the Genetic Code for Proteins."
7. Judson, *The Eighth Day of Creation,* 470.
8. Sydney Brenner, letter to Max Perutz, June 5, 1963, http://elegans.swmed.edu/Sydney.html.
9. Don Riddle, home page, http://www.biotech.missouri.edu/Dauer-World/People/Donmenu.html.
10. Paul Berg, "The Eighth Feodor Lynen Lecture: Biochemical Pastimes . . . and Future Times," in *Molecular Cloning of Recombinant DNA,* eds. W. A. Scott and R. Werner (New York: Academic Press, 1977).

11. Michael Rogers, *Biohazard* (New York: Knopf, 1977), 37.
12. Stanley N. Cohen et al., "Construction of Biologically Functional Bacterial Plasmids in Vitro," *Proceedings of the National Academy of Sciences* 70 (1973): 3240–44; John F. Morrow, Stanley N. Cohen, Annie C. Y. Chang, Herbert W. Boyer, Howard M. Goodman, and Robert B. Helling, "Replication and Transcription of Eukaryotic DNA in *Escherichia coli*," *Proceedings of the National Academy of Sciences* 71 (1974): 1743–47.
13. There were ten altogether, including luminaries such as David Baltimore and Norton Zinder.
14. Michael Rogers was the *Rolling Stone* journalist; he went on to write the book *Biohazard*.
15. Crick et al., "General Nature of the Genetic Code for Proteins."
16. See C. Thacker et al., "Functional Genomics in *Caenorhabditis elegans*: An Approach Involving Comparisons of Sequences from Related Nematodes," *Genome Research* 9 (1999): 348–59. For somewhat similar analysis of the human genome, see also The BAC Resource Consortium: V. G. Cheung et al., "Integration of Cytogenetic Landmarks into the Draft Sequence of the Human Genome," *Nature* 409 (2001): 953.

CHAPTER SEVEN: The Pig Farm of Dr. Moreau

1. Wallace Immen, "Milking genes for all they're worth," *Globe and Mail*, November 29, 2001, A3.
2. Alan McHughen, *Pandora's Picnic Basket: The Potential and Hazards of Genetically Modified Foods* (Oxford: Oxford University Press, 2000), 47–48.
3. Eyal Press and Jennifer Washburn, "The Kept University," *Atlantic Monthly*, March 2000.
4. Carolyn Abraham, "Cloning pig parts for people," *Globe and Mail*, January 3, 2002, A1; see also Abraham, "Stakes high in race to produce pig parts for human transplant," *Globe and Mail*, January 4, 2002, A1.
5. Jaenisch and Ian Wilmut had previously published on these developmental problems, resulting mainly from errors in epigenetic reprogramming, and argued that since there are so many problems with cloned animals, it was unsafe to clone humans; see Rudolf Jaenisch and Ian Wilmut, "Developmental Biology: Don't Clone Humans!" *Science* 291 (2001): 2552. See also National Academy of Sciences, National Academy of Engineering, and Institute of Medicine. Committee on Science, Engineering, and Public Policy (National Academies of Science), *Scientific and Medical Aspects of Human Reproductive Cloning* (Washington, DC: National Academy Press, January 18, 2002).
6. In fact, these pigs farrowed one day after my departure. See "Piglet clones elicit squeals of delight," *Toronto Star*, February 12, 2002, A11.
7. Wilmut, Campbell, and Tudge, *The Second Creation*, 137.
8. Ibid., 132. I found Willadsen working as an embryologist for the assisted reproduction unit of Saint Barnabas Medical Center in New York, under Jacques Cohen, a man whose *bone fides* in assisted human reproduction go right back to Edwards and Steptoe.
9. Emanuele Cozzi et al., "Characterization of Pigs Transgenic for Human Decay-Accelerating Factor," *Transplantation* 64 (1997): 1383–92; see also Dongwan Yoo

and Antonio Giulivi, "Xenotransplantation and the Potential Risk of Xenogeneic Transmission of Porcine Viruses," *Canadian Journal of Veterinary Research* 64 (2000): 193–203.

10. Geron, "Geron Announces Second Interference over Nuclear Transfer Patents." The first interference granted by the U.S. Patent Office showed that the Roslin-Geron patent application had been made before the application from Advanced Cell Technology. The second interference declared that the Roslin-Geron application was also senior to that made by Infigen, Inc. In an interference hearing, the onus is on the junior party to show that it made the invention before the senior party.

11. *Infigen, Inc., Plaintiff v. Advanced Cell Technology, Inc. and Steven L. Stice, Defendants,* First Amended Complaint, November 2, 1998; Answer to Amended Complaint and Demand for Jury Trial, December 7, 1998; Final Settlement Agreement; United States District Court, Western District of Wisconsin, No. 98C 0431C.

12. Ibid., Final Settlement Agreement, 4–6.

13. Wilmut, Campbell, and Tudge, *The Second Creation,* 148–49.

CHAPTER EIGHT: Rendezvous at the Fountain of Youth

1. Honey, "The man who aims to cheat death"; Carolyn Abraham, "Cloning research paying off, Texan says," *Globe and Mail,* December 1, 2001, A3.

2. Michael Ruse, "An appreciation: Stephen Jay Gould: He fought every 'ism' in the book," *Globe and Mail,* May 23, 2002, R9. Gould testified at the trial against creationism and in favor of the science of evolution. He was a profound critic of Darwin's gradualist conception of change. Gould thought species diverge rapidly after catastrophes, filling up freshly emptied environmental niches. He used paleontological evidence to support this conception, which he called "punctuated equilibrium."

3. See Greider, "Telomeres and Senescence."

4. Ibid.

5. Carol W. Greider and Elizabeth H. Blackburn, "Telomeres, Telomerase and Cancer," *Scientific American* 274, no. 2 (1996): 80–85.

6. Greider, "Telomeres and Senescence." In her review of the literature, Greider told the story this way: Olovnikov noticed that people did not get on at the very front of the train because there was an engine there; he thought maybe a similar thing happened with a replicating DNA strand.

7. *Tanakh,* Leviticus 16:27–29.

8. This paper actually appeared in 1986 as H. J. Cooke and B. A. Smith, "Variability at the Telomeres of Human X/Y Pseudoautosomal Region," *Cold Spring Harbor Symposia in Quantitative Biology* 82 (1986): 213–19.

9. C. B. Harley et al., "Telomerase, Cell Immortality and Cancer," *Cold Spring Harbor Symposia on Quantitative Biology* 59 (1994): 307–15.

10. T. M. Nakamura et al., "Telomerase Catalytic Subunit Homologs from Fission Yeast and Human," *Science* 277 (1997): 955–59.

11. Bodnar et al., "Extension of Life-Span by Introduction of Telomerase into Normal Human Cells."

12. Service, "'Fountain of Youth' Lifts Biotech Stock."

13. Statements of Changes in Beneficial Ownership, U.S. Securities and Exchange Commission, Form 4, Michael D. West, received December 10, 1997, and February 10, 1998.
14. Robert P. Lanza et al., "Extension of Cell Life-Span and Telomere Length in Animals Cloned from Senescent Somatic Cells," *Science* 288 (2000): 665–69.
15. Gretchen Vogel, "In Contrast to Dolly, Cloning Resets Telomere Clock in Cattle," *Science* 288 (2000): 586–87.
16. C. Ward Kischer, "The Corruption of the Science of Human Embryology," letter to the *American Bioethics Advisory Commission Quarterly*, Fall 2002, American Life League, http://www.all.org/abac/aq0203.htm.
17. *Roe et al. v. Wade*, District Attorney of Dallas County, Appeal from the United States District Court for the Northern District of Texas, No. 70–18; Argued December 13, 1971; Reargued October 11, 1972; Decided January 29, 1973.

CHAPTER NINE: Vested Ethics

1. Later, a U.S. congressman would claim that a government audit found that Advanced Cell Technology misused these grants, applying them to human cloning research, which the NIH cannot fund. In fact, the audit found that only $150,000 should be disallowed, but even that did not go to human embryo research. See Michael West, letter to Mark Souder, May 16, 2002, http://www.advancedcell.com/Letter_to_Souder.pdf.
2. Ronald M. Green, "The Ethical Considerations," *Scientific American* 286 (January 2002): 48–50.
3. Ibid.
4. Audrey R. Chapman, Mark S. Frankel and Michele S. Garfinkel, *Stem Cell Research and Applications: Monitoring the Frontiers of Biomedical Research*, American Association for the Advancement of Science and Institute for Civil Society, 1999, http://www.aaas.org/spp/sfrl/projects/stem/report.pdf, ix.
5. Lisa Belkin, "The Made-to-Order Savior," *New York Times Magazine*, July 1, 2001, 38. This story recounts how a researcher named Mark Hughes was hired by the NIH in 1994 to do research on preimplantation diagnosis, under guidelines developed by a federal advisory committee but Congress banned federal funding for all such research. Hughes had to find private funds.
6. One MP interested in the committee's work later showed me the contract that the Ottawa Hospital's Fertility Centre made its clients sign. The contract stipulates that if they abandoned their frozen embryos by, among other things, failing to pay for their maintenance, the program at the center would take over "ownership and control" and, depending on the client's indicated choice, put them in another woman, use them for research or dispose of them.
7. Bill C-13, Reprinted as amended by the Standing Committee on Health as a Working Copy for the use of the House of Commons at Report Stage and as Reported to the House on December 12, 2002.
8. Lisa Priest, "Embryo adoption program to offer infertile couples one last chance," *Globe and Mail*, March 4, 2002, A1.
9. National Academies of Science, *Scientific and Medical Aspects of Human Reproductive Cloning*.
10. The Nuremberg Code emerged from the war crimes trials following World War II. These trials exposed the cruelty and malevolent power exercised by Nazi

doctors over imprisoned soldiers, Jews, Roma and the mentally and physically incompetent. These doctors did research on these subjects without worrying that their procedures put their subjects at risk, without considering how their subjects might benefit, without even simple measures to reduce pain. The Nuremberg Code requires that research be systematic not capricious, that human subjects be protected from research that will put them at risk without the possibility of benefit and, above all, that research subjects must give informed and voluntary consent.

11. National Academies of Science, *Scientific and Medical Aspects of Human Cloning*, 2–5.
12. Aaron Zitner, "University to create human stem cells." *Toronto Star*, December 11, 2002, A4.

CHAPTER TEN: Dr. Watson, I Presume?

1. Barbara Ehrenreich, "Double Helix, Single Guy," review of *Genes, Girls, and Gamow*, by James D. Watson. *New York Times Book Review*, February 24, 2002, 6.
2. Jan Witkowski, personal correspondence, February 6, 2002.
3. See Return of Organization Exempt from Tax, Form 990, Cold Spring Harbor Laboratory, 1996, 1997, 1998, 1999, 2000; Robertson Research Fund, Inc., 1999. 2000; Cold Spring Harbor Laboratory Association, 2000.
4. See Browne, *Voyaging, and The Power of Place*, 1996, 2000.
5. Watson, *The Double Helix*, 166–67, 181.
6. Watson, *Genes, Girls, and Gamow*, 111.
7. Watson, *The Double Helix*, 69, 70.
8. Watson was quoted in the *New York Times* in 1993, the fortieth anniversary of the publication of the model of DNA, saying "I've had all these guilt feelings over the discovery. It's taken 40 years to feel I'm almost justified"; see Carol Strickland, "Watson relinquishes major role at lab," *New York Times*, March 21, 1993, LI1.
9. Ibid.
10. Comfort, *The Tangled Field*, 79.
11. Ibid., 66.

CHAPTER ELEVEN: Roots

1. Watson, *The Double Helix*, 4.
2. For much of this story of Darwin's life, I am indebted to Janet Browne and her magnificent two-volume work on Darwin's life, *Voyaging* and *The Power of Place*.
3. Browne, *Voyaging*, 80–82.
4. Ibid., 86–88.
5. Ibid., 88–90.
6. Ibid., 117–37.
7. Ibid., 209.
8. Ibid., 129.
9. Browne, *The Power of Place*, 23–27.
10. Browne, *Voyaging*, 227–29.
11. Ibid., 186–87.

12. Ibid., 336, 348.
13. Ibid., 288–95.
14. Ibid., 348.
15. Ibid., 349.
16. Browne, *The Power of Place*, 98.
17. Browne, *Voyaging*, 468.
18. Browne, *The Power of Place*, 32–38.
19. Browne, *Voyaging*, 502.
20. Adam Perkins, personal communication, February 14, 2002.
21. Charles Darwin to J. D. Hooker, May 9, 1856, Charles Darwin Collection, University Library, Cambridge, DAR 114 (2101).
22. Browne, *The Power of Place*, 14.
23. Darwin to Hooker, June 29, 1858, Charles Darwin Collection, DAR 114 (2290).
24. Darwin to Hooker, June 29, 1858, Charles Darwin Collection, DAR 114 (2298).
25. Browne, *The Power of Place*, 35.
26. Darwin to Hooker, July 5, 1858, Charles Darwin Collection, DAR 114 (2303).
27. Darwin to Hooker, July 13, 1858, Charles Darwin Collection, DAR 114 (2306).
28. Browne, *The Power of Place*, 44.
29. Ibid., 83.
30. Ibid., 111, 128.
31. Ibid., 111.
32. Darwin to Hooker, July 2, 1860, Charles Darwin Collection, DAR 115 (2853).
33. Darwin to Hooker, August 4, 1872, Charles Darwin Collection, DAR 94 (8449).
34. Francis Galton to Darwin, December 24, 1869, Charles Darwin Collection, DAR 105 (7034).
35. G. A. Gaskell to Darwin, November 13, 1878, Charles Darwin Collection, DAR 165 (11744).
36. G. A. Gaskell to Darwin, November 20, 1878, Charles Darwin Collection, DAR 165 (11752).

CHAPTER TWELVE: Descendants

1. Nicholas Wade, *Life Script: How the Human Genome Discoveries Will Transform Medicine and Enhance Your Health* (New York: Simon and Schuster, 2001): 15.
2. Frederick Sanger and Margaret Dowding, eds., *Selected Papers of Frederick Sanger: With Commentaries* (Singapore: World Scientific Publishing, 1996).
3. Ibid., xiii.
4. Sanger, "The Chemistry of Insulin," Nobel lecture, Dec. 1, 1968, *Selected Papers*, 161.
5. Susan Wright, "Recombinant DNA Technology and Its Social Transformation, 1972–1982," *Osiris*, 2nd ser., 2 (1986): 307.
6. Sanger, "The Chemistry of Insulin," 146.
7. DNA is a symmetrical molecule made up of two complementary strands. Organization of nucleotides along one sugar-phosphate backbone is complemented by the organization of nucleotides along the opposite backbone. Each backbone is made of alternating molecules of sugar (deoxyribose) and phosphate. The number of carbon atoms in a sugar molecule ring are numbered in chemical nomenclature from 1 to 5. The base pairs in the DNA helix are attached by hydrogen bonds to carbon atom number 1 in each sugar ring. The

phosphate is joined to carbon atom number 5. An OH molecule attaches to carbon atom number 3. The top of the left backbone has a phosphate group projecting off carbon atom number 5. At the bottom of this backbone there is an OH group projecting from carbon atom number 3. The partner backbone is organized in the reverse order. Sanger's polymerase always cut in the same direction—from the bottom.

8. Sanger, *Selected Papers*, 342.
9. Ibid., 344, Gilbert did reference Sanger's work in his Nobel lecture; Walter Gilbert, "DNA Sequencing and Gene Structure," Nobel e-Museum, http://www.noble.se/chemistry/laureates/1980/gilbert=lecture.html.
10. Sanger, *Selected Papers*, 346.
11. Ibid.
12. Ibid.
13. Ibid., 347.
14. Sulston and Ferry, *The Common Thread*, 91.
15. Ibid., 91–92.
16. Ibid., 60.
17. Ibid., 62.
18. Ibid., 60.
19. Ibid., 105–06.
20. Wade, *Life Script*, 35. See also Strickland, "Watson relinquishes major role at lab."
21. Sulston and Ferry, *The Common Thread*, 86.
22. Philip Hilts, "Head of Gene Map Threatens to Quit," *New York Times*, April 9, 1992, A26. See also Sulston and Ferry, *The Common Thread*, 89.
23. Wade, *Life Script*, 16.
24. Sulston and Ferry, *The Common Thread*, 93–98.
25. Ibid., 108.
26. Ibid., 144–48.
27. Wade, *Life Script*, 40.
28. Sulston and Ferry, *The Common Thread*, 162–64.
29. Ibid., 149–54.
30. Ibid., 221–22.
31. Ibid., 224.
32. Wade, *Life Script*, 67.
33. Brendan A. Maher, "Public-Private Genome Debate Resurfaces," *The Scientist* 16, no. 7 (April 1, 2002): 24.
34. Sulston and Ferry, *The Common Thread*, 224.
35. See The Wellcome Trust, Financial Framework, http://www.wellcome.ac.uk/en/1/awtviscorfra.html; Catalyst BioMedica Ltd, http://www.wellcome/ac.uk/en/1/awtviswhocat.html; Board of Governors, http://www.wellcome.ac.k/en/1/awtvtswhogov.html. For the rules governing the Trust, see The Companies Acts 1985 to 1989, Company Limited by Guarantee and Not Having a Share Capital, New Articles of Association of The Wellcome Trust Limited, 2000; The Companies Acts 1985–1989, Company Limited by Guarantee and Not Having a Share Capital, Memorandum of Association of The Wellcome Trust Limited, Incorporated 24 April, 1992, printed with changes made by resolutions dated 13 May, 1992, 1st June 1995, 10th February 1999, and 2000; The Charity Commissioners for England and Wales, Under the power given in the Charities

474 • ELAINE DEWAR

Act 1993 Order that from today, the 20th Feb. 2001, the following Scheme will
govern the charity formerly known as Will Charities of Sir Henry Wellcome (The
Research Undertaking Charity and the Museum & Library Charity) which shall
be known as the Wellcome Trust (210183), Case No. 82105.

CHAPTER THIRTEEN: Hello Dolly

1. David Hencke, "Treasury loses out on billions from cloning of Dolly," *The Guardian*, February 7, 2000, 3.
2. Roslin Institute, Annual Report, 1999/2000.
3. I. Wilmut, A. E. Schnicke, J. McWhir, A. J. Kind, and K. H. S. Campbell, "Viable Offspring Derived from Fetal and Adult Mammalian Cells," *Nature* 385 (1997): 810–13, as reproduced in Wilmut, Campbell, and Tudge, *The Second Creation*, 307.
4. Hencke, "Treasury loses out on billions from cloning of Dolly."
5. Ibid.
6. Geron Corporation, "Geron Acquires Roslin Bio-Med and Forms Research Collaboration with the Roslin Institute," May 4, 1999, http://www.geron.com.
7. Hencke, "Treasury loses out on billions from cloning of Dolly"; William Cash, personal interview, March 2002. The report was finally published as part of a larger inquiry in 2003. See note 13 below.
8. Wilmut, Campbell, and Tudge, *The Second Creation*, 112–21.
9. Ibid., 150.
10. Ibid., 132–35.
11. Jaenisch and Wilmut, "Don't Clone Humans!"
12. Wilmut, Campbell, and Tudge, 235–37.
13. Report by the Comptroller and Auditor General ordered by the House of Commons. *Reaping the Rewards of Agricultural Research*, January 20, 2003, 36.

CHAPTER FOURTEEN: Fresh from the New Garden

1. Http://www.sdv.fr/judaisme/israel/revel.htm. See also Haim Watzman and Susan Petersen Avitzour, "Israel's Great Expectations," *Nature Biotechnology* 17 (2001): 518–20; hadassaah.org/news/features/110501.htm.
2. Michel Revel, "Outright Condemnation of Cloning Research Is Premature," *The Scientist* 12 (January 19, 1998): 8.
3. A stem cell conference was held in Singapore in 2003 at Biopolis, a city within a city that is devoted to biotech.
4. Vogel, "In the Mideast, Pushing Back The Stem Cell Frontier."
5. Ibid.
6. Benjamin E. Reubinoff et al., "Neural Progenitors from Human Embryonic Stem Cells," *Nature Biotechnology* 19 (2001): 1134–40.
7. Vogel, "In the Mideast, Pushing Back the Stem Cell Frontier."
8. Izhak Kehat et al., "Human Embryonic Stem Cells Can Differentiate into Myocytes with Structural and Functional Properties of Cardiomyocytes," *Journal of Clinical Investigation* 108 (2001): 407–14.
9. Rappaport Family Institute for Research in the Medical Sciences, Institute Profile, 2002–2005, http://www.Techno.ac.Il/-rapinst.
10. K. L. Skorecki et al., "Y-Chromosomes of Jewish Priests," *Nature* 385 (1997): 32.

11. Suheir Assady et al., "Insulin Production by Human Embryonic Stem Cells," *Diabetes* 50 (2001): 1691–97.
12. Revel, "Outright Condemnation of Cloning Is Premature."
13. Israel Academy of Sciences and Humanities, Bioethics Advisory Committee, *The Use of Embryonic Stem Cells for Therapeutic Research*," August 2001, http://www.academy.ac.il/bioethics/index=e./html, 27.
14. Michel Revel, "Ethics and Genetics: Are Human Rights and Human Traditions Threatened by Scientific Progress?" *Here-Now4U Online Magazine*, http://www.here-now4u.ed/eng/ethics_and_genetics_are_human.htm.
15. Ibid.

CHAPTER FIFTEEN: Is It Safe?

1. Nick Kaplinsky et al., "Maize Transgene Results in Mexico Are Artefacts," *Nature Advance Online* (April 4, 2002); David Quist and Ignacio H. Chapela, Reply, *Nature Advance Online* (April 4, 2002).
2. Ibid.
3. David J. Sherratt, ed., *Mobile Genetic Elements* (Oxford: Oxford University Press, 1995), 1.
4. Ibid., Preface.
5. Zoltan Ivics et al., "Molecular Reconstruction of *Sleeping Beauty*, a *Tc1*-Like Transposon from Fish and Its Transposition in Human Cells," *Cell* 91 (1997): 501–10.
6. Press and Washburn, "The Kept University."
7. UC Berkeley/Novartis Agreement, as reproduced in Joint Hearing record of the Senate Committee on Natural Resources and Wildlife Senate and Select Committee on Higher Education, California Legislature, May 15, 2000.
8. McHughen, *Pandora's Picnic Basket*, 137–38.
9. See UC Berkeley/Novartis Agreement, Appendix A, List of Participating Faculty, 28.
10. Kaplinsky et al., "Maize Transgene Results In Mexico Are Artefacts."
11. Peter Rosset, Testimony, Joint Hearing, 77–80.
12. Joel Lexchin et al., "Pharmaceutical industry sponsorship and research outcome quality: systematic review," *British Medical Journal* 326 (2003): 1167.
13. Matthew Metz, Testimony, Joint Hearing, 27–32.
14. Gordon Rausser, Testimony, Joint Hearing, 43.
15. Metz, Written Testimony, Joint Hearing. 2.
16. Berg, "The Eighth Feodor Lynen Lecture."
17. Rogers, *Biohazard*, 34.
18. Ibid., 69–70.
19. Paul Berg et al., "Potential Biohazards of Recombinant DNA Molecules," *Science* 185 (1974): 303.
20. Paul Berg et al., "Asilomar Conference on Recombinant DNA Molecules," *Science* 188 (1975): 991–94.
21. Maxine F. Singer and Paul Berg, "Recombinant DNA: NIH Guidelines," *Science* 189 (1976): 186–88.
22. Kathleen Stratton, Donna A. Almario, and Marie C. McCormick, eds. *Immunization Safety Review: SV40 Contamination of Polio Vaccine and Cancer,*

Institute of Medicine of the National Academies, October 22, 2002, http://www.iom.edu/report.asp?id=4317.

CHAPTER SIXTEEN: Computing Life

1. Wade, *Life Script*, 66–67.
2. Sulston and Ferry, *The Common Thread*, 224.
3. Ibid.
4. W. James Kent and Alan M. Zahler, "Conservation, Regulation, Synteny, and Introns in a Large-Scale *C. briggsae-C. elegans* Genomic Alignment," *Genome Research* 10 (2000): 1115–25.
5. W. James Kent and David Haussler, "Assembly of the Working Draft of the Human Genome with GigAssembler," *Genome Research* (2001): 1541–48.
6. His paper on this subject was presented at a Cold Spring Harbor conference on Genome Sequencing and Biology in May 2000; see J. Kent et al., "Alternative Splicing of Human Genes," *Genome Sequencing and Biology*, Cold Spring Harbor, May 2000.
7. Stephen Wolfram, *A New Kind of Science* (Champaign, IL: Wolfram Media, 2002), 859.
8. Ibid., 1102.
9. Ibid., 1102, 1105.
10. See A. Krogh et al., "Hidden Markov Models in Computational Biology: Applications to Protein Modeling," *Journal of Molecular Biology* 235 (1994): 1501–31.

CHAPTER SEVENTEEN: The Sovereign State of Biology

1. Francoise Baylis, "CIHR Guidelines for Human Pluripotent Stem Cell Research," National Press Theatre, Ottawa, March 4, 2002, http://www.cihr.ca/president/speeches/20020304speech_e.shtml; CIHR, *Human Pluripotent Stem Cell Research: Guidelines for CIHR-Funded Research*, 2002, http://www.cihr.ca/about_cihr/ethics/stem_cell/stem_cell_guidelines_e.shtml.
2. Alan Bernstein, Testimony to the Standing Committee on Health, April 17, 2002, http://www.parl.gc.ca/inetimages/CresBlues.gif.
3. See J. E. Dick et al., "Introduction of a Selectable Gene into Primitive Stem Cells Capable of Long-Term Reconstitution of the Hematopoietic System of W/W/Mice," *Cell* 42 (1985): 71–79.
4. In 2001 there were thirteen institutes of health research within the CIHR: Aboriginal Peoples' Health; Cancer Research; Circulatory and Regulatory Health; Gender and Health; Genetics; Health Services and Policy Research; Healthy Aging; Human Development, Child and Youth Health; Infection and Immunity; Musculoskeletal Health and Arthritis; Neurosciences, Mental Health and Addiction; Nutrition, Metabolism and Diabetes; and Population and Public Health.
5. Campbell Clark, "Cloning of embryos considered: Federal Health Department may allow human cells to be grown for transplant," *Globe and Mail*, January 24, 2001, A5.
6. Bernstein, Testimony.

7. Health Canada, *Proposals for Legislation Governing Assisted Human Reproduction* (Ottawa: Health Canada, May 2001), sec. 35a, 35b.

8. Canada, House of Commons, *An Act to Establish the Canadian Institutes of Health Research, to Repeal the Medical Research Council Act and to Make Consequential Amendments to Other Acts*, 2000, sec. 6.

9. See Public Registry, http://www.strategis.ic.gc.ca/epic/internet/inoec=bce.nsf/vwGeneratedInterE/h_0e01270e.html.

10. Conflict of Interest Code, Public Registry.

11. Alan Bernstein, Declaration, May 30, 2001, Offices of the Ethics Councillor, http://strategis.gc.ca and curriculum vitae.

12. Timothy Caulfield et al., "MPs have the wrong focus," *Globe and Mail*, May 2, 2002, A21.

13. See the center's website at http://www.utoronto.ca/jcb/ source_funding.html.

14. Francoise Baylis, "Parliament's call for moratorium on stem cell research perplexing," *Hill Times*, May 6, 2002. See www.stemcellnetwork.ca/news/articles.php/id=47

15. Baylis, "A Reflection on the 'Place' of Bioethics."

16. See Canadian Institutes of Health Research, "Minister McLellan Announces an Investment of $88 Million to Train Next Generation of Health Researchers," http://www.cihr-irsc.gc.ca/news/press_releases/2002/pr-0217_e.shtml.

17. Elections Canada, Contributions and Expenses, 37th General Election, November 27, 2000, Brown, Bonnie, Oakville.

18. *An Act Respecting Assisted Human Reproduction*, sec. 5b.

19. Paul Szabo, *The Ethics and Science of Stem Cells* (Ottawa: Paul Szabo, 2002).

CHAPTER EIGHTEEN: A Noble Prize

1. Reuters, "Worm world wins Nobel," *Toronto Star*, October 8, 2002, A15.

2. Pall Corporation, Notice of Annual Meeting of Shareholders, 2002, 11.

3. Wade, *Life Script*, 33.

4. Ibid., 82.

5. Sydney Brenner, "The Tale of the Human Genome," *Nature* 416 (2002): 793–94.

6. Brenner, curriculum vitae. Http://www.nobels.se/medicine/laureates/2002/brenner-cv.html

LOOSE ENDS AND ACKNOWLEDGMENTS

1. Wendy Goldman Rohm, "Seven Days of Creation," *Wired*, January 2004, 120

2. Woo Suk Hwang, Young June Ryu, Jong Hyuk Park, Eul Soon Park, Eu Gene Lee, Ja Min Koo, Hyun Yun Jong Chun, Beyeong Chun Lee, Sung Keun Kang, Sun Jong Kim, Curie Ahn, Jung Hye Hwang, Ky Young Park, Jose B. Cibelli, Shin Yong Moon. "Evidence of a Pluripotent Human Embryonic Stem Cell Line Derived from a Cloned Blastocyst," *Scienceexpress*, www.sciencemag.org/12 February2004/Page1/10.1126/science.1094515

BIBLIOGRAPHY

BOOKS

Berg, Paul, and Maxine Singer. *Dealing with Genes: The Language of Heredity.* Mill Valley, CA: University Science Books, 1992.

Bowler, Peter J. *Evolution: The History of an Idea.* Berkeley: University of California Press, 1989.

Browne, Janet. *Charles Darwin: The Power of Place.* New York: Knopf, 2002.

———. *Charles Darwin: Voyaging.* New York: Knopf, 1995.

Carlson, Elof Axel, ed. *Man's Future Birthright: Essays on Science and Humanity,* by Hermann J. Muller. 1958. Albany: State University of New York Press, 1973.

Comfort, Nathaniel C. *The Tangled Field: Barbara McClintock's Search for the Patterns of Genetic Control.* Cambridge: Harvard University Press, 2001.

Crick, Francis. *The Astonishing Hypothesis: The Scientific Search for the Soul.* London: Touchstone, 1995.

———. *Life Itself: Its Origin and Nature.* New York: Simon and Schuster, 1981.

Davenport, Charles. *Heredity in Relation to Eugenics.* London: Williams and Norgate, 1912.

Davies, Kevin. *The Sequence: Inside the Race for the Human Genome.* London: Phoenix, 2002.

Edelson, Edward. *Francis Crick and James Watson and the Building Blocks of Life.* New York: Oxford University Press, 1998.

Galton, Francis. *English Men of Science: Their Nature and Nurture.* London: Macmillan, 1974.

———. *Hereditary Genius: An Inquiry into Its Laws and Consequences.* London: Macmillan, 1869.

———. *Inquiries into Human Faculty and Its Development.* London: Macmillan, 1883.

Gould, Stephen Jay. *The Lying Stones of Marrakech: Penultimate Reflections in Natural History.* New York: Harmony, 2000.

———. *Ontogeny and Phylogeny.* Cambridge, MA: Belknap, 1977.

Habermas, Jürgen. *The Future of Human Nature.* Cambridge, UK: Polity, 2003.

Hanson, Earl D., ed. *Recombinant DNA Research and the Human Prospect: A Sesquicentennial Symposium of Wesleyan University.* Washington, DC: American Chemical Society, 1983.

Henig, Robin Marantz. *A Monk and Two Peas: The Story of Gregor Mendel and the Discovery of Genetics.* London: Phoenix, 2001.

Judson, Horace Freeland. *The Eighth Day of Creation: Makers of the Revolution in Biology.* Exp. ed. Plainview, NY: Cold Spring Harbor Laboratory Press, 1996.

Kevles, Daniel J., and Leroy Hood, eds. *The Code of Codes: Scientific and Social Issues in the Human Genome Project.* Cambridge: Harvard University Press, 1992.

Kitcher, Philip. *Science, Truth, and Democracy.* Oxford: Oxford University Press, 2001.

Lee, Thomas F. *The Human Genome Project: Cracking the Genetic Code of Life.* New York: Plenum, 1991.

McGee, Glenn, ed. *The Human Cloning Debate.* Berkeley, CA: Berkeley Hills Books, 2000.

McHughen, Alan. *Pandora's Picnic Basket: The Potential and Hazards of Genetically Modified Foods.* Oxford: Oxford University Press, 2000.

Morange, Michel. *A History of Molecular Biology.* Translated by Matthew Cobb. Cambridge: Harvard University Press, 2000.

Nossal, G. J. V., and Ross L. Coppel. *Reshaping Life: Key Issues in Genetic Engineering* 2nd ed. Cambridge: Cambridge University Press, 1989.

Rogers, Michael. *Biohazard.* New York: Knopf, 1977.

Rossant, Janet, and Roger A. Pederson, eds. *Experimental Approaches to Mammalian Embryonic Development.* Cambridge: Cambridge University Press, 1986.

Sager, Ruth. *Cytoplasmic Genes and Organelles.* New York: Academic Press, 1972.

Sanger, Frederick, and Margaret Downing, eds. *Selected Papers of Frederick Sanger: With Commentaries.* Singapore: World Scientific Publishing, 1996.

Schrödinger, Erwin. *What Is Life? with Mind and Matter and Autobiographical Sketches.* Cambridge: Cambridge University Press, 2000.

Sherratt, David J., ed. *Mobile Genetic Elements.* Oxford: Oxford University Press, 1995.

Silver, Lee M. *Remaking Eden: How Genetic Engineering and Cloning Will Transform the American Family.* New York: Avon, 1997.

Singer, Peter. *Writings on an Ethical Life.* New York: Ecco, 2000.

Singh, Simon. *Fermat's Enigma: The Epic Quest to Solve the World's Greatest Mathematical Problem.* Toronto: Penguin, 1998.

Stock, Gregory, and John Campbell, eds. *Engineering the Human Germline: An Exploration of the Science and Ethics of Altering the Genes We Pass to Our Children.* New York: Oxford University Press, 2000.

Strathearn, Paul. *The Big Idea: Crick, Watson and DNA.* New York: Anchor, 1997.

Sulston, John, and Georgina Ferry. *The Common Thread: A Story of Science, Politics, Ethics and the Human Genome.* London: Bantam, 2002.

Sykes, Bryan. *The Seven Daughters of Eve.* New York: Norton, 2001.

Szabo, Paul. *The Ethics and Science of Stem Cells.* Ottawa: Paul Szabo, 2002.

Wade, Nicholas. *Life Script: How the Human Genome Discoveries Will Transform Medicine and Enhance Your Health.* New York: Simon and Schuster, 2001.

Watson, James D. *The Double Helix: A Personal Account of the Discovery of the Structure of DNA.* New York: Scribner, 1998.

———. *Genes Girls, and Gamow.* New York: Knopf, 2001.

Wilmut, Ian, Keith Campbell, and Colin Tudge. *The Second Creation: Dolly and the Age of Biological Control.* New York: Farrar, Straus and Giroux, 2000.

Wolfram, Stephen. *A New Kind of Science.* Champaign, IL: Wolfram Media, 2002.

ARTICLES

Adams, Mark D., Jenny M. Kelley, Jeannine D. Gocayne, Mark Cubnick, Michael H. Polymeropoulos, Hong Xiao, Carl. R. Merril, Andrew Wu, Bjorn Olde, Ruben F. Moeno, Anthony R. Kerlavage, W. Richard McCombie, and J. Craig Venter. "Complementary DNA Sequencing: Expressed Sequence Tags and Human Genome Project." *Science* 252 (1991): 1651–56.

Alexandre, Henri. "A History of Mammalian Embryological Research." *International Journal of Developmental Biology* 45 (2001): 457–67.

Allen, Garland E. "The Eugenics Record Office at Cold Spring Harbor, 1910–1940: An Essay in Institutional History." *Osiris*, 2nd ser., 2 (1986): 225–64.

Arney, Katherine L., Sylvia Erhardt, Robert A. Drewell, and M. Azim Surani. "Epigenetic Reprogramming of the Genome: From the Germ Line to the Embryo and Back Again." *International Journal of Developmental Biology* 45 (2001): 533–39.

Assady, Suheir, Gila Maor, Michal Amit, Joseph Itskovitz-Eldor, Karl L. Skorecki, and Maty Tzukerman. "Insulin Production by Human Embryonic Stem Cells." *Diabetes* 50 (2001): 1691–97.

Belcher, C. J. E., A. W. Tucker, J. A. Bell, W. McDonell, and D. G. Butler. "High Welfare Production of Gnotobiotic Pigs for Solid Organ Xenotransplantation Using a Modified Isolator Technique." *Animal Technology* 52, no. 2 (2001): 85–95.

Berg, Paul. "The Eighth Feodor Lynen Lecture: Biochemical Pastimes . . . And Future Times." In *Molecular Cloning of Recombinant DNA*, ed. W. A. Scott and R. Werner. Miami Winter Symposia 13. New York: Academic Press, 1977.

———. "Reflections on Asilomar 2 at Asilomar 3 Twenty-Five Years Later." *Perspectives in Biology and Medicine* 44, no. 2 (2001):183–85.

Bhardwaj, G., B. Murdoch, D. Wu, D. P. Baker, K. P. Williams, K. Chadwick, L. E. Ling, F. N. Karanu, and M. Bhatia. "Sonic Hedgehog Induces the Proliferation of Primitive Human Hematopoietic Cells via BMP Regulation." *Nature Immunology* 2, no. 2 (2001): 172–80.

Biggers, John D. "Research in the Canine Block." *International Journal of Developmental Biology* 45 (2001): 469–76.

Blackburn, Elizabeth H. "The End of the (DNA) Line." *Nature Structural Biology* 7 (2000): 847–49.

———. "Structure and Function of Telomeres." *Nature* 350 (1991): 569–73.

———. "Telomere States and Cell Fates." *Nature* 408 (2000): 53–56.

Blasco, Maria A., Han-Woong Lee, M. Prakash Hande, Enrique Samper, Peter M. Lansdorp, Ronald A. DePinho, and Carol W. Greider. "Telomere Shortening and Tumor Formation by Mouse Cells Lacking Telomerase RNA." *Cell* 91 (1997): 25–34.

Bodnar, Andrea G., Michel Ouellette, Maria Frolkis, Shawn E. Holt, Choy-Pik Chiu, Gregg B. Morin, Calvin B. Harley, Jerry W. Shay, Serge Lichsteiner, and Woodring E. Wright. "Extension of Life-Span by Introduction of Telomerase into Normal Human Cells." *Science* 179 (1998): 349–352.

Braude, Peter. "Preimplantation Genetic Diagnosis and Embryo Research: Human Developmental Biology in Clinical Practice." *International Journal of Developmental Biology* 45 (2001): 607–12.

Cantor, Charles R. "Orchestrating the Human Genome Project." *Science* 248 (1990): 49–51.

Chang, Annie C. Y., and Stanley N. Cohen. "Genome Construction between Bacterial Species in Vitro: Replication and Expression of Staphylococcus Plasmid Genes in *Escherichia coli*." *Proceedings of the National Academy of Sciences USA* 74 (1974): 1030–34.

Cheung, V. G., et. al. "Integration of Cytogenetic Landmarks into the Draft Sequence of the Human Genome." *Nature* 409 (2001): 953–58.

Cibelli, Jose B., Ann A. Kiessling, Kerrianne Cunniff, Charlotte Richards, Robert P. Lanza, and Michael D. West. "Somatic Cell Nuclear Transfer in Humans: Pronuclear and Early Embryonic Development." *e-biomed: The Journal of Regenerative Medicine* 2 (2001): 25–31.

Clarke, Ann G. "McLaren: A Tribute from Her Research Students." *International Journal of Developmental Biology* 45 (2001): 491–95.

Cohen, Stanley N., Annie C. Y. Chang, Herbert W. Boyer, and Robert B. Helling. "Construction of Biologically Functional Bacterial Plasmids in Vitro." *Proceedings of the National Academy of Sciences* 70 (1973): 3240–44.

Cohen, Stanley N., Felipe Cabello, Malcolm Casadaban, Annie C. Y. Chang, and Kenneth Timmis. "DNA Cloning and Plasmid Technology." In *Molecular Cloning of Recombinant DNA*, ed. W. A. Scott and R. Werner. Miami Winter Symposia 13. New York: Academic Press, 1977.

Cooke, H. J., and B. A. Smith. "Variability at the Telomeres of Human X/Y Pseudoautosomal Region." *Cold Spring Harbor Symposia in Quantitative Biology* 82 (1986): 213–19.

Corbo, Joseph, Anna Di Gregorio, and Michael Levine. "The Ascidian as a Model Organism in Developmental and Evolutionary Biology." *Cell* 106 (2001): 535–38.

Cosso, Giulio, Luciana de Angelis, Ugo Borello, Barbara Berarducci, Viviana Buffa, Claudia Sonnino, Marcello Coletta, Elisabeta Vivarelli, et al. "Determination, Diversification and Multipotency of Mammalian Myogenic Cells." *International Journal of Developmental Biology* 44 (2000): 699–706.

Counter, Christopher M., Ariel A. Avilion, Chatering E. LeFeuvre, Nancy G. Stewart, Carol W. Greider, Calvin B. Harley, and Silvia Bacchetti. "Telomere Shortening Associated with Chromosome Instability Is Arrested in Immortal Cells Which Express Telomerase Activity." *The EMBO Journal* 11 (1992): 1921–29.

Cozzi, Emanuele, Alexander W. Tucker, Gillian A. Langford, Gilda Pino-Chavez, Les Wright, Mary-Jane O'Connell, Vincent J. Young, et al. "Characterization of Pigs Transgenic for Human Decay-Accelerating Factor." *Transplantation* 64 (1997): 1383–92.

Crick, F. H. C., Leslie Barnett, S. Brenner, and R. J. Watts-Tobin. "General Nature of the Genetic Code for Proteins." *Nature* 192 (1963): 1227–32.

De Felici, Massimo. "Twenty Years of Research on Primordial Germ Cells." *International Journal of Developmental Biology* 45 (2001): 519–22.

De Felici, Massimo, and Gregorio Siracusa. "The Rise of Embryology in Italy: From the Renaissance to the Early 20th Century." *International Journal of Developmental Biology* 44 (2000): 515–21.

de Lange, Titia. "Telomeres and Senescence: Ending the Debate." *Science* 279 (1998): 334–35.

Dick, J. E., M. C. Magli, D. Huszar, R. A. Phillips, and A. Bernstein. "Introduction of a Selectable Gene into Primitive Stem Cells Capable of Long-Term

Reconstitution of the Hematopoietic System of W/W Mice." *Cell* 42 (1985): 71–79.

Donovan, Peter J. "High Oct-ane Fuel Powers the Stem Cell." *Nature Genetics* 29 (November 2001): 246–47.

Donovan, Peter J., Maira P. De Miguel, Masami P. Hirano, Melanie S. Parsons and A. Jeanine Lincoln. "Germ Cell Biology: From Generation to Generation." *International Journal of Developmental Biology* 45 (2001): 523–31.

Doolittle, W. Ford, and Carmen Sapienza. "Selfish Genes, the Phenotype Paradigm and Genome Evolution." *Nature* 284 (1980): 601–3.

Elliott, Carl. "Pharma Buys a Conscience." *The American Prospect* 12, no. 17 (2001).

Fantini, Bernardino. "The 'Stazione Zoological Anton Dohrn' and the History of Embryology." *International Journal of Developmental Biology* 44 (2000): 523–35.

Fraser, Claire M., and Malcolm R. Dando. "Genomics and Future Biological Weapons: The Need for Preventive Action by the Biomedical Community." *Nature Genetics* 29 (November 2001): 253–56.

Fredrickson, Donald S. "The First Twenty-Five Years after Asilomar." *Perspectives in Biology and Medicine* 44, no. 2 (2001): 170–82.

Furey, Terence S., Neilo Cristianini, Nigel Duffy, David W. Bednarski, Michel Schummer, and David Haussler. "Support Vector Machine Classification and Validation of Cancer Tissue Samples Using Microarray Expression Data." *Bioinformatics* 16 (2000): 906–14.

Giudice, Giovanni. "From a Home-Made Laboratory to the Nobel Prize: An Interview with Rita Levi-Montalcini." *International Journal of Developmental Biology* 44 (2000): 563–66.

Glass, Kathleen Cranley, and Trudo Lemmens. "Conflict of Interest and Commercialization of Biomedical Research." In *The Commercialization of Genetic Research: Ethical, Legal and Policy Issues,* edited by Timothy A. Caulfield and Bryn William-Jones. New York: Kluwer Academic/Plenum, 1999.

Goldberg, Robert B. "From Cot Curves to Genomics: How Gene Cloning Established New Concepts in Plant Biology." *Plant Physiology* 125 (January 2001): 4–8.

Greider, Carol W. "Telomerase Activation: One Step on the Road to Cancer." *Trends in Genetics* 15, no. 3 (1999): 109–12.

———. "Telomerase and Telomere-Length Regulation: Lessons from Small Eukaryotes to Mammals." *Cold Spring Harbor Symposia on Quantitative Biology* 58 (1993): 719–23.

———. "Telomeres and Senescence: The History, the Experiment, the Future." *Current Biology* 8 (1998): R178–81.

———. "Telomeres, Telomerase and Senescence." *BioEssays* 12 (1990): 363–69.

Greider, Carol W., and Elizabeth Blackburn. "Identification of a Specific Telomere Terminal Transferase Activity in Tetrahymena Extracts." *Cell* 43, part 1 (December 1985): 405–13.

———. "Telomeres, Telomerase and Cancer." *Scientific American* 274, no. 2 (1996): 80–85.

Hammer, Michael F., Karl Skorecki, Sara Selig, Shraga Blazer, Bruce Rappaport, Robert Bradman, Neil Bradman, P. J. Waburton, and Monic Ismajlowicz. "Y Chromosomes of Jewish Priests." *Nature* 385 (1997): 32.

Harley, Calvin B., A. Bruce Futcher, and Carol Greider. "Telomeres Shorten during Ageing of Human Fibroblasts." *Nature* 345 (1990): 458–60.

Harley, C. B., N. W. Kim, K. R. Prowse, S. L. Weinrich, K. S. Hirsch, M. D. West, S. Bacchetti, et al. "Telomerase, Cell Immortality and Cancer." *Cold Spring Harbor Symposia on Quantitative Biology* 59 (1994): 307–15.

Hemann, M. T., J. Hackett, A. Ijpma, and C. W. Greider. "Telomere Length, Telomere-Binding Proteins, and DNA Damage Signaling." *Cold Spring Harbor Symposia on Quantitative Biology* 65 (2000): 275–79.

Herndon, Laura A., Peter J. Schmeissner, Justyna M. Dudoranek, Paula A. Brown, Kristin M. Listner, Yuko Sakano, Marie C. Paupard, David H. Hall, and Monica Driscoll. "Stochastic and Genetic Factors Influence Tissue-Specific Decline in Ageing *C. elegans.*" *Nature* 419 (2002): 808–14.

Hogan, Brigid. "From Embryo to Ethics: A Career in Science and Social Responsibility: An Interview with Anne McLaren." *International Journal of Developmental Biology* 45 (2001): 477–82.

Hutchison, Clyde A., III, Sandra Phillips, Marshall H. Edgel, Shirley Gillam, Patricia Jahnke, and Michael Smith. "Mutagenesis at a Specific Position in a DNA Sequence." *Journal of Biological Chemistry* 253 (1978): 6551–60.

International Human Genome Mapping Consortium. "A Physical Map of the Human Genome." *Nature* 409 (2001): 934–41.

Irvine, S. H. "Contributions of Ability and Attainment Testing in Africa to a General Theory of Intellect." *Journal of Biosocial Science* suppl. 1, *Proceedings of the Fifth Annual Symposium of the Eugenics Society, London, September 1968* (1969): 91–102.

Irving, Dianne. "When Do Human Beings Begin? 'Scientific' Myths and Scientific Facts." *International Journal of Sociology and Social Policy* 19, nos. 3/4 (1999): 22–47.

Ivics, Zoltan, P. B. Hackett, R. H. Plasterk, and Z. Izsvak. "Molecular Reconstruction of *Sleeping Beauty,* a *Tc1*-Like Transposon from Fish and Its Transposition in Human Cells." *Cell* 91 (1997): 501–10.

Jackson, David A., Robert H. Symons, and Paul Berg. "Biochemical Method for Inserting New Genetic Information into DNA of Simian Virus 40: Circular SV40 DNA Molecules Containing Lambda Phage Genes and the Galactose Operon of *Escherichia coli.*" *Proceedings of the National Academy of Sciences USA* 60 (1972): 2004–2909.

Jessberger, R., and Paul Berg. "Repair of Deletions and Double-Strand Gaps by Homologous Recombination in a Mammalian in Vitro System." *Molecular Cell Biology* 11 (1991): 445–57.

Karanu, Frances N., Barbara Murdoch, Lisa Gallacher, Dongmei M. Wu, Masahide Koremoto, Seiji Sakano, and Mickie Bhatia. "The Notch Ligand Jagged-1 Represents a Novel Growth Factor of Human Hematopoietic Stem Cells." *Journal of Experimental Medicine* 192 (2000): 1365–72.

Kehat, Izhak, Dorit Kenyagin-Karsenti, Mirit Snir, Hana Segev, Michal Amit, Amira Gepstein, Erella Livne, Ofer Binah, Joseph Itskovitz-Eldor, and Lior Gepstein. "Human Embryonic Stem Cells Can Differentiate into Myocytes with Structural and Functional Properties of Cardiomyocytes." *Journal of Clinical Investigation* 108 (2001): 407–14.

Kent, W. James, and David Haussler. "Assembly of the Working Draft of the Human Genome with GigAssembler." *Genome Research* 11 (2001): 1541–48.

THE SECOND TREE · 485

Kent, W. James, and Alan M. Zahler. "Conservation, Regulation, Synteny, and Introns in a Large-Scale C. briggsae-C. elegans Genomic Alignment." *Genome Research* 10 (2000): 1115–25.

King, Mary-Claire, and A. C. Wilson. "Evolution at Two Levels in Humans and Chimpanzees." *Science* 188 (1975): 107–16.

Kirk, Karen E., Brian P. Harmon, Isabel K. Reichardt, John W. Sedat, and Elizabeth H. Blackburn. "Block in Anaphase Chromosome Separation Caused by a Telomerase Template Mutation." *Science* 275 (1997): 1478–81.

Koonin, Steven E. "An Independent Perspective on the Human Genome Project." *Science* 279 (1998): 36–37.

Koopman, Peter. "In Situ Hybridization to mRNA: From Black Art to Guiding Light." *International Journal of Developmental Biology* 45 (2001): 619–22.

Krogh, A., M. Brown, S. Mian, and K. Sjolander. "Hidden Markov Models in Computational Biology: Applications to Protein Modeling." *Journal of Molecular Biology* 235 (1994): 1501–31.

Lansdorp, Peter. "Self-Renewal of Stem Cells." *Biology of Blood and Marrow Transplantation* 3 (1997): 171–78.

Lanza, Robert P., Jose B. Cibelli, Catherine Blackwell, Vincent J. Christofalo, Mary Kay Francis, Gabriela M. Baerlocher, Jennifer Mak, et al. "Extension of Cell Life-Span and Telomere Length in Animals Cloned from Senescent Somatic Cells." *Science* 288 (2000): 665–69.

Lemmens, Trudo, and Benjamin Freedman. "Ethics Review for Sale? Conflict of Interest and Commercial Research Review Boards." *Millbank Quarterly* 78 (2000): 547–83.

Lemmens, Trudo, and Paul Miller. "Avoiding a Jekyll-and-Hyde Approach to the Ethics of Clinical Research and Practice." *American Journal of Bioethics* 2, no. 2 (2002): 14–17.

Lemmens, Trudo, and Alison Thompson. "Genetic Testing in the Workplace. 'What about Your Genes?' Ethical, Legal, and Policy Dimensions of Genetics in the Workplace." *Politics and the Life Sciences* 16 (March 1997): 57–75.

———. "Noninstitutional Commercial Review Boards in North America." *IRB: Ethics and Human Research* 23, no. 2 (2001): 1–12.

———. "Private Parties, Public Duties? The Shifting Role of Insurance Companies in the Genetics Era." *Genetic Information Acquisition, Access, and Control*, edited by Alison K. Thompson and Ruth F. Chadwick. New York: Kluwer Academic/Plenum, 1999.

———. "Selective Justice, Genetic Discrimination, and Insurance: Should We Single Out Genes in Our Laws?" *McGill Law Journal* 45 (2000): 347–412.

Lexchin, Joel, Lisa A. Bevo, Benjamin Djulbegovic, Otavio Clark. "Pharmacutical industry sponsorship and research outcome and quality: systematic review," *British Medical Journal* 326 (2003): 1167.

Merlo, Giorgio R., Barbara Zerega, Laura Paleari, Sonya Trombino, Stefan Mantero, and Giovanni Levi. "Multiple Functions of Dix Genes." *International Journal of Developmental Biology* 44 (2000): 619–26.

Meyerowitz, Elliot. "Prehistory and History of Arabidopsis Research." *Plant Physiology* 125 (2001): 15–19.

Monk, Marilyn. "Of Microbes, Mice and Man." *International Journal of Developmental Biology* 45 (2001): 497–508.

Morrow, John F., Stanley H. Cohen, Annie C. Y. Chang, Herbert W. Boyer, Howard M. Goodman, and Robert B. Helling. "Replication and Transcription of Eukaryotic DNA in *Escherichia coli.*" *Proceedings of the National Academy of Sciences* 71 (1974): 1743–47.

Nagy, Andras, and Janet Rossant. "Chimaeras and Mosaics for Dissecting Complex Mutant Phenotypes." *International Journal of Developmental Biology* 45 (2001): 577–82.

Nagy, Andras, Janet Rossant, Reka Nagy, Wanda Abramow-Newerly, and John C. Roder. "Derivation of Completely Cell Culture–Derived Mice from Early-Passage Embryonic Stem Cells." *Proceedings of the National Academy of Sciences* 90 (1993): 8424–28.

Nakamura, T. M., G. B. Morin, K. B. Chapman, S. L. Weinrich, W. H. Andrews, J. Lingner, C. B. Harley, and T. R. Cech. "Telomerase Catalytic Homologs from Fission Yeast and Human." *Science* 227 (1997): 955–59.

Nakatsuji, Norio, and Shinichiro Chuma. "Differentiation of Mouse Primordial Germ Cells into Female or Male Germ Cells." *International Journal of Developmental Biology* 45 (2001): 541–48.

Namsaraev, Eugeni A., and Paul Berg. "Rad51 Uses One Mechanism to Drive DNA Strand Exchange in Both Directions." *Journal of Biological Chemistry* 275 (2000): 3970–76.

Novitski, C. E. "Another Look at Some of Mendel's Results." *Journal of Heredity* 86, no. 1 (1995): 62–66.

Orgel, L. E., and F. H. C. Crick. "Selfish DNA: The Ultimate Parasite." *Nature* 284 (1980): 604–7.

Papaioannou, Virginia E. "The McLaren Effect: A Personal View." *International Journal of Developmental Biology* 45 (2001): 483–86.

Paria, Biblash C., Huengseok Song, and Sudhansev K. Dey. "Implantation: Molecular Basis of Embryo-Uterine Dialogue." *International Journal of Developmental Biology* 45 (2001): 597–606.

Reese, Martin G., David Kulp, Hari Tammana, and David Haussler. "Gene Finding in *Drosophila melanogaster.*" *Genome Research* 10 (2000): 529–38.

Reubinoff, Benjamin E., Pavel Itsykson, Tikva Turetsky, Martin F. Pera, Etti Reinhartz, Anna Itzik, and Tamir Ben-Hur. "Neural Progenitors from Human Embryonic Stem Cells." *Nature Biotechnology* 19 (2001): 1134–40.

Rex, John. "Race as a Social Category." *Journal of Biosocial Science,* suppl. 1, Proceedings of the Fifth Annual Symposium of the Eugenics Society, London, September 1968, edited by G. A. Harrison and John Peel (1969): 145–52.

Roberts, D. F. "Race, Genetics and Growth." *Journal of Biosocial Science,* suppl. 1, Proceedings of the Fifth Annual Symposium of the Eugenics Society, London, September 1968, edited by G. A. Harrison and John Peel (1969): 43–67.

Schechter, Alan N., and Robert L. Perlman. "Editors' Introduction to the Symposium on the 25th Anniversary of the Asilomar Conference." *Perspectives in Biology and Medicine* 44, no. 2 (2001): 159–61.

Sherwood, Peter. "The Yeast Genetics Course at Cold Spring Harbor Laboratory: Thirty Years and Counting." *Genetics* 157 (2001): 1399–1402.

Singer, Maxine. "What Did the Asilomar Exercise Accomplish, What Did It Leave Undone?" *Perspectives in Biology and Medicine* 44, no. 2 (2001): 186–91.

Singer, Peter A., and Abdallah S. Daar. "Harnessing Genomics and Biotechnology to Improve Global Health Equity." *Science* 294 (2001): 87–89.

Smith, Christopher D., and Elizabeth Blackburn. "Uncapping and Deregulation of Telomeres Lead to Detrimental Cellular Consequences in Yeast." *Journal of Cell Biology* 145, no. 2 (1999): 203–14.

Toma, Jean, Mahnaz Aknavan, Karl J. L. Fernandes, Fanie Barnabé-Heider, Abbas Sadikot, David R. Kaplan, and Freda D. Miller. "Isolation of Multipotent Adult Stem Cells from the Dermis of Mammalian Skin." *Nature Cell Biology* 3, no. 9 (2001): 778–84.

Tsang, Tania E., Poh-Lynn Khoo, Robyn V. Jamieson, Sheila X. Zhou, Siew-Lan Ang, Richard Behringer, and Patrick P. L. Tam. "The allocation and Differentiation of Mouse Primordial Germ Cells." *International Journal of Developmental Biology* 46 (2001): 540–55.

Venter, J. Craig, Mark D. Adams, Eugene W. Myers, Peter W. Li, Richard J. Mural, Granger G. Sutton, Hamilton O. Smith, et al. "The Sequence of the Human Genome." *Science* 291 (2001): 1304. http://www.sciencemag.org/cgi/content/-full/291/5507/1304.

Vercoutere, Wenonah, Stephen Winters-Hilt, Hugh Olsen, David Dearner, David Haussler, and Mark Akeson. "Rapid Discrimination among Individual DNA Hairpin Molecules at Single-Nucleotide Resolution Using an Ion Channel." *Nature Biotechnology* 19 (March 2001): 248–52.

Wakayama, Teruhiko, Viviane Tabar, Ivan Rodriguez, Anthony C. F. Perry, Lorenz Studer, and Peter Mombaerts. "Differentiation of Embryonic Stem Cell Lines Generated from Adult Somatic Cells by Nuclear Transfer." *Science* 292 (2001): 740–42.

Warnock, Mary. "Anne McLaren as Teacher." *International Journal of Developmental Biology* 45 (2001): 487–90.

Watson, James D. "The Human Genome Project: Past, Present and Future." *Science* 248 (1990): 44–48.

Watson, James D., and F. H. C. Crick. "Genetical Implications of the Structure of Deoxyribonucleic Acid." *Nature* 171 (1953): 964–67.

———. "Molecular Structure of Nucleic Acids: A Structure for Deoxyribose Nucleic Acid." *Nature* 171 (1953): 737–38.

Woo Suk Hwang, Young June Ryu, Jong Hyuk Park, Eul Soon Park, Eu Gene Lee, Ja Min Koo, Hyun Yong Chun, et al. "Evidence for a Pluripotent Human Embryonic Stem Cell Line Derived from a Cloned Blastocyst," Scienceexpress, www.scienceexpress.org/12February2004/Page1/10.1126/science.1094515

Wright, Susan. "Recombinant DNA Technology and Its Social Transformation, 1972–1982." *Osiris*, 2nd ser., 2 (1986): 303–60.

Yoo, Dongwan, and Antonio Giulivi. "Xenotransplantation and the Potential Risk of Xenogeneic Transmission of Procine Viruses." *Canadian Journal of Veterinary Research* 64 (2000): 193–203.

Zhu, I, He Wang, J. Michael Bishop, and Elizabeth Blackburn. "Telomerase Extends the Lifespan of Virus-Transformed Human Cells without Net Telomere Lengthening." *Proceedings of the National Academy of Sciences USA* 95 (1999): 3723–28.

ABSTRACTS

Barritt, Jason A., Carol A. Brenner, Henry E. Malter, and Jacques Cohen. "Mitochondria in Human Offspring Derived from Ooplasmic Transplantation: Brief Communication." *Human Reproduction* 16 (2001): 513.

Finn, Colin A. "Reproductive Ageing and the Menopause." *International Journal of Developmental Biology* 45 (2001): 613.

Grishok, A., A. E. Pasquinelli, D. Cote, N. Li, S. Parrish, I. Ha, D. L. Baillie, A. Fire, G. Ruvkun, C. C. Mello. "Genes and Mechanisms Related to RNA Interference Regulate Expression of the Small Temporal RNAs That Control *C. elegans* Developmental Timing." *Cell* 106 (2001): 23–24.

Ivanova, Natalia B., John T. Dimos, Christoph Schaniel, Jason A. Hackney, Kateria A. Moore, and Thor R. Lemischka. "A Stem Cell Molecular Signature." *Science* 298 (2002): 601–4.

Jaenisch, Rudolf, and Ian Wilmut. "Developmental Biology: Don't Clone Humans!" *Science* 291 (2001): 2552.

Jessberger R., V. Podust, U. Hubscher, and P. Berg. "A Mammalian Protein Complex That Repairs Double-Strand Breaks and Deletions by Recombination." *Journal of Biological Chemistry* 268, no. 20 (1993): 15070–15079.

Jiang, Yuehua, Bakrishna N. Jahagirdar, R. Lee Reinhardts, Robert E. Schwarts, C. Dirk Keene, Xilma R. Ortiz-Gonzalea, Maorayma Reyes, et al. "Pluripotency of Mesenchymal Stem Cells Derived from Adult Marrow." *Nature* 418 (2002): 41–49.

Jones, Steven J. M., Donald L. Riddle, Anatoli T. Pouzyrev, Victor E. Veculescu, LaDeanna Hillier, Sean R. Eddy, Shawn L. Stricklin, David L. Baillie, Robert Waterston, and Marco A. Marra. "Changes in Gene Expression Associated with Developmental Arrest and Longevity in *Caenorhabditis elegans*." *Genome Research* 11 (2001): 1346–52.

Legouis, R., A. Gansmuller, S. Sookhareea, J. M. Bosher, D. L. Baillie, and M. Labouesse. "LET-413 Is a Basolateral Protein Required for the Assembly of Adherens Junctions in *Caenorhabditis elegans*." *Nature Cell Biology* 2, no. 7 (2000): 415–22.

McKnight, T. D., M. S. Fitzgerald, and D. E. Shippen. "Plant Telomeres and Telomerases: A Review." *Biochemistry* (Moscow) 62 (1997): 1224. http://www.puma.protein.bio.msu.su/biokhimiya/contents/v62/abs/62111432.htm.

Ramalho-Santos, Miguel, Soonsang Yoon, Yumi Matsuzaki, Richard C. Mulligan, Douglas A. Melton. "'Stemness': Transcriptional Profiling of Embryonic and Adult Stem Cells." *Science* 298 (2002): 597–600.

Thacker, C., M. A. Marra, A. Jones, D. L. Baillie, and A. M. Rose. "Functional Genomics in *Caenorhabditis elegans*: An Approach Involving Comparisons of Sequences from Related Nematodes." *Genome Research* 9, no. 4 (1999): 348–59.

Waterston, Robert H., Eric S. Lander, and John E. Sulston. "On the sequencing of the human genome." *Proceedings of the National Academy of Sciences USA* 99 (2002): http://pnas.org/egi/doi/10.1073/pnas.042692499.

UNPUBLISHED MANUSCRIPTS

Irving, Dianne Nutwell. "Analysis: Parts I and II: Stem Cells That Become Embryos: Implications for the NIH Guidelines on Stem Cell Research, the NIH Stem Cell Report, Informed Consent and Patient Safety in Clinical Trials." Report commissioned by the Linacre Institute of the Catholic Medical Association and the International Federation of Catholic Medical Associations, 2001.
———. "Biomedical Research with 'Decisionally Incapacitated' Human Subjects: Legalization of a Defunct Normative Bioethics Theory."
———. "What Is 'Bioethics'? (Quid Est 'Bioethics?')."
Monk, Marilyn. "Embryonic Genes Expressed in Cancer Cells." 2001.

LETTERS, REVIEWS, VIEWS

Baylis, Francoise and Jocelyn Downie. "Ban cloning. Do you copy?" Globe and Mail, July 3, 2002, A15.
———. "Cloning for stem cell research unnecessary and dangerous." The Hill Times, February 3, 2003. http://www.stemcellnetwork.ca/news/articles.
———. "Parliament's call for moratorium on stem cell research perplexing." The Hill Times, June 5, 2002, http://www.stemcellnetwork.ca/news/article.
Bennett, Carolyn. "Assisted Human Reproduction Is on the Legislative Agenda." Update from Dr. Carolyn Bennett, MP (Spring 2003): 3.
Berg, Paul, David Baltimore, Herbert W. Boyer, Stanley N. Cohen, Ronald W. Davis, David S. Hogness, Daniel Nathans, et al. "Potential Biohazards of Recombinant DNA Molecules." Science 185 (1974): 303.
Berg, Paul, and Maxine Singer. "Regulating Human Cloning." Science 282 (1998): 413.
Bernstein, Alan. "Get ready for the next SARS." Globe and Mail, June 19, 2003, A15.
Brenner, Sydney. Letter to Max Perutz, June 5, 1963. http://www.elegans.swmed.edu/Sydney.html.
———. "The Tale of the Human Genome." Nature 416 (2002): 793–94.
Council for Biotechnology Information. Good Ideas Are Growing. Washington, DC: Council for Biotechnology Information, n.d. http://whybiotech.ca/html/pdf/Good_Ideas.pdf.
Council for Biotechnology Information and the Crop Protection Institute. Plant Biotechnology in Canada. n.d. http://www.croplife.ca/english/pdf/plantbiotechnology.pdf.
"Defining a New Bioethic" (editorial). Nature Genetics 28, no. 4 (2001): 297–98.
Dulbecco, Renato. "A Turning Point in Cancer Research: Sequencing the Human Genome." Science 231 (1986): 1055–56.
Ehrenreich, Barbara. "Double Helix, Single Guy." Review of Genes, Girls, and Gamow, by James D. Watson. New York Times Book Review, February 24, 2002, 6.
Hurlbut, William B. "The Ethics of Cloning." Scientific American 289 (November 2003): 14.
Ingram, Jay. "The story of the big clone story." Toronto Star, January 27, 2002, F8.

Johnson, Harriet McBryde. "Unspeakable Conversations, or How I Spent One Day as a Token Cripple at Princeton University." *New York Times Magazine,* February 16, 2003, 50.

Kaplinsky, Nick, David Braun, Damon Lisch, Angela Hay, Sarah Hake, and Michael Freeling. "Maize Transgene Results in Mexico Are Artefacts." *Nature Advance Online* (April 4, 2002).

Kischer, C. Ward. "The Corruption of the Science of Human Embryology." Letter to the *American Bioethics Advisory Commission Quarterly.* Fall 2002. American Life League. http://www.all.org/abac/aq0203.htm.

Krinos, Corinna M., Michael J. Coyne, Katja G. Weinacht, Arthur O. Tzianabos, Dennis L. Kasper, and Laurie E. Comstock. "Extensive Surface Diversity of a Commensal Microorganism by Multiple DNA Inversions." *Nature* 414 (2001): 555–38.

Luria, S. E. "Human Genome Program." *Science* (1989).

"NAS Ban on Plasmid Engineering." *Nature* 250 (1974): 175.

Pearson, Helen. "The Regeneration Gap." *Nature* 414 (2001): 388–90.

Petsko, Gregory. "The Father of Us All." *The Scientist* 16, no. 7 (2002): 13.

Quist, David, and Ignacio H. Chapela. "Transgenic DNA Introgressed into Traditional Maize Landraces in Oaxaca, Mexico." *Nature* 414 (2001): 541–43.

———. Reply. *Nature Advance Online* (April 4, 2002).

Revel, Michel. "Outright Condemnation of Cloning Research Is Premature." *The Scientist* 12, no. 2 (January 19, 1998): 8.

Wake, David B. "A Few Words about Evolution." *Nature* 416 (2002): 787–88.

Weissman, Irving L., and David Baltimore. "Editorial. Disappearing Stem Cells, Disappearing Science." *Science* 292 (2001): 601.

SELECTED POPULAR ARTICLES, OP/ED PIECES, NEWS STORIES

Abraham, Carolyn. "Chip may one day fix your failing memory." *Globe and Mail* March 14, 2003, A1.

———. "Cloning pig parts for people." *Globe and Mail,* January 3, 2002, A1.

———. "Worming our way to a longer life." *Globe and Mail,* July 13, 2001, A1.

———. "Cloning research paying off, Texan says." *Globe and Mail,* December 1, 2001, A3.

———. "Human genome decoder going non-profit." *Globe and Mail,* April 24, 2002, A8.

———. "Woman with faulty gene gets embryo without it." *Globe and Mail,* February 27, 2002, A1.

———. "Scientists look at creating a human-mouse embryo." *Globe and Mail,* November 28, 2002, A1.

———. "Stakes high in race to produce pig parts for human transplant." *Globe and Mail,* January 4, 2002, A1.

Abrams, Fran. "Inquiry into how public 'lost billions' over Dolly." *The Independent,* May 18, 2000, 5.

———. "Man of steel uses his will and electricity to beat odds." *Globe and Mail* September 13, 2002, A3.

Adams, Tim. "The clone arranger." *The Observer,* December 2, 2001. http://www.observer.co.uk/2001review/story/0,1590,624237,00.html.

Alphonso, Caroline. "Drug tests favour sponsor's product, study says." *Globe and Mail,* May 30, 2003, A8.

Andrews, Edmund L. "U.S. seeks patent on genetic codes, setting off furor." *New York Times,* October 21, 1991, A1.

Associated Press. "Scientists to attempt to create new form of life." *Globe and Mail,* November 22, 2002, A22.

Barrett, Amy. "Weird Science." *New York Times Magazine,* February 3, 2002, 9.

Basu, Arpon. "Montreal scientist clones calves using groundbreaking method." *Toronto Star,* December 10, 2001, A2.

Belkin, Lisa. "The Made-to-Order Savior." *New York Times Magazine,* July 1, 2001, 36.

Bonetta, Laura. "Getting Proteins into Cells." *The Scientist* 16, no. 7 (2002): 38

Branswell, Helen. "Chick study shows how faces could form." *Toronto Star,* December 20, 2001, A21.

Calamai, Peter. "Paper on biotech corn was flawed, journal reports." *Toronto Star,* April 6, 2002, A21.

Carr, David. "Networks say editor tried to sell clone 'exclusive.'" *New York Times,* January 5, 2003, 13.

Caulfield, Timothy. "I smell a cloned rat." *Globe and Mail,* January 4, 2003, A17.

Caulfield, Timothy, Abdallah S. Daar, Bartha M. Knoppers, and Peter A. Singer. "MPs have the wrong focus." *Globe and Mail,* May 2, 2002, A21.

Caulfield, Timothy, Abdallah Daar, Bartha Knoppers, Peter A. Singer, David Castle, and Ron Forbes. "Not all cloning is alike." *The Hill Times,* February 24, 2003. http://www.stemcellnetwork.ca/news/articles.php?id=128.

Cibelli, Jose B., Robert P. Lanza, and Michael D. West, with Carol Ezzell. "The First Human Cloned Embryo." *Scientific American,* November 24, 2001. http://www.sciam.com/article.cfm?articleID=0008B8F9=AC62=1C7S=9B8180 9EC588EF21.

———. "The First Human Cloned Embryo." *Scientific American* 286 (January 2002): 44–51.

Clark, Campbell. "Cloning of embryos considered: Federal Health Department may allow human cells to be grown for transplants." *Globe and Mail,* January 24, 2001, A5.

"Clue to chemistry of heredity found." *New York Times,* (June 13, 1953), L17.

Crick, F. H. C. "The Structure of Hereditary Material." *Scientific American* 191 (October 1954): 54.

Derbyshire, David. "Firm wins the right to clone billions from Dolly." *Daily Mail* (January 21, 2000), 41.

Doyle, Rodger. "Down with Evolution!" *Scientific American* 286 (March 2002): 30.

Elias, Paul. "Politics stalls stem-cell work." *Seattle Times,* November 18, 2002. http://www.seattletimes.nwsource.com.

Examiner News Service. "Drug makers lead the way as stock markets rally." *San Francisco Chronicle,* November 6, 1998.

Ezzell, Carol. "The $13-Billion Man." *Scientific American* 284 (January 2001): 29–30.

Fischer, Joannie. "The first clone." *U.S. News & World Report,* December 3, 2001, 30–63.

Galloway, Gloria. "Cloned baby Eve exists, Raelian says in court." *Globe and Mail,* January 30, 2003, A12.

———. "Leaders outraged by alleged cloning." *Globe and Mail,* December 28, 2003, A1.

————. "Lawyer's suit seeks charter of rights for cloned children." *Globe and Mail*, January 4, 2003, A1.

Gibbs, W. Wayt. "Biological Alchemy." *Scientific American* 284 (February 2001): 16–17.

————. "Shrinking to Enormity." *Scientific American* 284 (February 2001): 33–34.

————. "The Unseen Genome: Gems among the Junk." *Scientific American* 289 (November 2003): 46–53.

Gold, E. Richard, Timothy Caulfield, Bartha Maria Knoppers, Peter Bridge, Erika Duenas, and Lori Sheremeta. "Ottawa must act on DNA patents." *Montreal Gazette*, December 15, 2001, B7.

Goodstein, Laurie. "Priests' sex abuse in Boston causes ripples across U.S." *New York Times*, February 17, 2002, 16.

————. "Trail of pain in church crisis leads to nearly every diocese." *New York Times*, January 12, 2003, 1.

Graham, Sarah. "Journal Retracts Support for Claims of Invasive GM Corn." *Scientific American* (April 2002).

Green, Ronald M. "The Ethical Considerations." *Scientific American* 286 (January 2002): 48–50.

Greenaway, Norma. "Panel pushes stem cell bill closer to law." *Ottawa Citizen*, December 16, 2002. http://www.stemcellnetwork.ca/news/articles.php?id=114.

Halim, Nadia G. "Aftermath of Tragedy: Researchers, Government Officials Review Gene Therapy Trials." *The Scientist* 14, no. 1 (January 10, 2000): 6.

Hall, Carl T. "Biotech's fountain of hype." *San Francisco Chronicle*, January 15, 1998. http://sfgate.com/cgi=bin/article.cgi?file=/chronicle/archive/1998/01/15/BU270 37.DTL.

————. "Non-aging human cells created in lab: Bay firm's stock soars on hopes of medical advances." *San Francisco Chronicle*, (January 14, 1998. http://sf-te.com/cgi=bin/article.cgi?file=/chronicle/archive/1998/01/14/MN74909.DTL.

————. "Studies lift hopes for anti-aging technology: 'Immortality' enzyme doesn't cause cancer." *San Francisco Chronicle*, December 29, 1998. http://sf-gate.com/cgi=bin/article.cgi?file=/chronicle/archive/1998/12/29/MN48125 .DTL.

Harper, Tim. "New bill backs stem-cell research." *Toronto Star*, May 10, 2002, A3.

Hencke, David. "Treasury loses out on billions from cloning of Dolly." *The Guardian*, February 7, 2000, 3.

————. "Watchdog examines cloning 'giveaway.'" *The Guardian*, May 19, 2000, 11.

Hilts, Philip J. "Head of gene map threatens to quit." *New York Times*, April 9, 1992, A26.

Honey, Kim. "The man who aims to cheat death." *Globe and Mail*, November 27, 2001, A1.

————. "Sperm-less eggs used to create stem cells." *Globe and Mail*, February 1, 2002, A3.

Hong, Frances. "Drug money: The biotech business." *Examiner*, January 18, 1998. http://sfgate.com/cgi=bin/article.cgi?file=/examiner/archive/1998/01/18/ BUSINESS4747.dtl.

Hurst, Lynda. "Genome just a start for maverick scientist." *Toronto Star*, April 24, 2002, A2.

Immen, Wallace. "Milking genes for all they're worth." *Globe and Mail*, November 29, 2001, A3.

Jegalian, Karin, and Bruce T. Lahn. "Why the Y Is So Weird." *Scientific American* 284 (February 2001): 56–61.

Kline, Ronald M. "Whose Blood Is It, Anyway?" *Scientific American* 284 (April 2001): 42–49.

Kling, Jim. "The Complexity of Gene Silencing." *The Scientist* 16, no. 4 (February 18, 2002): 31.

Kolata, Gina. "In cloning, failure far exceeds success." *New York Times,* December 11, 2001, D1.

Koring, Paul. "Human cloning dangerous, unethical, experts say." *Globe and Mail,* March 29, 2001, A10.

Lawton, Valerie. "A long gestation." *Toronto Star,* April 5, 2003, G1.

Lemmens, Trudo. "Reading the Book of Life." *University of Toronto Bulletin,* February 16, 2001.

Lewin, Roger. "Do Jumping Genes Make Evolutionary Leaps?" *Science* 213 (1981): 634–36.

Lewis, Ricki. "A Case Too Soon for Genetic Testing?" *The Scientist* 16, no. 7 (April 1, 2002): 16.

———. "Debate over Stem Cell Origins Continues." *The Scientist* 16, no. 11 (May 27, 2002): 19.

Lu, Vanessa, and Karen Palmer. "Set stem cell rules now: MDs." *Toronto Star,* November 27, 2001, A1.

Maher, Brendan A. "Public-Private Genome Debate." *The Scientist* 16, no. 7 (April 1, 2002): 24.

Marx, Jean L. "Antibody Research Garners Nobel Prize." *Science* 218 (1987): 484–85.

McCarthy, Shawn. "MPs ask Rock to tighten rules on cloning." *Globe and Mail,* December 13, 2001.

McIlroy, Anne. "Canadians take big step in reproducing stem cells." *Globe and Mail,* January 30, 2001, A3.

———. "Cloning procedure may be the key to a second chance for dying species." *Globe and Mail,* March 4, 2004, A5.

———. "Olivieri affair was rife with errors, report says." *Globe and Mail,* October 24, 2002, A6.

———. "Stem cells trigger organ regeneration." *Globe and Mail,* June 23, 2003, A6.

Milroy, Steven. "Cloning con game." *Washington Times,* January 31, 2002.

Mitchell, Alanna. "Scientists clone human embryo." *Globe and Mail,* November 26, 2001, A1.

Munro, Margaret. "Embryonic stem cell research to get funding. 'A great pity' ethicist says: Agency is tired of waiting for Ottawa." *National Post,* May 14, 2003, A5.

———. "Stem cells coax body to heal itself: Bone marrow cells injected into mice quickly cure diabetes." *Ottawa Citizen,* June 23, 2003.

Munro, Neil. "Doctor Who? Scientists are treated as objective arbiters in the cloning debate. But most have serious skin in the game." *Washington Monthly,* 2002.

Nelson, Deborah, and Rick Weiss. "Gene test deaths not reported promptly." *Washington Post,* January 31, 2000, A1.

———. "Hasty decisions in race to a cure? Gene therapy study proceeded despite safety, ethics." *Washington Post,* November 21, 1999, A1.

Nelson, Marissa. "Raelian sect told to stop research on cloning." *Globe and Mail,* June 30, 2001, A11.

Osborne, Lawrence. "Got Silk." *New York Times Magazine,* June 16, 2002, 48.

Palevitz, Barry A. "Designing Science by Politics." *The Scientist* 16, no. 11 (May 27, 2002).

Papp, Leslie. "Umbilical cord cells could be brain-savers." *Toronto Star,* February 19, 2001, A14.

Pear, Robert. "U.S. to review research at hospitals for veterans." *New York Times,* April 13, 2003, A14.

Petit, Charles. "Regenerating gene cloned by scientists: New light shed by cancer cells." *San Francisco Chronicle,* August 16, 1997. http://sfgate.com/cgi=bin/article.cgi?file=/chronicle/archive/1997/08/15/MN29623.DTL.

Picard, Andre. "Montreal's 'Frankenstein.'" *Globe and Mail,* January 19, 2002, F3.

Press, Eyal, and Jennifer Washburn. "The Kept University." *Atlantic Monthly,* March 2000.

Priest, Lisa. "Cloning files snatched in Korean raid." *Globe and Mail,* December 31, 2003, A1.

———. "Embryo adoption program to offer infertile couples one last chance." *Globe and Mail,* March 4, 2002, A1.

"Raelians backpedal on proof of cloning." *Toronto Star,* January 4, 2003, A17.

Reaney, Patricia. "First cloned horse takes a bow in Italy." *Globe and Mail,* August 7, 2003, A2.

Reuters. "'Multiplicity' not the goal." *Toronto Star,* November 26, 2001, A16.

———. "Piglet clones elicit squeals of delight." *Toronto Star,* February 12, 2002, A11.

———. "Worm world wins Nobel." *Toronto Star,* October 8, 2002, A15.

Revel, Michel. "Ashes from a stolen frame." *Ha'aretz,* May 4, 2001, B11.

Roberts, Leslie. "NIH Gene Patents, Round Two." *Science* 255 (1992): 912–13.

———. "Why Watson Quit as Project Head." *Science* 256 (1992): 301–2.

Rohm, Wendy Goldman. "Seven Days of Creation." *Wired,* February 2004, 120.

Ruse, Michael. "An appreciation: Stephen Jay Gould: He fought every 'ism' in the book." *Globe and Mail,* May 23, 2002, R9.

Service, Robert F. "'Fountain of Youth' Lifts Biotech Stock." *Science* 279 (1998): 472.

Shreve, James. "The Secrets of the Gene." *National Geographic* 196 (October 1999): 42–75.

Skloot, Rebecca. "Sally Has Two Mommies and One Daddy." *Popular Science* (March 2003). http://www.popsci.com/popsci/medicine/article/0,12543,411770,00.html.

Smaglik, Paul. "Investigators Ponder What Went Wrong after Gene Therapy Death." *The Scientist* 13, no. 21 (October 25, 1999): 1.

Steed, Judy. "The promise of the future in a spider's gene." *Toronto Star,* June 9, 2002, C1.

Stern, Curt. "The Biology of the Negro." *Scientific American* 191 (October 1954): 80.

Stix, Gary. "What Clones?" *Scientific American* 286 (February 2002): 18–19.

———. "Who Owns You?" *Scientific American* 286 (March 2002): 35.

Stolberg, Sheryl Gay. "The Biotech Death of Jesse Gelsinger." *New York Times Magazine,* November 28, 1999, 137.

———. "Gene therapy ordered halted at university." *New York Times,* January 22, 2000, A1.

———. "Many approved stem-cell lines aren't ready to study, U.S. says." *New York Times,* September 6, 2001, 1.

————. "That scientific breakthrough thing." *New York Times,* December 9, 2001, 3.

Strickland, Carol. "Watson relinquishes major role at lab." *New York Times* (March 21, 1993, LI1.

Suhr, Jim. "Paralyzed Superman can move toes, fingers." *Toronto Star,* September 11, 2002, A3.

Talaga, Tanya. "Goodbye, Dolly: Cloned sheep buys the farm." *Toronto Star,* February 15, 2003, A23.

————. "Scientists pore over 'Book of Life' secrets." *Toronto Star,* February 12, 2001, A1.

Talbot, Margaret. "A Desire to Duplicate." *New York Times Magazine,* February 4, 2001, 40.

Toner, Robin. "Bush caught in the middle on research on stem cells." *New York Times,* February 18, 2001, 23.

Travers, Jim. "Calm debate is needed on cloning issue." *Toronto Star,* November 29, 2001, A31.

Tu Thanh Ha. "Cult's cloning claim in doubt as DNA review is called off." *Globe and Mail,* January 7, 2003, A1.

————. "More clones on way, sect says." *Globe and Mail,* January 6, 2003, A2.

————. "Transgenic goats help spin spider silk." *Globe and Mail,* January 18, 2002, A5.

Van Rijn, Nicholas. "Scientists clone human embryo." *Toronto Star,* November 26, 2001, A1.

Vogel, Gretchen. "In Contrast to Dolly, Cloning Resets Telomere Clock in Cattle." *Science* 288 (2000): 586–87.

————. "In the Mideast, Pushing Back the Stem Cell Frontier." *Science* 295 (2002): 1818–20.

Wade, Nicholas. "DNA in Africa: Group in Africa has Jewish roots, DNA indicates." *New York Times* (1999): www.biblemysteries.com/library/dna.

————. "Gene experiment comes close to crossing ethicists' line." *New York Times,* December 23, 2001, A22.

————. "Reading the Book of Life." *New York Times,* June 27, 2000, F5.

Walker, William. "Fetus is a 'child' Bush declares." *Toronto Star,* February 1, 2002, A1.

————. "'No prime suspect' in anthrax case, FBI says." *Toronto Star,* February 26, 2002, A10.

Watzman, Haim, and Susan Petersen Avitzour. "Israel's Great Expectation." *Nature Biotechnology* 17 (2001): 518–20.

Weiss, Rick. "Genetic code analysis yields clues to disease." *Toronto Star,* February 11, 2001, A1.

————. "In laboratory, ordinary cells are turned into eggs." *Washington Post,* May 2, 2003, A1.

————. "New stem cell source called possible." *New York Times,* February 1, 2002, A18.

————. "Scientists in U.S. grow 'kidneys' from cells of cloned cow embryos." *Toronto Star,* January 30, 2002, A3.

Weiss, Rick, and Deborah Nelson. "FDA lists violations by gene therapy director at U-Penn." *Washington Post,* March 4, 2000, A4.

Whitfield, John. "Cloned Cows in the Pink." *Nature,* November 23, 2001. http://www.nature.com/nsu/011129.011129=1.html.

Young, John A. T., and R. John Collier. "Attacking Anthrax." *Scientific American* 286 (March 2002): 48–59.

Zitner, Aaron. "University to create human stem cells." *Toronto Star,* December 11, 2002, A4.

REPORTS, PROCEEDINGS, NEWSLETTERS, HEARINGS, TESTIMONY

Allen, Grace. "The Families Whence High Intelligence Springs." Eugenics Record Office, *Bulletin,* no. 25. Cold Spring Harbor, NY: 1926.

American Society for Cell Biology. *Newsletter.* 25, no. 12 (December 2001).

———. *Press Book 2001.* 41st Annual Meeting, December 8–12, 2001.

Berg, Paul, David Baltimore, Sydney Brenner, Richard O. Roblin III, and Maxine Singer. "Asilomar Conference on Recombinant DNA Molecules." *Science* 188 (1975): 991–94.

Bernstein, Alan. Declaration, May 30, 2001. Offices of the Ethics Councillor. http://strategis.gc.ca.

———. Testimony before the Standing Committee on Health, House of Commons, April 17, 2002.

Canadian Bioethics Society. *Newsletter,* 8, no. 2 (August 2001).

Chenier, Nancy Miller. "The CIHR Embryonic Stem Cell Research Guidelines: Informed Consent and Privacy." Ottawa: Library of Parliament/Bibliotheque du Parlement, April 2002.

Cooperative Human Tissue Network. *Tissue Topics.* Newsletter. 5 (Winter 2001).

———. Annual Meeting Program. 41st annual meeting, December 8–12, 2001.

Davenport, Charles B. "The Family History Book." Eugenics Record Office, *Bulletin,* no. 7. Cold Spring Harbor, NY: 1912.

———. "The Trait Book." Eugenics Record Office, *Bulletin,* no. 4. Cold Spring Harbor, NY: 1912.

Elections Canada. Contributions and Expenses, 37th General Election, November 27, 2000, Brown, Bonnie, Oakville.

Gauthier, Michel J., ed. *Gene Transfers and Environment: Proceedings of the Third European Meeting on Bacterial Genetics and Ecology (BAGECO-3), 20–22 November 1991.* Berlin: Springer-Verlag, 1992.

Goddard, Henry H. "Heredity of Feeble-Mindedness." Eugenics Record Office, *Bulletin,* no. 1. Cold Spring Harbor, NY: 1911.

Israel Academy of Sciences and Humanities. Bioethics Advisory Committee. *The Use of Embryonic Stem Cells for Therapeutic Research.* August 2001. http://www.academy.ac.il/bioethics/index=e.html.

Juvenile Diabetes Research Foundation International. *JDRF Research E-Newsletter,* no. 9. http://www.jdrf.org/publications/enews.

Laughlin, Harry B. *Report No. 1.* Cold Spring Harbor, NY: Eugenics Record Office 1913.

McHughen, Alan. *Biotechnology and Food for Canadians.* Risk Controversy Series 2, edited by Laura Jones. Vancouver: Fraser Institute; New York: American Council on Science and Health, 2002.

National Academy of Sciences, National Academy of Engineering, and Institute of Medicine. Committee on Science, Engineering, and Public Policy. *Scientific and Medical Aspects of Human Reproductive Cloning.* Washington, DC: National Academy Press, January 18, 2002.

Report by the Comptroller and Auditor General. *Reaping the Rewards of Agricultural Research*. House of Commons, January 20, 2003.

Stratton, Kathleen, Donna A. Almario, and Marie C. McCormick, eds. *Immunization Safety Review: SV40 Contamination of Polio Vaccine and Cancer.* Institute of Medicine of the National Academies. October 22, 2002. http://www.iom.edu/report.asp?id=4317.

United States. Committee on Energy and Commerce. Subcommittee on Oversight and Investigations. "Issues Raised by Human Cloning Research." Testimony. http://www.energycommerce.house.gov/107/hearings/.

United States. Senate Committee on Natural Resources and Wildlife and Senate Select Committee on Higher Education. *Impacts of Genetic Engineering on California's Environment: Examining the Role of Research at Public Universities* (Novartis/UC Berkeley Agreement). State Capitol, Sacramento, California, May 15, 2000.

POLICY PAPERS, LEGISLATION, DRAFT LEGISLATION

Canada. House of Commons. Bill C-13. An Act to Establish the Canadian Institutes of Health Research, to Repeal the Medical Research Council Act and to Make Consequential Amendments to Other Acts. Tabled November 4, 1999.

Canadian Institutes of Health Research. *Human Pluripotent Stem Cell Research: Guidelines for CIHR-Funded Research.* March 4, 2002. http://www.cihr-irsc.gc.ca/e/publications/1487.shtml.

———. "Human Stem Cell Research: Opportunities for Health and Ethical Perspectives." Discussion paper of the Working Group (30 March 2001). http://www.cihr.ca/governing_council/ad_hoc_working_groups/ahwg_stem_cell_e.shgtml.

Health Canada. *Proposals for Legislation Governing Assisted Human Reproduction.* Ottawa, Health Canada: May 2001.

Minister of Health (Canada). Bill C-13. An Act Respecting Assisted Human Reproductive Technologies and Related Research. Reprinted as amended by the Standing Committee on Health as a working copy for the use of the House of Commons at Report Stage and as reported to the House on December 12, 2002.

Minister of Health (Canada). Bill C-13. An Act Respecting Assisted Human Reproduction. First reading: May 9, 2002.

Medical Research Council of Canada, Natural Science and Engineering Research Council of Canada, Social Sciences and Humanities Research Council of Canada. *Ethical Conduct for Research Involving Humans: A Tri-Council Policy Statement.* Ottawa: Medical Research Council of Canada, August 1998.

REAL Women of Canada. Brief on New Medical Technologies. November 2001.

United States Department of Energy. ELSI Retrospective: Human Genome Project Information. 1990–2001.

United States National Human Genome Research Institute. Ethical, Legal, and Social Implications (ELSI) Program, Project Descriptions, 1990–Present.

498 • ELAINE DEWAR

COURT FILINGS, PROCEEDINGS, RULINGS

Eve, by her guardian ad litem, Milton B. Fitzpatrick, Official Trustee v. Mrs. E., Respondent and Canadian Mental Health Association, Consumer Advisory Committee of the Canadian Association for the Mentally Retarded, the Public Trustee of Manitoba and Attorney General of Canada, Intervenors. Supreme Court Reports 1986.
Infigen, Inc. Plaintiff. v. Advanced Technology, Inc. and Steven L. Stice, Defendants. United States District Court, Western District of Wisconsin, case no. 98-C-0431-C. First Amended Complaint, November 2, 1998.
———. Answer to Amended Complaint, December 7, 1998.
———. Final Settlement Agreement.
McKay, Andrew W. Reasons for Judgement. Between: Monsanto Canada Inc. and Monsanto Company, Plaintiffs and Percy Schmeiser and Schmeiser Enterprises Ltd., Defendants. Docket: T-1593–98. Neutral Citation: 2001 FCT25a. March 29, 2001.
Muir v. Alberta, 32 ALTA. L. R. 3rd.
Roe et al. v. Wade. District Attorney of Dallas County. Appeal from the United States District Court for the Northern District of Texas. United States Supreme Court, Syllabus and Opinion of the Court, January 22, 1973.
Wisconsin Alumni Research Foundation Plaintiff v. Geron Corporation. United States District Court, Western District of Wisconsin. case no. 01-C-0459-C. Complaint for Declaratory Relief. August 13, 2001.

PRESS RELEASES, STATEMENTS

Alberts, Bruce. "U.S. Policy-Makers Should Ban Human Reproductive Cloning." National Academies News Conference. January 18, 2002.
Baylis, Francoise. "CIHR Guidelines for Human Pluripotent Stem Cell Research." National Press Theatre, Ottawa, March 4, 2002. http://www.cihr.ca/president/speeches/20020304speech_e.shtml.
———. "A Reflection on the 'Place' of Bioethics."
Bernstein, Alan. "I wish to reassure MPs no new funds for human embryonic stem cell research will flow to researchers until April 2003." Special to *The Hill Times*, May 6, 2002.
———. "President's Address." National Press Theatre, Ottawa, March 4, 2002.
———. Public Declaration of Declarable Assets. Conflict of Interest and Post-employment Code for Public Office Holders. May 30, 2001.
Canadian Institutes of Health Research. "Federal Funding Agencies to Launch New Panel on Ethics in Research Involving Humans." November 9, 2001. www.nserc.ca/programs/ethics/news-e.html.
———. "Minister McLellan Announces an Investment of $88 Million to Train Next Generation of Health Researchers." www.cihr-irsc.gc.ca/news/press_releases/2002/pr-0217_e.shtml.
Canadian Nurses Association. "Amendments Needed to Bill-C13 If CIHR Is to Live Up to Its Potential, Nursing Group Says." December 1999.

————. "Stem Cell Oversight Committee (SCOC): CIHR seeks Applications for Stem Cell Oversight Committee." May 22, 2003: www.cihr-irsc.gc.ca/about_cihr/organization/ethics/stem_cell/call_for_members_e.htm.

Colman, Alan. Director, PPL Therapeutics. "Comment on Correspondence Published in *Nature* on 27 May, 2001."

Geron Corporation. "Geron Acquires Roslin Bio-Med and Forms Research Collaboration with the Roslin Institute." May 4, 1999. http://www.geron.com.

————." Geron Announces Second Interference over Nuclear Transfer Patents." March 4, 2002. http://www.geron.com/pr_0304982.html.

————. "Geron Corporation Reports Publication of Research Supporting the Utility of Telomerase as a Universal Antigen for Cancer Immunotherapy." March 13, 2003. http://www.geron.com.

Greenpeace. "Greenpeace, World Scientists Call for Action to Save Mexico's Corn from Genetic Contamination." Media Center. November 28, 2001.

Health Canada. "Health Minister Launches Canadian Institutes of Health Research." June 7, 2000. www.hc-sc.gc.ca/english/media releases/2000/cihre.htm.

Hospital for Sick Children. "Sick Kids Attracts New Head of Cancer Research and Leading Stem Cell Researcher." July 31, 2002.

Infigen Inc. "Infigen Introduces Oldest Adult Holstein Clone in Canada, CrescentMead—a Margo II." November 19, 2001. http://www.infigen.com.

————. "Long-Term Infigen Study of Cloned Cattle and Pigs Finds That They Are Normal and Healthy." November 27, 2001. http://www.infigen.com

Institute for Human Gene Therapy. "Statement on the Death of Jesse Gelsinger." September 7, 2001. http://www.uphs.upenn.edu/ihgt/jesse.html.

International Institute for Human Gene Therapy. "Preliminary Findings Reported on the Death of Jesse Gelsinger." May 25, 2000. http://www.uphs.upenn.edu/ihgt/findings.html.

————. "Response to FDA." September 7, 2001. www.uphs.upenn.edu/ ihgt/resp2fda.html.

Joint Centre for Bioethics. "Statement on Sources of Funding." http://www.utoronto.ca/jcb/sources_funding.html.

PPL Therapeutics. "PPL Produces World's First Transgenic Pigs." http://www.ppl-therapeutics.com.

Rael. "Human Cloning Will Make Terrorist Attacks Inefficient and Will Allow the Judgements of the Perpetrators." Press release, September 14, 2001, http://www.rael.org/pres.

The National Academies. "U.S. Policy-makers should ban human reproductive cloning." January 18, 2002: www.national-academies.org.

West, Michael. Letter to Mark Souder. May 16, 2002. http://advanced cell.com/Letter_to_Souder.pdf.

CORPORATE DOCUMENTS

Canadian Institutes of Health Research. Annual Report, 2000–2001. Ottawa: Canadian Institutes of Health Research, 2001.

Cold Spring Harbor Laboratory. Return of Organization Exempt from Income Tax. 1996, 1997,1998, 1999, 2000.

Cold Spring Harbor Laboratory Association. Return of Organization Exempt from Income Tax. 2000.

CuraGen. Annual Report. 2000.

Pall Corporation. Annual Reports, 2001, 2002.

———. Annual report pursuant to Section 13 or 15 (d) of the Securities Exchange Act of 1934 for the fiscal year ended August 3, 2002.

———. Notices of annual meeting of shareholders, 2000, 2001, 2002.

The Rappaport Family Institute for Research in the Medical Sciences. Institute Profile. 2002–2005.

Robertson Research Fund, Inc. Return of Organization Exempt from Income Tax. 1999, 2000.

Roslin Institute. Annual Report, 1999/2000.

The Van Andel Research Institute. Scientific Report 2000–2001.

The Wellcome Trust. Scheme. The Charity Commissioners for England and Wales (210183). Sealing: 279S)01. Case no. 82105. February 20, 2001.

The Wellcome Trust Limited. Memorandum of Association. 2000.

———. New Articles of Association of the Wellcome Trust Limited. 2000.

WEB PUBLICATIONS

Andrology Institute of America. Dr. Zavos. curriculum vitae. 2001. www.aia-zavos.com/drz.htm.

———. Our Services. 2001. http://www.aia-zavos.com/aiaservices.htm.

———. Press release re. workshop on Human therapeutic cloning and press conference. 2001. http://www.aia-zavos.com/rome.htm.

———. Prices. 2001. http: http://www.aia-zavos.com/prices.htm.

Associated Press. "Science journal backs off corn report." MSNBC Science News, April 4, 2002.

BBC News. "Doctors defiant on cloning." March 9, 2001. http://news.bbc.co.uk/hi/english/sci/tech/1209716.stm.

———. "Dolly scientists play down US clone." November 25, 2001. http://news.bbc.co.uk/hi/english/sci/tech/1676037.stm.

———. "Europe rejects human cloning ban." November 29, 2001. http://news.bbc.co.uk/hi/english/sci/tech/1682591.stm.

CBC News Online. "Journal retracts article questioning GM crop safety." April 8, 2002. http://cbc.ca/stories/2002/04/08/Consumers/GMOstudy_020408.

Chapman, Audrey R., Mark S. Frankel, and Michele S. Garfinkel. Stem Cell Research and Applications: Monitoring the Frontiers of Biomedical Research. American Association for the Advancement of Science and Institute for Civil Society, November 1999. http://www.aaas.org/spp/sfrl/projects/stem/report.pdf.

CNN. "Scientists discover cellular 'fountain of youth.'" January 13, 1998. http://www.cnn.com/Health/9801/13/life.extention.

Ehrenstein, David. "Immortality Gene Discovered." Science 279 (1998): 177. http://www.sciencemag.org.

GE Food Alert Campaign Center. "Nature's Retraction Based on Concerns of Only One Referee." April 8, 2002. http://www.gefoodalert.org/News/news.cfm?News_ID=3291.

Gilbert, Walter. "DNA Sequencing and Gene Structure." Nobel 3-Museum. http://www.nobel.se/chemistry/laureates/1980/gilbert=lecture.html.

Kent, Saul. "Telomerase: The 'Immortalizing' Enzyme: Update on Geron Corporation." Life Extension Report 15, no. 1 (January 1, 1995). http://www.lef.org/anti-aging/telomer1.htm.

Kischer, C. Ward. "Cloning, Stem Cell Research and Some Historic Parallels." n.d. http://lifeissues.net/writers/kisc/kisc_02historicparallels.html.

National Academy of Sciences, National Academy of Engineering, and Institute of Medicine. Committee on Science, Engineering, and Public Policy. *Scientific and Medical Aspects of Human Cloning.* August 7, 2001. http://www.nationalacademies.org.

National Institutes of Health. "Stem Cells: A Primer." May 2000. http://www.nih.gov/news/stemcell/primer.htm.

Noble, Ivan. "Breakthrough for stem cell research." BBC News Online, November, 30, 2001. http://news.bbc.co.uk/1/hi/sci/tech/1683424.stm.

———. "Mexican study raises GM concern." BBC News Online. November 28, 2001. http://news.bbc.co.uk/1/hi/sci/tech/1680848.stm.

Orkin, Stuart H., and Sean J. Morrison. "Biomedicine: Stem-Cell Competition." *Nature* 418 (2002): 25–27. http://www.nature.com/cgi=taf/DynaPage.taf?file=/nature/journal/v418/n6893/full/418025a_fs.html.

Revel, Michel. "Ethics and Genetics: Are Human Rights and Human Traditions Threatened by Scientific Progress?" *Here-Now4U Online Magazine.* http://www.here-now4u.de/eng/ethics_and_genetics_are_human.htm.

Rifkin, Jeremy. "Dazzled by the Science: Biologists Who Dress Up Hi-tech Eugenics as New Art Form Are Dangerously Deluded." *The Guardian,* January 14, 2003. http://www.guardian.co. uk/comment/story/0,3604,874312,00.htm.

Steinberg, Jessica. "The Little Cells That Could." *Jewsweek,* (March 7, 2002). http://www.jewsweek.com/.

Vaughan, Christopher. "A Model Mathematician: David Haussler, Ph.D." Incyte Genomics. 2002. http://www.incyte.com/insidegenomics/int.

Wicker, Randolfe H. "Dr. Antinori Claims Successful Human Clone Pregnancy: Middle East Challenges West to Debate Medical Ethics and Cloning." The Reproductive Cloning Network. April 10, 2002. http://www.reproductive cloning.net/cgi-bin/ikonboard.

"What Is *Caenorhabditis elegans* and Why Work on It? An Introduction for Those Unfamiliar with 'The Worm.'" http://www.biotech.missouri.edu/Dauer-World/Wormintro.html.

Zavos, Panayiotis, and R. Moorgate. "Committee of Scientists for Safe and Responsible Therapeutic Human Cloning." The Reproductive Cloning Network. 2001. http://www.ReproductiveCloning.Net/Article/zavos.htm.

WEBSITES

Advanced Cell Technology. http://www.advancedcell.com.
Applera Corporation. http://www.pecorporation.com.
CLONAID. http://www.Clonaid.com.
Genome Canada. http://www.genomecanada.ca.
Geron Corporation. http://www.geron.com.
HumGen. http://www.humgen.umontreal.ca/en/.
Industry Canada. Public Registry. www.strategis.ic.gc.ca/SSG0e01048e.html.
Joint Centre for Bioethics. http://www.utoronto.ca/jcb.
Networks of Centres of Excellence. http://www.nce.gc.ca.
Nexia Biotechnologies. http://nexiabiotech.com.
PPL Therapeutics. http://www.ppl-therapeutics.com.
Reproductive Genetics Institute. http://www.reproductivegenetics.com.
Samuel Lunenfeld Research Institute. http:www.mshri.on.ca.

INDEX

Moyzis, Bob, 88
Muller, Hermann J., 54, 124
Myers, Eugene, 384–5
mystics/vitalists. *See under*
materialists
mythology, 173–4

Nägeli, Karl von, 25
Nagy, Andras, 10, 13, 52–3, 72
background, 58–65
Nakamura, Toru, 178
National Academy of Sciences
(U.S.), 46, 126, 344
and human cloning, 143
and regulation, 358–9
*Scientific and Medical Aspects of
Human Reproductive Cloning,*
204–5
National Bioethics Advisory
Commission (U.S.), 92–3
National Center for Biotechnology
Information, 372, 378, 394
National Human Genome Research
Institute (U.S.), 195
National Institutes of Health (U.S.),
32, 39, 85–6, 127, 359–62
biology as public, 262–3
and funding, 136
funding for Greider, 89–90
funding for West's company, 192
and patents, 272–3
and stem cell research, 107
National Science Foundation, 353–4,
358–9
Natural History Museum (London),
231, 233, 238–43
Neuberger, Albert, 258–9
Neural Regeneration Network, 103–4
Newton, Isaac, 29
Nexia, 112
Nobel Prize
Berg, Sanger and Gilbert, 355
T. Cech, 177
R. Franklin, 5
and the Gairdner, 76, 429–33
influence of, 66, 100, 124, 216, 226
G. Khorana, 119

B. McClintock, 69
H. Muller, 215
S. Brenner, 113, 448
F. Sanger, 255–7, 262–4
J. Watson, 56, 94, 213, 219–20, 223
M. West, 176
M. Wilkins, 213
Novartis, 148–9, 342, 345–8, 351–2,
354
NSERC (Natural Sciences and
Engineering Research Council
of Canada), 38
nuclear transplantation to produce
stem cells, 206
nucleic acids, 261–2
Nuremberg Code, 187, 205, 470n10
Nuremberg trials, 3
Nurse, Paul, 295, 453

Oko, Richard, 80
Olovniknov, Alexey, 173, 175
Olson, Maynard, 429, 434
Ontario Genomics Institute, 431
Ontario Research and Development
Challenge Fund, 415
organ transplants
of animals, 148–9
of cloned organs, 408
compared to tissue, 325
heart, 317–22
organ sales, 81
unavailable, 327–8
O'Toole, Tara, 79–80
Owen, Richard, 238–9, 249–51, 418

partition chromatography, 260
patented genes
business, 146, 363–5
of cloned sheep, 296–7
Cohen and Boyer, 126
competition over, 148–9, 160, 178,
273, 469n10
and enzyme telomerase, 8–9
of expressed sequence tag (EST),
272–3
governments, 300
and the Internet, 393

OLD
BLACK
MAGIC

OLD BLACK MAGIC

BLACK

MAGIC

by Jake Tanner

CROWN PUBLISHERS, INC.
NEW YORK

To loves lost . . .

Copyright © 1991 by Coal Creek Consortium

*All rights reserved. No part of this book may be
reproduced or transmitted in any form or by any means,
electronic or mechanical, including photocopying,
recording, or by any information storage and retrieval
system, without permission in writing from the
publisher.*

*Published by Crown Publishers, Inc., 201 East 50th Street,
New York, New York 10022. Member of the Crown
Publishing Group.*

CROWN is a trademark of Crown Publishers, Inc.

Manufactured in the United States of America

Library of Congress Cataloging-in-Publication Data

Tanner, Jake.
    Old black magic / Jake Tanner.—1st ed.
        p. cm.
    I. Title.
  PS3570.A553044   1991
813'.54—dc20   90-48208
        ISBN 0-517-57676-7

10 9 8 7 6 5 4 3 2 1
FIRST   EDITION

Prologue

The bar at Cabo San Lucas fits my body like a bandage. I slide into it when I'm tired and wounded and let the dark cool interior heal my internal injuries. Cabo has an unrelenting sun that beats down on the beaches and makes everything look surreal. The Two Yanks saloon is pure sanctuary.

The outside of the bar is nothing to write home about. It sits just off the beach in a cul-de-sac that was built for the turista cars in the 1970s. The exterior is painted cinder block topped with thatch, and the sign is a plank sculptured in traditional British pub motif. It was cannibalized from a fine old boat. Inside, there are exactly eight sea tables and an interesting assortment of substantial chairs. In a bar fight serious brawlers could probably throw the chairs, but if they could heft the tables, you'd want to make damned sure they were on your side. The tables are usually populated with tourists, hookers, and the vaguely criminal. Anyone with any sense sits at the bar. The bar is mahogany. Stryker had it imported. The real point of being at the bar is to be closer to Stryker Stephens.

Since mid-August I had looked at too many dead bodies. Each of them was different. All of them had died ultimately from a bullet, but each had been brutalized in other ways. The first body was ripped almost in half at the base of the spine. The others had their own ugly anomalies. The bodies were buried now except in my head.

So I had come to my second home for solace. In Denver, my residence of record, the temperature was beginning to get challenging. The first snows had fallen, covering the fourteeners, those fourteen-thousand-foot mountains that serve as every Colorado boy's rite of passage. The Rockies can be a struggle. Cabo is not. I come here for sun and ocean and the lazy conversations I can have with Stryker while we're trying to spear some unsuspecting fish below his split cabin ketch. And this time for solace.

It was four o'clock here, and an inconsequential rain was going about the business of wetting the sand so it would shine more brightly at sunset. Sunset was two hours and three margaritas away. Stryker slapped his rag against the bar and scowled at me.

"Stop thinking about it."

"Maybe later."

"Why is this misadventure so hard to get past? There've been worse."

Stryker's ability to ask incisive questions is known in the vernacular as cutting through the crap. He was right. God knows, I've seen enough dead people. Dead from war, homicide, and disease. Dead from ignorance, accident, or by their own hand.

Maybe my discomfort had to do with the fact that these had all been Buddhist corpses in Boulder, Colorado, where they had no business being in the first place. The investigation had conjured images from that part of the world where the Buddha once walked and taught. Except

2

that when I was there on the government's pay list, Buddha was nowhere to be found, and his monks were setting themselves on fire in the streets of Saigon.

Stryker already knew what was eating at me. He never comes out and tells me these things. He just leads me to answers by asking the questions. By the time I get to the answer, I'm convinced that I've come up with a brilliant original insight. Then I see the small wry smile on Stryker's face, and I realize I've been had. It's part of what makes him a great bartender. It was also what made him my mentor. He assumed that role at the Rand Corporation in the late sixties when an earnest bunch of us elected to enroll in his school for spooks instead of sporting regular dog tags. Ostensibly retired in this Mexican paradise, Stryker is still my first-line consultant in the intermittent investigations I can't seem to turn down.

"Stryker, gimme another one and get off my case."

"Another margarita?"

"*Por supuesto.*"

My Spanish always improves in Cabo. It's also the only place I can tolerate margaritas. In Denver they serve a watery version in tulip glasses made with grocery store limeade and a splash of Pancho Villa. In Cabo Stryker squeezes fresh *limónes* and doles out the Sauza Conmemorativa and Damiana in exacting proportions. It is a drink for the tongue with sweet and tart and salty. Stryker put the plain tumbler on a fresh napkin and leaned across the bar.

"I'm not looking forward to your brooding, Bullfrog. Perhaps you'd rather stay at the Twin Dolphin?"

"I'm being threatened with exile for being in a bad mood?"

"Worse. I'm threatening to kick your ass out of my bar."

It was an idle threat, provocatively designed to let me know that he was truly concerned about my blue funk. Stryker is built close to the ground and hefty. If you failed to note the intelligence in his eyes, you'd think he was the kind of cop who used to crack nappy heads in old Southern towns.

He has about fifteen years and thirty IQ points on me.

He wasn't asking me to tell him about the murders because he already knew as much as I did. What he was asking me to do was make personal sense of it. "Shed light, not heat," he used to say. That was harder.

There had been twenty-five regular members of the sangha, some in Denver, some in Boulder. A sangha is the Buddhist equivalent of a congregation. In America, like Protestants, Catholics, and Jews, they traced their specific heritage to historical branches of some great theological tree first grown in another country. American Buddhists seem less strange when you consider that all religion in the United States is imported except for the shamanism of the Native Americans.

This particular sect was Nyingma, and they subscribed to a form of Tibetan Buddhism. Most of them held regular jobs, and there wasn't an Asian in the bunch. They had no credentialed lama for their group. Instead they had what is known as a lay teacher. She had been anointed by the lay teacher before her, and the Tibetan hierarchy continued to tolerate the departure from tradition. Gretchen Reynolds was the sangha leader. Dr. Gretchen Reynolds. She had a Ph.D. in organizational psychology, and for a while I was convinced that she was either a deluded fool or an academic fraud. She was neither, but, nonetheless, her flock kept dying.

The first call had come from my friend, Mack Barnum, at the Boulder County Sheriff's Department. It was high August, and Mack would have preferred to be working a trout stream instead of a homicide. I can't say I was too excited about the prospect, either. But most of my work these days comes from old friends in tight corners to whom I can't say no.

But now it was November, and I looked down at the

once fresh napkin under my drink and noticed that I had ripped it into neat shreds. Stryker Stephens, who knows when to keep his mouth shut, waited calmly for some response.

"I'll make a deal with you."

"Proceed."

"Sauza Tres neat, grilled langostinos, real avocado, and after that I'll tell you everything you want to know. Not here. Your deck."

Stryker did a rare thing then. He grinned. I remembered secluded seminar rooms at Rand where we used to make book on whether we could make Professor Stephens do anything more than smirk. Stryker cleared my drink and called over his twenty-year-old *mozo* to take charge of the bar.

"I don't know how you've survived, Hopper. You're so goddamned easy."

Highway 36 connects Denver, Colorado, with Boulder, Colorado, in a diagonal running southeast to northwest. It's mostly winding flat road that runs through suburban sprawl at first, then through small towns that sit on old mine shafts and remember better days. For the last mile or so the road climbs imperceptibly, and drivers are lulled by plains on either side. It's an easy drive until you crest the top of the hill and in a split second you are staring at a bowlful of Rocky Mountains with a human settlement tucked into its lap. When mammals are startled the first thing they do is suck in more oxygen. That's how the view got to be known as breathtaking. The first-time drivers inevitably forget they're driving. They cross a couple of lanes and start to swerve to miss the concrete divider. It's a miracle that there haven't been more accidents. Any idiot knows there are mountains in Colorado, but it took a highway engineer with a sense of humor to grade the turnpike that way.

Mack Barnum had called four times on Sunday, and

it took a while to get around to calling him back. I know he called four times in the same day because I have the kind of answering machine that not only tells me who called, but when—and to the digital microsecond. Mack, who has been indentured to the Boulder County Sheriff's Department forever, looks like a cross between a rancher and an old hippie. Maybe forty years old and maybe more. I'd met him at a law enforcement conference on sting operations in the drug trade. We both decided about halfway through that most of what the *federales* were presenting was horseshit. We ran into each other on adjacent bar stools in the hotel lounge, and since then we've had dinner twice a year to examine the imponderables of the law.

I was surprised to be hearing from him and even more surprised that he was being so persistent. Nothing much rattles Mack. His voice sounded more strained than I'd ever heard it, so I decided to call him back and find out who or what was hassling him. But first I had to deal with the arcane ritual of "getting through" the switchboard.

"Boulder Sheriff's Department."

"Mack Barnum, please."

"May I ask who's calling?"

"May I ask who's asking?"

"One moment, please."

She clicked off, and I daydreamed about an electronic alternative to switchboard operators that would provide the same safety with less irritation.

"Mr. Barnum is out of the building. If this is law enforcement business, I'll be glad to put you through to one of his deputies."

"This is B. F. Hopper calling from a pay phone in Denver, and you tell Mack that I'll wait till he gets out of the can."

It works every time because people who answer phones for a living have to lie a lot, and they appreciate it when you understand what they're up against. She almost giggled.

"I'll put you through, Mr. Hopper."

7

Mack sounded pressured.

"Jesus Christ, I'm sorry, Hop. I hate talkin' on the box, and Frieda has strict instructions to pawn off all the calls from yo-yos."

"Barnum, you're a lousy public servant."

"Well, you ain't exactly a Boy Scout. Listen, I got a weird one. Homicide. I been fifteen years on the force in a drug-infested place, and I've never seen anything quite like this. Makes me wonder. I could sure use the benefit of your thinkin'."

"It's your expense account. I'd be glad to talk with you, but I prefer the shadows. This one already sounds a little too front page for me."

"I hear you. I'll need to get the department to authorize you to consult, and we'll have to pay you some kind of fee. For right now I just want to talk to you about it."

Mack is one of those people whose life appeared to have left him in a backwater when he could have done greater things. Those of us who know him agree that such an idea could not be further from the truth. I was inclined to help just because he was Mack, but I'm tired of murder and mayhem and I only do it when it's worth doing—and because Stryker says its part of the mission. Everybody worth knowing has one.

"Just give me the basics, Mack. Who, when, where, and how?"

"Name is Tom Burroughs. Comes from money, but he's been living what we call an 'alternative life-style' for the past ten years. Shares a house with a bunch of Buddhists in north Boulder. They found the body up close to the county line, where the hoot owls do it to the chickens. Early Sunday morning. It was ugly, Hopper."

When a Coloradoan says that something is ugly, it means something different from what it means elsewhere. Colorado was the last gulp that Americans took on their way to conquering a continent. When something is ugly

8

here, it means that it offends all of the civilization that the pioneers painstakingly dragged in, kicking and screaming, from Greece, Rome, Jerusalem, and the British Isles on their way to the far edge of the New World. All they ended up doing was creating Los Angeles, but they couldn't have known that then. I wasn't anxious to hear what came next.

"How ugly, Mack?"

"Small caliber down through the top of the skull."

"That's pretty kind. It's quick, at least. What's your problem?"

There was silence on the other end. It was unaccustomed, and I could see the thinning blond hair and the rangy body hunched over the desk, paunch in his lap, waiting for the right words.

"It was like someone was trying to divine his entrails. The coroner said that the first cut went in at the base of the tailbone and wound around until there were two separate parts of him. His rod and reel were on one side of the line and his guts were on the other. Belly button was gone altogether. I looked at the poor sonuvabitch and hoped like hell that the bullet went in first. Reminded me of those cattle mutilations a few years back...."

He got quiet again, and I knew that I was in the presence of a good man who knew violence and who drew the line at brutality.

"You better have a good reason for making me think about this, Mack."

"Well, I think I do. You wanna have breakfast? The Diner, maybe?"

So I was on my way in the four-wheel. Almost everyone in Colorado has more than one vehicle. It's a necessity. I'd gone to Denver to get parts for an aging well pump, leaving the old Mercedes at home. Now I was content to be in something that could take any road. Highway 36 was a piece of cake except for the top of the hill. I've seen the Himalayas and the Andes, but I'm not above losing my breath when it's warranted. I made it to 28th and Arapahoe and was digging into my omelet before Mack rounded the doorway of the dining room.

"Good to see ya, boy."

"Good mornin' to you, and I hope next time we have more pleasant things to talk about."

He grimaced and appropriated half of my healthy whole-grain toast. No one has served white bread in Boulder for years.

Mack ordered biscuits and redeye gravy, and I nearly choked at the sight of it. It's a heavy meal and displeases the cholesterol conscious. But I was born in Two Buttes, Colorado, on the eastern plains. The smell of it brought back mornings when I was too young to be finicky and my stomach could handle anything. In the 1950s in Two Buttes, the biscuits and gravy were delivered by somebody's mother in a modest pink waitress uniform. But this was Boulder in the last quarter of the twentieth century, and here they came via a coed on roller skates wearing a short skirt that revealed all the well-muscled thigh that 10K races can create.

In the middle of breakfast Mack tossed me the photos of the body and continued to devour his biscuits and gravy. You would have thought I was about to see the latest school pictures of his kids. He probably would have given me his first theory, but I don't think he had one. I reviewed the full-color glossies. Predictably, Mack handled this grotesque evisceration like a street cop writing another parking ticket. I pushed the official envelope back across the table. Mack looked up.

"I don't know what the dharma thumper was up to, but I can't believe that he deserved this. It's weird."

"What else do you know?"

"Nothin' much. The call came in from some hiker who was out communing with nature. Found the body tied up against a tree at the edge of a field. He'd been dead at least since midnight on Saturday. The only thing out of place was the ash under the blood with some clumps of dark residue, like wax. Looked like a small fresh campfire. Sent the waxy stuff to the lab."

"What do the people at his house say?"

"Oh, they didn't think it was particularly strange that he was gone all Saturday night, but they have some sort of service every Sunday morning and they were sort of surprised when he didn't show up for that."

"Any suspects among the roommates?"

"Don't think so. They can all account for each other pretty well past the likely time of death. S'pose it could be a conspiracy, but they looked like pretty gentle religious nuts to me."

"How many are there?"

"Twelve in the house. Eleven not counting Tom. There's another house somewhere in Denver with another thirteen members of the same sect."

"Did you call Denver's finest?"

"Sure, for all the good it'll do. They're working on it, but it didn't happen in their jurisdiction, so unless I can get someone personally interested, it'll get dropped."

"Does this group have a leader?"

"Yeah. Nice woman. Dr. Gretchen Reynolds. She's a professor at the university. Calls herself the president of the organization."

"You're sure they're Buddhist, Mack? Would you know the difference between a Buddhist and any of the other cryptomystics in this county?"

He set the coffee cup down in the saucer a little too hard. You never really notice Mack's eyes are blue until he turns them on you steadily. It's like looking down the steel blue double barrel of a twelve-gauge. He was offended. There are times when I have an attack of what Stryker calls "acidmouth," and whoever I'm with gets the benefit of just how sarcastic and cynical I can be. It's almost never personal—it's just the way I am. But not everybody has it in perspective.

"Well, Jesus Christ, Hopper, maybe I don't know the difference. Maybe I better ask one of the boys at the state to take over the case."

"Sorry, Mack. Figure out how long it's going to take to forgive me and let me get the check. I want to walk the scene, and it's getting late."

He bought it, but not without grunting and turning his eyes away to rest on a horizon that didn't exist in the restaurant.

In the sixties, downtown Denver didn't have any buildings over fifteen stories high. Now it has slick, cool structures of smoked glass and steel locked into a pitiful competition with the mountains. They would have looked good on Chamber of Commerce brochures for Omaha or Tulsa, but for my money, too many great old buildings had fallen to make room for them. They annoyed me.

I had agreed to meet Gretchen Reynolds at the Buddhist Center in Denver on Tuesday morning. I was now on the Boulder County Sheriff's Office discretionary payroll for a nominal fee. Mack and I had done it before. The sheriff always approved it, and I always gave the money to the Boys' Club. I was now free to poke around in the case with a little more status than a volunteer, and Dr. Reynolds had no way of refusing to see me.

Sometimes the best leads in a homicide come from knowing everything you can about the victim. I wasn't sure what to expect.

The center was an old Victorian in a seedier section of Capitol Hill. The Hill was like its counterparts elsewhere, full of fading nineteenth-century elegance and landlords that didn't care much who you were as long as you paid the rent and didn't piss in the hallways.

I rang the bell, and it was answered by a small, intense woman with a child on her hip. Didn't look like the leader of anything. In fact, she wasn't. She showed me to a communal living room to wait for Reynolds. I looked around and was reminded that living spaces occupied by unrelated individuals never have much cohesiveness. It seemed like everyone had contributed some piece of leftover furniture from their last rental unit. It was functional, cheerful, worn, and unaesthetic. On the other hand, someone had spent a bundle on the Buddha in the entryway. A good bronze. I recognized *zafus* and *zabutons*, the traditional cushions used for meditation. They were all in primary colors and scattered carelessly around the couches and chairs. I still wasn't convinced that Mack had actually found himself a real nest of dharma thumpers. This place bore little relationship to the Temple of the Emerald Buddha in Bangkok. It had no trace of the ascetic elegance I remembered.

A tall, slender woman entered the room. Her eyes were red and her face looked drawn, but everything else about her was formal and controlled. The suit was off the rack, but her jewelry was the real thing. She looked out of place and distant. I was prepared to dislike her.

"Thank you for coming, Mr. Hopper. We appreciate your help."

"Sorry about the circumstances."

Something flickered in her eyes, but she busied herself with all the little rituals that smokers have. She found the case and her lighter and lit one of those designer 100-mm low-tar wonders. There didn't seem to be anything particularly spiritual about her, and I was never to understand why she'd made the commitments she'd made in her life. I could see that chatter wasn't going to work as a preamble. I didn't even try.

"I wonder if we could talk about Tom. I only know what Mack Barnum tells me, and it's not much."

She drew a deep breath as if she were about to launch into a detailed narrative. Instead she asked me politely what it would be helpful for me to know. I decided on a tactic.

"I suppose I should ask about his movements on the day of the murder, but right now I'm more interested in who he was and how he got to be a member of your group."

She drew her legs up under her on the couch and relaxed a little. Stryker always said that interrogating a woman required that you ask about the meaning of events before you asked about the events themselves.

"Tom joined the group about two years ago while he was married to a woman who is a member. They've since divorced, so he lived in the Boulder house and she lives here."

I would have ignored the marital relationship except that ex-spouses are always suspects. I have only had one verifiable love in my life, and I have little curiosity about how the rest of the world manages such liaisons.

"Isn't it a little unusual for them both to hang around, given the circumstances?"

"Do you know much about Buddhism, Mr. Hopper?"

"I've read a little. I lived in Southeast Asia."

She looked startled, but she didn't miss a beat. "You may be familiar, then, with the concept of karma, which states that anything we do not resolve in this lifetime—"

"We'll have to deal with in the next."

"Exactly. Buddhist practice demands that we work with whatever pieces life presents."

"So why did they get a divorce? You'd think they'd have tried to patch it up so they wouldn't have to go through it again in some other lifetime."

My voice must have conveyed some disdain, but Reynolds just smiled at me indulgently as if she were about to deliver an adult explanation to a child.

"The right action in any given situation is not necessarily the obvious or traditional one."

I wanted out of the case right then. I foresaw personal aggravation for myself in this investigation. Most of all I was afraid that this woman was going to tell me that old Tom getting disemboweled in a field was part of his damned destiny. I decided to switch gears.

"So what did he do for a living?"

"He was an apprentice for a local printer. He also did a little counseling on the side."

"Versatile guy." She let that one pass, but I figured I was going to have to start being more diplomatic. "What sort of counseling?"

"Oh, mostly creative visualization. Maybe a little astrology. I hadn't talked with him about it recently. My impression is that he was seeing university students primarily. Kids who were just getting interested in Eastern traditions."

"Why not meditation?"

"Tom wasn't ready to make that kind of commitment to his own practice. You have to reach a certain level in your own meditation before you can teach."

I was disappointed. It would have been so convenient to have a routine psycho or an enraged wife in the case. Instead I had an ex-wife who didn't even hate Tom enough to get out of the same religious organization. That and a bunch of benign but flaky college students. There were probably twenty more questions that I should have asked Reynolds just to round out the interview, but I was too bored or too uneasy to do it.

"Mr. Hopper?" She leaned forward from the couch, and her features softened. "I'd really like to help you, but I don't think I am. Can you say more about what you're thinking or what you really want to know?"

"I want to know who killed him."

"I have no idea."

"Then I'll settle for what you thought of him."

She nodded as if I'd finally asked the right question.

"I think Tom was still struggling with some very old conflicts with his father. Mr. Burroughs is wealthy, aggressive, and opinionated. Tom is none of those things. The more his father pushed, the more Tom retreated. He finally retreated here."

"To do what?"

"To find a path that wasn't like his father's. To find some other way to be accomplished in the world. I know how strange this is going to sound, Mr. Hopper, but all I felt when I heard that Tom had been killed was sadness. I was not surprised. I don't think he believed he had a right to live."

"The coroner has ruled out suicide."

She smiled. "You miss the point, Mr. Hopper. There are many ways to allow yourself to die. Faced with a violent assailant, Tom probably didn't do much of anything to defend himself."

"Because he was a Buddhist?"

"No, because he was Tom."

3

When I came back to Colorado after 'Nam, I spent seventy-two hours in Two Buttes, and I haven't been back since. I spent enough time with my folks so they could see that I was in one piece. I hadn't lost either hazel eye, the hair on my chest, or the two inches that I had on my father. He always said he was six feet even but never really got over five ten. My parents argued about whether my hair was dark blond or light brown, and my old man opined that if I filled out some, I'd look just like Grandpa Pritchard.

I did the things one does in Two Buttes. I spit at the Nebraska border. I watched my epidermis start to flake in the hot, dry air. I walked a piece of Horse Creek on a windy day and remembered that the sirocco of the eastern plains can drive you as crazy as the Middle Eastern variety.

When I left Two Buttes after that weekend, there was no particular place I wanted to be. I settled into a house in Coal Creek Canyon, far enough back from the highway to miss the buzz of the traffic but close enough to feel the

---

occasional rumble of a righteous trucker jake-brakin'
down the slot with a strong wind at his back. The house
is twenty minutes from downtown Denver when I ignore
the bureaucratically derived speed limits and punch the
curves at twenty over the numbers on the yellow warning
signs.

I would argue that I have the best of several worlds.
There is a category of land use in the Rockies that is mountain suburban. Indoor plumbing, electricity, optional telephone service, a convenience store or two, and a county
government that plows the road in the winter within three
or four days of a storm. You pay taxes for this. The upside
is that when you walk out of your house in the morning,
there is nothing but quiet and mountains and smogless
air. On such mornings I could just as easily be a Ute or a
Comanche as a high-tech survivor of an aging twentieth
century.

On this morning I was sitting on my front porch reviewing every detail of Tom Burroughs's death as I knew
it. The preliminary interviews with the sangha members
had been unproductive. The follow-ups with Gretchen had
produced more background but no direction. The silent
field on the north county line hadn't yielded much, and I
could feel the trail blowing colder down the back of my
neck.

The phone bleeped at me. I didn't want to answer it.
Obviously I should not have elected the telephone option,
but I had always been deprived as a kid.

"This is Hopper."

"Barnum here. The old man is in town. He's all over
the department like freckles on a redhead. Get your glib
ass into Boulder. You talk prettier'n me."

"Burroughs's old man?"

"No question. There's even a resemblance."

"I'm not into public relations, Mack. What the hell do
you expect me to tell him? Try stonewalling it—maybe
he'll leave."

"If he leaves feelin' the way he feels now, we're gonna

have one-eyed jack-offs covering this bit like mama's green chili smothering a fair-to-middlin' burrito. Bail me out of this."

I got ornery. I always get ornery when I have a problem that I can't get quick and brilliant about. I didn't know who killed Tom Burroughs, and I didn't want to discuss that sore point with anyone.

"Mack, this is the last thing I'm interested in doing at the moment. I'll back you up for free, but I don't have to take orders."

"Awright, Hopper. Just see this guy, and if you want out after that, I'll buy the black Jack. All this fancy talk makes me feel like a trial lawyer making my closin' argument with a sock in my mouth."

It sounded fair. I promised to be down the mountain within the hour. I pulled on a pair of pants, and the phone went off again. It was a woman.

"Mr. Hopper?"

"Never heard of him."

"Uhm . . . is this 645–5826?"

"It certainly is, but what area code are you dialing?"

"Well, I only dialed a one because it's a toll call, but this is the number I was given for B. F. Hopper."

"So now that you've got him, what are you going to do with him?"

I heard a pause and then a sigh. By this time I knew who it was, and I could feel her reaching for spiritual patience with this thoroughly abrasive problem solver who nonetheless must have a soul.

"All right, Mr. Hopper. This is Gretchen Reynolds. I'm sorry to disturb you, but Mr. Burroughs is in town to transport Tom's remains back to Connecticut. We're having a Buddhist funeral service this afternoon. We've invited him, and he's agreed to come. You're also welcome."

The stress in her voice was subtle and as controlled as everything else about her. Welcome indeed. It was clear that she and Mack both wanted some magic poultice to slap over this tiger's wounds, and I was it. I like funerals almost as much as I liked eating dog meat in the Mekong, but Mack was a friend, Gretchen an enigma, and all funerals yield information.

"Gretchen, it's time for you and me to get off this 'Mr. Hopper' routine. 'Sir' will do just fine. Why is everybody in such a hurry for me to meet Tom's father?"

"I didn't mean to—"

"Yes, you did, and yes, I'll be there, but first I'm going to meet our East Coast envoy at the cop shop. Tell me what you know about Burroughs. Tell me why Mack Barnum, who can handle orangutans in heat, is suddenly so uptight. And while you're at it, tell me why you sound like you've been sucking lemons. Does the man have three heads or what?"

There was a short silence. I don't believe the lady guru was used to having people talk to her that way. Or maybe she was just deciding what to tell me. I didn't have to wait long. When the answer came it was concise and practical.

"John Burroughs is chairman of the board of a holding company that was started by his grandfather. It's old Catholic money. They own everything from high-tech R and D to commercial real estate. He is well respected in the business community. Tom once told me in confidence that his father's fortune was in excess of thirty million dollars, and that was two years ago. He has a wife and two daughters. Tom was his only son. John has no one in the family to groom for his position except a son-in-law he doesn't like. He's very angry right now, but he has only one head. One is all he needs."

"What time is the service?"

"Four o'clock."

"Thanks. I'll make it. With Burroughs in tow."

I turned my attention back to my closet. This was going to be a piece of theater whether I liked it or not, and I pondered what to wear to meet someone who had thirty million dollars and a dead son. I chose the only dark suit I own by a Hong Kong tailor, a monogrammed white shirt, and a conservative undertaker's tie, and added western boots—black ostrich. No sense letting Burroughs get too comfortable with my image.

On the way down the mountain I kept switching the radio between Waylon Jennings and Prokofiev. Two Buttes and Westchester. Board chairmen and Buddhists. I pulled into the parking lot at Mack's place of business by noon.

The receptionist admitted that Mack Barnum was expecting me and buzzed me through a security door to the back. Mack and John were holed up in one of the interrogation rooms, although it was anybody's guess who was interrogating whom. Odds were Burroughs would be more impressed with a purposeful stride than a saunter, but I finally decided I didn't give a damn whether he was impressed or not. I stood to the left of the door and looked through the bulletproof window. Mack caught my shadow out of the corner of his eye. Without looking at me, spilling a word, or interrupting his train of thought, he cracked the door and let me in.

An impeccable John Burroughs rose from a battered chair at a dingy baize table. The room got smaller, not because of his size, but because of his presence. I had sudden visions of eighteen-foot teak conference tables where corporate America demanded information about quarterly profits. This was the kind of man who presided over such tables with no wasted motion—or emotion, for that matter. A champion athlete in the profit-and-loss league. He looked like he'd been doing it forever.

I could see vestiges of his son's corpse in Burroughs's face, the same patrician angularity. He was twenty-five years older, shorter, heavier, but he was also the same. I wasn't able to listen to Mack making a feeble attempt at an introduction because John Burroughs had locked his eyes into mine, and as we shook hands I could feel him assessing the strength in my grip. It was a two-second acid test.

"Good afternoon, Mr. Hopper. I'm John Burroughs."

"Afternoon, sir. I guess Mack told you why I'd be here to meet you. You have to understand that Mack and I go back a ways. It's ex officio. I suppose you'd like to know what's going on. I wish I knew myself. I've been poking around where I can, but I have a limited role here. If Mack did it my way, he'd be out of a job."

His eyes narrowed, and he put one hand in a well-tailored pocket while he leaned the other, knuckles down, on the baize table. He measured his response carefully, then spoke.

"Hopper, B. F. Born in Two Buttes, Colorado. Educated at Stanford. The only Coloradoan admitted from outside of Denver for more than ten years. Bachelor's in electrical engineering, master's in computer science with a minor in theater. A.B.D. in international affairs—emphasis socialist political theory, a fellow at the Hoover Institute. An impressive list of postgraduate work. Some at the University of Hawaii, allegedly sponsored—and then Hong Kong, where you met your wife. Recruited by the Rand Corporation in the sixties. Counterintelligence in Cambodia and 'Nam. Lost your wife there and dropped out, but the Pentagon still keeps an eye on you....Let's dispense with the down-home preliminaries, Hopper. I know who you are."

He had my dossier. He didn't have all of it, but it was enough to indicate whom he knew and the circles in which he moved.

There are three things Stryker Stephens taught me to do when I was caught in this corner. The first was to stop behaving, to control my facial muscles, my breathing, and my skeletal alignment so that the my opponent read "No response." The second was to focus my own perceptions externally so that the other guy's face, breathing, and muscular movement became like a road map that I could follow to my destination. The third thing that he taught me was to use a verbal feint, something mundane that belied the importance of anything that had just happened.

I could see my poor friend Mack, six feet away, with his eyeballs widened and keeping his mouth scrupulously shut—so shut, I suspect his jaws ached. I turned toward the door.

"Mack, I'd like to take you to lunch, but I think that John and I have some things to discuss that would only compromise your official position."

I looked back over my shoulder, and Mack was looking straight at Burroughs and shrugging as if to say, "It's your call, Sarge." John nodded and followed me out the door, the grief beginning to show through the cracks in his reserve.

By the time Burroughs and I mazed our way to the front door, I'd decided that I might as well play this one straight. *Ex officio* means having nothing to lose.

"Your Excellency, I don't know what you think you were doing back there, but if I'd wanted every mother's uncle to know about my past, I would have taken out an ad in the paper. Pedigrees are private. Mack and I don't deal that game. I won't ask you where you got your information, but I'd like to know why you went to all the trouble."

When I got to my wheels I turned to face him. He looked mildly abashed.

"Where can we talk, Hopper?"

"There's a park just up the road. We can walk."

We cleared the sheriff's parking lot and hit the footbridge without having said jack shit. We just walked, and our strides matched just fine. I considered taking a stab at being human.

"I understand that this is a very difficult situation for you, sir."

He nodded. "Don't call me sir."

"I'm sorry you don't like 'sir,' but I only refer to my friends by their first names, and 'Mr. Burroughs' sounds a bit stuffy. I also think we have tougher problems to deal with for now."

"Do you have a first name?"

"Yessir, but my mother is the only one who uses it. Hopper's fine. I'm surprised you didn't get that information where you got the rest of it."

"No big mystery, Hopper. I have friends in D.C. and friends in Denver. When my Denver man reported that Boulder County had called you in, my people in Washington checked you out. Your name was hardly my greatest concern. Upon my arrival, Barnum had nothing to report except what I already knew. So I'm frustrated and planning to hire my own people. Sitting on my hands is not a talent I've ever cultivated. Mack and his bureaucracy be damned. I'm going to move this thing off dead center even if it shouts 'Tilt.'"

If there was a veiled criticism in what he said, I didn't detect it. He seemed like a man with no choice but to do what he'd always done, which was to take charge. He looked at me sideways, and I stared straight ahead.

"Are you available, Hopper?"

"Available for what?"

"I want to know who killed my son."

"Are you trying to hire me?"

"Yes."

"I'm not for hire. This case is a mess, and I haven't been able to come up with a goddamned thing in the last forty-eight hours. That's not much of a job application. Why me?"

"Do you abhor money?"

"Sometimes I resent it and the prostitution it supports. I'm a poor mercenary but a fair crusader. You can't buy the truth. Sometimes you can buy the facts you want to hear, or a piece of theater, or an operative, but in this kind of investigation—"

"I'm betting on the come. I know that. Don't patronize me, young man."

I'd touched the hem of the chairman's mantle and gotten my wrist slapped, but I was beginning to like him all the same.

"I'll take it on your terms only if we both win."

"Meaning what?"

"If, to your satisfaction, I close the lid on the worm that did this and you feel okay about it with or without an arrest, we'll discuss the terms with my agent. If I don't, you owe me nothing."

"That's more than fair. Any other terms?"

"No, but I don't take orders well."

"Yes, I heard that. Seems you do pretty well without them. What do we do first?"

"We've both been invited to the Buddhist service for Tom."

He winced. We turned around and headed back to my mature Mercedes, a buggy so old that Burroughs probably hadn't inhabited one since he was a kid trying to get some in the backseat.

Thirty minutes later I hooked a right onto High Street and hoped this gig wasn't going to be too hard on either of us. Burroughs seemed to be a proud and private man, and I wasn't sure he could see his son's other life without experiencing a lot of stiff-necked pain. Besides, I had no idea what a Buddhist funeral in Denver would be like. In Tibet the lamas sit on mountaintops to die and present themselves to vultures as carrion. It is part of their bodhisattva vow to serve every sentient being. Even vultures. It's simple.

We were admitted by a sangha member, but Gretchen was nowhere in sight. I don't know why that bothered me, but it did. We were led silently to an alcove set up with folding chairs, and Burroughs took possession of his chair like a foreign dignitary. We faced the larger room, where various people were clustered around something that looked like a round, flat altar. They sat on cushions, and our elevation seemed unnatural. We were the only ones in the alcove.

On the altar was a chair draped in a white scarf with

a picture of Tom on the seat. It looked like a yearbook picture. Burroughs looked at the photograph once, hard, and then turned his head to survey the passing scene. The sangha members looked more like a congregation when they were all sitting on those damned cushions. It still didn't look like Bangkok, but at least they looked like they belonged together.

I recognized most of them from the interrogations Mack and I had conducted. Some were tearful, some were stone-faced, and some looked like they were somewhere else. I guess they were meditating.

Gretchen finally arrived in a long traditional Tibetan gown. It didn't look that strange if you'd grown up in the sixties, but I was used to seeing her in business suits, so it required a second once-over.

She nodded at us, and I nodded back. Burroughs kept his neck rigid and folded his arms across his chest. Gretchen walked to the front of the makeshift altar and began a prayer. She told me later it was Sanskrit. Burroughs pursed his lips, narrowed his eyes, and watched her like a disgruntled panther.

They started a Tibetan chant, and I thought he was going to lose it. He leaned forward a fraction of an inch, and I caught the blood pounding through a vessel at his temple and the jaw muscles moving the bones in front of his ears in double time. Only one prayer, to the great Metreiya, had been translated into English. I don't know whether that was better or worse. At least all the foreign noise kept him ignorant of the content of his son's spirituality, but the Metreiya prayer left no doubt.

Gretchen's eulogy was a different matter. She did a nice job even by Methodist standards. As she talked about Tom's finer qualities, Burroughs's posture softened, his shoulders sagged. For a minute or so he was looking at her without rancor.

Then the sangha was invited to speak of their personal memories. I shifted on my chair so that I was shoulder to shoulder with Burroughs, and it was both instinctive and useless. They were going to say what they were going to say.

A woman named Patty spoke first. She was a long cool drink of water, and her voice only shook a little.

"I keep remembering the night that I asked Tom to help me with this personal problem and he was so gentle and right on. He talked about the three jewels of Buddhism, and I went to meditate. Tom always knew what to say. He could have been a teacher someday." There were tears streaming down her face.

Buddha, dharma, sangha. The teacher, the path, and the community. I wondered whether Burroughs could translate that into CEO, corporate objectives, and personnel. Could he see Tom as a middle manager in this eccentric organization? Or could he see only an atypical woman presenting his son as a stranger with foreign ideas? I couldn't tell.

A man named Billings was next.

"I remember Tom talking about failure. He said that failures were an opportunity to learn about illusion. He said he'd failed with his family, but he'd learned so much from it that he succeeded with us."

It went on and on until everyone had spoken. The themes were the same. Tom Burroughs never hurt anyone. If Tom Burroughs had ever had any vindictive or competitive urges toward his fellow human beings, they were nowhere in evidence.

Then the gift giving began. People dropped flowers and incense sticks on the altar and said their last good-byes. It was not unlike the throwing of carnations onto a Christian casket in a consecrated grave. Old man Burroughs was stone-faced.

I figured the worst of it was over. I had forgotten that Buddhism is a paradoxical religion and that funerals were serious business, not designed to provide comfort to the mourners, but to speed the soul on its journey away from the attachments of the lifetime most recently abandoned.

Gretchen took the picture of Tom out of the frame and

ripped it to shreds, stuffing it into a large brass incense burner. She lit a match, and fragrant smoke escaped from the vents of the receptacle.

Burroughs sucked in air through his nostrils with such force that I was convinced he would exhale fire and kill the Buddhist bitch on the spot with the heat of the flames. He was both the dragon and St. George, Catholic to the core and repulsed by this heathen desecration of his son's memory.

Gretchen heard the intake of breath and turned in our direction. She told me later that the burning was a ritual, symbolic of the sangha's desire to detach from the soul and to send it on a good journey to a next life. It may have been well intentioned in the tradition of the *Tibetan Book of the Dead*, but it was also a serpentine dagger twisting and turning in the old man's gut.

At the end of it, I understood how men create wars out of just such misunderstandings.

I drove Burroughs to Stapleton International Airport. It is the fifth busiest airport in the world, just behind London's Heathrow. Not bad for a cow town.

We were silent. Leaving the center seemed to take the wind out of the chairman of the board. He brooded, hunched in the passenger seat, and asked a few desultory questions about Denver's economy. He only had a briefcase, and we walked to the northwest gate looking for flight 692 to Westchester County Airport—nonstop. I sat with him until boarding. They called his flight, and we both got up on cue. The sun was over the yardarm.

"I'll be in touch, Hopper. Stay on it."

"No problem, sir."

He turned, and his mouth was hard—but his eyes were two pools. He put his hand on my shoulder and squeezed, and I don't know what the hell happened, but we ended up in a Latin *abrazo*, one of those rare displays of male affection that the South Americans are much better at than we are. For a moment I was Tom, I think, and he was my rangy rancher father in Two Buttes.

He picked up his briefcase and stalked to the gate. My teeth were clenched. Who needs this shit, I thought, as I wandered over to an airport minibar and ordered a double black Jack with one cube.

The overhead page droned on, and I heard my name come over the system in a sterile, familiar cadence.

"Mr. Hopper, Mr. B. F. Hopper, please come to a white paging telephone. Mr. Hopper, Mr. B. F. Hopper."

I was tempted to ignore it, but I downed the Jack and found a telephone.

"This is Hopper. You paged me."

"Uh, yes, Mr. Hopper. Stand by, please. I'll patch you in to the Boulder Sheriff's Department. Detective Barnum on the line."

It was all white noise for a moment, then Mack came on, cold as ice.

"I think the sonuvabitch got another one. Sweeney just called from the Capitol Hill precinct. He's got a fresh corpse named Childress. They found her in an alley two blocks from the center. Rape and homicide, he thinks, but the coroner won't confirm that for hours. Not more than ninety minutes ago."

"What makes you think—"

"She's one of Gretchen's dharma thumpers, asshole."

My brain completed the association between "Childress" and the woman named Patty who had spoken with such feeling at the funeral.

"Where's the body?"

"Cherokee Street, north door. I'll meet you there in twenty minutes. Move it."

I broke my own record for getting out of the airport and was speeding down Martin Luther King Boulevard before I stopped to notice how angry I was. Most people who commit murder do it once. They usually kill someone they know, and it's a crime of passionate impulse. Not so with the serial sickies. They affront the law by striking

33

again and again right under your nose. Embedded within each crime is a taunt. This miserable bastard had managed to kill a woman that I had been in the same room with maybe an hour or two before. He hadn't even waited till dark.

Having a P.I. license in the state of Colorado does not entitle you to a parking space at the downtown police administration building. I screwed around with the one-ways looking for a meter. I ended up a block away, but the walk gave me time to get cool again.

I asked for Sweeney at the front desk, and they waved me through without the usual I.D. procedures. Hardly anybody knows about the coroner's private slab at the downtown brig. It's a small, medically equipped room they use when everybody wants to avoid the zoo at Denver General. Mack was already there since he had the advantage of flashing lights to get him down the highway at eighty-five miles an hour.

I shook Sweeney's hand and listened to the coroner summarizing his findings in that flat, matter-of-fact voice with which all coroners are born.

"...and that the immediate cause of death was a gunshot wound from a small-caliber weapon. The bullet entered from the back of the neck just above the first spinal vertebra and severed the spinal cord on its way out through the trachea. The close range and precision of placement suggest that the victim was already unconscious from a sharp blow to the back of the head. There is a diamond-shaped flap of skin approximately one centimeter long missing from the middle of the forehead. The incision was likely made after the gunshot wound."

He clicked off the recorder and looked over at Sweeney.

"Off the record, Al, I'd say this clown knew something about anatomy. I think all the bruises and contusions happened before the shot. He may not have expected her to fight quite so hard, but he knew exactly how to position that bullet after he finally got her knocked out. Same thing with the incision on the forehead. Nice and neat."

"Thanks a lot. We'll look for suspects with M.D.'s."

I walked over to the body. I would later read the transcript of the coroner's tape several times, but just then I wanted to take my own look. Her face appeared to have taken two quick punches to the left jaw and right eye, which was swollen shut. The rest of the bruise patterns were typical of the kind of scuffle in which an assailant has to hang on hard to a thrashing victim. There was blood at the throat and the back of the head. The coroner noted that the abrasions on her back and buttocks probably came from being thrown against the brick wall in the alley a couple of times.

We continued to debrief. Despite the fact that her underwear was missing, there was no evidence of sexual assault. Forensics was still working her clothes over for human hair, but the coroner hadn't found any tissue or blood under her fingernails. He was pretty convinced that it was a male assailant who was shorter than the victim, but at that point it was all conjecture.

Sweeney, Barnum, and I repaired to a less oppressive office to chew on what we had.

I started with Sweeney.

"What'd you get from the scene?"

"Not much. The call came in from a local car that was flagged down by an old crazy who kept jabbering about the pretty lady in the alley. He's harmless except that he rolled the corpse before he went to look for help."

"Do you still have the barriers up?"

"Yeah. The neighborhood is freaked out. I'll keep two guys there until tomorrow morning."

I glanced at Mack and he was fidgeting.

"Let's go look, Mack. I want to get a feel for the ground between the center and the scene."

I left thinking about pretty Patty, who, unlike Tom, chose to fight like a banshee to stay incarnated in this lifetime. I wondered what Gretchen would say.

Police barriers are fragile psychological things. In Denver they consist of yellow plastic ribbon strung between anything at hand and a white sawhorse or two. The ribbon says CRIME SCENE— DENVER POLICE DEPARTMENT in black lettering. The barriers physically restrain no one except the law-abiding, who will come within a hair of the tape to gawk without ever quite touching it. Ribbons will do for pedestrian traffic, but to cut off vehicular traffic at both ends of an alley requires squad cars, white Ford LTDs with the universal police motto on the side in royal blue. To SERVE and PROTECT, it says. Most of the time they're better at protecting parades than dirty alleys where people are about to get murdered.

Vice has it easy. Those guys can survey, play at undercover, bust small-timers to get information. There's always something to do when you're working vice. Homicide is different. Homicide oozes into action after everyone who was supposed to serve and protect blows it. The DPD psychologist once told me that the vice cops are always in danger of becoming cynical and abusing due process, but the homicide detectives get suicidal on a regular basis.

We started at the scene. Mack flashed his sheriff's badge to a uniform, and we stepped over the barriers. The chalk lines that they'd drawn around the corpse were midway down the alley in a recess hidden by a Western Waste trash dumpster. It was well hidden from the side streets. A lot of privacy for violence. The block was pretty typical for the Hill, zoned B-8 with Victorians cut up into little office spaces. Early on a Sunday evening there were no office tenants around to see or hear very much.

Mack chatted with the crime lab guys, who would have been through hours ago except that they are always required to sift through the goddamned dumpsters looking for weapons. I paced around the twenty-square-foot area in which most of the action must have taken place. It was clean as a whistle. Whatever had been on the ground had been bagged and logged by the lab guys.

You could see where the head wound had been inflicted. There was blood splatter on the metal D&H Towing sign that hung low on the back wall and notified the public that this was twenty-four-hour private parking and offenders would be ticketed and towed. I knelt down to look at nothing in particular and concentrated on the line where the black asphalt met the red bricks of the building. It was nine o'clock, and the late-arriving August night had finally dropped. The back alley lights went on courtesy of Public Service.

Stryker Stephens used to teach a technique that he called "third-eye reconnaissance." At first we all thought he was eccentric, but the more we practiced, the better we got. So I squatted with my back to Mack and the lab guys and passively looked at nothing, moving my head 180 degrees from right to left and back again. "Scan without judging," Stryker would say. "Be a sensing device and not an average human being who compulsively makes meaning out of everything he sees. If you must make meaning, you're dependent on your previous experience and your concepts, which are limited as hell. You'll miss things that way."

There was a small dull gleam in the asphalt about six inches from the chalk outline of Patty's neck. It was tucked away in a little crevice. I touched the wall about a foot above the sparkle and ran my fingers lightly down the brick. I wanted to look like a meditative investigater contemplating the wall's navel. When my fingers reached the asphalt I scraped the little piece of metal into my hand and palmed it into my pocket. I squatted for another ten seconds, shifted my weight, and stood up. I walked back to where Mack was still palavering. He looked frustrated.

"Well, Hopper, they got about as much as you'd expect from a setup like this. A lot of stinking garbage and no evidence."

"Let's go back to Cherokee, Mack. I want to look at the statements Sweeney's men got from the sangha. Then I want to talk to Reynolds—and Tom Burroughs's wife."

"I thought you wanted to walk the neighborhood."

"Don't need to. He approached her somewhere between the center and alley, and she went with him willingly."

"Is this one of your intuitions, Hopper?"

"How else could it have come down, Mack? The first victim ends up in a rural backwash, miles away from where he's last seen, no signs of a struggle. There's not even any evidence that the murderer roughed him up before he started shooting and carving. The second victim fought like a she-cat, but not until she got into the alley. If she'd done that on the street, it would have been better than fifty–fifty that even in this godforsaken neighborhood somebody would have noticed. He's clearly not a hulking hoodlum because there is no elegance in the way he subdued her. He got lucky when he managed to back her into the wall. My guess is that she was almost as strong as he was. The point is that he didn't need to subdue her until real late in the game. And remember that the only thing that ties the two murders together is a small-caliber gun and the fact that the victims were both Buddhists. Let's assume that the man has an angle on luring Buddhists into dangerous places."

Mack snorted and stuffed his hands into his pants pockets. "Well, thank you, Professor Hopper. I certainly do appreciate your educating an old country sheriff like myself. Do you have any words of wisdom that you might share with me about motive in this case? Just anything at all will do, don'tcha know."

Trust Mack to poke me in the rib that hurt the most. I drove back to Cherokee, reviewed the statements, and made a phone call to arrange breakfast with Gretchen. The little piece of metal was burning a hole in my pocket.

Reading the statements from the sangha was about as useful as consulting a palm reader. Patty had left the center after the potluck, which seemed to be some arcane American Buddhist version of a wake. She went to take a walk to "clear her mind," she'd said to at least two people. It was only 6:30 P.M. in August and still broad daylight. Nobody tried to stop her. They were Buddhists, and I guess they thought that everything would be "perfect for her learning." The rest was from the police blotter.

By the time I got back to Coal Creek it was about to be Monday morning. I do some of my best thinking after midnight. I shoved the Saturday breakfast dishes into the sink to clear a space on the dining room table where the light is good. I grabbed a white sheet of paper and dumped the little piece of metal onto it, where it clinked once and settled. For all I knew it might have been a piece of gilded foil from a beer bottle or a slice of tinted aluminum from a can.

Instead, it was a gold crown, and whoever had lost it

probably had a sensitive exposed nerve or, at the very least, a stubby half tooth in their mouth. Might have been the corpse, who wouldn't have minded much. The blow to the head would have done it. Might have been some unlucky stranger who'd been sleeping it off in that same alley. It was a little bunged up, but it looked too clean to have been there very long.

It was an odd-looking piece of work. It had an unusual back ridge on it, and I couldn't tell what position it might have held in the average mouth. Somewhere between the incisor and the molar. Not something that would show except in a broad grin. I rolled it around gently, inviting it to speak to me, but it was just a little lump of mute dental magic. I tossed it back onto the white paper and picked up the phone to call Mack.

Then I cradled the receiver. I thought about it. I picked the phone up again. I repeated the sequence three times. I didn't want to call Mack. I didn't want to call anyone except maybe Dr. Abe, a genius in forensic dentistry who'd assisted with a few tooth clues in the past and taught me a lot in the process. I thought better of it. No need to disturb the old man's sleep. The phone call to Abe could wait until morning.

The crown probably didn't mean anything anyway. It was evidence only because it had come from the scene, and if I was withholding it, then I was doing so on behalf of my client, who wanted to get things out of the blocks as fast as he could. The argument probably wouldn't stand up in a court of law, but it was one o'clock in the mountain morning and there wasn't a jury in sight.

I wrapped the tooth in the paper with a well-executed pharmacist's fold that I'd learned at an evidence seminar. I stashed it before lowering my bones onto the couch to take a nap until it was a more decent hour to make phone calls. I slept till noon.

There oughta be a law about being awakened by the Bell system. I knocked the phone off the end table reaching for it and muttered some profanity that I don't even remember. It was the high priestess calling to take me to task for missing breakfast.

"Are you all right, Hopper?"

"Just dandy for someone who slept on a lumpy couch for a few hours. What time is it, anyway?"

"A little after twelve. I waited at the diner as long as I could, but I have to arrange for—"

Her voice broke, and I listened to her cry for a while. It's a terrible thing to wake up to the sound of a woman crying. Particularly if she's the type that doesn't cry very much. I closed my eyes and waited before I asked her to turn off the faucets. I figured she had a right.

"Listen to me, Gretchen. I need to see you this afternoon. I also want to see Tom's ex-wife, Antoinette. Are you aware of any special relationship between Patty and Tom? Say, a friendship that was turning into something a little more?"

She sniffled and took a deep breath. "No. Patty was happily married to Kevin. I can't imagine . . ."

"Yeah, well, have Kevin there, too. We may be overlooking the obvious."

"Hopper, are you trying to say that you think that someone in the group did this? That's ludicrous. The night Tom died, Antoinette was asleep in her bed, and Kevin stayed late at the potluck yesterday. We're Buddhists, not sorcerers. No one can be in two places at one time. There was no opportunity. What the hell kind of person are you, anyway?"

I'd made her mad, but she'd also stopped crying. I needed Gretchen Reynolds intact and on my side. I took a stab at soothing her.

"Look, you have a choice between me asking the routine questions or some clumsy sergeant from the force. I'm sure your folks are pure and innocent, but in these situations no cop is going to take that for granted. Do I get my audience or not?"

She sighed. "What time will you be here?"

"By three o'clock I want the whole congregation at the Denver Center."

"But some of them went to work and won't be back until—"

"They're your flock. Round them up. By the way, what are you doing for dinner?"

She groaned and slammed the receiver down, but she'd forgive me before the day was out. I called Mack, and he was grumpy. I briefed him on my plan for the dharma thumpers, but I didn't say anything about the gold crown. I asked him to bring a copy of the typed version of both coroners' reports and to wait for me to pick him up for the three o'clock soiree.

I reviewed the reports in the car while Mack drove. They didn't indicate any missing teeth. Just a missing flap of skin, a missing navel, and two missing lives.

They were all in the living room when we got there. Mack kept looking around like he was in some sort of museum. Antoinette Burroughs and Kevin Childress, the recently bereaved, sat together on the couch. They looked numb, incapable of violence beyond dental floss. My suspicion was deceitful to begin with. I'd just wanted the chance to get them in one room under the leadership of their teacher because what I had to say to them required a little drama and Gretchen's full support.

"Drama has its uses," Stryker used to say, "but most people end up getting addicted to it. If you must have the play, make sure you're the director." I paused at the archway to the living room and said nothing. The room buzzed, and the sound was like a beehive that had been unnecessarily disturbed. I focused on the queen bee.

"Gretchen, thank you for arranging this opportunity. We'll try to be brief so that those members of the group who must return to work can do that without too long a delay."

I stopped for a moment and looked at every face in the room, starting at one end and moving leisurely to the other, including in my search three children and two cats. By the time I finished the hive was quiet, attentive. I elected to remain standing in the archway. I folded my arms, looked down at the ground, and paced very slowly until I knew that all of the eyes in the room followed my movements. I returned to the center, stopped, and looked up abruptly.

"Mrs. Burroughs, Mr. Childress, I know this is painful for you, but I ask you to bear with me. Dr. Reynolds tells me that you all have a commitment to this group and to something that you call 'the path.' I don't pretend to understand all of that, but I'm going to take her word for it. What I understand very well is that two of your people are dead, and that it doesn't end there. Tom was butchered in a field, and Patty was battered before the bullet got her. Mack Barnum, who is standing back there, can tell you lots of stories about killers who like to thump, blast, and carve. This killer likes all three. Chief Barnum is in charge of this investigation and has several requests of you. Mack?"

Out of the corner of my eye, I saw one of the women reach over to put her arms around a nine-year-old boy whose eyes were wide and fast losing innocence. He resisted a little and then let himself fall back against her familiar shoulder. I could not protect him with anything but the truth.

I ceded my spot in the archway to Mack. He was gearing up to do a law enforcement speech with a few embellishments from the Stryker school of domination theatrics that I'd laid on him during the ride over. He cleared his throat and took off his hat.

"Those of us in the department who investigate death for a living look for patterns. I can tell you for sure that

these murders weren't committed at random. Even if Patty
and Tom had been members of some Protestant congregation, and killed with different weapons, we would start
with the theory that the murders were connected just because the victims knew each other. As it is, we're lookin'
at the same weapon and two victims that are part of a . . .
more unusual religious group. So it seems like someone is
stalking the members of this sect."

I'd fed him that line and coached him on taking a long
five count because I wanted everyone in the room to reach
his or her own understanding of being prey. "Check for
the effect," Stryker used to say, so I used the time to observe how many human beings Mack had transformed into
scared rabbits. I looked for twitching fingers, shifts in body
position, a narrowing of eyelids. I found it everywhere
except in Gretchen, who was staring at me with cool, clear
eyes, almost mocking in their steadiness.

"Now it's not the department's job to tell you how to
live your lives. You got a right to do what you think's best,
but I want you to have the information you need to figure
out what that is. You gotta keep in mind that any one of
you is a potential victim. Since Patty's murder, we've been
watching out for you as best we can, but don't get too
comfortable 'bout that. We're usually more effective at
trapping criminals after the fact than we are at preventing
murder. It's the nature of the beast. There's not a damn
thing anybody can do for you if you refuse to help us protect you."

He took one more beat—his own idea. The crowd was
divided between those members who were mentally putting on their armor and those who were looking confused.
Mack had folded his arms and was looking for gross revelations of guilt. He had one more line.

"If you're the murderer, we'll get you eventually. But
it's highly probable that the killer has picked one of you
as the next victim, and we can't help you without your
cooperation."

The hive exploded again, this time with shock and anger.

A hefty full-bearded man who looked like he should be cutting timber in the mountains stood to face Mack off. "What are you suggesting, Sheriff? That one of us did it? That we should all be looking at each other with distrust? You want us to do your job and put each other under surveillance?"

Mack was cool in these circumstances. He'd been dealing with university protesters and angry ranching communities for most of his years.

"Don't be an ass, Tarzan. If anyone in this group was a serious suspect, they'd be in custody by now."

I could see the lumberjack's fists clench. I didn't think he was really going to go after Mack, but a little good-guy PR to balance Mack's bad-guy bluster seemed called for.

"I think Mack's suggesting that you deal with simple realities like being more cautious." I girded my loins for a metaphysical discussion of reality. It wasn't to come from him because he sat down abruptly like a football player taken to task by his coach. The next challenge came from a woman in her early twenties, her jeans worn at the knees and other predictable places. She was a rabbit, but she was a brave rabbit. Her hair was badly cut, but her features were finely sculpted. She had the energy of a reformed cynic.

"I came here to learn about getting over fear. Buddhism teaches that we create our fears in our own minds out of our own karma. I don't appreciate your little manipulations. I trust everyone in this room. If I die now, it will be because that's what I need to learn in this lifetime. You're asking me to buy into some cops-and-robbers illusion. Do you understand what you're saying and who you're saying it to?"

I let Mack take that one.

"Look, if you have to trust everyone in this room to stay straight with your god, that's your business. In my opinion it's not your safest option. I'd prefer you to be less trusting with everyone outside this room, at least for the time being. I want you to stay close to people you know and not take unnecessary chances. If that's manipulation, I'm guilty, ma'am."

Mack stood steadfast and cased the room, allowing his eyes to light on me once for moral support. Precipitously, Gretchen stood up and joined us in the archway. When she spoke her voice was as smooth as glass and unhurried.

"I understand your distress. Please listen. What Sheriff Barnum has chosen to tell us is not easy to integrate. The beliefs that are the foundation of this sangha are far removed from murder, distrust, and, most of all, from the fear of death. I would remind you, however, that a cornerstone of Buddhism is the piercing of illusion. That a human being is either a murderer or not a murderer is a duality, an illusion. Keep in mind that we assume that, in a lesser karmic state, each one of us has murdered and has been murdered. Please attend, then. What Mack suggests is offensive to the ego but not to the teachings." I felt her hand slip into mine. It was cool and dry to the touch.

The eyes of the bearded man darted from Gretchen's face to mine. I couldn't quite read the question in his face, but it seemed now that his pique was directed at me and not Mack. Gretchen looked at him beatifically, and he started to weep with his face in his hands. The young woman also hung her head as if chastised, and suddenly I wished to be back in Two Buttes with this case and my history still in front of me.

Most of all, I wanted to quash the intuition that Gretchen Reynolds was a suspect. It was difficult for me to see her slender fingers at a trigger or around the shaft of a knife, but she was much too calculating in this situation, much too available to explain. And she had much too much power with these frightened people.

A slight internal shudder ran through me, as it always does when an ally becomes ambiguous. I scrambled for a good argument against adding her to the list and found none. She had no more of an alibi than most of her flock. The room got cold.

Mack ended his spiel. He gave them practical advice on what to do next, and I disengaged myself from Gretchen, avoiding her eyes. I declined to invite her to dinner. I needed time to test the new perception.

Nobody came up to make conversation as we left. Mack pushed open the front door, and I nearly ran into him because he just stopped. I could see enough to know that there was a woman standing in front of him on the porch. I heard him growl.

"Ah, shit, Kate."

"Hey, Mack."

I stepped out from behind him to the sight of a young woman, twenty-six or twenty-seven, tops. She looked like a kid with her hand in the cookie jar. Barnum was annoyed. We exited, and Mack shut the door behind him with a little more force than was absolutely necessary. She stood her ground and tossed her long brown hair behind her shoulder. She adjusted the horn-rims and looked like a graduate student in anthropology.

"I hope that whoever you got this from told you that you can't use it."

"They mentioned it."

"What do you want, Kate?"

"An exclusive . . . when it's time."

"Awright. Come see me tomorrow."

She nodded and turned to go. She looked innocuous, but some of the best investigative reporters I've ever known looked innocuous.

"Who's she, Barnum?"

"She's from the *Camera*. I'll handle it."

I have no use for most of the press, but if Mack said he'd handle it, there was no reason to doubt that he would. We retired to the Satire Lounge for chili rellenos smothered in green. It is blessedly dark in the Satire. It's a drug-dealing, shady-business, pickup bar frequented by Colfax locals, but it's also the kind of place where you can avoid any trouble that isn't yours. Safely seedy. After the third Corona and the fourth surly encounter with the waiter, I started to think about home.

"So, what's next, Hopper?"

"I gotta go see a man about an idea. I'll call you tomorrow. I also need more time with Reynolds."

"She spook you, too, friend?"

"Just a little. Just enough."

On the drive home my mind wasn't working well enough to have written a "wish you were here" postcard to Mom from Miami. Days are funny things. They are twenty-four-hour markers of how close you are to the goal line. Every day that I didn't have a coherent theory was another day that didn't end well. I kept seeing old man Burroughs's face when I'd left him at the airport, and it bothered me even more than the check that had arrived in the mail. The check embarrassed me. I'd sent it back.

I threw on the swamp cooler because not even the higher altitude could always mitigate the August heat. I found the tooth and, leaving it in its envelope, turned it over in my hand several times before I pivoted toward the phone.

I dialed the international operator and gave her the number. She spoke quick, accented English, and I determined that she was Cuban. Like bicycle riding, some skills never leave you.

"Is that Cabo San Lucas, sir, or Cabo—"

"San Lucas. Stryker Stephens."

It finally rang through the white noise.

"¡*Diga!*"

"Stryker?"

"Who's calling?"

"Jesus Christ, it's me...."

"Stand by, please."

The line went dead momentarily. When he came back, the transmission was clearer. I don't know why.

"Howdy, how're you?"

Nobody knew where the twang came from, maybe Missouri. It never left him entirely.

"I got a misery. No, worse—I got amoebic disintegration."

"What kind?"

"Two corpses. No explanation."

"We all die, Hopper."

"Not this way. The circumstances are too bizarre. Motive is unclear, leads slim to none. I'm stuck, I guess."

He sighed. He sighed like the respectful parent of a two-year-old. He sighed like a committed professor with a freshman naif.

"Data?"

"Some."

"Enough for a first theory?"

"No."

"Maybe you better call me back when you have one."

Such commands are usually followed by a peremptory click on the other end of the line, but he hesitated this time and there was nothing I could do but make myself more available.

"No first theory, but I'd like to debrief. Is that acceptable?"

"Try not to be effusive, Hopper. Just outline your current difficulties."

I did. It took three minutes. It should have taken two. Stryker Stephens used to set a stopwatch next to my ear and growl, "Report." I would do it again and again until the communication was telegraphic. He took all of it this time with no comment. The butchery, the tooth, the Buddhists, all of it.

"Tell me whether the incision on the first corpse was above or below the coccyx."

"Below."

"And the bullet on the second one invaded the throat?"

"Dead center."

"Any other injuries?"

"He was missing a belly button. She had a small diamond cut out of her forehead. The incisions were precise in both cases. I figure it was a little extra sadistic expression. He had all the privacy in the world when he killed Burroughs, but I'm surprised he took the time to play around that way in a public alley. Childish."

"I don't agree. Pursue it."

"Where? With whom?"

"Consider the strategic locations. Good luck."

"Thanks a lot," I said lamely. The connection went dead.

I speed-dialed Gretchen's number. I was pissed that Stryker had left it in my lap again, expecting originality.

"Hello, this is Gretchen."

"Hopper, here."

"Yes. Can I help?"

"Maybe."

"It was a difficult learning for us today."

"Why is that?"

"I need to explain it to you personally."

"My thought exactly. Make a suggestion."

"Uhmmm. Breakfast tomorrow?"

"Where?"

"My house. Park Hill: 2375 Elm. Eight o'clock."

"Done."

I withdrew into reverie to consider strategic locations.

Park Hill sits between the airport and Hilltop. Park Hill is where you live if you're professional, liberal, and into gentrification projects. It's also where you live when you're not yet rich enough to afford Hilltop, where the houses start at a quarter mil. Gretchen's bungalow was neatly porched and drivewayed and looked like any other middle-class home on the block until you got inside. She greeted me with slightly damp hair, wearing a long striped caftan. I could see the break-fast dishes on the dining room table.

"Mornin', Dr. Reynolds."

"Good morning. Would you like some coffee?"

I nodded and did a quick recon of the room while she disappeared into the kitchen. It was the kind of house that drives cleaning help screaming through the front door. It wasn't dirty in the least, but it was crammed to the hilt with the business of living: files, books, catalogs, plants, objets d'art. The couch and chairs were Thai, the stemware

Czechoslovakian, the grandfather clock vintage Midwest. The lady had done some traveling.

It was six steps from the living room to the dining room, and I took them slowly. Gretchen arrived with coffee, and we sat down to a New Age breakfast of fruit and yogurt sprinkled with some kind of seeds. Organic, no doubt.

I'd come only with the basic principle of rattling her composure. I'd heard her cry, seen her grieve, held her hand, but I hadn't any goddamned idea what she'd be like if I made her anxious.

"So what do you hear from the flock?"

"They're upset. We had a meeting after you left. It took two hours, but I finally got everyone settled down. They're doing what you asked. They set up a sort of buddy system and a way to do telephone checks."

"Who's the asshole with the beard?"

"I beg your pardon?"

"The asshole with the beard. Who is he?"

"If you mean the man who was being so defensive yesterday, his name is Dan Wyatt."

No dice on that one. She sat calmly scooping up strawberries. I decided to become more personal.

"Why do you do this?"

"Do what?"

"Hang out with a bunch of neurotic losers."

She put her spoon down and looked at me appraisingly. Then she smiled. "Well, strictly speaking there's no way I can hang out with neurotic losers unless I define them as losers. And I don't, although you're welcome to."

Stryker intervened from that place in my memory bank where I keep him for emergencies.

"A tactic perceived is no tactic at all, Bullfrog." The infuriating thing about Stryker is that he's right even when he's not around.

Gretchen rose to clear her dishes, and I followed her, returning my largely uneaten breakfast to the kitchen. The next move was part tactic, part impulse, but I didn't stop to evaluate it.

I caught her around the waist as she turned from the sink. I kissed her long enough to make sure I was welcome and to feel her hands sliding up my back. Now *that* made her anxious.

I pulled back and had about three seconds to look at her face before she recomposed it. The anxiety was there, but it was a simple open sexual agitation. I looked for fear, guilt, or the scheming gleam of eye that is diagnostic of the lunatic and the sociopath. I acquitted her on all counts and kissed her again.

I nibbled an earlobe absentmindedly. "Gretchen, tell me that everything you've ever said to me is true."

She pulled back to look at me, narrowing her eyes. "What are you investigating now, Hopper?"

"The probability that you're an insane religious fanatic who is systematically killing members of the congregation. I apologize if that offends your finer sensibilities, but I had to know."

I closed my eyes and steeled myself for the inevitable stiffening of muscle, hardening of stance, sensual rejection. Instead she tilted her head and looked at me quizzically.

"Interesting methodology, Hopper. How does it work?"

She was stroking the small of my back. Most men don't know it, but the upside to cerebral women is their disinterest in behaving the way they are supposed to behave. They're not necessarily good at getting dinner on the table on time, but they think it's perfectly reasonable that you've just exculpated them from a couple of brutal murders. She wasn't going to hold it against me.

I started to explain on the way to the bedroom.

Praise God that Buddhism has fewer sexual sanctions than any other religion save paganism. I had a lot left on my "to do" list, but I couldn't motivate myself out of her bed. It felt self-indulgent, so I decided to go back to work.

"Tell me about being a bodhisattva."

"What do you want to know?"

"How do you know you are one, for openers? Do you apply and wait to hear or what?"

"*Bodhisattva* means 'he who is brave enough to walk on the path of the bodhi.'"

"Darlin', I'll walk on the path of your body anytime."

"Pay attention, Hopper! The bodhi are the awakened ones."

"So the bodhisattva are enlightened folk."

"Not necessarily. It's not a requirement that the bodhisattva be fully awake, only that he or she be willing to walk the path and take the vow."

"What vow?"

"When you become a bodhisattva you promise to work for the enlightenment of all sentient beings."

"Sounds like taking the pledge to me. Does that include spiders and vultures? They're sentient."

"Why do you make fun of commitment?"

"Because most of what passes for commitment in the world is facile and phony. Besides, real commitment is too important to get serious about."

"You are a very cynical man. It comes from here." She traced a lazy circle around my solar plexus.

"That's the site of my incipient ulcer, not my cynicism."

"It's the site of the third chakra."

"The third who?"

"Chakra. A chakra is an energy center. There are seven of them in your body."

"Maybe in your body, lady. Not in mine. I just keep the usual stuff like kidneys and intestines in there."

"And some very powerful knots of energy."

"Whereabouts?"

"Well, chakras are actually part of the spiritual body, but western linear thinkers locate them physically. The first chakra is called the Kundalini, and it's just about here at the base of the spine."

"That's very nice. What does it do besides keep my ass in place?"

"It's the source of security and survival. Very basic and powerful. The second chakra is right about here."

She pointed to a spot on my abdomen. I tried to move her hand south, but the guru was teaching and refused to be distracted by other parts of my anatomy.

"Geshe Kelsang places it four finger widths below the navel. It's the center of sensation and emotion."

"Good for Geshe. Go on."

"The third chakra at the solar plexus is the center of action and power. It's where great social dramas are created. It's the easiest place to get stuck in your own ego."

"Are you insinuating that I'm egotistical?"

"No, I'm suggesting that the cynicism developed because you're blocked from expressing yourself in the fourth chakra."

"Which is?"

"The heart chakra."

"I thought I expressed myself rather well from the heart chakra. You didn't enjoy it?"

The guru giggled. "Sex is in the second chakra, and, yes, you express yourself very well from there. The heart chakra is the center of all the forms of love. Not just romance, but compassion, empathy. All of it."

What I'd learned in my detour through Buddhism was coming back to me.

"And numbers five, six, and seven?"

"Throat, crown, and the middle of the forehead where the third eye is. Don't you want to know what they do? . . . Hey, where are you going?"

I was disentangling myself from the bedcovers and pulling on my pants. I wasn't ready to talk to her about it just yet.

"I have an appointment at the medical center. I didn't realize what time it was. We'll finish this lesson later, and I'll work on massaging my solar plexus. I want you out of this house and into someplace safer as quickly as you can make that happen."

"But why?"

"Safety. You're not immune from the murderer, you know."

"I suppose I could move into the center for a while."

"I'd rather you didn't. Do you have any friends who aren't Buddhists?"

"Yes. But I can't just call them up and move in. Besides, I wouldn't want to put them in any danger."

"All right. Get a room at a hotel. Not in town. A busy one at the Tech Center. Try the Hyatt, the Sheraton, and the Marriott—in that order—and use my name. I'll try calling you at the Hyatt first in an hour and a half."

I was operating on pure instinct. She was confused but disinclined to argue. I was out the door five minutes later.

The University of Colorado Dental School is secluded in one corner of the medical center's campus. I made the sixteen blocks from Gretchen's house in four minutes and parked illegally. Doctor Abe was an old friend, but not the kind of man you keep waiting. The appointment was for eleven, and I broke the tape at ten fifty-eight.

Abe Silverman, D.D.S., is the best forensic dentist in Denver. His students get snapped up by other cities because by the time he's through with them they're indispensable to police departments. He was the only Jewish dentist invited to Uruguay on the Mengele case, and it wasn't because of his Yiddish accent.

"Thanks for finding the time, Abe. Have they got you covered up as usual?"

He rose from his desk and walked me to the couch in his office. He is a small, grandfatherly-looking man with a salt-and-pepper beard and horn-rimmed glasses. We'd shared a few experiences in the past and had come to respect one another's style.

"Not so bad. Besides, I'll always make time for you. You bring me interesting problems. In fact, I used the Adams County case in class the other day. I said to these baby dentists, 'When you find a cop type who's willing to learn the science, take the time to teach—because, once he learns, you have a representative in the streets.' Then I told them about you. So what did you bring me?"

I stopped being warmly nostalgic and withdrew the paper-wrapped crown from the inside pocket of my jacket.

"Hopper, this deserves a box. Gold is soft. It could get crushed. So what's the story?"

"I want to know what kind of mouth it came out of and who made the crown."

"Well, the second part is easy. See this little ridge in the back here? It's a Liggett ridge."

"Is that a technique?"

"No, it's the quirky practice of a very good dentist named Liggett. There are other things that work just as well, but he likes to do certain kinds of crowns just this way. I think it's ego. It's like initialing your work."

"Does anyone else use it? Maybe one of his students?"

"He's not a teaching dentist, he's a doing dentist. Until last year when he retired and gave it up for golf. As for the mouth this came from, I couldn't tell you, but Liggett could. What's the rap?"

"Murder."

"Are you trying to identify the victim or the perpetrator?"

"The latter."

"Got a court order to obtain records?"

"No."

"Get one."

"Can't."

Abe stopped fondling the crown and looked right through me. "Withholding evidence is a crime, Hopper."

"Only if it's adjudged to be evidence."

"Withholding a material fact is also a crime."

"It's temporary."

"What do you expect Liggett to do for you under the circumstances?"

"I don't expect him to do anything for me. I expect him to do it for you."

Abe frowned and shook his head. I was having second thoughts about not having covered my ass by at least telling Mack.

"Hopper, you're still getting forgiveness instead of permission."

"Abe, trust me on this one. This guy is on his second body. I have reason to believe there'll be a third. I don't know how much time I've got. I don't know if this crown even belongs to him. Just call Liggett and give me a glowing reference. If he doesn't want to play, I'll go back and do it the right way."

"And lose your license?"

"And lose my license if necessary."

"Give me the afternoon to reach him. I'll call you."

I thanked him and rose to leave.

"Hopper!"

I took my hand off the doorknob.

"One more thing. Jim's a nice man. Tell him you're trying to identify a victim. His feelings would be hurt if he thought one of his patients was a murderer."

"Got it. Thanks."

14

I returned to Coal Creek to wait for Abe's call and make a few of my own. I called the Hyatt from a pay phone before I left. I couldn't reach Gretchen, but the front desk assured me that she was safely ensconced in a room on the sixth floor. I left a message for her to call the mountains.

I pulled in the rutted driveway to my humble abode, and before I even made the turn toward the garage, I cut the engine. Something was wrong. I smelled it before I could plug in any evidence from my other senses.

"Sensing danger consists of the ability to smell the odorless," intoned Stryker-in-memory.

The garage door looked undisturbed, the security system hadn't been tripped, and the garbage cans were in the same place. Finally I located the source of my distress. There was a ceramic frog missing from the kitchen windowsill ten feet away from the garage door and visible from the driveway. The frog was trained to hold a plastic scouring pad in its mouth, but it usually held soap scum instead. It had been peacefully in place when I left. It is the kind of detail that one notices only when it exits a bigger picture.

I got out of the car without shutting the door and whispered up to the rear entrance of the house. I heard nothing but a lazy breeze in the pines. The door was locked, and there were no signs of tampering. If someone had gotten in, it was through the front door or a window. I unlocked the back door and slunk into my kitchen. It was a morass of smashed crockery and spilled food. The frog was in shards. I hugged the walls and made it into the living room, where the news wasn't much better. Still no sound. The furniture was overturned, and my books were strewn around the room, some of them with ripped pages.

I did it room by room until I had satisfied myself that I was the only human being on the premises. The phone cord had been sliced, and it took me a couple of minutes to find the spare. I still hadn't discovered the point of entry. I let out a short karate yell just to regain my dignity, then hit the phone. Dan Wyatt's bewhiskered face kept floating up in front of my mind's eye.

I called Mack first.

"Get a team up here to the house, Barnum. Our serial sickie is also a serious vandal."

"Which house? The Denver house or the Boulder house?"

"My house, jackass. The one without the contents insurance."

"Oh, sweet Jesus. He paid you a little visit, did he? I'm on my way."

I got on the line to the international operator again. I was righteously irritated at Stryker's insistence on the Socratic method, but I had to admit that his cognitive compass always pointed north.

It rang and it rang and it rang.

"Yes!"

It was Stryker's favorite greeting for people who disturbed him with phones that wouldn't stop ringing.

"Chakras!" I shouted back.

"Precisely. Now did you call me to show off your latest accomplishment, or is there something I can do for you?"

"What did I interrupt, Stryker? Your afternoon siesta?"

This is what is known as a private joke since Stryker barely sleeps at night, let alone the middle of the afternoon.

"Worse, Hopper. You interrupted Maggie."

I almost blushed. Maggie is the kind of hooker who hadn't always been one. There is a magnificent world-weariness to their friendship. I liked her, and I felt bad that their coitus was now interruptus.

"I apologize. I'll be quick. As you bloody well know, the ritual component of the murders is based on chakras. So far he's gone through the first, second, fifth, and sixth. That leaves the third, fourth, and seventh."

"On the contrary. He took care of the crown chakra, the seventh, with the bullet to Tom's brain. He needs only one more murder to complete this cycle. He is clearly obsessed, and he will strike next at the third and fourth chakras. He appears to think that he can gain chakra energy but carving them out of their home. Chakras are psychic realities and, as such, reside in the etheric rather than the mundane body. Since those two bodies don't always occupy the same space, they are physically connected only in the western mind. I believe you will have to move quickly. Tom's been dead over a week."

There it was. An alarm went off in my head.

"Tom? I don't recall that I told you his name."

"You didn't."

"Who did? What are you up to, Professor?"

"Mike Lawless told me, and I'm not up to anything. I was up to something, but you've destroyed the mood. Are you inclined toward one of the sangha members?"

"I am inclined toward whoever it was who destroyed half of my worldly possessions this morning. Now, explain the Lawless connection."

"Lawless is an acquaintance of the elder Burroughs."

"Why doesn't that surprise me?" I could see Mack's car bumping up the driveway. "Well, Stryker, I have to deal with local law enforcement for a bit. Will you be available for further speculation late this evening?"

"Provided that it is not idle."

"My love to Maggie."

"Humph." He hung up a trifle too emphatically.

Mack sauntered in, whistling in admiration for the vandal's thoroughness. "Did you a number, didn't he? How'd he get in?"

"Don't know. Your folks can figure that out. I got more important questions."

"Lay 'em on me."

I let Mack have about half of the current theory. Gretchen. Dan Wyatt. Chakras. Just enough to allay any suspicion that I was gunnysacking information. The phone rang. It was Abe Silverman. I scribbled a downtown address.

"Mack, the lab guy should be here soon. I need to get back to Denver. Can you hold the fort?"

"Sure. Check in at the department late this afternoon and we'll see what we've got."

I took the Mercedes because it was the only vehicle with good air-conditioning. Dr. James Liggett had consented to consult.

15

Living in Coal Creek means commitment to a kind of commuting that bears no resemblance to the East Coast variety. You have to know your car, and you have to pay attention. Daydream once too often and no chickenshit little railing is gonna save you from the momentum that a two-thousand-pound car gathers as it's falling off the side of a mountain.

On the other hand, once you get used to it, you can think yourself through a lot of trouble while you're screaming through a series of chicanes. It was only three o'clock in the afternoon and I'd already made love to a guru, corrupted a forensic dentist, had my house ransacked, and destroyed a romantic afternoon for my mentor. Some days, I'm magic.

Of all the things I could have chosen to be uneasy about, I chose to be uneasy about Lawless. The case was bizarre enough without his tangential participation.

I've only met Mike Lawless a few times face to face, and that's okay with me. I'll never entirely understand his relationship to Stryker, but local legend has it that they're the Two Yanks. It's the kind of relationship that last existed in the Korean conflict. By the time we all got to 'Nam, alliances based on a military experience were no longer possible. The nobility was gone.

We've had a few tense moments, Professor Stephens and I, when in a rare and exotic mood I referred to Lawless as that "tight-assed Jesuit Machiavellian elitist pimp." Calling Maggie a whore could get you killed. Putting down Lawless will get you a dour look of contempt. With Stryker, I never apologize, but it's always understood. I never allow myself to relax in Mike's presence.

Michael is what my Kansas City grandmother would have called a fine figure of a man. He lounges in tuxedos like they were bathrobes, and his thick, curly hair continues to radiate a perfect level of gray light at the temples.

Many years ago, in D.C., I ended up at a congressional picnic with him. I watched infants of both sexes coo at him in delight, and I saw matriarchs patting him on the knee, batting their eyelashes, reminiscing. And that doesn't begin to describe the vulturesque circling of women between twenty-five and forty-five toward whom he maintained a wholly genteel indifference.

I couldn't imagine how he'd met Burroughs, but I could imagine why. Lawless gravitates toward money and power. He was as slick as Burroughs was forthright, and I guess I was looking around the corner to see how Lawless could exploit Burroughs's grief. Stryker tells me I'm paranoid and parochial, but he won't discuss the substance of his attachment to Lawless. I've never been able to get him drunk enough to tell the story. I asked him once why the bar was called the Two Yanks, and he brushed me off, saying it was a joke from an old war movie.

I made a mental note to call Herr Burroughs and moved on to the next bit of business. The wild-eyed Wyatt. Something felt very right about the theory that he ransacked my house. It was the uncontrolled act of an overgrown kid who has tantrums in public. However, something was very wrong with the rest of the story. If he was looking for something, what was he looking for? If he wasn't the murderer, why was he looking for anything? And what could he have against me? No doubt it would have been effortless for him to get Tom and Patty into a field or an alley, but I couldn't conceive of his wielding the weapons with the degree of precision that was present in those murderous acts. He was the most attractive suspect to date, but it just didn't compute in the six inches of critical intestine that I use to assess such possibilities.

The last bit of business took me into the outer suburbs, where I had to deploy a bit more of my autonomic system to deal with the traffic on Sixth Avenue. I made it into town and parked on Wazee Street because it's the only sensible place to park. It has plain old-fashioned angle spaces without attendants or high-rise fees. I was on my way toward my best chance to make sense out of the case, and I was getting grumpy as hell.

I walked down the 16th Street mall. It used to be a four-lane downtown artery. Now it is laid with upscale paving stone imported from some out-of-state mine probably owned by the mayor's brother-in-law. This, despite the fact that there is plenty of rock as close as Boulder from a rock pit owned by an old Italian mining family.

The mall serves no traffic. It has park benches and trees. It is an imitation of village life. I consider it ersatz because Denver has no business being nostalgic about villages. The ghosts who walk the historic districts are rough customers from the mining camps. I suspect that they look with amusement or disdain at the Dove Bar carts and the New York hot dog stands. I do.

I was generally unsettled. A hundred human organisms within my direct line of sight were oppressing me with their dogged intentions. They were intent on getting from one place to another, controlling their children, avoiding the police, buying merchandise, negotiating deals on the street. It made me weary. I only wanted one of them. I wanted the one who intended to kill another bodhisattva.

The address Abe had given me was in a bank building midway down the mall. I thought it was an odd place for a dentist. When I got to the suite, it became clear that it was a doc-in-the-box outfit and I wasn't expected. An efficient-looking blonde challenged me with one of those "have a nice day" smiles.

"How can I help you, sir? Are you a walk-in?"

"No, I'm a shoo-in. I'm here to see Dr. Liggett."

She looked down at the book and frowned. "Well, Dr. Liggett is in the building this afternoon, but he doesn't see patients. I don't see how you can have an appointment with him. Are you sure it isn't Dr. Friedman or Dr. Evans?"

"Positive."

"Well, what sort of procedure were you going to have done, sir? If I could just have your name, I'm sure I could straighten this out."

I heard him before I saw him. He bellowed.

"Allison, my little rose blossom!"

He came striding into the waiting room from one of the inner sanctums. About fifty-eight, an Arizona tan, two inches taller than God, with hands like hams and a turquoise belt buckle as big as a midget's fist. He wore a bola tie, for chrissake. He clapped a hand on her shoulder in avuncular fashion.

"Rose Blossom, I got this guy comin' in named Hooper or Humper or something. Send him back when he gets here."

"Dr. Liggett?"

I stuck out my hand because that's what you do with the hearty types.

"I'm Hopper. I appreciate your taking the time."

"My pleasure. Abie tells me you know an old friend of mine."

If Abie wasn't the old friend that he was referring to, I was lost and surprised that Abie hadn't clued me in.

"Who's that?"

"A full gold crown. Did you bring it?"

I looked around the waiting room meaningfully. "Certainly. I wonder if we could..."

"Oh, sure. It's confidential and all. Come on back with me to the lab."

Someone had rehabbed the suite to franchise specifications. The rooms were done in dentist pastels with insipid art on the walls. It stood in pathetic contrast to the hallways, in which it had been less expensive for the developer to leave in place the Carrara marble and the worn brass fittings of banking. I don't like being in either setting particularly, but given a choice I would have stayed in the hallway.

The lab had all the usual dentist gizmos. We sat on stools. I handed him the crown, and for a moment he didn't even look at it.

"My crowns don't usually fall out, Mr. Hopper. What happened?"

"It's a highly unusual situation, as I hope Dr. Silverman explained. I'm not at liberty to discuss the details. For our purposes, it would be enough for you to identify this particular patient. Unofficially. If the case develops in the way we believe it will, I'll be back with all the *t*'s crossed. I apologize if this is awkward for you."

His brow furrowed, but he didn't say no. Then he shrugged and unwrapped the crown. He put it on a clean tray and grabbed a pair of dentist's pincers. He turned it around twice at eye level, squinting. He put it down a little too slowly, as if it had suddenly become a piece of a black hole, infinitely heavy. He turned to face me.

"Well, sir, if I'm going to give up the rules the way they're written in the ADA manual, then you're going to have to give up your rule about not discussing the details. Is this person dead, Hopper?"

"I don't know."

"In that case, you're looking to identify a suspect and not a corpse. Thought so. The coroners usually call about the corpses."

So much for Abe's request about sparing the good doc's feelings. I saw no reason to lie.

"That's partially true. The fact that your patient lost a crown at the scene of a crime does not automatically mean he or she is a suspect. Forensic methods get more sophisticated all the time, but no one can tell whether that crown came loose while a crime was being committed or the day before. If it was the day before, then your patient is home free. I'd simply like to talk to whoever it is. Do you recall that anyone made an appointment in the last few days to get a gold crown replaced? Or maybe someone called to complain about a crown coming loose?"

"First, tell me why you didn't go through channels."

The once jovial dentist was now tight-lipped and stern. His face didn't look like it wore that expression often. "Finesse is always preferable to pissing contests!" Stryker hissed in my head. I decided on a variation of the truth.

"It's the kind of case that prompts investigators to bend the rules. If I'd gone through channels, I'd be putting this person, who may be innocent, through the nightmare of a lifetime. If this person is not innocent . . . I want to do the spade work personally."

He looked back at the tooth on the tray and shook his head with more dismay than irritation. "At least tell me what he's supposed to have done."

"He?"

"Yes, he."

"You know the patient just by looking at the tooth?"

"Yes, goddamn it. I don't recognize every crown I ever made, but I've known this patient from the time he was a pup. He spent a lot of hours in my chair. The ridge is unusually pronounced on this tooth because he had a difficult problem with his bite. Now what is it you think he did?"

"Abe said you were the kind of dentist who had concern for his patients. Let me stress that I have no reason to believe that the presence of a lost crown is incriminating at this point. Is that what you're worried about?"

"No, sir. I'm worried about doing the right thing. Did you tell Abe that you were operating on the fringe, or did you con him into calling me?"

"Do you know Abe very well, Dr. Liggett?"

"I took a forensics course from Abe about ten years ago because I had one of my offices in the old section of Aurora where people got into serious trouble more often than at my uptown office. Besides, I always like to learn things from the pros. I know Abe well enough."

"Then, presumably, you know the answer to your question. He doesn't con."

I knew that I was going to win or lose on that point, and I stopped breathing to concentrate on the tiny muscles in his face that communicate "yes" or "no" far sooner than words. All I saw was struggle and an unqualified "maybe." Finally his head snapped sideways in frustration.

"My concern for this patient is not the problem, Hopper. I did excellent work on his mouth. That was my job. But out of hundreds of patients in thirty years there are only three or four that I never could warm up to, and he was one of them. He's a weird little guy. Fair is fair, and I gotta make sure I'm not cooperating with you because I happen to be of the opinion that he's probably capable of doing something awful. I don't want to . . ."

He trailed off then. He was an ethical type. He was practical, too, and I started to wonder just how much he was protecting himself from being dragged into a legal horror show, complete with ugly press negatively affecting his doc-in-the-box venture. I had another card to play.

"Well, let's suppose that I'm not investigating anything, but that I happen to get into a conversation with a dentist about one of his old patients who may or may not remain unnamed. Does that feel more comfortable? Truth is, I could come back with the right papers and then you'd have to tell me. Which way suits you better?"

He sat back down on his stool with his legs apart and his hands clasped in front of him. He smiled briefly. "Fair enough, Hopper. One more condition. I'll give you the story. You give me the nature of the crime. No names."

He was flat out, and I'd done as well as I was going to do under the circumstances. I gave him the honorable nod of a gentleman and pushed the record button of my brain.

"I used to have an office at Colfax and Yosemite. I sold the practice three years ago, but that's where I usually saw him. The first time he came, he was thirteen with a mouth full of lousy teeth. His mother showed up on the first visit, and I never saw her after that. He kept coming with the Medicaid forms. It was back in the days when I was still willing to do my bit for the poor, so I drilled and filled. He was the kind of kid who always buttoned the top button of his shirt and wore long sleeves in the summer. Sort of short and puny. History of asthma. He chipped his front tooth once, and some social worker showed up wanting to know if I thought the chip was consistent with having your face smashed into a door frame. They dropped the charges against his mother's boyfriend, but I always looked for suspicious injuries after that."

"How long did you see him altogether?"

"Years. By the time he was seventeen we'd done the lion's share of the basic repairs. After that he always showed once a year with cash to get his teeth cleaned or maybe some little cavity filled. He disappeared for about four years in the late sixties or early seventies. Can't say I was all that unhappy to see him go."

"When did you do the crown?"

"Just before I retired. He had a tooth that we'd been saving for a long time. It just didn't have any mileage left, so I ground it down and crowned it. His permanent incisors were too long and misplaced for the rest of his mouth, and the crown corrected for that."

74

"So far he sounds like lots of people. Why do you remember him so well?"

Liggett shook his head and grunted. He shifted on the stool and looked me in the eye. "He was a spooky kid. I have a son about his age, and so at first I treated him like I treated Jim, Junior, but I learned pretty quick that if I touched him with anything but the drill, he would flinch. Now the drill didn't seem to bother him much at all. I have always tried very hard to keep things as painless as possible, but sometimes it's just not possible. The mouth is a tender place in the body. Teeth are rock hard, but the tissues are sensitive. Well, there were procedures that I had to do that hurt. I couldn't make them not hurt, and he would lie there and stare into the light and get this intense look in his eyes like he was...feeding on the pain. Then there was the aquarium trouble."

"Aquarium?"

"I kept an aquarium in the waiting room. Some marketing consultant told me it was comforting. Just some goldfish, black mollies, and a plant or two. Twenty-gallon tank. Nothing special. One day, when he was sixteen, Jeff came in for a routine appointment. I was the only one in the office. It was a pretty slow day, but there was a root canal in front of him. He had to wait about ten minutes, and there was no one after him. He got in the chair and I was examining him when my receptionist came in and whispered something to my assistant. Seems all the fish were dead and floating on top of the water. The receptionist was sure that he'd done it. None of my staff ever liked him, so I tried to be especially impartial, but when I confronted him about it he started to hyperventilate. He got hysterical. I calmed him down. I suspected a chemical, so I frisked his jacket and there was a tin can of drain cleaner in his inside pocket. Drano or Red Devil or something. I asked him why he did it, and he said that some things didn't deserve to live. By then he was pale and shaking. I tried to call his mother to pick him up, but nobody an-

swered the phone at his home. I told him he was going to
have to pay for the fish out of his allowance. I don't know
how to say this, but he sort of leered at me with this crazy
smile and said that he'd done me a favor. That was over
sixteen years ago, and I can still feel my fist clenching
because I just wanted to hit him."

"You didn't do anything else?"

"I continued to treat him. He seemed to grow out of
it. He seemed almost normal. Paid me for the fish and
apologized. And then he disappeared. When he got back,
he was different. He said he'd been in the army and it had
done him a lot of good. Taught him discipline and commitment. I was glad to hear it. I don't think I ever wanted
to know much about him personally, but I assume his life
had been pretty crummy to that point. He still wore long
sleeves in the summer, but he was a little less high strung
and I didn't lose any more fish. The last time he came to
see me ... oh, hell, it's nothin'."

"Maybe it's nothing, but tell me anyway."

"When I put the crown in I was having trouble with
getting enough leverage and pressure, so I leaned my right
arm against his chest to apply more force, and I felt this
metal necklace or something under his shirt. It was thick
like a piece of good Santa Fe silver, and it felt like a cross.
When we were all through with the procedure I asked him
about it. He seemed pleased that I'd asked, and he dug
into his shirt. He came up with a sculptured crucifix on a
thick chain. Gorgeous piece of work. His eyes got sort of
glassy, and he smiled at me the same way he smiled about
the dead goldfish. Then he was out of the chair and gone.
I never saw him again."

The good doc was looking two feet above my left shoulder. His voice had not changed in pitch or timbre, but his
speech was slower.

"So?" I said with a complementary lack of speed.

"I thought about it later, and some little thing bothered me ... but I couldn't put my finger on it. Something

76

about it didn't seem natural—just like he never seemed natural."

"Did you ever figure it out?"

"Yeah. I figured it out. Not that it makes much sense."

"Try me."

"Dentistry, besides being a medical specialty, is part sculpture and part engineering. It requires a good sense of spatial relationships. When he reached into his shirt to bring up the crucifix, I recall that he rotated it in a half-circle."

"Meaning what?"

"Meaning that the crucifix must have hung around his neck upside down."

I was careful about my response. The good doc was not an enthusiastic informant, but I tried one last time.

"Will you tell me his name?"

His concentration broke then, and he looked at me like I was some kind of irritating insect buzzing around his face.

"I don't think that there's anything else I can tell you under the circumstances. If anybody asks—you came to ask me some questions, and I didn't know the answers."

"I've agreed to tell you the nature of the crime. If I told you that the crime involved murder, would you reconsider?"

"Come back when you can legally make me tell you his name. I've told you enough."

"Thank you, sir, I'll be back. And thank you for your help."

White male. Small. Early thirties. First name Jeff. Last seen at an office at Colfax and Yosemite. I had more than enough.

olfax Avenue is the four-lane business loop of State Highway 40 running east and west. Stay on it long enough and it will get you out of town in either direction. It was named after Schuyler Colfax, a New Yorker who was the vice-president of the United States when the settlers began moving into Denver in 1870. It has none of the graciousness of Speer Boulevard, none of the glamour of 17th Avenue Parkway. It just runs east and west like a faithful old dog and is home to all manner of retail and commercial eyesores. Every once in a while some rehabber gets a bank to lend him some money for restoration, but Colfax is indomitable. If you slice out the sleaze in one place, it will surface again down the road a piece.

At the west end it turns into a feeder for Interstate 70 and is lined with respectable small businesses. Downtown it deteriorates just past a four-star restaurant that sits at the edge of the Capitol building. It turns into pawnshops, porno palaces, liquor stores, and the like. Continuing east,

it shapes up at major intersections, thanks to the rehab-
bers, but in between it disintegrates into fast-food joints
and the kinds of bars that specialize in wet T-shirt contests
for urban cowboys. The visual agony ends just past Fitz-
simmons Army Medical Center after a few blocks of cheap
motels.

I was poking along in the right lane thinking about
how to approach the problem of locating the suspect. I
never drive Colfax if I can help it. Traffic signals are closer
together than telephone poles to a crop duster and syn-
chronized by some kind of sadist. Yosemite is just about
at the city limit where Denver slides imperceptibly into
Aurora. All I could visualize at that corner was a fried
chicken place and a Burger King. I had no idea what kind
of structure I was looking for, but I knew that I'd recognize
it when I got there.

A block away from the intersection I started to scan
the area in earnest. The bank was off in the distance and
dwarfed everything else on the street. Nothing that looked
like a dentist's office existed to my left, so I concentrated
on the south side of the street. There were two candidates,
two nondescript square little buildings of tan brick probably built in the mid–1950s.

As I got closer, they sorted themselves easily. The first
was an insurance office, and the second bore a small sign
that said LIGGETT, SCHUSTER AND SMITH, P.C., DENTISTS. I
suppose it was false advertising since Liggett hadn't been
there for a couple of years, but it saved me some time and
I was grateful for their sloppiness.

I parked the car in a two-hour space on the street,
intending to be there for one-eighth of that time. I dug out
an old briefcase that was filled with a yellow legal pad, a
calculator, and other small props. I walked into the fried
chicken place and ordered a large coffee, inhaling the
grease that had emulsified itself into the air. Choosing a
seat that allowed me to watch the dental office, I reviewed
cover stories.

It's not easy to get into a private home these days

without a good-looking fake I.D., but businesses are easier. The personnel feel safe because there are other people around. I needed to get a look at the access points for the office and the location of the records. I decided on the public servant ploy and extracted a business card from my wallet that said that I was indispensable to the city of Aurora. I abandoned the coffee, which tasted more like French fries than brewed beans. I had my game plan down.

It took fifteen seconds to get to the doorway, and I almost ran into the old lady who was also coming up the walk. I opened the door for her gallantly, relishing the opportunity she would give me to look around before I had to talk to anyone. She toddled up to the receptionist's window. It was one of those frosted sliding glass affairs that separated the peon patients from the molar moguls.

The waiting room was plain, with middle-of-the-road taste. There was an unexplained door at the far end of the room and another that clearly led back to the treatment rooms. The receptionist was sort of amidships. The old lady finished announcing herself and sat down to review the latest *Better Homes and Gardens*. With a brief glance at the twenty-gallon aquarium in the corner, I took my turn at the window. The receptionist was very different from Allison the rose blossom. This one looked like she'd been there for years, and she probably had a name like Thelma or Rose. She was matronly and neutral in her greeting, attired in a sensible tweed skirt and a blouse with a bow. I estimated that she tipped the chronological scale at about fifty.

"May I help you?" she said.

"I hope so, ma'am. I'm with the City Water Department, and we're concerned with water pressure in this neighborhood, particularly for the higher-volume users like yourself. I wonder if there's someone available who could escort me through your offices to test pressures? We're having trouble with a main."

I thrust the business card at her, and she examined it

like it was an overdue bill. "If you'll wait a moment, I'll speak with Dr. Schuster."

I waited while she disappeared into the back. The filing was arranged horizontally in a cabinet at one end of her desk. There were two five-foot shelves of apparently active records. The lower shelf was the only one with the cabinet cover closed, and I guessed that it held about two-hundred inactive records. He was in there somewhere. Sorting two-hundred records is a formidable task without a last name, but it gets faster if you can eliminate all women and all men not named Jeff or born in the wrong decade. "Even gross details discriminate," I heard from Stryker in my head. The receptionist returned with a dental hygienist and said that the little blond slip of a girl would tour me around.

"Please be quick. There are patients in all but one room," she said.

"Thank you so much. I certainly will." I turned to the blond tooth scraper and smiled as engagingly as I could. She was solemn. The treatment rooms filled out the oblong building and branched off from a long hall. I extracted the yellow legal pad and assiduously wrote down notes on each sink, turned each of them on briefly, and made up some fancy numbers expressing their water pressure. In each treatment room the hygienist apologized to the occupants and said I was a city inspector. The patients, with their mouths agape and already violated, could not have cared less, and the dentists and other hygienists were busy being compulsive.

The windows in the treatment rooms were long and thin, sealed from the great outdoors, and would not admit an adult male comfortably. The last room on the corridor, where the bookkeeper sat, had a full door with a hopefully pickable lock and a security chain that my rubber band trick might work on. I was finished.

I thanked the receptionist briefly, returned to my car, and drove back downtown using 14th Street, which is faster and has fewer stoplights.

I stopped at the Manhattan Cafe, ordered a bourbon at the dark wood bar, and pondered. Mostly I was worried about the fried chicken place, which stayed open until two A.M. and would attract drunks, vagrants, and even cops looking for a free cup of French-fried coffee. I drove back to Coal Creek to get darker clothes.

I thought about calling Mack. I decided to keep him innocent until I needed him. So far I'd only had to lie a little on this case. Now I was upping the ante to breaking and entering. And theft.

When I left Coal Creek, I was wearing denim and a black leather flight jacket against the cooling September night. I dug out a pair of flat-heeled soft leather shoes. Stryker calls them my ninja boots. They make no sound on concrete and less on grass. The pick set and rubber bands were tucked into a pocket that had been sewn into the lining to hold such tools along with a pry bar and bolt cutters in case I got desperate or hurried. I stuffed rubber surgical gloves into my hip pocket.

The break-in would be easy, but being on the border of Denver and Aurora meant I had two sets of late-night patrol cars to worry about on a street that police have learned to keep an eye on. I waited until three A.M. so that even the guy who closed the fried chicken place would be gone.

I dumped the car in an alley a block east off 14th Street and walked. Casing a street is different from casing a building. It requires all your peripheral vision and eyes in the back of your head. I moved casually, trying to fit in.

There was no parking lot in back of the building, just the door. I checked for alarm wires and found none. The lock picked just fine. I took a deep breath and slid the door open a couple of inches, listening for other trouble, like a guard dog. The chain lock was right for the elastic whiz bang—but I was about to make my first noise, and I looked around for headlights or even lights in the houses behind me. It isn't likely that a sleeping person will be awakened by the sound of a chain rattling, but you think about it at that hour of the morning.

The rubber bands did their job relatively noiselessly, and I waited only a few seconds before entering. No response from the neighborhood. I closed the door and made my way to the receptionist's station without light. It was no challenge to get through the lock on her door handle even in the dark. I used the gloves. The next problem was the file cabinet. For that, I extracted a small flashlight with an iris that allowed me to dial a tight beam or a broader one. I chose a tight beam and the right pick, and the file cabinet lock was history.

I squatted to reach the lower shelf that held the inactive files. They were in numbered sequence with no predictable relationship to dates. The rest was up to luck. There might be one Jeff born in the right decade in the two-hundred records, or there might be ten.

Some tasks are better accomplished with your brain off. I started at the beginning and pulled every Jeffrey, Jeffry, and Geoffry in the bunch, one at a time. I eliminated a sixty-five-year-old, a twelve-year-old, and a girl whose parents apparently had wanted a son.

Fifteen minutes later the pool of candidates pared itself down to three. Jeff Dugan, aged thirty-two. I almost eliminated him because he was five feet ten—which is not that short, but who knows how Liggett, at his altitude, perceived someone at five ten. Jeff Gottmacher, also aged thirty-two and five six. Jeff Rotheker, aged thirty-one and five four. All Jeffs, all Medicaid patients in their youth, all the right age. I spent no time looking at the arcane markings on the little pictures of teeth. I'd learned something about dentistry from Abe, but I didn't want to spend minutes I didn't have in a dark dental office at three-thirty in the morning making clumsy diagnoses of little teeth pictures. I swiped them all.

I relocked the file cabinet the same way I'd opened it. I tucked the files in the back of my pants and covered them with the back of my jacket. I went to the back of the office and searched the area through one of those skinny windows. No impediments in sight.

I made the car in two minutes.

could have taken a direct route back to Coal Creek. I should have taken a direct route back to Coal Creek. Instead I went south on Havana and hooked up with I-225. There were three times the truckers on the road in comparison with cars and precious few of either kind of vehicle. I hit the junction with I-25 just north of Belleview and stopped to consider that I was crazy to be doing this at four A.M. I was on my way to the Hyatt knowing she was asleep and knowing that I would not be welcome in the elegant lobby. I was dressed for Colfax. There is, however, no law against going to a house phone and dialing a room number. If she didn't answer, I'd head for the barn.

I pulled into the sleeping parking lot and walked into the lobby like I had a very clear destination. There was one drowsy desk clerk behind the counter, who roused himself long enough to check out my demeanor. I waved and walked toward the elevators. I doubled back and found a house phone.

It rang six times before she answered.

" 'Lo?" She was dead asleep, but she was otherwise alive.

"Gretchen, it's Hopper. Can you wake up?"

"Hnnmm...yeah. Who...are you?"

"A pilgrim. A penitent. Wake up, kid. It's morning. Pay attention here. I'm going to come up to the room and knock on your door."

"Door?" I could hear her struggling to figure out what the door had to do with the conversation.

"The door, Gretchen, the door. I'm going to knock and you're going to let me in."

"Hopper?"

Consciousness dawned. Now I had to keep it alive.

"Yep. Hopper. That's spelled H-o-p-p-e-r. Are you awake now?"

"Yes."

She was not only awake, but she sounded very anxious, as if my presence at four in the morning was threatening—but then I guess most people view wee-hour intrusions with some apprehension. It was the damnedest thing. The woman didn't allow for very much protectiveness, but she made me feel protective nonetheless. As if Buddha weren't quite doing a good enough job for her.

I hit the right floor and got what I hoped was the right door and knocked quietly. There was a sound of water in the bathroom and it took a minute, but when she answered she was clear-eyed with a freshly washed face. I kissed her forehead and would have taken her back to bed, but the dental records wouldn't wait.

I headed for the little table they always put in hotel rooms. Files in hand, I sat down to do a second review.

"Dr. Reynolds, see if you can rustle us up some coffee." She dialed, still silent but accepting.

Dugan, Gottmacher, and Rotheker. He had a name now. It was one of those. I opened all three records at once. I looked at the marks that Liggett had penned onto the templates of the three mouths. It looked like a lot of hen scratches. All three of them had had extensive work. Gretchen appeared at my right shoulder.

"What are those?"

"Dental records. I stole them."

"I don't understand."

I was saved by room service with a pot of coffee and a *Denver Post*. I had no objection to telling her, but I didn't want to take the time until I had it figured out.

Dugan's file was first up for more careful attention. He had the least hen scratches and seemed to have appeared on Liggett's doorstep when he was only nine. Besides, five ten still seemed too tall. I trusted Liggett's memory enough to eliminate Dugan provisionally.

Gottmacher was twelve plus a few months when he entered Liggett's practice. Unlike Dugan, he could have been perceived as a thirteen-year-old. Lived in the neighborhood. Single-parent family. Dentists make very sketchy notes, and I read through the list of procedures with dates attached without seeing anything that indicated a Liggett ridge. It didn't surprise me. Even the proud papa of the procedure probably would have called it something multisyllabic and technical. The notes revealed three crowns, one of them in a reasonable location within the mouth. The trail of dates ended at about the right time. Gottmacher was a suspect.

Rotheker had the most hen scratches. His history revealed extractions, crowns, fillings, bridgework. He had a mouth like a war zone. The patient had shown up when he was fourteen, but if he was as puny as Liggett remembered him, he could have looked a year younger. Rotheker was a suspect.

It had been a long, wearing night. I had two suspects now and twice that many phone calls to make. It was five A.M. in Denver and seven A.M. in Westchester. I needed to wait for the time zones to catch up with my next steps. I closed the file and looked at Gretchen, who was sitting quietly on the bed. I wondered how long it would take to get the nightgown off. I suspected it would depend on how curious she was.

I closed my eyes, pressing against my right eyeball with my thumb and my left eyeball with my middle finger. The poor man's acupressure. Hazily at first and then more clearly, I saw my wife's face and then her body. I considered opening my eyes to dispel the images, but my eyelids were welded shut. It used to happen a lot when I was tired and I couldn't keep the mental barriers up against it. This time the scene was a small hotel bed in a much smaller hotel room. I could feel Caitlin align her tiny softness with the full length of my body. My hands twitched thinking about sliding them down her back to stroke her ass. With a lazy, relaxed concentration, we could take on any problem at all. It was a game of sorts. Sentences were interrupted by kisses. Solutions were pursued until such time as sensuality turned to urgency, and all problems were deferred save for how we would love each other that moment.

My eyes snapped open, and it was the Hyatt again and Gretchen was still waiting patiently.

"You look troubled."

"Not troubled. Just impatient."

I reviewed my progress with her. She asked thoughtful, perceptive questions and chose not to comment on my lack of methodological orthodoxy. She was becoming too much of a friend to be a therapeutic roll in the hay, and, as I had painfully reminded myself, she was not my partner. I imagined we would have other moments, Gretchen and I, but not on this lonely morning. Like I said, I should have taken the direct route back to Coal Creek.

I got up to leave. Gretchen rose, and I held her quietly. I had the uncanny feeling that she knew everything there was to know about my internal struggle and forgave me. It was something in the way she held her body. She was spooky sometimes. I promised to call with any news and headed for the elevator.

Much of what would happen next depended on Abe and my persuasive powers. Like Damon Runyon, I figure that everything in life is six to five against, but if I could get Abe to pick Gottmacher or Rotheker, I would save the time it would take to find both of them.

I got back to my pathetic, debris-strewn house, and there was a note from Mack saying that they'd gotten no prints but that access had been through an apparently unlocked back door. He noted that it was okay to clean up the mess now. There was also a message on my tape asking where the hell I was and why hadn't I called and the whole godammit. I wasn't ready to talk to Mack yet, so I called him where I knew he wasn't and left a message that things were okay and I'd get with him that evening.

be Silverman's office informed me at eight A.M. that he was in class until nine A.M. I luxuriated in a shower and a thirty-minute catnap. It's not really sleep, of course, but a technique that renders your mind nonpresent for a little while. I am informed by Stryker that two or three of those a day are all the human body really needs. We sleep because we've never given up the slothful habits of our mammalian predecessors.

I called again at nine and nine-twenty, refusing to leave messages because I would lose control of the process. Once you leave a message, receptionists expect you to quit bugging them and wait for a call back. So I kept saying I wasn't anywhere I could be reached and that I would call back. I got him at nine-fifty.

"This is Dr. Silverman."

"Hi, Abe. It's Hopper. Got a few minutes this morning? I have something new."

"Hopper, my friend, you'll be a long time making up

for this one. I talked with Jim. He didn't appreciate your methods, and neither do I. I'm not interested in knowing what you have that's new until you show up with something official. My skin is already itching because I feel complicit. Someday we'll speak of it again, but not now. Leave me alone."

"But, Abe—" The line was dead. I was mildly puzzled. I hadn't even had time to take the risk of telling him about the theft, so he was reacting to something else in the situation, and I was shit out of luck.

I couldn't go back to Liggett with stolen records. I couldn't go back until I could get Abe to identify a Liggett ridge on Rotheker or Gottmacher. I could get Liggett to confirm it if his respected colleague identified it first, but I didn't think I could budge him without that.

My choices were coming clean with Mack or coming up with something that would move Silverman off his position. The trouble with using Mack was that we would end up dealing with the boys on the bench. I knew I'd already screwed up any normal investigative procedure that would warrant a warrant.

I supposed that I could start on locating both Gottmacher and Rotheker and hope like hell that I could find something solid enough to get the system moving. I cringed at the additional time it might take. There was a murderous clock ticking loudly in my head. I didn't think I had much time. I didn't think the killer would stop killing while I looked for him.

I could feel every neuron in my head stretching toward an answer. Each bit of data arranged and rearranged itself in fanciful structures, not necessarily logical. I poured three fingers of bourbon, took a sip, and remembered it was before noon. I tossed the rest of it in the sink. I reached for the phone and placed a long-distance call to Cabo. It rang too long and I ended up with the *abuela* who cleans Stryker's digs and helps out on occasion at the bar. Her eyesight is failing, so Stryker usually ends up redoing the crystal, but she has a job there for life. She calls me Señor Rana—Mr. Frog—because that's the closest she wants to get to pronouncing Hopper.

She said he was at the bar. There is no phone at the Two Yanks, only a ham radio in the back room that crackles with communiqués from all over the world. I didn't leave a message because Lucia doesn't do messages.

It was midmorning in Cabo. The bar would get busy about noon. I needed a long-distance messenger, and Lucia wouldn't do. In a fit of inspiration I called the only cab company in Cabo and asked for Guillermo. Guillermo dispatched when he wasn't driving. There wasn't much to dispatch. The whole company consisted of three rattly broken-down chebbies that spent most of their time shuttling tourists from the airport at San José del Cabo to the hotels.

Guillermo was delighted to hear from me, so delighted that I had to listen to the latest story about his *novia*, the poor woman he's been engaged to for ten years. She will be arthritic by the time she gets to the altar, but he speaks of her with grand passion. I explained my dilemma. I asked him to go to the bar, deliver the message, and collect his tariff from Stryker.

"No hay problema. Ya voy."

"¿Immediatamente? ¿Seguro?"

"En seguida, señor. Te lo prometo."

Guillermo's promises were honorable, but one had to allow for the slowness of pace that is Cabo. He'd get there, but it could be an hour.

I was antsy. I called Directory Assistance, too impatient to go looking for the White Pages.

"Mountain Bell. What city, please?"

"Aurora. Do you have a listing for a Jeffrey Gottmacher at 1732 Syracuse?"

I could hear her clicking away on the CRT. It took a while, which made me think there was nothing to find.

"I have no listing on Syracuse. I only show a J. Gottmacher on Niagara in Denver."

"I'll take it." Syracuse was the last address on the dental record, but Niagara was a slightly better neighborhood. It was the kind of move an Aurora blue-collar worker would make happily when he got a few bucks together. I hung on until the number was repeated and the recording ran its course—I got another operator.

"I'd like the number for Jeff Rotheker on Beeler in Aurora." I spelled it.

"One moment, please." This time I got the electronic voice immediately. I jotted down the phone number. This one hadn't moved from 1365 Beeler, which was his last known address. I wanted to jump in the truck and take a look at the residences, but I wanted to talk to Stryker more.

I had a sudden annoying thought. What if Lucia got it wrong? What if he wasn't exactly at the bar? Down the beach from the bar is a walled adobe fortress known affectionately as "the compound." Those windowless walls are all that can be seen from the outside. There is a door with an American buzzer and an intercom. There is no address and no way to get in unless someone wants you there. The only people who could possibly want you there are Stryker or Lawless. If Professor Stephens was engaged in other business at the compound, no one would tell Guillermo, and neither Guillermo nor anyone else in town would pursue Stryker at the compound without an invitation.

The first time I saw the compound was in 1972. The war was over and my wife was dead.

I was still deciding whether I wanted to live. I sat in my house in Coal Creek Canyon and called every person I knew who had ever known Stryker Stephens. Rand wouldn't return my calls. Most everyone else had lost track of him as I had just after the bombing of Cambodia. None of my highly placed contacts was worth a damn, and I decided that Stryker was also dead or didn't want to be found. The results would have been the same. In a moment of inspiration, I called the woman who had been his secretary at Rand. My palms were sweaty and my breath was sour from the latest three-day binge, but I managed to get the phone to work, and with considerable warmth and some hesitation she told me about Cabo and hoped that he wouldn't be angry at her since no one was supposed to know. It made sense. Julie had always known where he was and knew exactly whom to tell and whom not to tell.

I hopped a Mexicana turista flight to Mazatlán and took the car ferry across the inlet. It was the fastest way to get there in those days. Cabo is not a big place. He is hard to miss. They pronounced his name "Strai-care," but they knew who I meant. An old checker cab conveyed me to the Two Yanks.

It was one in the afternoon and the sun was blinding, but when I walked in the door, it could have been midnight in the bar. It had one tiny window and in the middle of the day was still lit with candles and small hurricane lamps. I was exhausted, my stomach ached, and my vision was blurred.

I stood at the door waiting for my pupils to dilate to let in more light. I was looking for a waiter and rehearsing my question about Stryker's whereabouts in the pidgin Spanish that I knew then. I focused on the handsome mahogany bar, and my eyes drifted upward to rest on the hands of the bartender serving a Lone Star long-neck to a blond woman.

These hands were more like paws with broad palms and stubby fingers. I was twenty feet away, but my eyes were adjusted and I could see the thick wrists that drove the quick, precise movements. I had seen these hands before as they pointed to things on blackboards, punched keys at computer terminals, and lifted double bourbons in great saloons that nobody else knew about.

I heard the beer bottle hit the bar, and my head jerked up to look at the face. For a split second I was terrified that I had been imagining those hands and that they didn't belong to the missing person I had hired myself to find.

Stryker Stephens attended to wiping off the bar and

washing glasses, lost in the same kind of self-possessed reverie I remembered from California. His hair was very gray for someone his age, and he'd gained a few pounds. His eyes, behind the same aviator lenses, were still dark and impenetrable even at ten paces.

I would have stood there longer, expressionless, not breathing, except that this goddamned parrot sitting on a perch behind the bar leapt on Stryker's shoulder and he looked up.

He stared at me with my two-day stubble, sour clothes, and a face that had frozen into Western stoicism. He looked to the right and to the left. He placed the parrot back on its perch. He motioned for one of his employees to replace him at the bar and walked out to meet me.

Some knot inside me untied itself.

"Hopper."

"Yeah."

He reached up and put his hand on my shoulder, which is no mean trick since I have about a foot on him. He wheeled me around like a Detroit bouncer and we were suddenly outside blinking at each other.

There are men who consider crying to be a form of vomiting, and they will avoid it at any cost. So we did, casually wiping our eyes against the sun as we walked toward the compound.

"So what are you doing here?" I said, my voice back and my gut peaceful.

"Exactly what I want and nothing else. What are you doing here?"

I couldn't answer him, and he didn't press. We reached the compound, and he took a large latchkey from his pocket and inserted it in the wrought-iron lock.

"How did you find me?"

"Julie."

He grunted and turned the key. The heavy mission oak door swung open, and we walked through. The deception of Mexican architecture is benign. It hides beauty from the streets. We walked into a courtyard that might have been in another world with cactus flower, bougainvillea, and a small pool. Tiny sculptures were tucked into shady recesses.

"Do you live here?"

"No."

We crossed the courtyard to an archway that admitted one to a circular hall and to doors of leaded glass. We took the main entrance into the cool interior of a living room. It was set in masculine comfort, half Mexican, half stateside.

He tossed the key onto a massive coffee table. "Welcome to the hermitage. I'll show you to your room."

"Stryker?"

"Yes."

"I came to talk."

"I don't talk to recalcitrant students who are too tired or drunk to stand up. Go to sleep."

I didn't argue. On the way to my room I passed a library with big leather chairs and good art. I passed a conservatory filled with enough audio equipment to put most studios to shame, a grand piano, and a full set of drums. Adjacent was a replica of his old data analysis station at Rand. It was "clean room" clean—stark and home to a minicomputer with a modem, a fax machine, a telex—and a scientist in Mexico's answer to "muthaless" Bell—sophisticated and powerful radio equipment. The little compound is much bigger than it looks from the outside—another illusion. It is filled with the toys of grown boys.

I had no luggage to unpack. Stryker left as soon as he had shown me my room and bath. There was a phone next to the commode. I slept in clean linen sheets and imagined that once he looked in on me. When I woke the next morning there was an Oaxacan blanket on the bed that hadn't been there the night before.

I'd had sweet dreams tangled with nightmares of Caitlin, Stryker, Two Buttes, and Southeast Asia. I awoke fourteen hours later. I felt like I'd made it through the looking glass.

A slight Mexican child, no more than twelve, brought me breakfast. It consisted of fresh-squeezed orange juice, chorizo, corn tortillas, and fruit. The coffee was laced with Kahlúa.

I was groggy and lifting myself slowly off my back as he set the tray down at the bedside table.

"Buenos días, señor."

"Bonaz dee-az," I mumbled at the bright smile and reedy voice.

I heard Stryker's basso profundo from the doorway.

"Vete, 'jito. Tu mamá te necesita á la cocina."

He was the cook's son, and he bobbed his small head reverentially as he left. Stryker does that to people. Now the boy is a grown man and manages the charter fishing business for the Two Yanks. That morning, his was the first face I'd looked at without bitterness for years.

The second was Stryker's.

Stryker sat at the edge of the bed, stuck his paws under his armpits, and surveyed the debris in my face.

I remember it as the day I started to heal. I hadn't completed the process yet, but I considered that I owed the man my life. Another few days of booze and sleeplessness, and I would have put the barrel of a gun in my mouth and kissed my life good-bye.

As if on cue, the phone rang and jolted my brain fast-forward to real-time Coal Creek Canyon to answer it.

"I am mildly intrigued by the possibility that this is important. Guillermo is certainly honored to have been your courier. What the hell's going on?"

"I'm close, but of course that only counts in hand grenades, as you would say. If I can't turn my technical expert around, I'm going to have to waste time going after two suspects at once."

"Where is the obstacle? Be precise."

I explained. He concurred that Silverman was the most efficient linchpin in the investigative mechanism.

"This will probably take two hours to implement. I suspect that Silverman may not get back to you until late in the afternoon. Get some sleep."

"Stryker, what are you going to do? Brief me."

"I only brief people when they are collaborators in the event. You have nothing to do but get some sleep and wait for a call from Silverman. Stop wasting time. I have to place a transatlantic call."

The wizard was about to do a little magic, and I wasn't in any position to argue if he didn't want to give away the secret to the trick.

"Keep me posted," I said, feeling left out and a little like Mack Barnum must feel when I do the same thing to him.

At three fifty-five in the afternoon, I woke from a dreamless sleep to take a call from Dr. Abe.

"Hopper!"

"Yessir!"

"Today I received a call about your problem. I will see you as soon as you can get here. My office. Can you make it before five o'clock? There'll be traffic."

There was still reluctance in his voice. He was going to help me somewhat grudgingly, and he wanted to get it over before dinnertime. I assured him that I could make it in thirty-five minutes.

"Abe?"

"Yes?"

"Don't tell me anything that I'm not supposed to know, but where did that call come from?"

"Tel Aviv. You didn't know?"

Sonuvabitch if the portly bartender hadn't reached way back into Abe's life to get it done. Stryker raised the meaning of the word *network* to an awesome level.

I got to Abe's office at the appointed time. He flinched when he saw the purloined records, but he set up the Gottmacher and Rotheker files side by side on his desk, never looking at me.

It took two minutes. His face remained impassive as his eyes darted between one and the other. Then one eyebrow raised itself slightly. He double-checked something earlier in both records. Then he slammed the Rotheker file shut and shoved it across the desk.

"A Liggett ridge on a gold crown installed to correct a bad bite problem. Please leave now. Take the other file with you as well. Find Rotheker."

He took his glasses off and rubbed his eyes. I was dismissed. On the one hand, the hair on the back of my neck was standing straight up because I was closing in on the quarry. On the other hand, I was dismayed at Abe's anger and disappointment.

"Abe, I'm sorry. If there had been another way . . ."

He didn't respond. I shut the door to his office quietly. There was no point in being provocative and asking to use the phone in the office. I found a pay phone in the hall. It was going to be difficult to get through to Liggett, but I had to try. Rose Blossom answered. She made the usual obfuscating noises about the doctor being in conference.

"Please tell him that Jeff Rotheker is on the line. I think he'll want to talk to me."

Apparently she performed as instructed. He was on the line immediately, and although he limited himself to saying, "This is Dr. Liggett," the apprehension in his voice was the only corroboration I needed just then. I hung up.

I dialed Jeffrey Rotheker's number. No answer. I knew very little about the man. Did he work? Did he get home at five o'clock? Could I stake out 1365 Beeler without arousing suspicion? I didn't have any answers, but I left the building and took 17th to Aurora to have a look around in broad daylight.

The 1300 block of Beeler had never seen glorious times, but it had seen better days. It was loaded with seven-hundred-square-foot frame-and-stucco houses on small lots with skinny driveways. Some of them hadn't seen paint for a long time. It was the kind of neighborhood that housed retail clerks and gas station attendants going to school at night to become mechanics. Some people drank too much here on weekend nights when they had the money. Abandoned wives with small children found rents they could afford on welfare or minuscule salaries. Retired folks worked tiny gardens and canned tomatoes. It wasn't the kind of street you could sit on without being noticed by someone. I needed a vehicle with an excuse. The Mercedes wouldn't do.

20

I have friends in Denver. Good friends. The kind who will give you the shirt off their back or their vehicles or any bit of information you might need, no questions asked. I called a friend with a converted van, the kind of thing you can park in a seedy neighborhood, crawl into the back of, and have a private view through the one-way black windows. I called another friend at the state motor vehicle department, and he ran the name and address for me. Rotheker would show up in a 1978 dark blue Valiant if he showed up at all.

I was in position by five-thirty. The garage door was open and the garage empty, so I hadn't missed him. Other folks came home from work to meals of Kentucky Fried Chicken or pizza or burgers. I could see televisions through picture windows. I could see fights starting between husbands and wives, brothers and sisters. No Rotheker. It was almost seven.

Finally a noisy blue Valiant came up the street from

the Colfax end and putzed its way up the skinny driveway at 1365. I was about to get my first look.

It wasn't much. He was five feet four all right, dark-haired and slight. I couldn't make out much of the face. He used the back door, and it was already dark. Lights went on. I could see him shucking his jacket and cap and moving into the kitchen.

I had no fancy stuff with me. No telescopic infrared spotting scopes. No night cameras. No bugs to plant under windows. No cellular phone. If the guy was going to have a quiet evening at home, it was going to be a long god-damned night. I was going to have to wait him out till morning staring at a dark house. If I'd done it by the book, there'd be a plainclothes guy from Aurora pulling stake-out duty with backup and the luxury of a two-way radio. It's not that I think I'm too good for that kind of job; it's just that I'm not well suited.

When I was getting ready for my tour of duty with the spook patrol in Southeast Asia, Stryker taught me to breathe without moving my chest or making a sound. Just enough to get the oxygen into my lungs, but not enough to constitute respiration. I can conceal myself behind a rice paper wall for hours without moving, totally intent. But put me in a comfortable van on a back street in Aurora and my foot starts to tap, my fingers drum on my knees, and I find the silence distracting.

By ten P.M. I was doing isometric exercises to keep myself from getting van fever. I was working on my right bicep when all of the lights went out, one after another, and I groaned. The little bastard was going to go to bed with two murders to his credit and a lone wolf parked in front of his house chewing on the leg that got caught in this soporific trap.

I was about to go back to my isometrics when his back door slammed shut and he emerged in clothes so dark I could see nothing but movement. He was lugging some kind of case, too small for a suitcase and too large for a briefcase. He opened the door to the Valiant and set the case on the front seat. The engine started like a hog with a bad case of gas.

He pulled out, and I had about thirty seconds to make a decision. I could follow him, or I could break into his house. If I guessed wrong and he came back prematurely, I'd have to render him unconscious and my hand would be tipped. If I followed him, I might be preventing murder three.

I compromised. I let him get a block ahead of me and pulled out slowly. I followed him at a comfortable distance. He drove all the way east to Gun Club Road and pulled into a mobile-home park. I just kept going.

It was ambiguous. He might have been dropping off his little case and driving back home. Drugs, maybe. Or he might have been visiting his girlfriend and the mysterious valise just contained a change of clothes. I was reasonably sure that whoever was in the trailer was not a bodhisattva. More decisions.

I did a 180 and made it back to Beeler in ten. It was easy. One of the back windows had a flimsy lock. The neighbors were half-asleep, and I was up and in before they noticed.

I palmed my flashlight. I was in a bathroom. It was sparse, furnished with a couple of towels and a roll of colored toilet paper. The house was cool, as if the thermostat were always set to minimize the utility bill.

The bathroom led to a hall, which in turn had three doorways—one to the kitchen, one to the living room, and another to the bedroom. There was a linen closet and a clothes chute and nothing much else to comment on. I took a look at the kitchen, cracked the refrigerator door and a cupboard or two. Nothing stood out. There was peanut butter and jelly and white bread, but nothing that vice would find of interest. I moved into the living room and drew the curtains. I didn't want to risk turning on the lamp, but I wanted a little privacy.

I hit the flashlight and found a radio, a couch from the Salvation Army, an old tuxedo chair, and the kind of magazine rack that you made in wood shop in junior high. It was all neatly arranged, but it was very depriving. If the house burned down tomorrow, he couldn't have honestly collected more than fifty bucks for the contents. There were no pictures on the wall, nothing comfortable or homelike. It was spare and depressing as hell.

I moved to the back bedroom, expecting to see a monk's cell. Instead I wandered into a museum of sorts. There was a cloying smell in the room like bad incense combined with unwashed laundry. I pulled the shade and risked the light because the shadows and textures coming from the walls were richer than all the rest of the house put together.

The bed itself was, in fact, monklike. A single twin with dingy white sheets. But the walls were hung with talismans of some sort. Masks and gargoyles and demons. There was a small antique dressing table in the corner, and it was set like an altar with black candles, half-burned and oddly reflected in the long narrow mirror.

On the third wall was a small bookcase, and I stooped to browse. Anton LaVey, H. P. Lovecraft, *The Enochian Keys*, Nietsche, Goethe's *Faust* and, inexplicably, an entire set of Dune. Above the bookcase was a framed piece of parchment with a list of names beautifully wrought in calligraphy under a stylized pentagram.

Beelzebub. Belial. Leviathan. Behemoth. Abaddon. Then *Lucifer* in larger letters. Hanging from the ceiling, dangling next to the parchment, was an inverted crucifix. Liggett hadn't been imagining things. The kid was Satanic. My stomach, generally of cast iron, flopped once in revulsion. I could feel the bile crawling up my throat.

I was in the right place and this was the right guy, but so far all I knew was that he believed in the Devil. The Constitution guarantees the right to do that unless you hurt somebody, and every cell in my body knew that he had. So far I couldn't prove it.

The art wasn't going to tell me anything that I didn't already know, so I turned back to the bookcase. Fifteen minutes had elapsed. If he'd just stopped to drop something off at the trailer court, he'd be back soon, and I still had no material evidence.

The first two shelves were filled with the classical texts, including Goethe's *Faust*. The last shelf was an occult supermarket. There was a traditional tarot and a New Age tarot. Volumes of astrology, numerology, martial arts, t'ai chi, and finally an edition of *The Clear Light of Bliss*. A Buddhist text. I plucked it and found the chapter on chakras. It was dog-eared and underlined. I would have taken it, but I wanted him to know nothing of my visit.

I needed more than circumstance. The knife or the gun would do—something that would connect the lunatic room with the reality of two deaths. I shrugged and moved to the closet. Members of the American Satanic Church, if he was one, probably didn't do things so differently from other hoodlums. They probably hid things in closets.

I switched off the light and listened for the sound of cars. Hearing none, I flipped the flashlight to full beam and entered to search the tiny closet. There wasn't much to see. A couple of pairs of pants. A jacket. Some hats and lots of denim. And a shoebox from Payless Shoe Stores.

I grabbed it with no enthusiasm. It was too heavy. Inside, it held some letters from correspondents in Laramie, Los Angeles, and as far away as Miami. I didn't stop to read them. Underneath the letters, unsheathed and casual, were a stiletto and a small-caliber firearm. I replaced the box carefully. I was in no position to take them. My job was to get out of there and bring Mack Barnum back to find them.

I left the house at ten-thirty and parked the van in the same place on the street. I waited for him. He came home at two in the morning. His arrival was uneventful. He carried the same valise back into the house. The lights were on, and then they were off. Satan's child slept and found nothing amiss.

Forty-five minutes later I was back in Coal Creek with an urgent need to do something—damn near anything. I dialed a number in Connecticut. It was about four in the morning there, and I should have chided myself for my lack of consideration.

The voice that answered was awake.

"Burroughs here."

"Hopper calling from Denver. How are you, sir?"

"What's happened?"

"I found him. Now what do you want me to do with him?"

I don't know why I asked. I knew what I was bound by law to do. But I'd made a deal with Burroughs. With or without an arrest. He had to decide. There were other means that he had at his disposal, and it was his bloodline that had been interrupted, not mine.

Burroughs understood the question. I hung on to the phone loosely and tried not to care about the answer. If we were going to take this through the criminal justice system, I was going to have to conserve my patience. I would need it just to get the collar to happen. If we were going to do something else, I would need all my energy to finish the job.

The loss of sleep was beginning to wear on me. I can't remember having a preference at that moment. I was indifferent. I would do what I had to do. Not think about it too much. I waited in that state of suspension that I had learned slowly but quite well. His answer was swift and emanated from an old Jesuit tradition. Never be obvious with your vengeance.

"Do you have enough to prove it?"

"I have enough."

"Get him arrested."

"I'll do my best."

"What's the problem?"

"It took a fair amount of illegal activity to get to the truth. I don't know if I can find a judge who'll ignore it. Mack will have to help with the window dressing."

"You need a warrant?"

"I need a small miracle."

"I've got one or more owed. Wait to hear from me."

Everyone was a friggin' magician except ol' Hopper, who was just a burned-out spook from Two Buttes. Jesus, I was tired. I couldn't worry about it. He was out there. I didn't think he'd kill anyone before tomorrow afternoon, but I was too spent to do anything about it if he tried.

My sleep was disturbed by dreams I refused to remember.

I slept for a couple of hours—just enough to remind my body of what it was missing, but not enough to satisfy the need. At 6:30 A.M. I called Mack at home and he was already mean and ugly.

"Where the fuck have you been, Hopper? Do you think I'm some jack-legged dog-ass you can tether to a telephone pole and leave till Sunday?"

"Shut up, Mack. And drop the wounded dog crap because now I can give you all of it and I'm going to need your help. Big time."

"Jeez Christ. What have you done now?"

"I found him. But to do it I broke a rule or two. Burroughs said he could help us through the rough spots—and to wait. I don't know who Burroughs has in his back pocket, but he sounds confident. Right now, I need your collaboration."

Mack sighed, and I could see him wiping his ham hand across the back of his neck. I'd managed to put a lot of important friendships on the line for this one. Mack included.

"Gimme what you got."

So I gave. No censorship. Everything but Stryker and the call to Tel Aviv, which was outside Mack's need to know. Hell, Stryker's call to Tel Aviv was even beyond *my* need to know.

"So what's my case for a warrant? A story from my good friend Hopper, who lies to his friends and lets himself into dental offices without a key. How'm I supposed to explain to a judge how I know what I know? You want me to say it's a hunch? You want me to buy off a psychic to testify that she had a vision about 1365 Beeler and we'd like to check it out? Think that would suit him?"

I let him carry on. He had a right.

"Mack, run a make on Rotheker. Do all the things you have a right to do without entering his private property. Put a man on him. Give Burroughs some time. Give me some time. I'm not doing this because it improves my love life, ya know."

There was a silence on the other end of the phone that signaled the kind of flexible frontier reasoning of which Mack is capable. You can't always find it in Washington or New York, but, God willing, you can find it in Colorado.

"All right, Hopper. I'll see what I can do. But it's the last goddamned time. You hear me? The last motherfucking goddamned time."

I called Gretchen, and there was no answer. I called the center and found everyone present or accounted for at the moment.

I wanted to be out on the street. I wanted to be welded to Jeff Rotheker's hip. I wanted to tie him up and put him in a cage until I could make the legal system move.

I made another call to Cabo.

"¡Diga!"

"Stryker."

"Is it done?"

"Not quite. Burroughs thinks he has a judge who can get Mack into the house."

"Whose house?"

"Belongs to a guy by the name of Jeff Rotheker. Appears to believe in chakras and Satan."

"Have you established motive?"

"Don't need to. If I can get Mack into the house in exactly the right way, he'll have weapons and a wild-eyed suspect. I'm pretty sure Rotheker is part of the American Satanic Church. It won't play well in Denver. He did it. I have no question."

"I see. So until Liggett is subpoenaed on the dental records and Mack runs the gun through ballistics you've got nothing but circumstantial woo-woo. Tell me, can this case be tried anywhere but in the newspapers?"

He was the consummate devil's advocate.

He was also correct, and I resisted it. I was tired and I'd worked hard. With any luck it was about to be the prosecutor's problem. Grudgingly, I had to admit that a lot of people had pulled a lot of strings for me in this situation, and I was no closer to explaining motive than I had been when I rifled the bookcase for clues. I didn't know what Satanism had to do with Buddhism or what Tel Aviv had to do with dentistry. I didn't even know for a fact that the ballistics of the gun would match the bullets taken from the victims. I was a little weary of everyone reminding me of what I didn't know.

"Hopper?"

"I'm here."

"Precisely the problem. You are *there*. Step back. Consider the data. You've done a remarkable job in a short time. That doesn't make it enough."

I was used to Stephens breathing down my neck like the master of the dojo, but the cumulative hits I'd taken on this case were reaching dangerous proportions. I hadn't quite recovered from the unbidden memory of Caitlin, either. If I didn't get some sleep, hallucination was a real possibility.

"No—not enough yet—there's more. But I hope not a lot more." I hissed it, resentful of the demand that I was making on myself.

"Hopper. Take care."

"Kiss Maggie for me."

"With pleasure."

I cut the connection to Cabo, and I concentrated on the very short term.

It was the first opportunity I'd had to deal with the chaos in my house. I put the large things back in place first. Clothes back in drawers, books back on shelves. The crockery in the kitchen took half an hour to sweep and discard. The ceramic frog was indeed dead.

The vandalism still held little interest—murder is very distracting that way—but it was a loose end that dangled annoyingly in a case that still had too many loose ends for my taste. I was certain that the visit to my house was not paid by Rotheker. Betting on Wyatt was pure hunch. I had no motive there, either.

I put the finishing touches on the living room. It wasn't perfect, but at least it didn't look like the inside of a dumpster anymore. There was a knock at my door, and through the window I saw Kevin Childress and Dan Wyatt, as if on cue. Childress looked stoic and determined. Wyatt hung his head like a guilty puppy. I smelled the possibility of some answers, so I opened the door.

"Gentlemen?" I did not invite them in. Childress had been appointed spokesperson.

"Mr. Hopper, we'd like to talk with you."

I shrugged and walked casually into the living room, keeping them both within peripheral vision. They followed and remained standing. I sat on the couch and stretched out my legs.

"Have a seat. Start talking whenever you're ready."

"Dan has some information that we think might be important."

"About?"

"Well . . . about this friend of his who's been asking a lot of questions about Buddhism and the teachings."

I turned to Wyatt, who still had his eyes glued to the floor.

"Seems to me that Mack and I have given everybody sufficient opportunity to make statements. Why am I hearing about this now?"

He finally picked his head up and looked at me. All I could read was fear and shame. Rotheker's status as prime suspect was secured. Wyatt was a jerk, but he was no murderer.

"I . . . um . . . got a little crazy about this whole thing. It happened so fast, and you were . . ."

"I was what?"

"Not someone I could trust, at first."

"Wyatt, I am not trying to make friends here. This is a murder investigation. You don't have to trust me. In fact, I would suggest that you not trust me at all. You never know when I might ask Mack to arrest you. I find it real interesting that you know my address. Makes me think you might have been here before."

He gulped but continued to look at me. Kevin put his hand on Wyatt's shoulder and shook it a little.

"Tell him, Dan. Just get it over with."

It was like Childress had flipped a switch on a talking machine. Pretty simple story, really. Wyatt had it bad for the lady guru, who persisted in relating to him like a teacher and an authority figure and not a karmic lover. I got tapped for the role of evil Lothario bent on stealing her away from him. It was pretty pathetic.

"So you were looking for evidence that I was a bad guy. You had to break dishes and tear books?"

"I'm sorry. Like I said, I was a little crazy."

I grunted. There wasn't anything else to say. "So what does this friend of yours have to do with all this?"

Wyatt sighed as if relieved to get on to less personal topics. "I met him about a year ago. He came to an introductory lecture that Gretchen was giving at the center. He seemed really interested, and we got to talking. He asked for some books he could read, so I gave him some titles. I also lent him my copies of basic stuff."

"Get to the point, Wyatt."

"Well, maybe it's nothing, but this guy seemed to be watching us. He asked a lot of personal questions about what life was like at the house. Wanted to know who had studied the longest, where they worked, which issues in Buddhism gave us the most trouble. He came to every public event we had for about a year, and then he just disappeared."

"How long ago was that?"

"About three months ago."

"You haven't heard anything from him since?"

"Not a word. I just blew it off until Tom got killed. Then I thought about him. I don't know why."

"Is there anything that connects him with Patty or Tom?"

"Not particularly. He was curious about everybody and everything, like I said. Jeff was pretty intense even when he was just making conversation."

The name fell into my lap with a thud. I was cool. I'm almost always cool. Drives women crazy and makes my mama proud.

"Do you have any reason to believe that he's involved in this? So far all you've told me is that he's an intense guy who's a little nosy."

Childress was fidgeting in his chair. "Tell him about the portal theory, Dan."

"Yeah, I was just getting to that. See, Mr. Hopper, just before Jeff disappeared we got into this long rap about chakras. He said he wondered whether chakras weren't the access points to another dimension. He thought if you could control chakras, you'd be able to harness immense metaphysical power. It was pretty distorted, weird stuff. He kept talking about chakras as if they were actual physical things. The throat chakra is not the throat and the heart chakra is not the heart, but I couldn't get him to understand that. Anyway, he kept talking about them being the seven portals to control the soul."

I was having a few first theories of my own. I had no desire to share them. The phone saved me. I took the call in the kitchen and let the two bodhisattvas stew for a bit.

"Hello."

"Hopper, I swear to Christ you lead a charmed life. I just got a call from Judge Fastabend. Said he 'heard' we were getting close to a suspect. Said he was going to dinner at the Red Lion Inn tonight, but he'd be available if we needed him to issue a warrant."

My, my. Old Catholic Eastern money has tentacles that reach all the way to the West.

"Mack, get the warrants first—one for the guy and one for the house—and don't go without me. I think we are going to get to use them both. I've got to finish up with a couple of Buddhists who are telling me all sorts of interesting things about our man. Seems he attended every public event that the sangha held for the last year. They've all seen him, Jake. Unless we lose this guy to a mugger, an accident, or a bad Miranda, it oughta be like old times."

"Stay put, Hopper. I'll call you when I've got papers in fist."

I hung up and sauntered back into the living room.

"Wyatt, I'm not going to press charges on the break-in, but in exchange for my kindness, you're to remain at the center until I say you can leave, or I'll track you down, stomp a mud hole in your ass, and fill it with cactus. I need to be able to reach you at a moment's notice. Go back to the center and start thinking about everything this guy ever said to you. We may need you to recite it. Do you know anything unmetaphysical about him, like where he works, where he lives, his last name?"

Wyatt's brow furrowed, and his eyes rolled up and to the right. "I don't remember that he ever told me his last name. He did say he was an orderly at a hospital, but I don't know which one. Don't know where he lives. I only saw him at the center."

"Okay. I'll be in touch."

Childress seemed irritated. "What are you going to do, Mr. Hopper? My wife died two days ago, and it seems to me this is at least a lead."

"Perhaps, Childress, perhaps. Now I believe you two were on your way out?"

He shot me a sour look, and they both left. I wandered back into the kitchen, having begun to notice this empty space in my stomach. I grabbed eggs and a can of hash and put away a pot of coffee before Mack called again to invite me back to Boulder.

25

The Boulder County Sheriff's Department is a pre-fab concrete pink-speckled job. Supposed to match the flatirons, I guess. Probably some out-of-town architect's idea. Police buildings get more effete all the time. Bothers me. I think they ought to look gray and foreboding so that people who have dealings there come in with fear in their heart and leave remembering that they've screwed up. The politicos should put the law enforcement community relations department somewhere else and make it pink-speckled prefab with rose gardens and other posies.

Mack had his size fourteens up on his desk—and a file in his lap.

" 'Bout time, Hopper. We haven't located him yet, but we've got all three jurisdictions involved and some information on him. The juvenile records, if there were any, are expunged, but he's clean after the age of eighteen. No criminal record in the army, either. He's an orderly at St. Luke's Presbyterian. Good work habits, rotating shifts.

He's working afternoons for the rest of the week. You wanna search first and pop him at the hospital, or should we do it at the house?"

"Who's coming to this party?"

"It's an intimate little gathering of three—you, me, and an Aurora detective—professional courtesy, their jurisdiction."

"We've got more control at the house."

"Yeah, but if he's there and I can't find the hardware, we'll have to leave, and he'll know he's been fingered. We can't arrest him for having black candles and books about the Devil, y'know."

"Is somebody watching the house now?"

"Yeah. He's not home."

"All right. We'll go in after we're sure he's pushing gurneys."

I wasn't looking forward to spending another two hours sitting on my hands. The closer you get to the goal line, the more time extends itself—kind of like the dream where the tiger chases you and the harder you try, the slower you run.

"Mack, what do you say we get to Denver now? If I have to sit here with my thumb up my ass, I'm gonna get cranky. I'll buy your lunch."

"Right behind you."

We took my rig and Mack's handset with the inter-agency channel. We weren't going to need a squad car until we moved into the hospital. It was probably the first time since the beginning that Mack and I had had more than a few minutes together. We didn't talk much until we got to Federal Boulevard, where you start to understand that you're entering a city.

"So, Hopper, how come you didn't tell me that you're a war hero?"

"Because I'm not."

"Hmmph."

"Where'd you spend the last war, Mack?"

"Bein' a too-old-to-be-drafted law enforcement official. Playing games with riled-up students on the campus.
Sometimes it seemed they were too scared to face real
bullets, but they still needed to be part of the war. They
ran their own little war games, and we got to be the Viet
Cong. It was all very polite compared to what was happening in 'Nam, but it was an imitation of it. A civil dis-
order, know what I mean?"

"Just part of the chess game, Mack."

"So what was it like to be one of the pieces, Hopper?"

"It was a job."

"I'm...sorry about your wife."

"Thanks. Me too."

The exchange was over before we got to the interstate,
and I knew we wouldn't speak of it again. Women frequently complain that men don't know how to talk to each
other. Maybe they're right. Doesn't mean we don't communicate. Mack and I had just "talked" about the confusion in his own life, his envy of my adventures, and his
empathy for my loss. I responded with my own fatigue
and sorrow with the way the world works and my gratitude for his concern and friendship. I'm not saying it's the
best way to do it; I'm just saying that that's how it's done.
There isn't one woman in twenty who would have understood the nuances if she'd been in the backseat eaves-
dropping.

We hit the White Spot on Colfax. Food's lousy, but the
floor show's great at just about any time of the day. There
are certain places in cities that are like watering holes in
the Serengeti. Doesn't matter if you're the hunter or the
hunted. Everyone has the right to drink bad coffee and eat
greasy burgers in peace. Drug addicts, runaways, teenaged
hookers, punkers with purple hair, discouraged workers
who can't bear to look at another want ad. Crazy people
who mutter into their cups. Old men and women who can
handle the prices and enjoy being part of a scene. And
cops. Uniforms, mostly. The restaurants like 'em. They

make it all seem less dangerous and crazy. Free coffee and doughnuts is a small price to pay to get a cop to take his break in one of your booths.

It was two-thirty and a little late for lunch, but Mack ordered two chili dogs and a cherry Coke. I stuck with a milk shake because they're hard to ruin and my stomach felt like an acid vat.

The food came and Mack attacked it like it was his last meal. I surveyed the passing scene. To my right at the next table was the Queen of old Aurora. Not the new Aurora with its upscale office parks, but old Aurora with its seven-hundred-square-foot frame houses and junk cars in the yard.

She'd waddled in by herself and was now stressing your standard restaurant chair to the max. Two hundred eighty pounds without breathing hard. Sparkly blue polyester pants suit with snags, eleven chins, and three-month-old nail polish. Her hair was bottle red and frizzy, pinkie finger sporting a diamond the size of a gnat's eye. It was the only thing small about her.

Some excesses can't be contained. I watched her thighs tumble over the edges of the seat and her breasts flow onto the tabletop. She ate fast and brushed the crumbs into her crotch. Her arms were the size of an elephant rump roast. She had a little fastidious mouth that was countersunk into the folds of her fleshy cheeks. It was voracious. It consumed.

Gretchen would have said that the fat lady was a sentient being. The bodhisattva would have proceeded to work for her enlightenment. Mack was committed to serve and protect an abstraction, the "community," the Queen of old Aurora included. Stryker, in a cynical mood, would have said that it ain't over till the fat lady sings. I wondered if she sang. I was repelled and fascinated by her all at the same time, just as I was repelled and fascinated by Rotheker. I was about to try to hog-tie a sick murderer as much for her as for Burroughs or Reynolds or Stephens. I didn't like her much. She was Mammon in America, but she was also somehow part of the mission.

"Time to roll, Hop."

I drained the milk shake, sucking up the last sickly-sweet chocolate slurp like a kid. Mack left to pop for the tab. There was a red plastic rose stuck in a cheap glass vase sitting next to the ketchup on the table. I plucked it and laid it next to the fat lady's plate on my way out. She looked up at me for a moment with watery blue eyes that were suddenly more alert and intelligent than I would have expected. I bowed slightly, and she blushed like a young girl.

Like the man said, it was time to roll.

26

It was quiet on Beeler Street. The Aurora cop was in position halfway down the block. Jake used the radio to say that we'd arrived and that we were going in. Rotheker hadn't been home since Aurora had taken position at ten A.M. Mack asked that Aurora PD coordinate with Denver to locate the suspect at St. Luke's.

We knocked. That's a rule. There was no answer. I picked the lock for Mack, and the door gave way to the same ascetic room that I remembered. It was no better in the daytime. I took off to the right and circled through the kitchen. He took off to the left and went into the hall, where we met before breaching the bedroom.

The door to the bedroom was closed, and there was something ominous for me in that—something from another sensory level. We didn't say anything, but Mack drew his gun and put his hand on the doorknob, his back to the door frame. He opened it slowly.

There was a lump in the monk's bed the size of a

person. I heard Mack suck in a sharp breath. A sheet was neatly tucked over it, hiding the face. My second sense exploded in my face.

Mack's frustration demanded expression. It was no less than mine.

"Hop, I don't think we should have stopped for warrants and coffee. Fuck the system. Damn the sickos it protects. Screw the whole goddamned asshole this world has become."

I guess he wasn't as hardened to it as I had thought.

"Unless you expect surprises, the world will always have the advantage." Stryker in my head again. Insistent and embedded.

Mack and I looked at each other, reaching for a decision. I motioned my head toward the closet, and he moved to open it. The shoebox was there. I pointed to it.

"How am I supposed to use all of my sleuth training if you already know where everything is?"

"I don't know a thing, Mack. But I've always been good at Easter egg hunts."

The letters were still there. The gun was still there. The stiletto was gone. Didn't matter anymore. There was a dead body in his room. Less need for hardware. Less need for a tight argument. The presence of a corpse should be enough to beat the inertia of our illustrious criminal justice system.

Mack bagged the gun and turned to me. "Time to pull the sheet, Hopper. I'll get Aurora's man in here. We'll need all the corroboration we can muster."

He punched the radio, barked a half order, half request, and looked sadly back at me. I walked reluctantly to the front door to let our liaison in. There was something heavy in my chest, an unnamed fear. I was wound too tight on this one.

Mack came into the living room, nodded at the Aurora guy, and asked him to check out the lump, make a report, and call the coroner.

I looked at Mack, perplexed, and started for the bedroom. The next thing I knew I had a meat hook on my tricep.

"What the hell, Barnum?"

"Stay where you are, Hopper. You don't need it."

The Aurora guy came out of the bedroom, and to his credit he was only white in the face and not green at the gills.

"Where's the phone?"

"Over there."

"Shit, Barnum. You could've warned me."

He called for the lab and the coroner. Mack didn't move a hairbreadth, watching me like I was the mongoose and he was the snake.

Something was very wrong here. This was not Mack. Whatever or whoever was in the bedroom was something he didn't want me to see.

"Mack. Quit being an asshole. You can't shelter me. I've seen it all."

He stared at me for a few seconds, but it seemed like longer. Then he shrugged, his career and his body approaching middle-aged burnout. He moved out of the doorway and let go of my arm.

The Aurora detective had left the sheet at half-mast. It was grisly. The chest cavity gaped, and the solar plexus was a bloody mess. The heart had been extracted. The third and fourth chakras had been literally removed with psychotic determination and quasi-surgical skill.

Gretchen Reynolds lay on the monk's pallet with the ghost of a smile on her face. I had no more understanding of why she had come here or why she had died than I had of chakras and satanic intention.

The gall of failure rose in my throat like a slow-acting poison. She was partially clothed, and my impulse was to cover her, to protect her in death somehow better than I had protected her in life. In truth, it didn't matter. She was the prime portal to Rotheker's fantasy dimension. I lifted the sheet over her body and fought against the memory of its brief, uncompleted pleasure.

I walked back into the living room, and Mack kept his eyes masked.

"Coroner's on his way. You can skip the hospital if you want to. We can handle it."

"I'm sure you can, Mack."

"I...uhm..."

"Shut the fuck up, Barnum. Do we know whether he showed up at work?"

"Not yet. We've got his supervisor looking for him."

I looked at my watch. It was 3:40 P.M.

"I thought you said he had good work habits."

Mack grunted and asked the Aurora guy to route the coroner's report and keep the house under unobtrusive surveillance. We got in the car and drove to Cherokee Street.

From the car we dispatched squad cars to the Denver and Boulder houses. From Cherokee we put out an all-points with the license number and description of the old blue Valiant. We also put security at St. Luke's on alert. We told them that a uniform unit was on the way, code 3, that he was very dangerous, and that they were to observe but not try to apprehend. We left the press out of it entirely. Mack called his office to report on our progress and the stalemate.

I paced like a cat. It was four-thirty. If Rotheker were going to show up on the job, he would have done so by now. Where was he? Halfway to Montana, where you could get lost forever? Somewhere on the way to Nebraska? West to Los Angeles in his on-its-last-legs car?

At seven P.M. I dialed Westchester.

"Burroughs, here."

"This is Hopper. The trail's getting cooler."

There was silence on the other end of the line. Interminable silence. "I see," he said finally.

"There's been a third murder."

He sighed. "I can see that doing it by the book was costly; it usually is. I appreciate your keeping me informed. Carry on. And Hopper...let me know if I can speed this up. I might have other resources to put at your disposal that you could manage much more efficiently than the authorities." Then a click without the customary protocol.

From a police perspective, my procedural sins were all forgiven. Through Mack I could now mobilize the arm of the law—atrophied as it was. But we needed it to influence ordinary people. I started with Silverman and Liggett.

They got an informal call from the chief of police and the district attorney. They showed up in good suits. I didn't have to explain how I got the dental records. I didn't have to be in the same room with them. Mack turned in the tooth. Liggett identified the patient, and Silverman concurred. Before Gretchen's death I would have been political, polite, and forgiving with the good doctors. Now I had no time for it. They were advised that they would be subpoenaed in due course. Rotheker was officially a wanted man. Suspicion of homicide.

St. Luke's called in every day for the next three days to report that they hadn't seen him. Aurora PD checked out every trailer in the Gun Club trailer court, and nobody claimed to know him. One trailer pad had been vacated since the night he'd been there, but they hadn't left any forwarding address. They talked to the neighbors, but they couldn't find anyone who would admit to being a friend. They rounded up all the mystics and the wicca and occult weirdos they could find and questioned them with a light hand.

Somehow another Sunday afternoon rolled around. That made it three weeks since Mack had called me about Tom's murder. Three weeks. Three bodies. One of them Gretchen's. Seven chakras and one missing loony tune.

I tried to imagine a life in which no one died a violent death. I couldn't. The last time I had thought I knew of such a life, I was in high school in Two Buttes, dreaming stupidly of better worlds. Then four of my classmates—and in Two Buttes that's a big percentage—drank too many beers and drove off a bridge.

There is a particular fire drill one goes through when the fox has flown the chicken coop. Mack went through it. He notified everyone on God's green earth that Colorado had a problem named Rotheker. Every local, every adjacent out-of-state, and finally the feds. Each step of each remaining bodhisattva was dogged by the department. The path they were on was suddenly crowded.

The hospital officially terminated Rotheker's employment. Conversations with Burroughs were curt. Finally I called Cabo.

"¡Diga!"

"It's me."

"And who are you . . . exactly?"

He knew. There is no need to tell Stryker much. He probably got it from Lawless, who got it from Burroughs. It didn't matter. He knew.

"I am a man who is pissed, frustrated, and sick and tired of due process. I need something to do while I'm waiting for Satan to win."

"Cease and desist. That's infantile nonsense. At first you spent all your time identifying him. Now that he's identified, you must spend all your time finding him. What do you want from me? Ask clearly and quit pretending that I'm a mind reader."

My ears had been cuffed. "I need a new idea."

"There are no new ideas. Just old ideas that work and can be adapted."

"So gimme a clue."

"Give me a break."

"Stryker, this town is not like New York City. In Manhattan you can get lost because there's too many people for anyone to give a shit. But in Colorado you can get lost in the wilderness. He may be gone for a long time."

I could see Stryker pursing his lips, trying to decide which lever to flip in my psyche.

"It won't do."

"Agreed. I'll be in touch."

But not without Rotheker's scalp in my back pocket.

I abandoned Coal Creek with what Stryker called "cellular determination," all capacities ready to be focused on a single absorbing task. It was a universe of two now, Rotheker and me. I needed to squeeze digressions into an infinitely small space to energize the mystic magnet.

I assumed Rotheker knew that the police were on to him. I hoped he didn't know about me. That gave me the edge of anonymity and surprise. He might not know enough to watch for me, and I was going to be very hard to see even if he did. I predicted victory for myself on that basis alone if somehow I could put myself inside his head and find myself nose to nose with the little bastard. Now I had to worm my way into and through the maze of life that he could get lost in if he chose.

I checked into the Valli-Hi Motor Hotel at I-25 and Pecos—satellite TV, direct-dial phones, centrally located, government rates, all major credit cards. It wasn't convenient to anything unless you were a serious trucker on

your way from Montana to New Mexico. The good urban
bunker at $19.95 a night, single occupancy. How'd the song
go? "No pool, no pets ..."

There are two million people who live on the Colorado
front range. I hoped I didn't have to run all of them through
the sieve to find my nugget. And he wasn't necessarily on
the front range. The issue was a little more problematic
than it had been when I was still looking for him among
a few hundred dusty dental files. The statistics were meant
for a computer. So much for science—this one was going
to require a bit more reach. But I was beginning to know
him better by the minute. I was beginning to intuit the
patterns that might narrow the search.

There are places in Denver that have become pockets
of fringy people and their problems. Five Points shelters
burglers, gangs, and drugs. Lower downtown is Denver's
Greenwich Village, with artists in loft spaces elbowing
their way through the winos who sleep in the dumpsters
and doorways. But satanic cults? There's no address of
preference for that kind of evil in Denver.

There was a time in the seventies when Boulder was
home to a little bit of everything off-center. Most of
those people hung out on the Pearl Street Mall and slept
under the stars on the courthouse lawn. They were a network unto themselves, and they could often deliver news
from the ends of the earth. Most of them have grown up
and drifted on to be replaced by a little less interesting
group of drifters. The Hare Krishnas were now ensconced in a large Spanish stucco in Park Hill. Castle
Rising, the most comprehensive witchcraft shop in the
Rocky Mountains, had closed. But the shop hadn't been
much to start with, catering mostly to the Dungeons and
Dragons crowd.

The times and the pressures that came with them dic-
tated that perverse religious customs be conducted in secret. I only knew a couple of people in town who might
know about these secret places, and I started with them.

"Hellooo."

Mary never changed. She always answered the phone in the same questioning, breathy voice.

"Mary, my Wyoming princess, it's Hopper. And for the record, you haven't heard from me in years."

"How are you? What fun. It really has been forever."

"Where's your colleague in crime?"

"In the basement office. Do you need us both? Are you getting kinky on me?"

She giggled and I wanted to laugh with her, but laughter didn't come so easy these days. It had been squeezed into that infinitely small space with all the other stuff that made life normal for most folks.

"I need a consult, my friend. In the interest of efficiency, could you put her on the extension?"

"Uhm . . . certainly. Just a moment."

Mary Lou was all business by the time she got to the phone.

"Hopper, what's the matter?"

"I've got a strange set of circumstances, and I need to know whatever you know about local satanic cults. Where do I go? Who do I talk to? Who can I find to give me an instant seminar?"

There was dead silence on the line. I was pleased. They were telepathic, these two, and I imagined that they were exchanging their memories of all the fringes they'd worked on that involved persons who worshiped the red guy.

They'd both been tilling the fields of the Lord for a long time. Mary Lou had done a stint as a medical social worker and a theology student, and Mary was employed by the state to protect children from marauding parents. In the meantime they grew crystals, kept cats, studied everything that even hinted of the metaphysical, and grew more psychic as time went on. They knew the community in a way few other people knew it. Stryker would have called them an "information node."

The node delivered. For a while it was as if I weren't on the line. They chatted with each other.

"Well, M'Lou, he could start with Ruben in Lafayette.

And what was the name of that woman who lived off Federal Boulevard?"

"Yeah, that's what I was thinking. Ruben might be worth the time. Monica, too. And she had a sister in Aurora, remember?"

I interrupted. "What about Aurora, Mary Lou?"

"There was a case we followed for years in the courts. Protective Services never quite had enough evidence, but Monica's sister kept turning up with children who were oddly abused. Monica has a mental health history. Most people didn't believe her ranting and raving about the Devil, but I never felt right in that house. Anyway, we had reason to believe that she and her sister were involved in some sort of organized group. And I think the kids were used for some of the rituals. The department never caught up with them, and the police never had enough evidence to pursue anything but the dependency and neglect cases."

"So who's Ruben?"

Mary answered somewhat diffidently. "Ruben is one of those people who seems to live on air in a small house in Lafayette. He's part scholar and part lunatic. He seems to know the entire religious history of the area. About forty years old, I think. Straddles Hispanic and Anglo culture and a lot of subcultures as well. Very intuitive."

"How do I find these people?"

"You can find Ruben at 221 Tenth Street. At least that's where he was a year ago. M'Lou, do you know whether Monica is still where she was?"

"I think so. It's subsidized housing on Raritan. Call me tomorrow, Hopper, and I'll have the address. Are you ...all right?"

"I'll let you know as things progress. My thanks to you both. Any more hot tips on looking for Lucifer?"

"Yeah," Mary Lou said. "Be careful, he speaks with forked tail."

29

nless one has a very distinctive appearance, it's not difficult to change it. I didn't need to change my appearance as much as I needed to blend into a variety of woodworks. I slipped into jeans and a T-shirt. I added glasses. I looked okay for Lafayette and the dozen other places I might be that day.

I don't do guilt, although I imagined that I'd left a few disgruntled folks in my wake. I wasn't fully underground, but I wasn't about to check in. Mack was angry by now and puzzled. Burroughs was wondering whether he'd made the right decision in relying on my judgment. And Stryker? Who the hell knew what Stryker thought?

I picked up a 1979 Pinto at the Rent-a-Heap on South Sante Fe, a no-nonsense place used to renting cars to the horse track folks before they tore Centennial Raceway down to make way for "progress."

I went to Lafayette first. The blond lady who answered Ruben Ocala's door said he was giving a seminar at a free school in Boulder and would be back from class in the late afternoon. I asked what class he was teaching. She shrugged and said it was about shamans "or somethin'." I figured him for a Castaneda clone, an ersatz *brujo* keeping college dropouts entranced and this little piece of easy blonde in place. Long hair, a dime-store headband, and faded jeans came to mind. I told the live-in that I'd check back. I figured I had found my crash-course professor.

I went to a pay phone and dialed up M'Lou.

"I found Ocala and he's not home. What's the story on this Monica person?"

"Wait just a minute. I found her address in some old files. Here it is. Monica DeVries, 833 Raritan. I'm not sure she'll talk to you, Hop."

"I'm more persuasive with crazy people than you might think. I used to debate with the military, remember. Thanks for the address."

"You're looking for the guy who killed all those Buddhists, aren't you?"

"What makes you ask?"

"The *Boulder Camera* says the murderer is missing in action. Some guy named Inspector Barnum refused to comment. It didn't mention you, but I just figured . . ."

"Thanks for the information, M'Lou."

God bless the media. It could have been worse. Rotheker now had verification someone was looking for him. At least they didn't have my picture to run on the front page. I was too irritated to remember that the cublet reporter was just doing her job. This changed the game a little. I needed to become someone else entirely.

I headed off to north Denver. The house on Raritan was in a neighborhood not much different from the one in Aurora. Monica DeVries didn't appear to be home. The curtains were drawn, the door was shut tight, and there were no cars in the vicinity. I saw no lights despite the overcast day. But I had no time to stake it out. Besides, I was now a bandito. I took the direct approach and knocked on the door.

I waited a few moments and then tried the doorbell.

There were no cracks in the curtains to glance through and still no signs of life. I turned to go. Then I heard a scrabbling near the door. It started, then stopped. It was enough to wait for. The door had one of those little peep-holes that magnify whoever's standing on the front porch. I wasn't sure whether to concentrate on looking harmless or to concentrate on looking authoritative. M'Lou proba-bly could have told me volumes about Monica's psychi-atric travails, but pressed by time and cockiness, I'd neglected to ask.

Inevitably with the very young or the very old, the front door opens a crack with a chain across it and you're lucky if you can catch the pair of eyes that peer out at whatever kind of danger you might be.

Monica DeVries did not peer. She stuck her small face right into the crack and stared. There wasn't a single ten-tative verbal clue, not a "Yes?" or "May I help you?" She just stared. I guessed that she was five to eight years younger than the fifty-five she looked. A little puffy around the face, wispy hair, blond, dirty, uncombed. Her eyes were brown, a strange almond brown that, combined with their depth, was unsettling somehow.

I didn't know what I had here, but I wasn't selling Girl Scout cookies to the average suburban housewife. The only way to get in the door was to earn my way in. I waited for clues.

The stand-off lasted a long time. She stared, appraised. Finally she shouted at me.

"Well, what the hell do you want?"

"I want to talk to you!" I shouted back instinctively.

"Talk, talk, talk. I don't want to talk if you're going to stand on the porch." She was petulant now and closing the door. The next thing I heard was the chain being drawn. I walked into the house, and she scuttled sideways like a crab. There was a mild stench of garbage and stale cigarette smoke.

"Are you from SSI?"

"No."

She screwed her face up and looked scared. "Who are you? Why are you in my house? This is my house. You can't be here anymore. Go away."

She put her hands over her ears although I hadn't said a word and backed into the kitchen. I followed her. There were several pill bottles on the counter next to the stove. She slammed a few dishes around aimlessly and then sat down at the kitchen table.

I moved very slowly to the counter to take a look at the prescription names. She made no protest. Navane. Cogentin. Vitamins. Someone thought she was schizophrenic.

Turning around, I found Monica staring at me again, this time vacantly. Disenfranchised. I was frustrated that the preliminary sniffing around was this much trouble. There was nothing to be gained from talking to a lady this crazy. Rotheker was a million miles away. The best I could do was disengage gracefully.

"I'm sorry that I bothered you, Monica. It's the wrong house, and I shouldn't be here. I'll go now." I bowed slightly as Stryker stirred in my head: "Look through it, Hopper. Everything is information."

"Who are you? I want to know before you go in case anyone asks me who you are."

"My name is . . . Stan. I came to ask you about your sister."

"Oh, well, you should have said so. My sister is a very important person. She's a vessel. She's very pretty, and she had two children. Now she has one. Why do you want to know about her? Are you one of them? I mean, one of us? Praise Lucifer. I'm tired. I've always been tired. Do you ever get tired?"

"Yes, Monica, I get very tired. Where is your sister?"

"I don't know. She won't talk to me now. I made her mad. I made her boyfriend mad. I made everyone in the coven mad. Madder and madder and madder. . . ."

She drifted off into singsong, autistic for the moment.

"Monica?"

Her head snapped, and she stared at me like a cat poised and drawn tight for the attack.

"Monica, where is your church?"

"My church is St. John's Cathedral. I take the bus except when it snows. Are you from the church?"

"No, Monica, but I'm interested in churches. Do you have any more churches?"

"I had another church once, but now they don't want me anymore. Only my sister."

"And where is that church, Monica?"

"I'm not supposed to tell, unless..."

"Unless what?"

"Unless the person is seeking the Beast of Revelation."

"And if a person is seeking, Monica, then what?"

"Then I could tell them."

"And if I were to tell you that that was my desire, would you tell me, Monica? It's very important that I join those who worship the Beast."

"Hmph! What do you think? That they take out ads in the Yellow Pages? I don't know where they are. I know where they used to be, but they probably moved—they have to move all the time, you know—and they won't tell me where they are because they're mad at me. You're a very stupid man."

I moved closer to her and took the other chair at the table. She tolerated it. I looked her in the eye, and she froze like a deer in the headlights.

"Why did they get so mad at you, Monica?"

She hung her head and mumbled into her sweater collar. "I was in the hospital. In Fort Logan. I told my doctors some things I shouldn't have told."

"What happened after you told?"

"They took my sister's kid away. But they couldn't find the coven because I can keep some secrets. I just can't keep them all."

She started to rock back and forth, and I figured I didn't have much time. Her face was caving in from the stress.

"Monica, it's all right to tell me where they used to be because they're not there anymore, right? It's not a secret. I promise you they won't get mad. They might even invite you back in for finding a hardworking new member."

She kept rocking and staring. It was a slim lead, but it was all I had at the moment. I decided that patience was the order of the day. Finally she spoke. Her voice was thin, but the words were well articulated and clear.

"Arise now and wrap thyself in the mantle of darkness, wherein all secrets abide." She looked at me, appraisingly again and with some dignity that resided in the set of her shoulders.

"If you are really seeking, you will find them. If you are not, you will not. So I'll tell you. It was a house in Boulder. A big house where Alpine dead-ends at the foothills."

I knew the area. They were all big houses.

"How will I know which one it is, Monica?"

She giggled and was gone again. "You'll know, you'll know, you'll know, but they're not there. Only the priest is there. Only the priest."

I left her rocking and babbling. I marveled at a culture that had enough room in it for the Queen of old Aurora, Monica DeVries, and my lady guru.

was tempted to go to Boulder immediately, but Lafayette was on the way, and with any luck Ruben could give me some background. I made it in twenty, and this time the door was opened by a big bear of a man, very Hispanic, Zapata mustache, graying at the temples. He looked at me with mild curiosity and a certain friendliness. So much for assumptions.

"Mr. Ocala?"

"Yes?"

"My name is Stan Altman. I'm sorry to disturb you this way, but I'm doing some research on alternative theologies and I was referred to you by a friend of a friend. I would have written you first, but I'm passing through on my way back to California. This was easier than making a separate trip. I wonder if I could buy you a beer and pick your brain."

He took it all in calmly, as if people had been knocking on his door for years looking for some kind of information.

"And the name of your friend?"

I made one up and apologized that I didn't know the name of his friend, who, in turn, knew Ruben Ocala. But Ruben was not a suspicious type, and I knew I'd get my interview. He shrugged.

"I can meet with you now for about an hour. If you need more time, it'll have to wait until tomorrow. Come in. I have Tecate—will that do?"

"That'll do just fine, but, please, if I'm to intrude without notice, I would at least like to buy the beer."

"No need."

I stepped into his house. The blonde was nowhere in sight. The furniture was old and comfortable, and there was nothing in the living room to suggest that religion was important to this man.

"I understand that you teach a course on shamanism. Are you affiliated with the University of Colorado?"

"The university only teaches such things in cultural anthropology courses. My students wish to apply spiritual tenets to their daily lives. What exactly are you studying?"

"Conversion to non-Christian or charismatic worship. I'm interviewing people in places like Berkeley, Ann Arbor, and Boulder since they seem to attract many of the nonconformists who leave their family churches."

"You're a theology student?"

"No, I'm a writer."

"If you're going to write one of those sensational books about weird cults that steal middle-class children, I probably can't help you very much. I have no interest in them."

I wondered if he meant the books, the cults, or the children, but I concentrated on looking earnest and engaging. I was on thin ice if I couldn't convince him that I wouldn't abuse his knowledge.

"No, actually I'm interested in people who have made a commitment to one of the organized alternatives. Buddhism, Science of Mind, even Satanism."

"And have you spoken with such people?"

Match point. I could either keep it broad and vague or start to zero in.

"Actually, I've spent a good bit of time with a sangha in Michigan, and I'm in regular contact with a number of metaphysical groups in the Bay Area. Predictably, I'm having the most trouble finding people who will admit to belonging to the American Satanic Church or to knowing anything about it. Anyway, I'd be grateful if you'd share any perceptions you have about unorthodox religions here. I understand you're the local expert on the topic."

He sighed, not with annoyance, but with a weariness born of repetition. Kind of like that of a university professor who is about to teach an introductory class to a gaggle of freshmen for the thirtieth time.

"I was born in La Plata County in Durango. My family lived close to El Rio de Las Animas Perdidas. Lost souls. I suppose I became interested first in the Indian and Catholic practices of my own culture. When I moved to Boulder County in the seventies, I became a student of many religions. We have them all here. Even the Satanists and the Manichees."

Pay dirt, maybe, but I didn't want to appear too anxious. I took a pull at my beer and hoped he would continue of his own accord. He did not. I put a toe in the water.

"Isn't it a little unusual to have the darker religions represented here? I mean, Denver is sort of a New Age capital, and the Manichees are more historical than contemporary, aren't they?"

"Not so unusual. From my perspective, both groups are a response to medieval Christianity. Compared with Eastern religions and Native American practices, they are relatively young. You'll find them anywhere in the Christian world that provides a community tolerant of deviance."

"Well, I'm still sort of surprised that the Devil's in

Boulder. I'd assumed I'd have to look for that alternative on one of the coasts. Any chance of meeting with representatives from those two groups?"

"Perhaps, although the Manichees are more often just individuals who espouse the philosophy. They worship God and the Devil equally, and they're not joiners. The Satanists have more structure. There's been a small coven in north Boulder for years. They're harmless for the most part. They attracted some genuinely disturbed people a few years ago, and there were allegations of sexual abuse. Word has it that they expelled those members. They're run by an attorney who keeps a low profile because of his practice. I know nothing about what might exist in Denver."

"Do you think it would be possible to meet with the lawyer? I only have a couple more days before I have to head back to San Francisco. I'd hate to miss the opportunity."

"I don't know. His name is Allen Rushmore, and he's in the phone book. I would prefer that you didn't use my name. I have little use for the spirit they serve."

"Ruben, excuse my curiosity, but do you subscribe to any particular religious persuasion?"

"I was a Catholic priest for ten years. Now I meditate. I share the shaman's belief that all parts of creation have spirit. But I suspect that on my deathbed I will still say the confiteor. In Latin."

The hour had gone rapidly. I rose to leave. "Always remember to thank the accidental ally," I heard from Stryker-in-my-head. I extended my hand, and Ocala gripped it firmly.

"Thanks for your help. I appreciate it."

He smiled and bobbed his shaggy head. "You're welcome. And whoever you really are and whatever you're really doing, I wish you Godspeed."

Intuitive, Mary had said.

I got to the car and punched up my favorite news/talk show station on the radio. The newscaster said that the

Boulder Camera reported that the suspect in the recent murders of Buddhists in Boulder and Denver counties had fled the state and was thought to be headed for the Black Hills in the Dakotas. I remembered that Mack had said he would handle the *Camera,* and he had. I was covered after all.

I had no idea whether I was getting any closer to Rotheker, but I was getting closer to the local nest of Satanism. I bet myself two bits that Rushmore had inhabited a house on Alpine at one time or another and that he was Monica's "priest." I found him in the phone book under General Practice. It was no effort to convince his secretary that I was looking for a divorce attorney, and I soon had an appointment for the following day at ten A.M. as Gary Taylor from Boulder.

I drove back to my temporary digs. There was a lot of homework to do in a short time. I had to turn myself into a prospective acolyte in less than twenty-four hours. I plowed through most of the Anton LaVey in short order, skipping the parts in French and German and sticking to the English translations. I didn't have to be an expert, but it behooved me to know enough to seem like a player at this point in the game.

Sleep was fitful. I dreamed of the coffins that Anton said were an indispensable part of the death and rebirth ritual in his church. *Le Messe Noir* played like a porno-graphic video in my head. I awoke ready and alert at five A.M. and finished my study.

I chose a black conservative suit and a red tie. Rush-more would appreciate the colors. The drive to Boulder was my last rehearsal. By the time I reached the office on Walnut I was in role and ready to convince LaVey himself that I was seeking the services of Shaitan.

Entering the suite, I saw trendy gray-and-mauve ap-pointments worthy of an expensive but uncreative interior decorator. I had formed no hypotheses about Rushmore, preferring to experience him with as few preconceptions as possible. I had met the Devil himself in the fields of Southeast Asia, but I'd never met a registered rep.

His secretary looked normal enough. She was a bright-eyed redhead with freckles, and the Norman Rockwell im-age was appropriately ironic. I accepted the cup of coffee.

At ten I was ushered into a pleasant office, equally trendy, and got my first look at the clandestine coven leader. Allen Rushmore didn't look anything like Ralph Bellamy in *Rosemary's Baby*. He looked like an aging prep-pie, a little too old for yuppiedom but large-boned, clean-cut, and classic. He was very professional, and there were few preliminaries.

"I understand that you're looking for someone to help you with a divorce, Mr. Taylor. My fees are one hundred and thirty-five dollars an hour, and if you'll answer a few questions, I'll be able to tell you approximately how much time it will take. Has your wife retained counsel?"

"I have no wife, Mr. Rushmore."

"Excuse me?"

"I have no wife."

He looked at me more closely and narrowed his eyes. "Mr. Taylor, if this is some kind of joke, I'm not amused. If you're in trouble, I'll be glad to refer you to one of my colleagues who handles criminal cases. But let's not play games. Who are you?"

My eyes bored into his for effect, and neither of us

blinked. "I am the King that magnifies himself, and all the riches of creation are at my bidding."

It was right out of the Black Book, the *Al-Jilwah*. To his credit he still didn't blink but responded in kind, his eyes still locked on mine.

" 'I have made known unto you, O people, some of my ways. So saith Shaitan.' Very well, Mr. Whoever-you-are. You read old books. What do you want?"

"I wish to become a celebrant in your . . . group. I apologize for the deception, but I could hardly explain it to your secretary."

He got up from his desk and turned to the window that opened onto a view of the Rockies. He clasped his hands behind his back with the attitude of a man surveying his polo fields or his new Porsche. Very self-contained and cool. No raving maniac here.

"Mr. Taylor, who sent you here?"

"Mr. Rushmore, I am not at liberty to say. I've been a student of the scripture for some time. I'm ready to make a commitment. It doesn't matter how I found you. Must I go elsewhere?"

He turned around slowly and got cagey. "Mr. Taylor, I am known to a select few as a scholar in the field of which you speak. Now and again some law enforcement agency with nothing better to do sends some undercover cop to investigate my hobby. You must understand that I am cautious about these things by nature. I am afraid that if you come without references, there's nothing I can do for you."

I apologized silently to Ruben Ocala and trusted that the ends justified the means. "I come to you by way of Ruben Ocala, and I found Ruben by way of a friend at Duke University who is an acknowledged expert in parapsychology. Neither of them know of my real interest. I can't give you any other references except for Satan himself."

Rushmore threw his head back and laughed. "Believe me, I'll check that reference. Is Taylor your real name?"

"Yes."

"And what do you do for a living, Taylor?"

"I'm a free-lance advertising copywriter. I think you'll agree there's no better way to celebrate the material world."

I remembered an exercise that Stryker used to call "lying in truth." Our seminar group would practice creating a consistent, interwoven pack of lies that worked because they were merely variations of the truth. I could lie to Rushmore all night long if that's what it took, and I would betray no anxiety because the core of what I said to him would be true.

We talked for another half hour. I was subtly quizzed on my knowledge of the ancient lore, my intentions, and my ambition. I invented a seamless identity and offered various ways for him to corroborate the lies I was telling him. The upshot was that Rushmore invited me to an "orientation" that was to be held two days later. A moderate payoff for my trouble.

"We have several new aspirants. They're all very high quality people. I will be stepping down from my role as invocator in favor of one of them. He is very special. I think you'll be impressed." He was careful to explain that there would be no ritual during the orientation and that each aspirant would be considered for inclusion on his own merit. I had been warned in the texts that the Satanic church allowed no "sincerely interested" visitors at their masses. It ruined the energy flow. I had to make it into the club. The ritual would follow the orientation. Perhaps I would be invited, perhaps not.

He gave me an address, and we did a little mumbo jumbo at the end. The whole thing reminded me of the parting of two Masons or fraternity brothers in the days when that was hip. On the way out of his office, I took several deep breaths and tried to kill the images of Gretchen Reynolds on Rotheker's bed of death.

The next two days were not easy. I struggled with myself about whether I ought to contact Mack to check on the department's progress on locating Rotheker. I didn't like the idea of wasting my time with a Boulder coven if he had been spotted in some other place, but I wanted no complications or distractions. Mack would have to wait. I thought about calling Stryker and rejected the idea. The mission was my job now. And I had to cram for my finals.

I checked out the neighborhood that Monica had described. She was right. I knew the house when I saw it. It had subtle but distinctive features. It was an old Victorian with glossy black trim and gargoyles in the garden. The mailbox said Rushmore, and I got impatient to get on with it.

The universe of two persisted for forty-eight hours more, and then I left to go to my orientation.

There were eighteen people in the living room of a house on 5th Street not far from the location Monica had given me. There was something reminiscent of the house on High Street, but these were not Buddhists. The faces were equally earnest, but not as open.

The aspirants, I was quickly to learn, were not allowed to mingle. Rather, we met with the coven as individuals. I didn't see the others. Apparently we had been cloistered in various parts of this rambling house until our presence was requested by the initiates.

It started at seven, and I wasn't called until nine. The waiting was onerous. I put my mind on passive hold, blocking out all but my purpose. I worked my body into a respectable state of readiness and tested each of my senses for acuity. I needed to know whatever there was to be learned here.

I recognized Rushmore, and there was a blond woman at his side who looked a bit like Monica DeVries. Younger, prettier, and more passive, but part of the same family. There were no children, although a few members looked to be in their early twenties. The women wore good clothes, and the men looked reasonably prosperous. It might have been a cocktail party.

My examination lasted for forty minutes, after which I was led back to my room to wait for an invitation or the proverbial black ball. The men had asked most of the questions. There was nothing particularly ominous about it. It was kind of like a civil service exam. I had sounded motivated and knowledgeable enough for a neophyte.

At eleven o'clock a young woman showed up bearing a black robe of heavy cotton twill with a hood and a soft belt of red twine. She was honey blond and thin. Mid-twenties, I guessed. She looked like a teacher or a nurse. She said that not everyone would go through *l'air épais* this evening, but that the robe was mine to keep until my turn, probably sometime in the next few days. She kissed me then with her tongue halfway down my throat. When she came up for air, she said she would be pleased to give me birth.

The "ritual of stifling air," *l'air épais*, is a ceremony in which the weak, inadequate self dies and is reborn in the service of Satan. This young woman would be in the coffin of my dreams as part of the initiation of the celebrant. Maybe there was nothing more to satanic cults than another excuse to get laid.

The getup was not hard to get on, and she was already wearing hers. It felt like Halloween. We walked down the stairs to the living room, and the lights were out. There were several more people lining up to descend the basement stairs. The women bore candles. It was impossible to recognize anyone.

My guide, her honey-blond hair now covered by a hood, led me to the stairs. Three or four people ahead of me was a short hooded figure who struck a visual chord. A memory with no words. Next to him was someone whose build looked much like Rushmore's even under all the material.

There were twenty-two steps, and the last step led into a mirrored hall that continued for twelve feet until it entered a mirrored chamber. We were herded into a circle around an altar. The incense was enough to make me gag.

Rushmore made his way to the altar with all the casualness of a Protestant minister about to start his service. He beckoned to a smaller figure, who joined him on the macabre stage. At first I thought it was a woman.

Rushmore threw his hood off to reveal a head that was clean-shaven. Whatever hair I had seen in his office was a damned good rug. He didn't look preppie anymore.

Then the short figure threw his hood back as well, and I got a hard look at Rotheker's face for the first time. Time changed for me. Perception changed. My muscle tone changed. My right hand reached through the voluminous sleeves of the robe and touched the tool I had chosen for just this occasion. It was an easily concealable, small-caliber single-shot with a meticulously machined and incredibly effective silencer. An assassin's weapon. No one had checked. Apparently none of these satanic wonders were terribly paranoid. The Beast would surely protect them.

My arm hadn't even begun the motion, but my eye had already drawn a bead on his seventh chakra. Like a pro practicing an inner game of tennis, I visualized the shot to his forehead, and I could see the small-caliber red hole silently appear in the exact center while his body crumpled to the altar and his eyes widened in surprise. I had maneuvered a position close to the hall, and I would be down it and up the stairs before the rest of them had a chance to react. And it would be done. Someone else— Mack, maybe—would clean out this nest of vipers later, but at the moment I had a clean shot at getting the most important part done. For Tom, for his father, for Patty, for Gretchen.

My right arm started to slide into position, and I saw myself in slow motion bringing the weapon on point. Before the gun breached my sleeve, I suddenly had a sense of Stryker. It arrested the movement. At first I thought it was simply the omnipresent Stryker in my head, saying, "No, Hopper, not that way!" but the presence was stronger than that and eerie. I kept my eyes motionless on the altar and my body coiled to strike.

It was surreal. Rushmore started to speak about the passing of the baton to a new generation, and I could feel myself concentrating backward to the strange presence. Very carefully, I started to move the weapon back to its holster. As the gun shifted back into place, so did my task. It was time to record the coven ritual in photographic detail, to look for an opportunity that could be used later to take Rotheker down and deliver him. It was time, for example, to notice that there were now twenty-two people in the room and that four of us had made the grade. One of us would be initiated that night. Depending on the order of the remaining initiations, I was looking at very little time or a few weeks.

The order of the initiations was also important. If either Rotheker or I were to be the center of attention, it would be damned difficult to get anything done. It would be handier if they initiated someone else the following week, leaving Rotheker and me free to complete our fate. But I'd managed to find him in the first place I looked, and I wasn't expecting any more favors from the universe. I tuned back into Rushmore.

"It has been my pleasure to watch the growth and success of our group over the last ten years. I believe that our devotion has been rewarded by Satan in the form of the young man who stands at my right. Some of you have met with him and some have not. Over the next few weeks all of you will have ample time to get to know him better. Allow me to say that he brings with him the knowledge of new portals to the soul. With his help and Satan's power, we can regain dominion on this earth."

Whatever Rotheker had sold Rushmore and the coven elders, they'd bought full retail. Or Rotheker had sold it to Rushmore, who'd sold it to his deacons. If Rushmore knew about the murders, then I had an attorney I could fry. It didn't displease me. Rushmore continued.

"I've asked Mr. Rotheker to say a few words prior to our welcoming Mr...."

Rushmore stopped and looked at the woman I believed to be Monica's sister. He was like any other attorney looking at his legal secretary for help.

"Fisher." She said it in a stage whisper.

"Welcoming Mr. Fisher into the group. Since there are so many of you at this time, and since I know you are all anxious to officially become a part of the coven, we are going to accelerate our schedule. *L'air épais* will be conducted tomorrow evening for Mr. Daily and the following evening for Mr. Taylor. Since Mr. Rotheker comes to us from another respected coven, the traditional initiation will not be necessary. Mr. Rotheker will assume his role as priest and invocator beginning with Mr. Daily's initiation. This accelerated schedule will allow us to quickly create the core of a new satanic movement. Jeff, please introduce yourself."

Rotheker took a step forward. The deadly bastard looked relaxed, and I had to apply more psychological muscle to keep my rage at bay. He spoke softly and intensely.

"I am pleased to be here. In the coming months, I look forward to working with you and the many new celebrants we can expect. I am confident that Satan's time has come and that together we will usher in his dark reign. *Domine Satanas.*" He inclined his head and took a step back.

I looked at his small stature, listened to the mild hypnotics in his speech, and suddenly it wasn't hard to imagine Tom or Patty agreeing to walk with him right into their deathtraps. That didn't explain Gretchen, but that question would have to wait. The ritual was beginning.

They played it right out of the book, although in English instead of the original French. Rushmore, the scholar, probably could have told me exactly how the modern version varied from the thirteenth-century version. The ritual was a gift from the Knights of the Templars, and it represented the transition from self-denial to self-indulgence. There were only seven roles to be played, and the rest of us were audience. It took an hour and ten minutes. After that people filed up the stairs, discarded their robes, and toasted Mr. Fisher with champagne. Rushmore and Rotheker were engrossed in conversation in a corner of the living room.

I chatted idly with the honey blonde, wondering what it would be like to be fornicating in a closed coffin with her. I would never know. My head was spinning wildly with the argument between the urge to take Rotheker out now and the more rational and professional options that I should create. I turned to once again size up my prey.

The son of a bitch was gone! The decision had been stolen from me. My heart sank. Where the hell was he? Was he on to me? I had more than a hundred unanswered questions.

I made some lame excuse to the blonde and the others and got my ass out of there. Now I really had to use everything I could muster to pull this one out. And I had had him in the sights. For the moment, there was no justice.

I moved quickly out the door because I had about forty-five seconds of concentrated decision making to do. One option was to throw myself into finding Rotheker with dispatch. I had two objections. The first was the possibility that Rotheker was simply in some other part of the coven house or that he was staying at Rushmore's for the night. Neither made a capture impossible, just more complicated. Second was the improbable possibility that if I found Rotheker alone anywhere, I would do something else besides deliver him to Mack. Improbable, but so tempting that I was motivated to be unusually careful.

Another option was to assume that the world was safe from Rotheker while he was busy taking over the coven. Assumptions are dangerous things, but human beings would be unable to think without them. The last time I had made an assumption, I had been terribly wrong, and it had cost me Gretchen.

I elected option two as I inserted my key in the ignition

and left to ponder methods. "Gary Taylor" had a reservation at a mom-and-pop motel outside of the city. After the Valli-Hi I would have preferred the Boulderado and its historic comfort, but too many knew me there.

When I got to the motel room, I turned down the tangerine bedspread, increased the air-conditioning, kicked off my shoes, and assumed a thinking posture on the bed. It was not an easy problem. Somehow I had to spirit away the new high priest in view of a gaggle of witnesses and make it look natural. One phone call to Mack, of course, would address all the tactical issues. The ritual started at eight and lasted seventy minutes. Mack and his boys could bust the door down, corner everybody downstairs, and I could just point the finger. It was easy. It was logical. It was a damned shame that I couldn't bring myself to do it.

Rotheker was *mine* now. I wouldn't even trust Barnum with the final takedown. I wanted complete control and my hands on the man. I had precise requirements and no stomach for some dramatic SWAT team event. Rotheker would leave my hands only when I could deliver him directly into Mack's with no interference and no danger of procedural screw-ups.

So I was back to the problem of removing him in some natural way. Any kind of muscle was probably out. There was no verbal ploy I could think of that would compete with his priestly duties well enough to get him outside of the house and away from the group. Shooting his legs out from under him and standing off the coven seemed unnecessarily risky and a little too John Wayne. I needed something unobtrusive, and I needed control of the timing. Since Rotheker would act as invocator that evening, whatever it was had to take place before or after the ritual itself. Slipping him a Mickey during the champagne celebration required that he drink champagne, and he had declined after Mr. Fisher's rebirth. Curare darts and blowpipes were not something I could pick up the next day at McGuckin's Hardware.

I closed my eyes and got myself into a light trance

because I was starting to think in circles. Absent the brain, the mind does strange and wondrous things.

When I opened my eyes, I had a shopping list. A spray can of Tough Skin, Band-Aids, DMSO, and a couple of interesting potions I had learned about from a witch doctor during one of the fuzzy parts of my past.

That accomplished, I undressed and slid between the cool sheets for four hours of dreamless sleep.

T he success of the plan hinged on my being able to make physical contact with Rotheker prior to the beginning of the ritual. The living room would do. The mirrored room would also do if I could get to him before he stepped onto the altar.

The beauty of it was that it didn't matter where I touched him. All that was necessary was that the Band-Aid, applied to the palm of my hand, make contact with exposed skin for about five seconds. Drugging a human being is usually done orally or by injection. Fortunately the space program has made giant steps in the field of drug infusion. Using a carrier chemical, drugs are easily transmitted through the skin at rather predictable rates. The method was originally used with antinausea drugs to combat motion sickness. I had something else in mind. DMSO would do just fine and was the most easily attainable.

The Tough Skin would cover my palm to prevent my absorbing the toxin. The next layer would be the saturated Band-Aid. It would carry the chemical inconspicuously. I could grab his arm, shake his hand, or lay my hand on his neck in a sign of camaraderie. Then it would just be a matter of time. I had called up my chemistry, wrestled with Rotheker's weight and susceptibility, and adjusted the concentration accordingly. I figured thirty minutes— plus or minus a few. It was time for a moment of silent prayer.

It wasn't foolproof. It was possible that Rotheker would be cloistered elsewhere in the house before the ceremony, but it seemed unlikely. After all, he had to establish rapport with his celebrants. They were to become the core of his revolution. It was also possible that through some slip-up, I would get myself.

I was ready by noon, and that included a quick trip to a veterinary supply store and an industrial chemical concern in Commerce City. Waiting out the afternoon and early evening was irritating, but I made it to the house on 5th Street at seven forty-five.

The curtain was at eight. The first good news was that Rotheker was sitting on the couch with the honey blonde, intensely involved in their conversation. I would have said a "Hallelujah," but it seemed out of place. Rushmore was nowhere to be seen, and the other coven members were straggling in slowly.

The couch was one of two that faced each other in the center of the room. Rotheker sat at one end, and the honey blonde, enraptured, sat next to him. Coven members were knotting up in front of him to listen to him talk about his theory of satanic materialism. I positioned myself about five feet behind the back of the couch and looked at the amount of skin between his hairline and his collar. It was enough to get my hand around, but if I came at him from behind, the startle response might prevent the length of contact I needed. Liggett had said that Rotheker flinched when touched, and I believed him. I needed something compelling to say to him to offset the inappropriateness of what I was about to do. If I could get to him now, the deed was done, and he would soon become dizzy as hell, too dizzy to be conducting rituals. I hadn't yet figured out what to do with the confusion this would create, but that was the improvisational part of the whole thing.

My body made the decision for me. I took three long steps to the back of the couch and laid my hand on the side of his neck, then leaned over and looked into his eyes. He flinched all right, but not enough to break the hold.

"Jeff, please forgive me for interrupting you, but it's getting close to eight and I wondered if I could have a word with you privately before we start the ceremony?"

The clock in my brain was ticking off "One thousand one, one thousand two, one thousand three," and finally I got to the five-second mark. Rotheker was looking at his watch, and the honey blonde was clearly annoyed that I had disrupted her rapture.

"It's getting late, and I think we should prepare to go downstairs. But I'll be able to speak with you after the ceremony."

I took my hand off his neck and took a step back, hoping to look like an embarrassed kid who had interrupted his teacher at the wrong time.

"Sure, that's fine. I'm sorry."

"No problem."

The maneuver had not only infused the drug, but it also broke up Rotheker's audience. They were now busily getting into their robes. Rushmore showed up to dim the lights and help his deacons light the candles for the ladies. It was showtime.

We shuffled down the stairs, and the next thirty minutes seemed to stretch endlessly. I stood as close as I could to the altar so I could watch him and get to him quickly when the drug took effect.

He was a much more charismatic invocator than Rushmore. At the fifteen-minute mark the entire coven was watching him like he was Olivier doing Othello. A piece of symphonic funeral music by Berlioz was playing in the background.

At the twenty-minute mark, something unsettled me.

When you spend a lifetime being in tight spots, there's a visceral instinct about threats to your backside. There was something or someone in the room that didn't belong there, and it or they were at my back.

I heard someone clear his throat, and that seemed out of place as well. I looked at Rotheker and he was still doing fine, so I decided to investigate the menace. The throat-clearing person, judging from the sound, was standing behind me.

I took one step to the right and one step back, and the figure was at my left now. Rotheker continued his drama. I glanced to the side, and at that moment, the hooded figure raised his hand to his face and brushed back just enough of his hood to reveal a profile.

The hand would have been presumptive evidence, but the profile was proof of a different sort.

Cabo was missing a bartender. How he'd arrived at my starboard elbow at just this moment was a jigsaw, but Mack would be relieved. I had cover.

However much I appreciated the comfort zone, I was aggravated. I could visualize Stryker sitting in his Eames chair in Cabo San Lucas sifting through the incomplete information that he had and becoming irritated that I hadn't called to report. When it became overwhelming, he no doubt had summoned Sergio to pack his bags. It was an unwelcome precedent. I wanted to know how he'd gotten to Rushmore in a fraction of the time it had taken me. I presumed I'd get that later. For the moment I just accepted it as a nice piece of counterintelligence from the master. I was half-mad with no time to think about my anger.

The chameleon nodded as if to say that the next steps were up to me and he was in backup mode. I turned my focused attention back to Rotheker.

Rotheker had begun the section of the ritual called the Denunciation. About midway through, he raised his hand to signal to a deacon that the Wine of Bitterness be offered to the celebrant. His fingers were trembling, and now I listened and watched him like a starving cat stalking a fat mouse. His breathing became more shallow. The celebrant was busy drinking his last drink before rebirth. I noted a slight quaver in Rotheker's voice.

"And now at last authentic word I bring, witnessed by every dead and living thing."

He reached up and rubbed his eyes, and I knew his vision was doubling. He began to sway on the altar, and the coven began to notice that something was wrong. They started chittering, and the celebrant looked around wildly for direction. I watched Rotheker's knees, and when the first one buckled, I was up on the altar—the epitome of the competent and concerned new member of the coven.

"Are you all right?" My voice projected to the back row.

"I . . . ahh . . . I—"

The nausea hit him, and the second leg buckled. Rushmore was precipitously at my side, looking distressed. I glanced quickly at him man to man, as if I were on his team and there to help. I lowered my voice to a conspiratorial whisper.

"Allen, he needs some air. I'm trained as an EMT. Let me take him upstairs. Can you finish?"

"Of course I can finish. Oh, Jesus, what's wrong with him?"

"Dunno, but finish with Mr. Daily. If the air doesn't help, I'll ditch the robe and drive him to Memorial. I'll handle it."

"Thanks, Taylor."

The coven looked stricken as I draped Rotheker's arm around my shoulder and got him off the altar. Rushmore took charge. Out of the corner of my eye, I saw Stryker near the doorway, and when Rotheker's body started to sag on the other side he moved in like a ballerina and we started up the stairs. It was not time to speak yet. The play had to be perfect. I could hear Rushmore saying to the coven that Rotheker seemed to have been taken ill. He took up the Denunciation, where Rotheker had left off.

We breached the door with Rotheker's sagging body between us. As we crossed the dark living room and left the house, we took turns holding him up while we shucked the robes.

"Vehicle?" Stryker barked like a motivated Doberman. "We don't have much time."

There were footsteps behind us. Stryker grabbed Rotheker and pushed him to the porch step, forcing Rotheker's head between his knees. A weak-chinned Mr. Fisher stood at the door, glaring, and then came out onto the porch.

"Is he all right?"

"We don't know yet. We're going to let him get his breath and then we'll see."

Fisher looked at Stryker and, like the newly initiated coven member that he was, became suspicious.

"Excuse me, I don't believe I've seen you before. Are you a new member?"

I preempted Stryker's answer by delivering a short, noiseless chop to Fisher's solar plexus and another to exactly the right nerve in his neck. He went down without a sound, and I dumped him over the porch rail behind a bush that at the time seemed to have been planted there for just that purpose.

Rotheker's grogginess made him awkward to handle, but we got him onto the backseat of the rented car and were out of Rushmore's sphere of influence in a minute and a half.

I took a route into the foothills.

"All right, Bullfrog. This isn't the way to the Boulder County Sheriff's Department. I know. I checked its location. What now?" This wasn't Stryker in my head. It was Stryker in my face.

My knuckles were white on the steering wheel. I turned onto a back road and drove the heap into a foothills meadow and stopped the car. I got out without saying a word. Rotheker was helpless, and Stryker left him on the backseat and walked a ways to join me as I gazed at the red outcroppings of rock that bordered the meadow in the twilight.

"Why should I give him to Boulder County, Professor? So they can try him? So one of Rushmore's cronies can get him off on an insanity plea?" I stuffed my hands in my pocket and paced the field. Stryker stood still and watched.

I paced my way back to him and got up close. His was an immobile face. He spoke softly.

"This is not wartime, Hopper. You have no authority to execute."

"It wasn't wartime when he executed three people. The system is going to let him walk because he's a disadvantaged kid who fought in Vietnam. It's the same old shit, Stryker. No one has to be accountable to anyone for anything as long he can find an attorney who's good enough. Some barrister in a sharkskin suit is going to make the jury weep for this poor lost soul. I predict the worst that will happen to him is that he'll do a few years in the loony bin in Pueblo!"

I was shouting, and it seemed to me that Stryker was absorbing my rage and containing it. It was so like him that my anger began to dissipate in the face of his calm loyalty.

"Remember, Hopper, that every human being on earth would like more control over the events in their lives. It's simply not possible. I'm sorry. The mission is to deliver him to Mack and report to Burroughs."

I exhaled, and the last vestiges of my rage flowed out with that forceful breath.

"Done."

We both walked back to the car. Rotheker had passed out. The dosage was set so that he'd clear in about an hour and the county personnel could read him his rights. I called Mack from Dixie's bar across from the flats while Stryker ordered a burrito. I got a draw.

We'd left Rotheker in the car, hands tied, feet bound. It was not unheard of for a worker from the nuclear plant across the street to be sleeping it off on the backseat of a car in front of Dixie's place. No one noticed.

Stryker tore into the burrito smothered in green. He asked pertinent questions in between bites.

"Where did you learn your new tricks, Bullfrog?"

"What new tricks?"

"I recognized the rabbit punch to that guy's gut back at the house, but the little bit of business you used on his neck is a new one. Brought him down quick. Looked like a Vulcan nerve pinch from *Star Trek*. Where'd you learn it?"

"Hah!" I stared at my beer. The truth was that I wasn't sure where or when I'd learned it. The eighteen months after Caitlin's death were a bit of a blur to me. I did it all on automatic pilot. I met ninjas in Southeast Asia, I did drugs, I learned from witch doctors. Total re-

call was not available for that period of time, only fragments—some of them fragments of my imagination, I
was sure.

"Let's just say that I learned it sometime between
working for the government and working for myself."

"Fair enough. I'd like to know what else you
know."

"Someday, Stryker...someday."

My head felt like a fifty-pound longshoreman's hook
belting itself into a sturdy box. I had pressing questions.

"How did you find me, Stryker? Magic?"

"No magic, just logic. I have the advantage, Bullfrog.
I trained you. I know how you approach a problem, how
you research a question, and how you cover your tracks.
I knew, for example, that you would look for a credible
expert to help get you up to speed on occult activity in the
area. Boulder County is not a large place. It took me forty-
eight hours to find Ruben Ocala. He's one of the good guys.
After that, it was easy."

"Of course. Stan Altman..."

"Was an alias you used in Bangkok in 1971. I may be
an old fart, Hopper, but there's nothing wrong with my
memory."

I shook my head. I was a little embarrassed that it
had been so easy.

"That explains 'how,' but it doesn't explain 'why.'
Mack would have been pleased with your backup, but your
appearance makes me feel like a snot-nosed kid who can't
be trusted to get a job done. You're probably the only
person in the world who could inspire me to feel that way.
Why now, Professor?"

Stryker pushed his plate away with stubby fingers. "I
learned of Gretchen's death from Burroughs. I was reminded of Caitlin. I came because..." He paused, and his
shrug would have been imperceptible to the average observer.

"You came because you thought I'd go off the deep
end?"

He grunted, shaking his head and regaining control of his voice. "I came because I think there are some things no man should be forced to go through alone if alternatives are available. No disrespect intended."

We cleared the tab just as the Boulder County car was sliding to a stop in the gravel parking lot.

Mack removed Rotheker from my Rent-a-Heap and put him into a sheriff's car. Rotheker was coming to. The collar wasn't exactly by the book, but I knew Mack would make it righteous enough for any county judge. Stryker asked for the car keys, said the car would be in short-term parking at Stapleton. He had always hated mopping up. Too much paperwork. It didn't surprise me much. I wished that I could go back to Cabo with him and park my behind on the ketch, but I had other tasks to accomplish.

Mack chewed my ass pretty good, but the fact was that he was more relieved than angry. Eventually they deposed me.

The star cub reporter did a sterling job with the exclusive that she'd extracted from Mack. The deal was based on her willingness to run a cover story for us while I was underground. In the article, she was disciplined and rational and never mentioned my name. She speculated on Rotheker's peculiar personal logic, but she never overstepped her bounds. Avoiding trial in the press and all that. I liked the piece. I liked her. If Caitlin and Gretchen had not been straddling my memory like a dual-headed Medusa, I think I would have devoted a weekend to getting to know her. Instead I tucked her into my contingent future.

By that time I was free to be done with it, but, in my own mind, I had one more chore to do. I pulled rank as a consultant to the county and asked to have an audience with Rotheker. There were too many loose ends for my taste, too many things I didn't know. I wanted it from the horse's mouth.

They were holding him in the county jail before arraignment, and I asserted my way into the rigid institutional schedule. I didn't dress up for the interview.

Sometimes I feel so damn naive. Now that he was exposed I guess I expected him to turn into some saturnine monster with spittle at the corners of his mouth, leering wildly. Instead he sat there looking much younger than his years in a button-down blue shirt and corduroy jeans. His hands were folded prayerfully in front of him. He didn't turn his head when I entered the room, but he greeted me politely.

"Good morning, Mr. Hopper. Congratulations."

"Congratulations for what?"

"For succeeding. Leaving that gold crown behind was very sloppy of me. It was all you needed. I felt it drop. I didn't think I had time to look. It was a lapse of judgment. It's not like I don't know about dental records." He smiled like a kid caught in a peccadillo. This cat was one for Professor Stephens because I was damned if I could make sense of any of it.

"Barnum tells me you confessed and asked for a lawyer. That right?"

"Yes. A public defender has already been assigned. I will also get a court-appointed psychiatrist on the state's tab. They think I'm crazy."

"Are you?"

"We have an old prayer in my church, Mr. Hopper. It says that we are 'standing at the gates of Hell to summon Lucifer, that he might rise and show himself as the harbinger of balance and truth to a world grown heavy with the spawn of holy lies.' Is that crazy?"

My gut rolled over, and for a few seconds I lost my equilibrium. It's funny how a phrase can do that to you. The spawn of holy lies. I saw my wife's body in a Vietnamese field again. I saw her face and breasts eaten away by American napalm. The "eyes only" records said that she was a counterinsurgent caught in friendly fire. The grotesque spawn of holy Pentagon lies in a world with no balance or truth.

I didn't know who I wanted to kill first—the little satanic shit in front of me, General Westmoreland's minions, or myself. It had been a long time since I'd killed anyone, and the violent impulse was not welcome.

I eased myself back into the present. Stryker used to admonish me to "stay with the eagle." I repeated it to myself once as an antidote, breathing deeply of fetid jail air. I never knew what it meant, but it always worked.

The creep was staring at me with more animation because he knew he'd hit some little piece of pay dirt. I decided I was interested in his lunatic ministry.

"Why pick on the Buddhists? If you wanted a little press for the Devil, why not just shoot a bunch of people from the top of the May D&F tower?"

"That's an excellent question. I thought about it a lot and decided that it was superfluous violence. I am convinced that Christianity is waning. It is a reactive, insecure, one-dimensional force. The fight against Lucifer has probably kept it alive longer than warranted. Buddhism is different."

"How so?"

"The Buddhists are not so foolish as to denigrate Satan. Instead they confuse people with all their talk of demons and wrathfuls and preaching about illusion. They are dangerously seductive. They relegate Lucifer to the limited hell-worlds or to the worldly sphere of samsara. As if any of us could escape him. It is disrespectful from our point of view. The problem is that they are taking hold of the strong ones who are abandoning Christianity and who would come into Lucifer's service were it not for the fairy tales that Buddhists tell."

"Did you think that knocking off three or four of them would do the trick?"

There was a vein in his neck that stuck out like a sewer pipe. It was his turn to be agitated.

"I talked with each of them about joining us and giving up their fairy tales, but none of them would budge. I wanted them to contribute the power of their chakras, and none of them would agree. So I took them."

I pitched my voice a bit lower and got a little more menacing. "I think I understand what you did with Patty and Tom, but how did you get to Gretchen?"

He leaned back in the standard county-issue plastic chair and looked thoughtful. "She was the easiest of them all."

My chest was constricted, and I was fearful of what I was about to know. "How so?"

"She was a genuine bodhisattva. Committed to the enlightenment of all sentient beings. So I left a message at the Denver center. She called back. I asked her for help. She couldn't say no. I met her at a restaurant. I explained what I wanted and she refused. Refused. I killed her in the parking lot and drove her to my house. Understand, Mr. Hopper, that I would have preferred that Gretchen, Tom, and Patty join my efforts. In lieu of that, I took the power they had."

He got weird-looking then, and maybe there was a little spittle at the corners of his mouth.

"And if Satan is so powerful, why didn't you keep working at recruiting them . . . and their chakras? Surely he could have delivered them eventually."

"I am the servant of the Prince of Darkness and do his bidding. There was no other way. It was timely."

My panic was gone. Satanism is a cheap card trick. It is an adolescent religion that wishes things were different from the way they are. Its members wish it so desperately that they become knee-jerk oppositional. White is black. Day is night. Good is evil. You have to destroy a village to save it.

I was done with him and with the whole lousy issue. My curiosity was well-nigh dead, but I decided to ask one more question.

"Where'd you learn about killing, kid?"

"The Ia Drang valley, sir."

"Regular army?"

"Yessir."

"Combat infantry?"

"No, sir. A medic."

I stared at him, and he stared back. I didn't want to know anything else about him. Case closed, the dead buried.

Epilogue

It was time for Cabo. I called my favorite travel agent, and four hours later I was on a Mexicana flight. No super-saver, but a quick seat. Before I left I called Mack to let him know that he could reach me long distance. Dockets were heavy, and it would take months to get to trial.

I took a carry-on bag with a change of underwear, a clean pair of jeans, two shirts, one pair of shorts, a tank suit, and a razor.

I was on the last flight into the airport, and I grabbed a cab driven by one of Guillermo's *compadres*. He recognized me. He knew the address.

Dusk had fallen on Cabo, and the ocean was a color that had no name in English. There were fewer rosy clouds here because there were fewer pollutants to turn the sun's rays into dramatic pinks and reds. Instead, sundown was gently purple and blue.

Lucia let me into Stryker's house and put me into the guest bedroom. I slept like a piece of granite.

It is impossible to detect Stryker's presence when he doesn't want to be detected. It was ten in the morning on the next day, and the sun had a head start at warming this little tip of earth, including my adobe bedroom, which was heating up with the morning. When I awoke, Stryker was sitting on a chair at the side of the bed, staring at the ceiling.

"Good morning. It's about time."

I groaned. The "Good morning" was less of a greeting than a challenge. I'd come for R&R, and I would get it— but not without debriefing.

"Damn you, Stryker! You're so goddamned intrusive. It's bad enough that you never leave my mind. Bad enough that you felt you had to hold my hand in Denver. I don't need a portly mentor as a mother. What's your problem? Don't you trust me anymore?"

"Cut the crap, Bullfrog. I learned a great deal in Denver, but there will be no repeat performances. Get your ass out of bed. Lucia has breakfast ready. Meet me at the compound in half an hour. Can you move that fast?"

It was a rhetorical question, and he was gone before I could answer it.

I took a thirty-second shower and dressed. Lucia stuffed me full of rich strong coffee, fresh pulpy juice, refritos, and eggs. She told me I was too skinny to pass up the extra tortillas on the plate.

The compound is down the road a piece. I loped along the beach, nearly running, and I allowed myself the luxury of wallowing for a moment in the grief and rage that had started at Beeler Street.

I went inland and passed the Two Yanks, which showed no outward signs of life but which catered, even at this hour, to its own unique congregation. I continued for a few yards and rang the bell at the door of the compound. The door buzzer sounded instantly. I was expected.

Sergio Cordero was waiting for me in the foyer. I never know how I feel about Sergio. He's a good-looking rich kid from a Mexican oil family. He has a degree from UCLA, and the most he's been able to do with his life is to be concierge for the Two Yanks. An overqualified gofer. Stryker's omnipresent diplomatic assistant. Lawless's lackey.

"*Muy buenas días, Cordero.*"

"*Bienvenido, Hopper. El profesor te espera. Ven conmigo.*"

I followed. We ended up in the high-tech study. Stryker was hunched over a CRT punching up screens full of coded data.

"Stand by, Hopper. I'm almost through."

In any other circumstance I would have busied myself with visual reconnaissance, but this was Cabo and he was Stryker, so I stared at his machinations hoping to learn something. He swiveled to face me.

"Sit down, please. You, too, Sergio. John Burroughs will be here in a moment. I sent Mike to get him."

"Why is he here?" I was startled, and it came out a bit more abruptly than I'd intended.

Stryker was unruffled. "He arrived with Sergio yesterday."

Something was up, and I settled in to watch the show. Lawless and Burroughs walked through the door. Burroughs looked less imposing without his boardroom costume, but he was still natty in a muted madras shirt, sport jacket, and crisply pressed tan slacks. We nodded. Lawless took a seat to the right of Stryker, Sergio to the left, and Burroughs next to me. I was looking at my tribunal.

Burroughs's eyes floated around the room, taking it all in. I waited for Stryker to call the meeting to order, but it was Lawless who spoke first.

"John, let me thank you again for coming. We have a few matters to discuss. You once asked whether you could make a contribution to our work. It occurs to me that we have not ... perhaps ... defined our work clearly. I cannot ask you to contribute to our efforts until you know what they are. I cannot explain it adequately without all the players you see before you."

Burroughs crossed his arms like an old deal maker and looked at Lawless like at an animal he had seen before. Lawless didn't miss a beat.

"Stryker, Hopper, Sergio, and myself constitute a cadre. We are attached by choice in a common cause. We are an ad hoc quartet, and I like to think that we operate with intelligence and discretion. There is no easy way to explain to you which projects we undertake and which we avoid or, for that matter, why....We operate under a rather strict code of ethics. There are certain occasions and events which require intervention on the basis of concern, equity, or justice. You've asked me what you can contribute to our efforts, and I've gathered the cadre because the answer must come from all of us."

Burroughs uncrossed his arms and leaned forward. "My son is dead, and I can't change that. His killer is in custody. I'm pleased with what you've been able to accomplish without appearing to be too far outside the confines of the system. Tell me what I can do for you."

"I'm not certain yet. You've mentioned money. I'm inclined to ask you for a line of credit instead."

Burroughs furrowed his brow, and I almost hoped he was going to say no, because whatever was taking place at this meeting seemed irrevocable.

"Tell me what you do. Each of you."

Lawless moved in like a good maître d'. "I toil in the fields of government. I'm a lobbyist of sorts. Mostly I make friends for my friends. You know what I do, John."

Burroughs nodded in my direction and swiveled in his chair. "And we know what you do, Hopper. You're the operative. You get things done one way or the other. I thank you for that."

I looked him straight in the eye and saluted him because old John had my number. He turned his questioning gaze to Sergio.

"And you, sir? You made my plane reservations. Are you the secretary of this organization?"

"On the contrary, señor. I am the ambassador. I know

about protocol, and I, too, know how to get things done. I am the concierge of the Two Yanks, but I am also more than that."

I was staring hard at Sergio and reevaluating his position. Whatever he did for Stryker and Lawless, it was more than I understood at the moment.

Burroughs stood up and walked to the fax machine. He ran his fingers lightly across the controls. The silence in the room was deafening. Burroughs whirled in a half-circle. His eyes landed on Stryker's face like a 747 making an emergency landing on a carrier deck.

"And you are the pivotal member of this cadre, the fulcrum, the power...aren't you, Dr. Stephens?"

Dr. Stephens raked his hand down the length of his face, and his body transformed itself into what I've come to call the "come to Jesus" posture.

"I am the tactician, John. Do not confuse that with 'leader.' I have the kind of mind that reduces most problems to rubble and I'm proud of that, but I have no stomach left for the kind of wheeling and dealing that Lawless does. I am too impatient to exercise the superb diplomatic skills of a Sergio Cordero. I have installed many of my own skills into Hopper's repertoire, and most of them he is able to execute more quickly and effectively because of his youth and physical prowess. Lawless called us a cadre. He is correct. Each one of us contributes a unique element to any operation.

"John, there are always critical situations in any country that cannot be resolved without sufficient capital and the ability to work by a slightly different set of rules. Informally and covertly, we can move quickly and quietly to help people who cannot help themselves or who need specific expertise—as you did in this instance. All I can do is assure you that these people merit our assistance, but, for the most part, you will not know who they are. That is as much for your protection as ours. Therefore, I'm uncomfortable with a 'donation' from you, but a line of credit could certainly speed our response time and expand our range."

I might have been irritated at Stryker's tacit assumption that I wanted to play this game, but I wasn't. Whatever deal was being cut here was not only irrevocable, it was probably inevitable. I didn't have all the pieces yet, but I got the big picture.

There was no way to know what sort of strategy Burroughs used to make decisions, but I figured it wouldn't take long, and I was right.

"How much can you use in the near term, where do you want to set up the account, and what do we call this entity?"

Lawless moved back in. "How about a quarter of a million in an offshore account, John?"

"I'd prefer to make it half a million to start. We can increase it later. I have a bank in the Dutch Antilles that is trustworthy and circumspect. Will that do?"

Lawless nodded. Burroughs continued.

"For my own business purposes I would like to attach a name to the account. What's suitable?"

We were silent for a moment. To name the cadre was to give it a new dimension of reality. It was fitting that it was Stryker who spoke.

"Call it the Key Foundation, John."

"Fine. Consider it done. I'll confirm with Lawless."

Stryker walked to the window, and I saw him smile. I suspect that I was the only one in the room who got the joke. For Stryker had really named us the Ki Foundation. "Ki" is a martial arts word that means "invisible life force."

My insides quivered. I hoped that this fork in the road would bring something less emotionally arduous than what I'd just experienced. It didn't seem likely.

Two days later, Stryker and I sat on his deck, the promised langostinos delivered. I delivered, too. Everything he wanted to know. And when I was done venting my spleen about the uselessness of most human endeavor, he looked at me noncommittally.

"It's a matter of rolling with the punches, Hopper. Practice. You have a natural talent ... and now you have a context ... and partners."

Ki. Invisible. Life force.